An Introduction to Analysis

An Introduction to Analysis

Robert C. Gunning

PRINCETON UNIVERSITY PRESS

Princeton and Oxford

ISBN 978-0-691-17879-0

Library of Congress Control Number: 2017963672

British Library Cataloging-in-Publication Data is available

This book has been composed in ITC Stone Serif and Avenir

Printed on acid-free paper. ∞

Typeset by Nova Techset Pvt Ltd, Bangalore, India

Printed in the United States of America

1   3   5   7   9   10   8   6   4   2

To Wanda

Contents

Preface

The course notes on which this book is based served as the text for the accelerated honors analysis course that I have been teaching at Princeton University. The traditional honors mathematics program at Princeton consists of three one-semester courses, covering calculus in one variable in the first semester, linear algebra in the second semester, and calculus in several variables in the third semester, and was usually taken by undergraduates in their first three semesters. In recent years undergraduates have entered the university with more experience in rigorous and abstract mathematics, and have been particularly eager to pass on to more advanced mathematics courses as quickly as possible. Consequently as an experiment we tried condensing the contents of the three semesters into a one-year course, with a fair bit of additional material as well, called the accelerated honors analysis course. About a third of the students taking the honors sequence have chosen the accelerated version, and generally they have done quite well indeed. The course begins rather abstractly with a background in set theory and algebra, treats differentiation and integration in several dimensions from the beginning, treats some standard material on linear algebra of particular interest in analysis, and ends with differential forms and Stokes's theorems in n dimensions, with the classical cases as examples. The course covers a good deal of material and proceeds quite rapidly. It meets twice weekly, for ninety minute sessions, but in addition there are tutorial sessions three times a week in the evening, run by student tutors to help the students read and understand the somewhat densely written course notes and to provide guidance in approaching the problems. Providing these extra sessions is possibly the only way to cover the material at a not too unreasonable pace. Those students who feel that the lectures are rather rushed do turn up quite regularly for the tutorial sessions.

During the 2007–2008 academic year Lillian Pierce and I cotaught the third term of the regular honors course, while she was a graduate student here, and we reorganized the presentation of the material and the problem assignments; her suggestions were substantial and she played a significant part in working through the revisions that year. The goal was to continue the rather abstract treatment of the underlying mathematical topics following Michael Spivak's now classical treatment, but with somewhat more emphasis on differentiation and somewhat less emphasis on the abstract treatment of differential forms. The problems were divided into two categories: a first group of problems testing a basic understanding of the essential topics discussed and their applications, problems that almost all serious students should be able to solve without too much difficulty, and a second group of problems covering more theoretical aspects and tougher calculations, challenging the students but still not particularly difficult. The temptation to include a third category of optional very challenging problems, to introduce interested students to a wider range of other theoretical and practical aspects, was frustrated by a resounding lack of interest on the part of the students. When the accelerated course was set up in the 2010–2011 academic year I used the notes we had developed for the third term of the regular course sequence and added additional material at the beginning, including more linear algebra.

I should like to express here my sincere thanks to the students who compiled the lecture notes that were the basis for the first version of the course notes: Robert Haraway, Adam Hesterberg, Jay Holt, and Alex Schiller; to the graduate students who have cotaught the course and have done the major share of grading the assignments; and to the students who have taken the course these past few years, for a number of corrections and suggestions for greater clarity. I would particularly like to thank my colleagues Lillian Pierce, for her many suggestions and very great help in reorganizing the course, and Tadahiro Oh, for a very thorough review of the notes and a great many very valuable suggestions and corrections. I owe a particular debt of gratitude to Kenneth Tam for a very close reading of the last versions of the notes. He pointed out and helped correct some errors that had slipped into previous revisions; he was an excellent critic on matters of consistency and clarity; and he suggested some very significant improvements and additions. The remaining errors and confusion are my own responsibility though.

An Introduction to Analysis

1

Algebraic Fundamentals

1.1 Sets and Numbers

Perhaps the basic concept in mathematics is that of a **set**, an unambiguously determined collection of mathematical entities. A set can be specified by listing its contents, as for instance the set $A = \{w, x, y, z\}$, or by describing its contents in any other way so long as the members of the set are fully determined. Among all sets the strangest is no doubt the **empty set**, traditionally denoted by $\emptyset$, the set that contains nothing at all, so the set $\emptyset = \{\ \ \}$; this set should be approached with some caution, since it can readily sneak up unexpectedly as an exception or counterexample to some mathematical statement if that statement is not carefully phrased. There is a generally accepted standard notation dealing with sets, and students should learn it and use it freely. That a belongs to or is a member of the set A is indicated by writing $a \in A$; that b does not belong to the set A is indicated by writing $b \notin A$. A set A is a **subset** of a set B if whenever $a \in A$ then also $a \in B$, and the condition that A is a subset of B is indicated by writing $A \subset B$ or $B \supset A$. There is some variation in the usage though. With the definition adopted here, $A \subset B$ includes the case that $A = B$ as well as the case that there are some $b \in B$ for which $b \notin A$; in the latter case the inclusion $A \subset B$ is said to be a **proper inclusion** and it is denoted by $A \subsetneq B$ or $B \supsetneq A$. In particular $\emptyset \subset A$ for any set A; for since there is nothing in the empty set $\emptyset$ the condition that anything contained in $\emptyset$ is also contained in A holds vacuously. Sets A and B are said to be **equal**, denoted by $A = B$ or $B = A$, if they contain exactly the same elements, or equivalently if both $A \subset B$ and $B \subset A$.

The **intersection** of sets A and B, denoted by $A \cap B$, consists of those elements that belong to both A and B, and the **union** of sets A and B, denoted by $A \cup B$, consists of those elements that belong to either A or B or both. More generally, for any collection of sets $\{A_\alpha\}$, the intersection and union of this collection of sets are defined by

$$\bigcap_\alpha A_\alpha = \{\, a \mid a \in A_\alpha \ \text{for all } A_\alpha \,\},$$
$$\bigcup_\alpha A_\alpha = \{\, a \mid a \in A_\alpha \ \text{for some } A_\alpha \,\}. \tag{1.1}$$

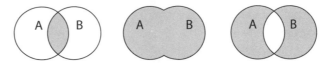

Figure 1.1. Venn diagrams illustrating the sets $A \cap B$, $A \cup B$, $A \triangle B$, respectively

Two sets A and B are **disjoint** if $A \cap B = \emptyset$. The intersection and union operations on sets are related by

$$A \cap (B \cup C) = (A \cap B) \cup (A \cap C),$$
$$A \cup (B \cap C) = (A \cup B) \cap (A \cup C). \tag{1.2}$$

The **difference** of two sets A and B, denoted by $A \sim B$, consists of those $a \in A$ such that $a \notin B$, whether B is a subset of A or not; for clarity $\emptyset \sim \emptyset = \emptyset$, a result that might not be altogether clear from the definition of the difference. Obviously the order in which the two sets are listed is critical in this case; but there is also the **symmetric difference** between these two sets, defined by

$$A \triangle B = B \triangle A = (A \sim B) \cup (B \sim A) = (A \cup B) \sim (A \cap B). \tag{1.3}$$

The symmetric difference, as well as the intersection and union, can be illustrated by **Venn diagrams**, as in the accompanying Figure 1.1. It is clear from the definitions that

$$A \sim (B \cup C) = (A \sim B) \cap (A \sim C),$$
$$A \sim (B \cap C) = (A \sim B) \cup (A \sim C). \tag{1.4}$$

If $A, B, C, \ldots$ are all viewed as subsets of a set E, and that is understood in the discussion of these subsets, then a difference such as $E \sim A$ often is denoted just by $\sim A$ and is called the **complement** of the subset A; with this understanding (1.4) takes the form

$$\sim (A \cup B) = (\sim A) \cap (\sim B),$$
$$\sim (A \cap B) = (\sim A) \cup (\sim B). \tag{1.5}$$

It is worth writing out the proofs of (1.2) and (1.4) in detail if these equations do not seem evident.

A **mapping** $f : A \longrightarrow B$ from a set A to a set B associates to each $a \in A$ its image $f(a) \in B$. The set A often is called the **domain** of the mapping and the set B the **range** of the mapping. The mapping $f : A \longrightarrow B$ is said to be **injective** if $f(a_1) \neq f(a_2)$ whenever $a_1 \neq a_2$; it is said to be **surjective** if for every $b \in B$ there is $a \in A$ such that $b = f(a)$; and it is said to be **bijective** if it is both injective and surjective. A bijective mapping $f : A \longrightarrow B$ is said to establish a **one-to-one correspondence** between the sets A and B, since it associates to each point $a \in A$ a unique point $b \in B$, and each point of B is associated in this way to a unique point of A. A mapping $f : A \longrightarrow B$ is bijective if and only if there is a mapping $g : B \longrightarrow A$ such that $g(f(a)) = a$ for every $a \in A$ and $f(g(b)) = b$ for every $b \in B$. Indeed if f is bijective then it is surjective, so each point $b \in B$ is the image $b = f(a)$ of a point $a \in A$, and f is also injective, so the point a is uniquely determined by b and consequently can be viewed as the image $a = g(b)$ of a well-defined mapping $g : B \longrightarrow A$, for which $b = f(g(b))$; and for any $a \in A$ substituting $b = f(a)$ in the preceding formula shows that $f(a) = f(g(f(a)))$, so since f is injective it follows that $a = g(f(a))$. Conversely if there is a mapping $g : B \longrightarrow A$ such that $g(f(a)) = a$ for every $a \in A$ and $f(g(b)) = b$ for every $b \in B$ then the mapping f is clearly both injective and surjective so it is bijective. The mapping g is called the **inverse mapping** to f; it is usually denoted by $g = f^{-1}$, and it is a bijective mapping from B to A. For any mapping $f : A \longrightarrow B$, not necessarily bijective or injective or surjective, there can be associated to any subset $X \subset A$ its **image** $f(X) \subset B$, defined by

$$f(X) = \left\{ f(a) \in B \mid a \in X \right\}, \tag{1.6}$$

and to any subset $Y \subset B$ its **inverse image** $f^{-1}(Y) \subset A$, defined by

$$f^{-1}(Y) = \left\{ a \in A \mid f(a) \in Y \right\}. \tag{1.7}$$

The image $f(X)$ may or may not coincide with the range of the mapping f. If $f : A \longrightarrow B$ is injective then clearly it determines a bijective mapping from any subset $X \subset A$ to its image $f(X) \subset B$, so in a sense it describes an injection of the set X into B. If $f : A \longrightarrow B$ is bijective the inverse image $f^{-1}(Y)$ of a subset $Y \subset B$ is just the image of that subset Y under the inverse mapping f^{-1}; it should be kept clearly in mind though that $f^{-1}(Y)$ is well defined even for mappings $f : A \longrightarrow B$ that are not bijective, so mappings for which the inverse mapping f^{-1} is not even defined. Thus the notation $f^{-1}(Y)$ really has two different meanings, and some care must be taken to distinguish them carefully to avoid confusion and errors. For any mapping $f : A \longrightarrow B$ and any subsets $X_1, X_2 \subset A$ and $Y_1, Y_2 \subset B$ it is fairly easy to see that

$$\begin{aligned}
f(X_1 \cup X_2) &= f(X_1) \cup f(X_2), \\
f(X_1 \cap X_2) &\subset f(X_1) \cap f(X_2) \text{ but it may be a proper inclusion,} \\
f^{-1}(Y_1 \cup Y_2) &= f^{-1}(Y_1) \cup f^{-1}(Y_2), \\
f^{-1}(Y_1 \cap Y_2) &= f^{-1}(Y_1) \cap f^{-1}(Y_2);
\end{aligned} \tag{1.8}$$

this is an instance in which the inverse images of sets are better behaved than the images of sets. To any mappings $f : A \longrightarrow B$ and $g : B \longrightarrow C$ there can be associated the **composite** mapping $g \circ f : A \longrightarrow C$ defined by

$$(g \circ f)(a) = g(f(a)) \quad \text{for all } a \in A. \tag{1.9}$$

The order in which the composite is written should be kept carefully in mind: $g \circ f$ is the operation that results from first applying the mapping f and then the mapping g.

To any set A there can be associated other sets, for example, the set consisting of all subsets of A, sometimes called the **power set** of A and denoted by $\mathfrak{P}(A)$. The power set $\mathfrak{P}(\emptyset)$ of the empty set is the set that consists of a single element, since the only subset of the empty set is the empty set itself; the power set of the set $\{a\}$ consisting of a single point has two elements, $\mathfrak{P}(\{a\}) = \{\{a\}, \emptyset\}$; and the power set of the set $\{a, b\}$ consisting of two points has four elements, $\mathfrak{P}(\{a, b\}) = \{\emptyset, \{a\}, \{b\}, \{a, b\}\}$. To any two sets A, B there can be associated the set

$$B^A = \{ f : A \longrightarrow B \} \tag{1.10}$$

consisting of all mappings $f : A \longrightarrow B$. If $A = \{a\}$ consists of a single point there is the natural bijection $\phi : B^{\{a\}} \longrightarrow B$ that associates to any mapping $f \in B^{\{a\}}$ its image $f(a) \in B$, since the mapping f is fully determined by its image and any point $b \in B$ is the image of the mapping f for which $f(a) = b$. On the other hand there is the natural bijection $\phi : \{a\}^B \longrightarrow \{a\}$ since there is a single mapping $f \in \{a\}^B$, the mapping for which $a = f(b)$ for every $b \in B$. If $B = \mathbb{F}_2 = \{0, 1\}$ is the set consisting of two points 0 and 1 there is the natural bijection $\psi : (\mathbb{F}_2)^A \longrightarrow \mathfrak{P}(A)$ that associates to each mapping $f : A \longrightarrow \mathbb{F}_2$ the subset $E_f = f^{-1}(1) \subset A$; for the mapping f is determined uniquely by the subset $E_f \subset A$, since $f(x) = 1$ if $x \in E_f$ and $f(x) = 0$ if $x \in A \sim E_f$, and for any subset $E \subset A$ the **characteristic function** χ_E of the set E, the mapping $\chi_E : A \longrightarrow \mathbb{F}_2$ defined by

$$\chi_E(a) = \begin{cases} 1 & \text{if } a \in E, \\ 0 & \text{if } a \notin E, \end{cases} \tag{1.11}$$

has the property that $\chi_E^{-1}(1) = E$. The bijective mapping ψ can be viewed as the identification

$$\mathfrak{P}(A) = (\mathbb{F}_2)^A; \tag{1.12}$$

sometimes the power set $\mathfrak{P}(A)$ of a set A is denoted just by $\mathfrak{P}(A) = 2^A$.

Yet a different way of associating to a set A another set is through an **equivalence relation** on the set A, a relation between some pairs of elements $a_1, a_2 \in A$ which is denoted by $a_1 \asymp a_2$ and is characterized by the following three properties:

(i) *reflexivity*, $a \asymp a$ for any $a \in A$;
(ii) *symmetry*, if $a_1 \asymp a_2$ then $a_2 \asymp a_1$; and
(iii) *transitivity*, if $a_1 \asymp a_2$ and $a_2 \asymp a_3$ then $a_1 \asymp a_3$.

If $\asymp$ is an equivalence relation on a set A then to any $a \in A$ there can be associated the set of all $x \in A$ that are equivalent to a, the subset

$$A_a = \left\{ x \in A \mid x \asymp a \right\} \subset A, \tag{1.13}$$

which by reflexivity includes in particular a itself. The set A_a is called the **equivalence class** of a with respect to the equivalence relation $\asymp$. Similarly to any $b \in A$ there can be associated the set A_b of all $y \in A$ that are equivalent to b. If $c \in A_a \cap A_b$ for an element $c \in A$ then $c \asymp a$ and $c \asymp b$ so by symmetry and transitivity $a \asymp b$; then by symmetry and transitivity again whenever $x \in A_b$ then $x \asymp b \asymp a$ so $x \asymp a$ and consequently $x \in A_a$, hence $A_b \subset A_a$, and correspondingly $A_a \subset A_b$ so that actually $A_b = A_a$. Thus the set A is naturally decomposed into a collection of disjoint equivalence classes; the set consisting of these various equivalence classes is another set, called the **quotient** of the set A under this equivalence relation and denoted by $A/\asymp$. This is a construction that arises remarkably frequently in mathematics. For example, to any mapping $f : A \longrightarrow B$ between two sets A and B there can be associated an equivalence relation on the set A by setting $a_1 \asymp_f a_2$ for two points $a_1, a_2 \in A$ whenever $f(a_1) = f(a_2)$; it is quite clear that this is indeed an equivalence relation and that the quotient $A/\asymp_f$ can be identified with the image $f(A) \subset B$.

The notion of equivalence also is used in a slightly different way, as a relation among sets rather than as a relation between the elements of a particular set. For instance the equality $A = B$ of two sets clearly satisfies the three conditions for an equivalence relation; it is not phrased as an equivalence relation between elements of a given set but rather as an equivalence relation among various sets, to avoid any involvement with the paradoxical notion of the set of all sets. As another example, two sets A, B are simply said to be **equivalent** if there is a bijective mapping $f : A \longrightarrow B$; that these two sets are equivalent is indicated by writing $A \leftrightarrow B$. This notion also clearly satisfies the three conditions for an equivalence relation among sets. Equivalent sets intuitively are those with the same "number of elements," whatever that may mean. One possibility for giving that notion a definite meaning is merely to use the equivalence relation itself as a proxy for the "number of elements" in a set, that is, to define the

cardinality of a set A as the equivalence class of that set,

$$\#(A) = \Big\{ X \mid X \text{ is a set for which } X \leftrightarrow A \Big\}. \qquad (1.14)$$

With this definition $\#(A) = \#(B)$ just means that the sets A and B determine the same equivalence class, that is, that $A \leftrightarrow B$ so that there is a bijective mapping $f : A \longrightarrow B$.

For sufficiently small sets this notion of cardinality can be made somewhat more explicit. Consider formal symbols $\{/, /, \ldots, /\}$, which are constructed beginning with the symbol $\{/\}$, followed by its **successor** $\{/, /\}$ obtained by adjoining a stroke, followed in turn by its successor $\{/, /, /\}$ obtained by adjoining another stroke, and so on; at each stage of the construction the successor of one of these symbols is obtained by adjoining another stroke to that symbol. The initial symbol $\{/\}$ represents the cardinality or equivalence class of sets consisting of the sets that can be obtained by replacing the stroke $/$ by an element of any set; its successor $\{/, /\}$ represents the cardinality or equivalence class of sets consisting of the sets that can be obtained by replacing the strokes $/, /$ by distinct elements of any set, and so on. The cardinalities thus constructed form a set $\mathbb{N}$ called the set of **natural numbers**. The natural number that is the cardinality represented by the symbol $\{/\}$ is usually denoted by 1, so that $1 = \#(X)$ for any set X consisting of a single element; and if $n \in \mathbb{N}$ is the cardinality described by one of the symbols in this construction the cardinality represented by its successor symbol is denoted by n' and is called the **successor** of n. As might be expected, a customary notation is $2 = 1'$, $3 = 2'$, and so on. It is evident from this construction that the set $\mathbb{N}$ of natural numbers satisfies the **Peano axioms**:

(i) *origin*, there is a specified element $1 \in \mathbb{N}$;

(ii) *succession*, to any element $a \in \mathbb{N}$ there is associated an element $a' \in \mathbb{N}$, called the *successor* to a, such that

(ii') $a' \neq a$,

(ii'') $a' = b'$ if and only if $a = b$, and

(ii''') 1 is not the successor to any element of $\mathbb{N}$;

(iii) *induction*, if $E \subset \mathbb{N}$ is any subset such that $1 \in E$ and that $a' \in E$ whenever $a \in E$ then $E = \mathbb{N}$.

The axiom of induction is merely a restatement of the fact that the set $\mathbb{N}$ is constructed by the process of considering the successors of cardinalities already in $\mathbb{N}$. A simple consequence of the axiom of induction is that every element $a \in \mathbb{N}$ except 1 is the successor of another element of $\mathbb{N}$; indeed if

$$E = 1 \cup \Big\{ x \in N \mid x = y' \text{ for some } y \in \mathbb{N} \Big\}$$

it is clear that $1 \in E$ and that if $x \in E$ then $x' \in E$ so by the induction axiom $E = \mathbb{N}$. The set $\mathbb{N}$ of natural numbers is customarily defined by the Peano axioms; for it is evident that any set satisfying the Peano axioms can be identified with the set of natural numbers, by yet another application of the axiom of induction.

Note particularly that it is not asserted or assumed that the cardinality of any set actually is a natural number. The sets with cardinalities that are natural numbers are called **finite sets**, while the sets that are not finite sets are called **infinite sets**. That a set S is infinite is indicated by writing $\#(S) = \infty$, a slight abuse of notation since it does not mean that ∞ is the cardinality of the set S but just that S is not a finite set. The set $\mathbb{N}$ itself is an infinite set. Indeed it is evident from the definition of the natural numbers that a finite set cannot be equivalent to a proper subset of itself; but the mapping $f : \mathbb{N} \longrightarrow \mathbb{N} \sim \{1\}$ that associates to each natural number n its successor n' is injective by succession and surjective by the preceding discussion, so $\mathbb{N}$ is equivalent to $\mathbb{N} \sim \{1\}$ hence $\mathbb{N}$ cannot be finite. It is traditional to set $\#(\mathbb{N}) = \aleph_0$; so if A is any set that is equivalent to $\mathbb{N}$ then $\#(A) = \aleph_0$ as well. A set A such that $\#(A) = \aleph_0$ is said to be a **countably infinite** set, while a set that is either finite or countably infinite, is called a **countable set**.[1]

Induction is also key to the technique of proof by **mathematical induction**: If $T(n)$ is a mathematical statement depending on the natural number $n \in \mathbb{N}$, if $T(1)$ is true and if $T(n')$ is true whenever $T(n)$ is true, then $T(n)$ is true for all $n \in \mathbb{N}$; indeed if E is the set of those $n \in \mathbb{N}$ for which $T(n)$ is true then by hypothesis $1 \in E$ and $n' \in E$ whenever $n \in E$, so by the induction axiom $E = \mathbb{N}$. Many examples of the application of this method of proof will occur in the subsequent discussion.

There is a natural comparison of cardinalities of sets defined by setting

$$\#(A) \leq \#(B) \text{ if there is an injective mapping } f : A \longrightarrow B. \qquad (1.15)$$

This is well defined, since if $A' \leftrightarrow A, B' \leftrightarrow B$, and $\#(A) \leq \#(B)$ there are a bijective mapping $g : A \longrightarrow A'$, a bijective mapping $h : B \longrightarrow B'$, and an injective mapping $f : A \longrightarrow B$; and the composite mapping

$$f' = h \circ f \circ g^{-1} : A' \longrightarrow B'$$

is an injective mapping hence $\#(A') \leq \#(B')$. As far as the notation is concerned, it is customary to consider $\#(B) \geq \#(A)$ as an alternative way of writing $\#(A) \leq \#(B)$, and to write $\#(A) < \#(B)$ or $\#(B) > \#(A)$ to indicate that $\#(A) \leq \#(B)$ but $\#(A) \neq \#(B)$. Since any finite set can be represented as a subset of $\mathbb{N}$ but does

---

[1]This terminology is not universally accepted, so some caution is necessary when comparing discussions of these topics; it is not uncommon to use "at most countable" in place of countable and "countable" in place of countably infinite.

not admit a bijective mapping to $\mathbb{N}$ it follows that $n < \aleph_0$ for any $n \in \mathbb{N}$. For any infinite subset $E \subset \mathbb{N}$ it is the case that $\#(E) = \#(\mathbb{N}) = \aleph_0$, even if E is a proper subset of $\mathbb{N}$; indeed each element of E is in particular a natural number n, so when these natural numbers are arranged in increasing order $n_1 < n_2 < n_3$ then the mapping $f : E \longrightarrow \mathbb{N}$ that associates to $n_i \in E$ the natural number $i \in \mathbb{N}$ is a bijective mapping and consequently $\#(E) = \#(\mathbb{N})$.[2] It is somewhat counterintuitive that any infinite proper subset of $\mathbb{N}$ has the same cardinality, or the same number of elements, as $\mathbb{N}$; if $A \subset B$ is a proper subset and both A and B are finite then $\#(A) < \#(B)$, but that is not necessarily the case for infinite sets. The further examination of inequalities among cardinalities of sets rests upon the following result, which is evident for finite sets but not obvious for infinite sets.

Theorem 1.1 (Cantor-Bernstein Theorem). *If there are injective mappings* $f : A \longrightarrow B$ *and* $g : B \longrightarrow A$ *between two sets A and B then there is a bijective mapping* $h : A \longrightarrow B$.

Proof: The mapping $f : A \longrightarrow B$ is injective, but its image is not necessarily all of B; the mapping g is injective so its inverse $g^{-1} : g(B) \longrightarrow B$ is an injective mapping onto B, but it is defined only on the subset $g(B) \subset A$. That suggests considering the mapping $h_0 : A \longrightarrow B$ defined by setting $h_0(a) = f(a)$ for all points $a \in A \sim g(B)$ and $h_0(a) = g^{-1}(a)$ for all points $a \in g(B)$. This is indeed a well-defined surjective mapping $h_0 : A \longrightarrow B$ that restricts to injective mappings on $A \sim g(B)$ and $g(B)$. However there may be, indeed actually are, points $a_1 \in A \sim g(B)$ and $a_2 \in g(B)$ for which $(g \circ f)(a_1) = a_2$ so $f(a_1) = g^{-1}(a_2)$ and the mapping h_0 fails to be injective; the problem thus lies in points

$$a_1 \in (g \circ f)(A \sim g(B)) \subset g(B).$$

If all such points are moved into the domain of the mapping f rather than the domain of the mapping g^{-1} that solves that problem for such points; but the mapping h_0 may still fail to be injective for a similar reason. The solution is to introduce the set

$$C = (A \sim g(B)) \cup \bigcup_{n \in \mathbb{N}} (g \circ f)^n (A \sim g(B)) \subset A;$$

---

[2]Actually $\mathbb{N}$ is the smallest infinite set, in the sense that if S is any infinite set then $\#(\mathbb{N}) \leq \#(S)$. The demonstration though does require the use of the axiom of choice; see the discussion in the *Princeton Companion to Mathematics*. Indeed by the axiom of choice it is possible to choose one of the points of S and to label it x_1; since S is infinite there are points in S other than x_1, so by the axiom of choice again it is possible to choose a point in S other than x_1 and to label it x_2; and inductively if $x_1, \ldots, x_\nu$ are labeled there remain other points in S, since S is infinite, so by the axiom of choice again choose another point and label it $x_{\nu+1} = x_{\nu'}$. That establishes the existence of a subset of S indexed by a subset $X = \{\nu\} \subset \mathbb{N}$ of the natural numbers, where X includes 1 and includes ν' for any $\nu \in X$ so by the induction axiom $X = \mathbb{N}$, as desired.

since $(A \sim g(B)) \subset C$ then $(A \sim C) \subset g(B)$. In terms of this set introduce the mapping $h : A \longrightarrow B$ defined by

$$h(x) = \begin{cases} f(x) & \text{if } x \in C \subset A, \\ g^{-1}(x) & \text{if } x \in (A \sim C) \subset A, \end{cases}$$

where $g^{-1}(x)$ is well defined on the subset $(A \sim C) \subset g(B)$ since g is injective. The theorem will be demonstrated by showing that the mapping $h : A \longrightarrow B$ just defined is bijective.

To demonstrate first that h is injective consider any two distinct points $x_1, x_2 \in A$ and suppose to the contrary that $h(x_1) = h(x_2)$. If $x_1, x_2 \in C$ then by definition $h(x_1) = f(x_1)$ and $h(x_2) = f(x_2)$ so $f(x_1) = f(x_2)$, a contradiction since f is injective. If $x_1, x_2 \in (A \sim C) \subset g(B)$ then $x_1 = g(y_1)$ and $x_2 = g(y_2)$ for uniquely determined distinct points $y_1, y_2 \in B$ since g is injective, and by definition $h(x_1) = y_1$ and $h(x_2) = y_2$ so $y_1 = y_2$, a contradiction. If $x_1 \in C$ and $x_2 \in (A \sim C)$ then by definition $h(x_1) = f(x_1)$ and $h(x_2) = y_2$ where $y_2 \in B$ is the uniquely determined point for which $g(y_2) = x_2$. Since $h(x_1) = h(x_2)$ it follows that $f(x_1) = y_2$ hence $x_2 = g(y_2) = (g \circ f)(x_1)$. Since $x_1 \in C$ then by the definition of the set C either $x_1 \in (A \sim g(B))$ or there is some $n \in \mathbb{N}$ for which $x_1 \in (g \circ f)^n (A \sim g(B))$; and in either case $x_2 = (g \circ f)(x_1) \in (g \circ f)^k (A \sim g(B)) \subset C$ for some k, a contradiction since $x_2 \in (A \sim C)$. That shows that h is injective.

To demonstrate next that h is surjective, consider a point $y \in B$ and let $x = g(y) \in A$. If $x \in (A \sim C)$ then by definition $h(x) = g^{-1}(x) = y$. If $x \in C$ it cannot be the case that $x \in (A \sim g(B))$ since $x = g(y) \in g(B)$, so it must be the case that there is some $n \in \mathbb{N}$ for which $x \in (g \circ f)^n (A \sim g(B))$; consequently $x = (g \circ f)(x_1)$ for some point $x_1 \in C$. Since $g(f(x_1)) = x = g(y)$ and g is injective it must be the case that $y = f(x_1)$; and since $x_1 \in C$ then by definition $h(x_1) = f(x_1) = y$. That shows that h is surjective and thereby concludes the proof.

The Cantor-Bernstein Theorem shows that the inequality (1.15) among cardinalities of sets satisfies some simple natural conditions; these conditions arise in other contexts as well, so they will be described here a bit more generally. A **partial order** on a set S is defined as a relation $x \leq y$ among elements $x, y \in S$ of that set with the following properties:

(i) *reflexivity*, $x \leq x$ for any $x \in S$;
(ii) *antisymmetry*, if $x \leq y$ and $y \leq x$ then $x = y$ for any $x, y \in S$; and
(iii) *transitivity*, if $x \leq y$ and $y \leq z$ then $x \leq z$ for any $x, y, z \in S$.

For example, the set of all subsets of a set S is clearly a partially ordered set if $A \leq B$ means that $A \subset B$ for any subsets $A, B \subset S$; and the set $\mathbb{N}$ of natural numbers clearly also is a partially ordered set if $a \leq b$ is defined as in (1.15). As in the case of equivalence relations, the notion of a partial ordering can be applied to relations among sets as well as to relation among elements of a particular set, so in particular it can be applied to the relation $\#(A) \leq \#(B)$.

Corollary 1.2. *The relation (1.15) is a partial order on the cardinalities of sets.*

Proof: The reflexivity of the relation (1.15) is clear since the identity mapping $\iota : A \longrightarrow A$ that associates to any $a \in A$ the same element $\iota(a) = a \in A$ is clearly injective; and transitivity is also clear since if $f : A \longrightarrow B$ and $g : B \longrightarrow C$ are injective mappings then the composition $g \circ f : A \longrightarrow C$ is also injective. Anti-symmetry on the other hand is far from clear, but is an immediate consequence of the Cantor-Bernstein Theorem; and that is sufficient for the proof.

Theorem 1.3 (Cantor's Theorem). *The power set $\mathfrak{P}(A)$ of any set A has strictly greater cardinality than A, that is,*

$$\#(\mathfrak{P}(A)) > \#(A) \quad \text{for any set A.} \tag{1.16}$$

Proof: The mapping that sends each element $a \in A$ to the subset $\{a\} \subset A$ is an injective mapping $g : A \longrightarrow \mathfrak{P}(A)$, hence $\#(A) \leq \#(\mathfrak{P}(A))$. Suppose though that there is a bijective mapping $f : A \longrightarrow \mathfrak{P}(A)$. The subset

$$E = \left\{ x \in A \mid x \notin f(x) \right\}$$

is a well-defined set, possibly the empty set; so since the mapping f is assumed to be surjective the subset E must be the image $E = f(a)$ of some $a \in A$. However if $a \in E$ then from the definition of the set E it follows that $a \notin f(a) = E$, while if $a \notin E = f(a)$ then from the definition of the set E it also follows that $a \in E$; this contradictory situation shows that there cannot be a bijective mapping $f : A \longrightarrow \mathfrak{P}(A)$ and thereby concludes the proof.

In particular $\#(\mathbb{N}) < \#(\mathfrak{P}(\mathbb{N}))$, so the set of all subsets of the set $\mathbb{N}$ of natural numbers is not a countable set but is a strictly larger set, a set with a larger cardinality. The set of all subsets of that set is properly larger still, so there is no end to the size of possible sets. Nonetheless there are many sets that appear to be considerably larger than $\mathbb{N}$ but nonetheless are still countable.

Theorem 1.4 (Cantor's Diagonalization Theorem). *If to each natural number $n \in \mathbb{N}$ there is associated a countable set E_n, then the union $E = \bigcup_{n \in \mathbb{N}} E_n$ is a countable set.*

Proof: Since each set E_n is countable the elements of E_n can be labeled $x_{n,1}, x_{n,2}$ and so on; the elements of the union E then can be arranged in an array

$$\begin{array}{ccccc}
x_{1,1} & x_{1,2} & x_{1,3} & x_{1,4} & \cdots \\
x_{2,1} & x_{2,2} & x_{2,3} & x_{2,4} & \cdots \\
x_{3,1} & x_{3,2} & x_{3,3} & x_{3,4} & \cdots \\
\cdots & \cdots & \cdots & \cdots & \cdots ,
\end{array}$$

where row n consists of all the elements $x_{n,i}$ ordered by the natural number $i \in \mathbb{N}$. Imagine these elements now rearranged in order by starting with $x_{1,1}$ then proceeding along the increasing diagonal with $x_{2,1}$, $x_{1,2}$, then in the next increasing diagonal $x_{3,1}$, $x_{2,2}$, $x_{1,3}$ and so on; if row n is finite the places that would be filled by some terms $x_{n,i}$ are blank so are just ignored in this ordering. When all of the elements in E are written out in order as

$$x_{1,1}, \ x_{2,1}, \ x_{1,2}, \ x_{3,1}, \ x_{2,2}, \ x_{1,3}, \cdots$$

and all the duplicated elements are eliminated there is the obvious injection $E \longrightarrow \mathbb{N}$, showing that E is countable.

Corollary 1.5. *The set of all finite subsets of $\mathbb{N}$ is countable.*

Proof: First it follows by induction on n that for any natural number $n \in \mathbb{N}$ the set E_n of all subsets $A \subset \mathbb{N}$ for which $\#(A) = n$ is countable; that is clearly the case for $n = 1$, and if E_n is countable then since

$$E_{n+1} \subset \bigcup_{i \in \mathbb{N}} \left\{ \{i\} \cup A \,\middle|\, A \in E_n \right\}$$

it follows that E_{n+1} is countable by Cantor's Diagonalization Theorem and the observation that a subset of a countable set is countable. Then since all the sets E_n are countable it follows again from Cantor's Diagonalization Theorem that the set $E = \bigcup_{n \in \mathbb{N}} E_n$ is countable, which suffices for the proof.

Problems, Group I

1. Write out the proofs of equations (1.4) and (1.8).

2. Show that $A \cup B = A \cap B$ if and only if $A = B$. Is it true that $A \cup B = A$ and $A \cap B = B$ if and only if $A = B$? Why?

3. The formula $A \cup B \cap C \cup D$ does not have a well-defined meaning unless parentheses are added. What are all possible sets this formula can describe for possible placements of parentheses? Illustrate your assertions with suitable Venn diagrams.

4. For which pairs of sets A, B is $A \triangle B = A \cup B$? For which pairs of sets is $A \triangle B = A \cap B$? For which pairs of sets is $A \triangle B = A$?

5. If $f \in B^A$, if $X \subset A$ and if $Y \subset B$ show that $f(X \cap f^{-1}(Y)) = f(X) \cap Y$.

6. If $f \in B^A$ is injective and if X_1, X_2 are subsets of A show that $f(X_1 \cap X_2) = f(X_1) \cap f(X_2)$.

7. Show that the set of all polynomials with rational coefficients is countable.

8. Is the set of all monotonically increasing sequences of natural numbers (sequences of natural numbers a_ν such that $a_\nu \leq a_{\nu+1}$) countable or not? Why?

Problems, Group II

9. Show that the set consisting of all countably long sequences $(a_1, a_2, a_3, \ldots)$, where $a_n = 0$ or 1, is not a countable set. Show though that the set consisting of all countably long sequences $(a_1, a_2, a_3, \ldots)$, where $a_n = 0$ or 1 but $a_n = 0$ for all but finitely many values of n, is countable. Is the set of all sequences $(a_1, a_2, a_3, \ldots)$, where $a_n = 0$ or 1 and where the sequences are periodic in the sense that $a_{n+N} = a_n$ for all n and for some finite number N depending on the sequence, countable or not? Why?

10. Is there an infinite set S for which the power set $\mathfrak{P}(S)$ is countable? Why?

11. For any mapping $f \in B^A$ set $a_1 \asymp a_2$ if $a_1, a_2 \in A$ and $f(a_1) = f(a_2)$. Show that this defines an equivalence relation on the set A. Show that the mapping $f : A \longrightarrow B$ induces an injective mapping $F : A/\asymp \longrightarrow B$.

12. For any given countable collection of sets A_m let

$$A' = \bigcup_{n=1}^{\infty} \bigcap_{m=n}^{\infty} A_m \quad \text{and} \quad A'' = \bigcap_{n=1}^{\infty} \bigcup_{m=n}^{\infty} A_m.$$

Describe the sets A' and A'', and show that $A' \subset A''$.

13. A partition of a set A is a representation of A as a union $A = \bigcup_\alpha A_\alpha$ of pairwise disjoint subsets $A_\alpha \subset A$. It was shown that to any equivalence relation $a_1 \asymp a_2$ on A there can be associated a partition of A into equivalence classes. Show that conversely for any partition $A = \bigcup_\alpha A_\alpha$ of a set A there is an equivalence relation on A so that the sets A_α are the equivalence classes. (Your proof shows that an equivalence relation on a set is equivalent to a partiton of the set, a useful observation.)

14. To each mapping $f \in B^A$ associate the mapping $f^* \in \mathfrak{P}(B)^{\mathfrak{P}(A)}$ that associates to any subset $X \subset A$ the image $f(X) \subset B$; the mapping $f \longrightarrow f^*$ thus is a mapping $\phi : B^A \longrightarrow \mathfrak{P}(B)^{\mathfrak{P}(A)}$. Is the mapping ϕ injective? Surjective? Why?

1.2 Groups, Rings, and Fields

It is possible to introduce algebraic operations on the cardinalities of sets, and for this purpose two further constructions on sets are relevant. The union $A \cup B$ of two sets was considered in detail in the preceding section; the **disjoint union** $A \bigsqcup B$ is the union of these two sets when they are viewed as being formally disjoint, so that if $a \in A \cap B$ then in the disjoint union the element a is viewed as an element $a_A \in A$ and a distinct element $a_B \in B$. Of course if the two sets A and B are actually disjoint then $A \bigsqcup B = A \cup B$. The **Cartesian product** of the two sets A and B is the set defined by

$$A \times B = \Big\{ (x,y) \big| x \in A, \ y \in B \Big\}. \tag{1.17}$$

In these terms the **sum** and **product** of the cardinalities of sets A and B are defined by

$$\#(A) + \#(B) = \#(A \bigsqcup B) \quad \text{and} \quad \#(A) \cdot \#(B) = \#(A \times B). \tag{1.18}$$

It is fairly clear that these operations are well defined, in the sense that if $A' \leftrightarrow A$ and $B' \leftrightarrow B$ then $(A' \bigsqcup B') \leftrightarrow (A \bigsqcup B)$ and $(A' \times B') \leftrightarrow (A \times B)$. In particular these operations are defined on the natural numbers if the sets are finite and on the number $\aleph_0$ if the sets are countable. They satisfy the following laws:

   (i)   the **associative law** for addition: $a + (b + c) = (a + b) + c$;
   (ii) the **commutative law** for addition: $a + b = b + a$;
   (iii) the **associative law** for multiplication: $a \cdot (b \cdot c) = (a \cdot b) \cdot c$;
   (iv) the **commutative law** for multiplication: $a \cdot b = b \cdot a$; and
   (v)   the **distributive law**: $(a + b) \cdot c = a \cdot c + b \cdot c$.

To verify that these laws are satisfied, if $a = \#(A)$, $b = \#(B)$, and $c = \#(C)$ then $b + c = \#(B \bigsqcup C)$ so $a + (b + c) = \#(A \bigsqcup (B \bigsqcup C)) = \#(A \bigsqcup B \bigsqcup C)$ while $a + b = \#(A \bigsqcup B)$ so $(a + b) + c = \#((A \bigsqcup B) \bigsqcup C) = \#(A \bigsqcup B \bigsqcup C)$, showing that $a + (b + c) = (a + b) + c$. Moreover $a + b = \#(A \bigsqcup B) = \#(B \bigsqcup A) = b + a$. Then for multiplication $b \cdot c = \#(B \times C)$ so $a \cdot (b \cdot c) = \#(A \times (B \times C))$ where $B \times C = \{(b,c)\}$ so $A \times (B \times C) = \{(a,(b,c))\} = \{(a,b,c)\}$; correspondingly $(a \cdot b) \cdot c = \#((A \times B) \times C)$ where $(A \times B) \times C = \{((a,b),c)\} = \{(a,b,c)\}$, and consequently $a \cdot (b \cdot c) = (a \cdot b) \cdot c$. Moreover $a \cdot b = \#(A \times B)$ and $b \cdot a = \#(B \times A)$; but the mapping that sends $(x,y) \in A \times B$ to $(y,x) \in B \times A$ is a bijective mapping so $\#(A \times B) = \#(B \times A)$ and consequently $a \cdot b = b \cdot a$. Finally $(a + b) \cdot c = \#((A \bigsqcup B) \times C)$ where $(A \bigsqcup B) \times C = (A \times C) \bigsqcup (B \times C)$ so $\#((A \bigsqcup B) \times C) = \#((A \times C) \bigsqcup (B \times C)) = a \cdot c + b \cdot c$ and consequently $(a + b) \cdot c = a \cdot c + b \cdot c$.

The algebraic notation is usually simplified by writing ab in place of $a \cdot b$ and by dropping parentheses when the associative laws indicate that they are not needed to specify the result of the operation uniquely, so by writing $a + b + c$

in place of $(a + b) + c$ and abc in place of $(a \cdot b) \cdot c$. Some care must be taken, though, since there are expressions in which parentheses are necessary. The expression $a \cdot b + c$ could stand for either $a \cdot (b + c)$ or $(a \cdot b) + c$ and these can be quite different. The informal convention is to give the product priority, in the sense that products are grouped together first; so normally $a \cdot b + c$ is interpreted as $(ab) + c$. If there are any doubts, though, it is safer to insert parentheses.

The addition of natural numbers is closely related to other operations on natural numbers. First, the successor to a natural number $a = \#(A)$, where A is a finite collection of strokes, is identified with the natural number $a' = \#(A')$, where A' is derived from A by adding another stroke; thus $A' = A \bigsqcup \{/\}$ hence $\#(A') = \#(A) + \#(\{/\})$ so

$$a' = a + 1 \quad \text{for any } a \in \mathbb{N}. \tag{1.19}$$

This provides another interpretation of the successor operation on the natural numbers and thereby fits that operation into the standard algebraic machinery. Second, there is the **cancellation law** of the natural numbers:

$$a + n = b + n \quad \text{for } a, b, n \in \mathbb{N} \text{ if and only if } a = b. \tag{1.20}$$

Indeed it is clear that $a + n = b + n$ if $a = b$, and the converse can be established by induction on n. For that purpose note that $a + 1 = b + 1$ is equivalent to $a' = b'$, which by succession implies that $a = b$; thus if $a + n + 1 = b + n + 1$ for some $n \in \mathbb{N}$ then $(a + n)' = (b + n)'$ so by succession $a + n = b + n$ and then by induction $a = b$. The cancellation law plays a critical role in extending the algebraic operations on the natural numbers. There is no analogue of the cancellation law for infinite cardinals; indeed it is easy to see that $\aleph_0 = \aleph_0 + 1 = \aleph_0 + 2$ and so on, so the cancellation law does not hold in this case. Third, the order relation $a < b$ for natural numbers, where $a = \#(A)$ and $b = \#(B)$ for sets A and B consisting of collections of strokes, indicates that there is an injective mapping $f : A \longrightarrow B$ but that there is not a bijective mapping $g : A \longrightarrow B$; the collection of strokes B can be viewed as a collection of strokes bijective to A together with some additional strokes C, hence $B = A \bigsqcup C$ and therefore $\#(B) = \#(A) + \#(C)$ so

$$\text{if} \quad a, b \in \mathbb{N} \text{ then } a < b \text{ if and only if } b = a + c \text{ for some } c \in \mathbb{N}. \tag{1.21}$$

That expresses the order relation for the natural numbers in terms of the group operations on the natural numbers. There is not a similar characterization of the order relation $a \leq b$ since there is no natural number that expresses the equivalence class of the empty set; that is one reason for seeking an extension of the natural numbers.

The basic reason for extending the natural numbers though is to introduce inverses of the algebraic operations of addition and multiplication as far as possible. To describe the algebraic properties of the extended sets it may be clearest first to describe these algebraic structures more abstractly. A **group** is defined to be a set G with a specified element $1 \in G$ and with an operation that associates to any elements $a, b \in G$ another element $a \cdot b \in G$ satisfying

    (i) the **associative law**: $(a \cdot b) \cdot c = a \cdot (b \cdot c)$ for any $a, b, c \in G$;

    (ii) the **identity law**: $1 \cdot a = a \cdot 1 = a$ for any $a \in G$;

    (iii) the **inverse law**: to each $a \in G$ there is associated a unique $a^{-1} \in G$ such that $a \cdot a^{-1} = a^{-1} \cdot a = 1$.

The element $1 \in G$ is called the **identity** in the group, and the element $a^{-1} \in G$ is called the **inverse** of the element $a \in G$. A group G is said to be an **abelian group**, or equivalently a **commutative group**, if it also satisfies

    (iv) the **commutative law**: $a \cdot b = b \cdot a$ for all $a, b \in G$.

A group may consist of either finitely many or infinitely many elements; for a finite group the cardinality of the group is customarily called the **order** of the group. An element a of a group is said to have **order** n if $a^n = 1$, the identity element of the group. Just as in the case of the natural numbers, it is customary to simplify the notation by writing ab in place of $a \cdot b$ and dropping parentheses when the meaning is clear, so by writing abc in place of $(ab)c$. The **cancellation law** holds in any group: if $a \cdot b = a \cdot c$ for some $a, b, c \in G$ then $b = (a^{-1} \cdot a) \cdot b = a^{-1} \cdot (a \cdot b) = a^{-1} \cdot (a \cdot c) = (a^{-1} \cdot a) \cdot c = c$. Although the multiplicative notation for the group operation is most common in general, there are cases in which the additive notation is used. In the additive notation the group operation associates to any elements $a, b \in G$ another element $a + b \in G$; the associative law takes the form $(a + b) + c = a + (b + c)$; the identity element is denoted by 0 and the identity law takes the form $a + 0 = 0 + a = a$; and the inverse law associates to each $a \in G$ a unique element $-a \in G$ such that $a + (-a) = (-a) + a = 0$.

The simplest group consists of just a single element 1 with the group operation $1 \cdot 1 = 1$. A more interesting example of a group is the **symmetric group** $S(A)$ on a set A, the set of all bijective mappings $f : A \longrightarrow A$, where the product $f \cdot g$ of two bijections f and g is the composition $f \circ g$, the identity element of $S(A)$ is the identity mapping $\iota : A \longrightarrow A$ for which $\iota(a) = a$ for all $a \in A$, and the inverse of a mapping f is the usual inverse mapping f^{-1}. It is a straightforward matter to verify that $S(A)$ does satisfy the group laws. Even more interesting though are the groups of bijective mappings of a set A that preserve some additional structures on the set A, often called the groups of **symmetries** of these sets. For instance, assuming some familiarity with elementary plane geometry, suppose that A is a plane rectangle as in the accompanying Figure 1.2

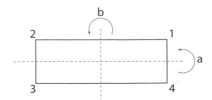

Figure 1.2. Symmetry group of a rectangle

and consider the Euclidean motions of three-space that are bijective mappings of the rectangle to itself. One motion is the flip a of the plane around the horizontal axis of the rectangle, a mapping that interchanges the vertices 1 and 4 and interchanges the vertices 2 and 3. Another motion is the flip b around the vertical axis of the rectangle, a mapping that interchanges the vertices 1 and 2 and interchanges the vertices 3 and 4. Repeating each of these mappings leaves the rectangle unchanged, so $a^2 = b^2 = 1$, the identity mapping. Performing first the flip a and then the flip b interchanges the vertices 1 and 3 and interchanges the vertices 2 and 4, so it amounts to a rotation $c = ba$ of the plane around the center of the rectangle through an angle of π radians or 180 degrees; and repeating this rotation leaves the rectangle unchanged, so $c^2 = 1$ as well. On the other hand performing the two flips in the other order really amounts to the same rotation, so $c = ab$ as well. These symmetries thus form a group of order 4, consisting of the identity mapping 1, the flips a and b and the rotation c. As for other finite groups, the algebraic operations in this group can be written as a multiplication table for the group, often called the **Cayley table** for the group, as in Table 1.1. The elements of the group are listed in the top row and the left-hand column; the entry in the table in the row associated to an element x of the group and the column associated to an element y of the group is the product $x \cdot y$ in the group. Evidently then the group is abelian if and only if the Cayley table is symmetric about the main diagonal, as is the case in Table 1.1. Each row and each column of the Cayley table for any group must be a list of all the elements of the group with no repetitions; for the multiplication of the elements of a group G by a fixed element $a \in G$ is a bijective mapping from the group to itself. In Table 1.1 the entries along the principal diagonal are all 1 reflecting that the elements of the group have order 2, so that $x^2 = 1$ for all elements x of the group. This group $\mathfrak{V}$ of symmetries of a rectangle thus is a group of order 4 in which each element has order 2; it is called **Klein's Vierergruppe**.[3]

Another group of order 4 is the group of rotations that preserve a square; that group consists of the identity mapping, a counterclockwise rotation a of

---

[3]Alternatively it is called Klein's four group; it was an example examined illustratively by Felix Klein in his classic book *Vorlesungen über das Ikosaeder und die Auflösung der Gleichungen vom fünften Grade* (Lectures on the icosahedron and the solution of equations of the fifth degree) in 1884.

TABLE 1.1.
Cayley table for the group of symmetries of a rectangle

	1	a	b	c
1	1	a	b	c
a	a	1	c	b
b	b	c	1	a
c	c	b	a	1

TABLE 1.2.
Cayley table for the group of rotations of a square

	1	a	b	c
1	1	a	b	c
a	a	b	c	1
b	b	c	1	a
c	c	1	a	b

$\pi/2$ radians or 90 degrees, a counterclockwise rotation b of π radians or 180 degrees, and a counterclockwise rotation c of $3\pi/2$ radians or 270 degrees. These operations clearly satisfy $a^2 = b$, $a^3 = c$, $a^4 = 1$, so the group can be viewed as consisting of the products a, a^2, a^3, $a^4 = 1$. This group $\mathbb{Z}_4$ is called the **cyclic** group of order 4, and its Cayley table is Table 1.2. It is clear from comparing Tables 1.1 and 1.2 that the two groups described by these tables are quite distinct; every element of the first group is of order 2 but that is not the case for the second group.

If G is a group and $H \subset G$ is a subset such that $1 \in H$, that $a \cdot b \in H$ whenever $a, b \in H$, and that $a^{-1} \in H$ whenever $a \in H$, then it is clear that the subset H also has the structure of a group with the group operation of G; such a subset is called a **subgroup** of G. A mapping $\phi : G \longrightarrow H$ from a group G to a group H is called a **homomorphism** of groups if $\phi(a \cdot b) = \phi(a) \cdot \phi(b)$ for all elements $a, b \in G$. A group homomorphism preserves other aspects of the group automatically. For instance if 1_G denotes the identity element of G and 1_H denotes the identity element of H then since $\phi(1_G) = \phi(1_G \cdot 1_G) = \phi(1_G) \cdot \phi(1_G)$ it follows from cancellation in H that $\phi(1_G) = 1_H$; moreover since $\phi(a) \cdot \phi(a^{-1}) = \phi(a \cdot a^{-1}) = \phi(1_G) = 1_H$ it follows that $\phi(a^{-1}) = (\phi(a))^{-1}$. A group homomorphism that is a bijective mapping is called a group **isomorphism**; the inverse mapping is clearly also a group isomorphism $\phi^{-1} : H \longrightarrow G$, and the groups G and H can be identified through this isomorphism. For example, it is easy to see that any group of order 4 must be isomorphic either to Klein's Vierergruppe $\mathfrak{V}$ or to the cyclic group $\mathbb{Z}_4$, so up to isomorphism there are just two groups of order 4, both of which are abelian. The smallest nonabelian group is one of order 6.

Groups involve a single algebraic operation; but the natural numbers and other sets involve two separate algebraic operations. The basic structure of interest in this context is that of a **ring**,[4] defined as a set R with distinct specified elements $0, 1 \in R$ and with operations that associate to any two elements $a, b \in R$ their **sum** $a + b \in R$ and their **product** $a \cdot b \in R$ such that

(i) R is an abelian group under the sum operation with the identity element 0;

(ii) the product operation is associative and commutative and has the identity element $1 \in R$, so that $(a \cdot b) \cdot c = a \cdot (b \cdot c)$ and $a \cdot b = b \cdot a$ and $1 \cdot a = a \cdot 1 = a$ for any elements $a, b, c \in R$; and

(iii) the operations are distributive, in the sense that $a \cdot (b + c) = a \cdot b + a \cdot c$ for any elements $a, b, c \in R$.

The additive inverse of an element $a \in R$ is denoted by $-a \in R$. The operation $a + b$ alternatively is called **addition** and the operation $a \cdot b$ alternatively is called **multiplication**. The additive identity 0 plays a special role in multiplication; for the distributive law implies that $a \cdot 0 = 0$ for any $a \in R$, since $a = a \cdot 1 = a \cdot (1 + 0) = a \cdot 1 + a \cdot 0 = a + a \cdot 0$ hence by cancellation $a \cdot 0 = 0$. Another consequence of the distributive law is that $(-a) \cdot b = -(a \cdot b)$, since $a \cdot b + (-a) \cdot b = (a + (-a)) \cdot b = 0 \cdot b = 0$; therefore of course $(-a) \cdot (-b) = a \cdot b$ as well, and in particular $(-1)^2 = 1$. A **field** is defined to be a ring F for which the set $F^{\times}$ of nonzero elements of F is a group under multiplication; thus a field is a ring with the additional property that whenever $a \in F$ and $a \neq 0$ then there is an element $a^{-1} \in F$ for which $a\,a^{-1} = a^{-1}a = 1$. For both rings and fields the notation is customarily simplified by writing ab in place of $a \cdot b$, by dropping parentheses if the meaning is clear, so by writing $a + b + c$ in place of $a + (b + c)$ and abc in place of $a(bc)$, and by writing $a - b$ in place of $a + (-b)$. If R is a ring a subset $S \subset R$ is called a **subring** if S itself is a ring under the operations of R, and correspondingly a **subfield** for fields. A subring $S \subset R$ thus must contain the elements $0, 1$, the sum $a + b$ and product ab of any elements $a, b \in S$, and the additive inverse $-a$ of any element $a \in S$; and in the case of fields S must contain the multiplicative inverse a^{-1} of any nonzero element $a \in S$. A mapping $\phi : R \longrightarrow S$ between two rings R, S is called a **homomorphism** of rings if

(i) $\phi(a + b) = \phi(a) + \phi(b)$ and $\phi(a \cdot b) = \phi(a) \cdot \phi(b)$ for any $a, b \in R$;

(ii) $\phi(1_R) = 1_S$ for the multiplicative identities $1_R, 1_S$ of these two rings.

---

[4]What is called a ring here sometimes is called a commutative ring with an identity, since there are more general algebraic structures similar to rings but without the assumptions that multiplication is commutative and that there is a multiplicative identity element.

Since a ring is a group under addition it follows as for general groups that $\phi(0_R) = 0_S$ for the additive identities and $\phi(-a) = -\phi(a)$ for any $a \in R$; but since R is not a group under multiplication it is necessary to assume that a homomorphism preserves the multiplicative identity elements. A **homomorphism** of fields has the same definition. A bijective ring or field homomorphism is called a ring or field **isomorphism**.

The natural numbers can be extended to a ring by adding enough elements to provide inverses under addition; however the additional elements must be chosen so that the extension is actually a ring, that is, so that the associative, commutative, and distributive laws continue to hold for the extended elements. A convenient way of ensuring this is to construct the extension directly in terms of the natural numbers and the operations of addition and multiplication of the natural numbers. Thus consider the Cartesian product $\mathbb{N} \times \mathbb{N}$ consisting of all pairs $\{(a, b)\}$ of natural numbers $a, b \in \mathbb{N}$ and define operations[5] on such pairs by

$$(a, b) + (c, d) = (a + c, b + d) \quad \text{and} \quad (a, b) \cdot (c, d) = (ac + bd, ad + bc). \quad (1.22)$$

It is a straightforward matter to verify that these two operations satisfy the same associative, commutative, and distributive laws as the natural numbers, as a simple consequence of those laws for the natural numbers. The only slightly complicated cases are those of the associative law for multiplication and the distributive law, where a calculation leads to

$$((a, b)(c, d))(e, f) = (ace + adf + bcf + bed, \ ade + acf + bce + bdf)$$
$$= (a, b)((c, d)(e, f))$$

and

$$((a, b) + (c, d))(e, f) = (ae + bf + ce + df, \ af + be + cf + de)$$
$$= (a, b)(e, f) + (c, d)(e, f).$$

Next introduce the equivalence relation on the set $\mathbb{N} \times \mathbb{N}$ defined by

$$(a_1, b_1) \asymp (a_2, b_2) \quad \text{if and only if} \quad a_1 + b_2 = a_2 + b_1. \quad (1.23)$$

It is easy to see that this actually is an equivalence relation; indeed reflexivity and symmetry are obvious, and if $(a, b) \asymp (c, d)$ and $(c, d) \asymp (e, f)$ then $a + d = b + c$ and $c + f = d + e$ so $(a + f) + c = a + (f + c) = a + (d + e) = (a + d) + e = (b + c) + e = (b + e) + c$ from which it follows from the cancellation law of the natural numbers, which must be used in the construction of the extension,

---

[5]The underlying idea is to view pairs (a, b) as differences $a - b$ of natural numbers, to define the ring operations as on these differences, and to introduce an equivalence relation among pairs that describe the same difference $a - b$.

that $a + f = b + e$ hence $(a, b) \asymp (e, f)$. It is an immediate consequence of the definition of equivalence that

$$(a, b) \asymp (a + c, b + c) \quad \text{for any } a, b, c \in \mathbb{N}, \tag{1.24}$$

a very useful observation in dealing with this equivalence relation. It is another straightforward matter to verify that the group operations on $\mathbb{N} \times \mathbb{N}$ preserve equivalence classes, in the sense that if $(a_1, b_1) \asymp (a_2, b_2)$ then

$$((a_1, b_1) + (c, d)) \asymp ((a_2, b_2) + (c, d)) \quad \text{and} \tag{1.25}$$
$$((a_1, b_1) \cdot (c, d)) \asymp ((a_2, b_2) \cdot (c, d)).$$

The quotient $(\mathbb{N} \times \mathbb{N})/\asymp$ of the set $\mathbb{N} \times \mathbb{N}$ by this equivalence relation is then a well-defined set, denoted by $\mathbb{Z}$ and called the set of **integers**; the operations of addition and multiplication are well defined among equivalence classes of elements of $\mathbb{N} \times \mathbb{N}$ so they are well defined on the set $\mathbb{Z}$, and these algebraic operations satisfy the same associative, commutative, and distributive laws as the natural numbers. In view of (1.24) for any $(a, b) \in \mathbb{N} \times \mathbb{N}$

$$(a, b) + (1, 1) = (a + 1, b + 1) \asymp (a, b) \tag{1.26}$$

so the equivalence class of $(1, 1)$ in the quotient $\mathbb{Z}$ satisfies the group identity law; and in addition for any $(a, b) \in \mathbb{N} \times \mathbb{N}$

$$(a, b) + (b, a) = (a + b, a + b) \asymp (1, 1) \tag{1.27}$$

so the equivalence class of (b, a) in the quotient $\mathbb{Z}$ acts as the additive inverse of the equivalence class of (a, b) and therefore $\mathbb{Z}$ is a well-defined ring.

The mapping $\phi : \mathbb{N} \longrightarrow \mathbb{Z}$ that associates to any $n \in \mathbb{N}$ the equivalence class in $\mathbb{Z}$ of the pair $(n + 1, 1) \in \mathbb{N} \times \mathbb{N}$ is an injective mapping, since if $(m + 1, 1) \asymp (n + 1, 1)$ then $m + 2 = n + 2$ so $m = n$ by the cancellation law of the natural numbers, which is used again here; therefore the mapping ϕ identifies the natural numbers with a subset $\phi(\mathbb{N}) \subset \mathbb{Z}$. Furthermore $\phi(m + n) \in \mathbb{Z}$ is the equivalence class of the pair $(m + n + 1, 1) \asymp (m + n + 1 + 1, 1 + 1) = ((m + 1) + (n + 1), 1 + 1) = (m + 1, 1) + (n + 1, 1)$ while $\phi(m) + \phi(n) \in \mathbb{Z}$ is the equivalence class of the pair $(m + 1, 1) + (n + 1, 1)$, so $\phi(m + n) = \phi(m) + \phi(n)$; on the other hand since $\phi(m \cdot n)$ is the equivalence class of the pair $(m \cdot n + 1, 1)$ while $\phi(m) \cdot \phi(n)$ is the equivalence class of the pair $(m + 1, 1) \cdot (n + 1, 1) = ((m + 1) \cdot (n + 1) + 1 \cdot 1, (m + 1) \cdot 1 + (n + 1) \cdot 1) = (mn + 1 + m + n + 1, 1 + m + n + 1) \asymp (mn + 1, 1)$ in view of (1.24) so $\phi(m \cdot n) = \phi(m) \cdot \phi(n)$. Thus under the imbedding $\phi : \mathbb{N} \longrightarrow \mathbb{Z}$ the algebraic operations on $\mathbb{N}$ are just the restriction of the algebraic operations on $\mathbb{Z}$, so $\mathbb{Z}$ is an extension of the set $\mathbb{N}$ with its algebraic operations to a ring. To see what is contained in the larger set $\mathbb{Z}$ that is not contained in the subset $\mathbb{N}$, it is evident from (1.24) that any pair $(m, n) \in \mathbb{N} \times \mathbb{N}$ is equivalent to either $(1, 1)$ or $(k + 1, 1)$ or $(1, k + 1)$ where $k \in \mathbb{N}$; the equivalence class of $(k + 1, 1)$ is in the image $\phi(\mathbb{N})$ of the natural number k, the equivalence class of $(1, 1)$ is the additive identity in $\mathbb{Z}$ but is not contained in $\phi(\mathbb{N})$, and the equivalence class of $(1, k + 1)$ is the inverse $-\phi(k)$ but is not contained in $\phi(\mathbb{N})$; so the only elements in $\mathbb{Z}$ not contained in the image $\phi(\mathbb{N})$ are precisely the elements needed to extend the natural numbers to a ring, in the sense that $\mathbb{Z} = \mathbb{N} \sqcup \{0\} \sqcup -\mathbb{N}$. Whenever $(m, n) \in \mathbb{N} \times \mathbb{N}$ then $(m, n) \asymp (m + 1 + 1, n + 1 + 1) = (m + 1, 1) + (1, n + 1) = \phi(m) - \phi(n)$, so the elements of the ring $\mathbb{Z}$ can be identified with differences of elements in the image $\phi(\mathbb{N})$ of the natural numbers in the ring $\mathbb{Z}$.

A particularly interesting class of rings are the **ordered rings**, defined as rings R with a distinguished subset $P \subset R$ such that

(i) for any element $a \in R$ either $a = 0$ or $a \in P$ or $-a \in P$, and only one of these possibilities can occur, and

(ii) if $a, b \in P$ then $a + b \in P$ and $a \cdot b \in P$.

The set P is called the set of **positive elements** of the ring R; and associated to this set P is the partial order defined by setting $a \leq b$ if either $b = a$ or $b - a \in P$. That this is indeed a partial order is clear: it is reflexive since $a \leq a$; it is transitive since if $a \leq b$ and $b \leq c$ then $b - a \in P$ and $c - b \in P$ so $c - a = (c - b) + (b - a) \in P$; and it is antisymmetric since if $a \leq b$ then either $b = a$ or $b - a \in P$ and if $b \leq a$ then either $b = a$ or $a - b \in P$, and since it cannot be the case that both $b - a \in P$ and $a - b \in P$ it must be the case that $a = b$. Actually it is a rather special partial order, a **linear order**, meaning that for any $a, b \in R$ either $b - a = 0$ or $b - a \in P$ or $a - b \in P$ and only one of these possibilities can arise; and these possibilities correspond to the relations $a = b$, $a < b$ and $b < a$ respectively. The relation of set inclusion $A \subset B$ is an example of a partial order that is not a linear order. As usual for an order $a < b$ is taken to mean that $a \leq b$ but $a \neq b$, and $b \geq a$ is equivalent to $a \leq b$ while $b > a$ is equivalent to $a < b$. Since $a = a - 0$ it is clear that $a > 0$ if and only if $a \in P$, so the positive elements of the ordered ring are precisely those elements $a \in R$ for which $a > 0$; note particularly that the additive identity 0 is not considered a positive element. It may be useful to list a few standard properties of the order in an ordered ring.

(i) If $a \in R$ and $a \neq 0$ then $a^2 > 0$; for if $a \in P$ then $a^2 \in P$, while if $a \notin P$ then $-a \in P$ and $a^2 = (-a)^2 \in P$.

(ii) The multiplicative identity is positive, that is $1 > 0$, since $1 = 1^2$.

(iii) If $a \leq b$ and $c > 0$ then $ca \leq cb$, for if $b - a \in P$ and $c \in P$ then $c \cdot (b - a) \in P$.

(iv) If $a \leq b$ and $c < 0$ then $ca \geq cb$, for if $b - a \in P$ and $c < 0$ then $-c \in P$ and $c(a - b) = (-c)(b - a) \in P$.

(v) If $a \leq b$ then $-a \geq -b$, for $(-a) - (-b) = b - a \in P$.

For the special case of the ring $\mathbb{Z}$ it is clear that the subset $\phi(\mathbb{N})$ satisfies the condition to be the set of positive elements defining the order on $\mathbb{Z}$; these elements are called the **positive integers**, noting again that with this convention 0 is not considered a positive integer.

Some interesting additional examples of rings can be derived from the ring $\mathbb{Z}$ by considering for any $n \in \mathbb{N}$ the equivalence relation defined by

$$a \equiv b \pmod{n} \quad \text{if and only if } a - b = nx \text{ for some } x \in \mathbb{Z}. \tag{1.28}$$

It is easy to see that if $a_1 \equiv a_2 \pmod{n}$ and $b_1 \equiv b_2 \pmod{n}$ then $(a_1 + b_1) \equiv (a_2 + b_2) \pmod{n}$ and $(a_1 \cdot b_1) \equiv (a_2 \cdot b_2) \pmod{n}$; hence the operations of addition and multiplication can be defined on equivalence classes, providing the natural structure of a ring on the quotient; that quotient ring is often denoted by $\mathbb{Z}/n\mathbb{Z}$. This ring is not an ordered ring; for if it were an ordered ring then $1 + 1 + \cdots + 1 > 0$ for any such sum since $1 > 0$, but the sum of n copies of 1 is 0. It is amusing and instructive to write out the detailed addition and multiplication tables for some small values of n.

To any ring R it is possible to associate another ring $R[x]$ called the **polynomial ring** over the ring R. The elements $p(x) \in R[x]$ are formal polynomials in a variable x, expressions of the form

$$p(x) = a_0 + a_1 x + a_2 x^2 + \cdots + a_n x^n \tag{1.29}$$

with coefficients $a_j \in R$; they can be written more succinctly and more conveniently in the form $p(x) = \sum_{j=0}^{n} a_j x^j$. If $a_n \neq 0$ the polynomial is said to have **degree** n, the term $a_n x^n$ is the **leading term** of the polynomial and the coefficient a_n is the **leading coefficient**. The ring R is naturally imbedded as a subset of $R[x]$ by viewing any $a \in R$ as a polynomial of degree 0. A polynomial really is just a string $(a_0, a_1, \ldots, a_n)$ of finitely many elements $a_i \in R$ in the ring, written out as in (1.29) for convenience in defining the algebraic operations. Two polynomials always can be written in the form $p(x) = \sum_{j=0}^{n} a_j x^j$ and $q(x) = \sum_{k=0}^{n} b_k x^k$ for the same integer n, since some of the coefficients a_j or b_k can be taken to be zero. The sum of these polynomials is defined by

$$p(x) + q(x) = \sum_{j=0}^{n} (a_j + b_j) x^j$$

so is just the sum of the corresponding coefficients; and their product is defined by

$$p(x) \cdot q(x) = \sum_{j,k=0}^{n} a_j b_k x^{j+k}$$

so is a collection of products of a coefficient of $p(x)$ and a coefficient of $q(x)$ grouped by powers of the variable x. It is a straightforward matter to verify that $R[x]$ is a ring with these operations, where the additive identity is the identity 0 of the ring R and the multiplicative identity is the multiplicative identity 1 of the ring R.

In rings such as $\mathbb{Z}/6\mathbb{Z}$ there are nonzero elements such as the equivalence class a of the integer 2 and the equivalence class b of the integer 3 for which $a \cdot b = 0$. Such elements cannot possibly have multiplicative inverses; for that would imply that $b = (a^{-1} \cdot a) \cdot b = a^{-1} \cdot (a \cdot b) = a^{-1} \cdot 0 = 0$, a contradiction. A ring for which $a \cdot b = 0$ implies that either $a = 0$ or $b = 0$ is called an **integral domain**. It is clear that any ordered ring R is an integral domain; for if R is an ordered ring, if $a, b \in R$ and neither of these two elements is zero, then $\pm a \in P$ and $\pm b \in P$ for some choice of the signs and consequently $\pm(a \cdot b) = (\pm a) \cdot (\pm b) \in P$ for the appropriate sign so $a \cdot b \neq 0$. In an integral domain the cancellation law holds: if $a \cdot c = b \cdot c$ where $c \neq 0$ then $0 = bc - ac = (b - a) \cdot c$ and since $c \neq 0$ it follows that $b - a = 0$ so $b = a$. Any field obviously is an integral domain; but there are integral domains, such as the ring of integers $\mathbb{Z}$, that are not fields. However any integral domain R actually can be extended to a field. To see this, consider the Cartesian product $R \times R^\times$ consisting of all pairs (a, b) where $a, b \in R$ and $b \neq 0$; and define[6] operations of addition and multiplication of elements of $R \times R^\times$ by

$$(a, b) + (c, d) = (ad + bc, bd) \quad \text{and} \quad (a, b) \cdot (c, d) = (ac, bd). \tag{1.30}$$

Introduce the equivalence relation on the set $R \times R^\times$ defined by

$$(a_1, b_1) \asymp (a_2, b_2) \quad \text{if and only if } a_1 \cdot b_2 = a_2 \cdot b_1. \tag{1.31}$$

It is easy to see that this actually is an equivalence relation. Indeed reflexivity and symmetry are trivial; and for transitivity if $(a_1, b_1) \asymp (a_2, b_2)$ and $(a_2, b_2) \asymp (a_3, b_3)$ then $a_1 b_2 = a_2 b_1$ and $a_2 b_3 = a_3 b_2$ so $a_1 b_2 b_3 = b_1 a_2 b_3 = b_1 a_3 b_2$ and consequently $(a_1 b_3 - a_3 b_1) b_2 = 0$ so $a_1 b_3 - a_3 b_1 = 0$ by the cancellation law. It is an immediate consequence of the definition of equivalence that

$$(a, b) \asymp (a \cdot c, \ b \cdot c) \quad \text{if} \quad c \neq 0, \tag{1.32}$$

a very useful observation in dealing with this equivalence relation. It is a straightforward matter to verify that addition and multiplication on $R \times R^\times$

---

[6]The idea of this construction is to introduce formally the quotients a/b of pairs of elements of the ring integers for which $b \neq 0$; and the definitions of the sum and product of two pairs are defined with this in mind.

preserve equivalence classes, in the sense that if $(a_1, b_1) \asymp (a_2, b_2)$ so that $a_1 b_2 = a_2 b_1$ then

$$((a_1, b_1) + (c, d)) \asymp ((a_2, b_2) + (c, d)) \quad \text{and} \tag{1.33}$$
$$((a_1, b_1) \cdot (c, d)) \asymp ((a_2, b_2) \cdot (c, d)).$$

Therefore these operations can be defined on the set of equivalence classes and it is readily verified that this set is a ring under these operations, with the additive identity element the equivalence class of $(0, 1)$, the multiplicative identity element the equivalence class of $(1, 1)$, and the additive inverse $-(a, b)$ the equivalence class of $(-a, b)$. The only slightly complicated case is that of the associative law of addition, where a calculation leads to

$$(a, b) + ((c, d) + (e, f)) = (adf + bcf + bde, bdf) = ((a, b) + (c, d)) + (e, f).$$

The quotient $F = (R \times R^\times)/\asymp$ of the set $R \times R^\times$ by this equivalence relation is then a well-defined ring. If $a \neq 0$ and $b \neq 0$ then $(b, a) \cdot (a, b) = (a \cdot b, a \cdot b) \asymp (1, 1)$, since the product $a \cdot b$ is nonzero in an integral domain; thus (b, a) is the multiplicative inverse of (a, b), and consequently F is a field. This field is called the **field of quotients** of the ring R. The mapping $\phi : R \longrightarrow F$ that sends an element $a \in R$ to the equivalence class $\phi(a) \in F$ of the element $(a, 1) \in R \times R^\times$ is clearly an injective mapping for which $\phi(a_1 + a_2) = \phi(a_1) + \phi(a_2)$ and $\phi(a_1 \cdot a_2) = \phi(a_1) \cdot \phi(a_2)$; in this way R can be realized as a subset of F such that the algebraic operations of R are compatible with those of F.

In particular since the ring $\mathbb{Z}$ is an ordered ring, hence an integral domain, it has a well-defined field of quotients, denoted by $\mathbb{Q}$ and called the field of **rational numbers**. Any rational number in $\mathbb{Q}$ can be represented by the equivalence class of a pair $(a, b) \in \mathbb{Z} \times \mathbb{Z}^\times$; and since $(a, b) = (a, 1) \cdot (1, b)$ then if $\phi(a)$ is the equivalence class of $(a, 1)$ and $\phi(b)^{-1}$ is the equivalence class of $(1, b)$ it follows that $\phi(a) \cdot \phi(b)^{-1}$ is the equivalence class of (a,b), so any rational can be represented by the product $a \cdot b^{-1}$ for some integers a, b, and as in (1.32) that rational also can be represented by the product $(a \cdot c)(b \cdot c)^{-1}$ for any nonzero integer c. It is customary to simplify the notation by setting $a \cdot b^{-1} = a/b$, so that rationals can be represented by quotients a/b for integers a, b. The field $\mathbb{Q}$ is an ordered field, where the set P of positive elements is defined as the set of rationals a/b where $a > 0$ and $b > 0$. To verify that, any rational other than 0 can be represented by a quotient a/b of integers; if the integers are of the same sign then after multiplying both by -1 if necessary both will be positive so $a/b \in P$; but if they are of opposite signs the quotient a/b can never be represented by a quotient of two positive integers so $a/b \notin P$. It is clear that if a/b and c/d are both in P then so are $(a/b) + (c/d)$ and $(a/b) \cdot (c/d)$, so P can serve as the positive elements in the field $\mathbb{Q}$. The natural numbers are a countably infinite set, as already noted. Since the integers can be decomposed as the union $\mathbb{Z} = \{0\} \sqcup \mathbb{N} \sqcup -\mathbb{N}$ of countable sets, where $\mathbb{N}$ is identified with the positive integers, then the ring $\mathbb{Z}$ is also a countably infinite set. The set of all pairs of integers is also countably infinite, by Cantor's Diagonalization Theorem; and since the rationals can be imbedded in the set of pairs $\mathbb{Z} \times \mathbb{Z}$ by selecting a representative pair for each equivalence class in $\mathbb{Q}$ it follows that the field $\mathbb{Q}$ is also a countably infinite set. This is yet another example of sets $\mathbb{N} \subset \mathbb{Z} \subset \mathbb{Q}$ where the inclusions are strict inclusions but nonetheless $\#(\mathbb{N}) = \#(\mathbb{Z}) = \#(\mathbb{Q}) = \aleph_0$.

The rationals $\mathbb{Q}$ form an ordered field, but one that is still incomplete for many purposes. For instance although there are rational numbers x such as $x = 1$ for which $x^2 < 2$ and rational numbers such as $x = 2$ for which $x^2 > 2$, there is no rational number x for which $x^2 = 2$, a very old observation.[7] However the rationals can be extended to a larger field $\mathbb{R}$, the field of real numbers, which does include a number x for which $x^2 = 2$, among many other delightful properties. This is a rather more difficult extension to describe than those from the natural numbers to the integers or from the integers to the rationals; indeed in a sense it is not a purely algebraic extension. Hence the discussion here will begin with an axiomatic definition of the field $\mathbb{R}$ of real numbers; the actual construction of this field as an extension of the rationals, which amounts to a verification that the field $\mathbb{R}$ described axiomatically actually exists, and the demonstration of its uniqueness, will be deferred to the Appendix to Section 2.2. To begin more generally, suppose that F is an ordered field, where the order is described by a subset $P \subset F$ of positive elements. A nonempty subset $E \subset F$ is **bounded above** by $a \in F$ if $x \leq a$ for all $x \in E$, and correspondingly it is **bounded below** by $b \in F$ if $x \geq b$ for all $x \in E$. That E is bounded above by a is indicated by writing $E \leq a$, and that E is bounded below by b is indicated by writing $E \geq b$; more generally $E_1 \leq E_2$ indicates that $x_1 \leq x_2$ for any $x_1 \in E_1$ and $x_2 \in E_2$. A subset is just said to be bounded above if it is bounded above by some element of the field F, and correspondingly for bounded below. If $E \subset F$ is bounded above, an element $a \in F$ is said to be the **least upper bound** or **supremum** of the set E if

   (i) $E \leq a$, and
   (ii) if $E \leq x$ then $x \geq a$;

---

[7] This is often attributed to the school of Pythagoras, an almost mythical mathematician who lived in the sixth century BCE in Greece; see for instance the *Princeton Companion to Mathematics*. Suppose that $a, b \in \mathbb{N}$ are any natural numbers for which $(a/b)^2 = 2$, or equivalently $a^2 = 2b^2$, where it can be assumed that not both a and b are multiples of 2. Since a^2 is a multiple of 2 then a itself must be a multiple of 2, so $a = 2c$; but then $4c^2 = (2c)^2 = a^2 = 2b^2$ so b must also be a multiple of 2, a contradiction.

the least upper bound of a set E is denoted by sup(E). Correspondingly if $E \subset F$ is bounded below, an element $b \in F$ is said to be the **greatest lower bound** or **infimum** of the set E if

(i) $E \geq b$, and
(ii) if $E \geq y$ then $y \leq b$;

the greatest lower upper bound of a set E is denoted by inf(E). It is evident that if $-E = \{ -x \in F \mid x \in E \}$ then sup$(-E) = -$inf(E).

For a general ordered field it is not necessarily the case that a set that is bounded above has a least upper bound, or that a set that is bounded below has a greatest lower bound; as noted earlier, the set of rational numbers $x \in \mathbb{Q}$ satisfying $x^2 < 2$ is bounded above but has no least upper bound. An ordered field with the property that any nonempty set that is bounded above has a least upper bound is called a **complete ordered field**. For any such field it is automatically the case that any nonempty set that is bounded below has a greatest lower bound, since sup$(-E) = -$inf(E) hence inf$(E) = -$sup$(-E)$.

The set $\mathbb{R}$ of the **real numbers** is defined to be a complete ordered field. That there exists such a field and that it is uniquely defined up to isomorphism remain to be demonstrated; so for the present just assume that $\mathbb{R}$ is a complete ordered field. There is a natural mapping $\phi : \mathbb{N} \longrightarrow \mathbb{R}$ defined inductively by $\phi(1) = 1$ and $\phi(n+1) = \phi(n) + 1$, where 1 denotes either the identity $1 \in \mathbb{N}$ or the multiplicative identity $1 \in \mathbb{R}$ as appropriate and $n + 1 = n'$ is the successor to n in $\mathbb{N}$. Since $\phi(1) = 1 > 0$ in $\mathbb{R}$ then $\phi(2) = \phi(1) + \phi(1) > 0$ in $\mathbb{R}$ as well, and by induction it follows that $\phi(k) > 0$ for every natural number $k \in \mathbb{N}$. Moreover $\phi(n+k) = \phi(n) + \phi(k)$ for all $n, k \in \mathbb{N}$, by induction on k, so $\phi(n+k) > \phi(n)$ for any $n, k \in \mathbb{N}$; consequently the mapping ϕ is injective. By yet another induction on k it follows that $\phi(n \cdot k) = \phi(n) \cdot \phi(k)$, so the mapping ϕ preserves both of the algebraic operations. The ring $\mathbb{Z}$ of integers was constructed in terms of pairs (a, b) of natural numbers. If the mapping ϕ is extended to a mapping $\phi : \mathbb{N}^2 \longrightarrow \mathbb{R}$ by setting $\phi(a, b) = \phi(a) - \phi(b)$ it is a straightforward calculation to verify that $\phi((a,b) + (c,d)) = \phi(a,b) + \phi(c,d)$ and $\phi((a,b) \cdot (c,d)) = \phi(a,b) \cdot \phi(c,d)$ in terms of the algebraic operations on $\mathbb{N}^2$ as defined in (1.22), and that $\phi(a,b) = \phi(c,d)$ whenever $(a,b) \times (c,d)$ for the equivalence relation (1.23); this extension thus determines a mapping $\phi : \mathbb{Z} \longrightarrow \mathbb{R}$ that is a homomorphism of rings. If $\phi(m) = \phi(n)$ for some $m, n \in \mathbb{Z}$ then $\phi(m+k) = \phi(n+k)$ for any integer k; there is an integer k so that $m + k$ and $n + k$ are contained in the subset $\mathbb{N} \subset \mathbb{Z}$, and since the mapping ϕ is injective on that subset it must be the case that $m + k = n + k$ hence that $m = n$. That shows that the ring homomorphism $\phi : \mathbb{Z} \longrightarrow \mathbb{R}$ is also injective. The ring $\mathbb{Z}$ is an integral domain, and the field $\mathbb{Q}$ of rational numbers is constructed in terms of pairs (a, b) of integers by the quotient field construction for any integral domain. If the mapping $\phi : \mathbb{Z} \longrightarrow \mathbb{R}$ is extended to a mapping $\phi : \mathbb{Z} \times \mathbb{Z}^\times \longrightarrow \mathbb{R}$ by setting $\phi(a, b) = \phi(a) \cdot \phi(b)^{-1}$ it is also a straightforward calculation to verify that $\phi((a, b) + (c, d)) = \phi(a, b) + \phi(c, d)$ and $\phi((a, b) \cdot (c, d)) = \phi(a, b) \cdot \phi(c, d)$ in terms of the algebraic operations on $\mathbb{Z} \times \mathbb{Z}^\times$ as defined in (1.30), and that $\phi(a, b) = \phi(c, d)$ whenever $(a, b) \asymp (c, d)$ for the equivalence relation (1.31); this extension thus determines a mapping $\phi : \mathbb{Q} \longrightarrow \mathbb{R}$ which is a homomorphism of fields, and it is readily seen to be an injective homomorphism by the analogue of the argument used to show that the ring homomorphism $\phi : \mathbb{Z} \longrightarrow \mathbb{R}$ is injective. Thus the field $\mathbb{Q}$ of rationals can be viewed as a subfield of the field $\mathbb{R}$ of real numbers, and will be so viewed subsequently; each rational number p/q can be identified with a well-defined real number.

Theorem 1.6. *For any real number $a > 0$ there is an integer n with the property that $1/n < a < n$; and for any real numbers $a < b$ there is a rational number r with the property that $a < r < b$.*

Proof: If it is not true that there is an integer n_1 such that $n_1 > a$ then $n \le a$ for all integers n so the subset $\mathbb{N} \subset \mathbb{R}$ is a nonempty set that is bounded above; and since the real numbers form a complete ordered field the set $\mathbb{N}$ must have a least upper bound b. The real number $b - 1$ then is not an upper bound for the set $\mathbb{N}$, so there is some integer $n > b - 1$; but in that case $n + 1 > b$, so that b cannot be an upper bound of the set $\mathbb{N}$, a contradiction. It follows that the assumption that there is not an integer $n_1 > a$ is false, hence there is an integer $n_1 > a$. In particular there is also an integer $n_2 > \frac{1}{a}$, and then $0 < \frac{1}{n_2} < a$. If $n \ge n_1$ and $n \ge n_2$ then $1/n < a < n$.

If $a < b$ then by what has just been shown there is an integer n_0 such that $\frac{1}{n_0} < (b - a)$. The set $S \subset \mathbb{R}$ of rational numbers $\frac{m}{n_0}$ for all $m \in \mathbb{Z}$ for which $\frac{m}{n_0} \le a$ has the upper bound a, hence it has a least upper bound $r \in \mathbb{R}$. Since $r - \frac{1}{2n_0}$ is not an upper bound of S there must be some rational number $\frac{m_0}{n_0} \in S$ for which $r - \frac{1}{2n_0} < \frac{m_0}{n_0}$, and $\frac{m_0}{n_0} \le a$ since $\frac{m_0}{n_0} \in S$. On the other hand $\frac{m_0+1}{n_0} = \frac{m_0}{n_0} + \frac{1}{n_0} \ge r + \frac{1}{2n_0} > r$ and consequently $\frac{m_0+1}{n_0} \notin S$ so $\frac{m_0+1}{n_0} > a$; and $\frac{m_0+1}{n_0} \le a + \frac{1}{n_0} < b$ so that $a < \frac{m_0+1}{n_0} < b$, which suffices for the proof.

In particular for any real number $r \in \mathbb{R}$ for which $r > 0$ there are integers $n \in \mathbb{Z}$ such that $n > r$ and $\frac{1}{n} < r$; an ordered field with this property is called an **archimedean field**. Furthermore for any two real numbers $r_1 < r_2$ there are rational numbers $\frac{p}{q}$ such that $r_1 < \frac{p}{q} < r_2$; so any real number can be approximated as closely as desired by rational numbers. On the other hand the field $\mathbb{R}$ as defined axiomatically is sufficiently complete that, for example, any positive real number has a unique positive square root. To demonstrate that, for any $r \in \mathbb{R}$ for which $r > 0$ introduce the sets of real numbers

$$X = \left\{ x \in \mathbb{R} \mid x > 0, x^2 < r \right\} \quad \text{and} \quad Y = \left\{ y \in \mathbb{R} \mid y > 0, y^2 > r \right\}.$$

These sets are clearly nonempty and $x < y$ for any $x \in X$ and $y \in Y$; so by the completeness property of the real numbers there are real numbers x_0, y_0 for which $x_0 = \sup(X)$ and $y_0 = \inf(Y)$. Any $y \in Y$ is an upper bound for X so it must be greater than the least upper bound of X, thus $x_0 \leq y$; and x_0 is a lower bound for Y so it must be less than the greatest lower bound of Y, thus $x_0 \leq y_0$. If $x_0^2 < r$ then $x_0^2 = r - \epsilon$ for some $\epsilon > 0$, and if $a = 2x_0 + 1$ and h is a real number for which $0 < h < \min(1, \epsilon/a)$ then $2x_0 h + h^2 < (2x_0 + 1)h = ah$ so

$$(x_0 + h)^2 = x_0^2 + 2x_0 h + h^2 < r - \epsilon + ah < r;$$

thus $(x_0 + h) \in X$, which is a contradiction since $x_0 = \sup(X)$, and consequently $x_0^2 \geq r$. On the other hand if $y_0^2 > r$ then $y_0^2 = r + \epsilon$ for some $\epsilon > 0$, and if $0 < h < \min(y_0, \epsilon/2y_0)$ then

$$(y_0 - h)^2 = y_0^2 - 2y_0 h + h^2 > r + \epsilon - 2y_0 h > r;$$

thus $(y_0 - h) \in Y$, which is a contradiction since $y_0 = \inf(Y)$, and consequently $y_0^2 \leq r$. Altogether then $r \leq x_0^2 \leq y_0^2 \leq r$ hence $x_0^2 = y_0^2 = r$. The uniqueness of the square root of course is clear.

Problems, Group I

1. Verify that the operations on pairs of natural numbers defined in equation (1.22) satisfy the same associative, commutative and distributive laws as do the corresponding operations on the natural numbers, and that these operations are compatible with the equivalence relation on these pairs, in the sense that equation (1.25) holds.

2. i) Show that the power set $\mathfrak{P}(A)$ of a set A is an abelian group, where the group operation is the symmetric difference $A \triangle B$ of sets.
   ii) What is the order of this group?
   iii) Write out the multiplication table for this group in the special case that $\#(A) = 2$.

3. Write out the Cayley table for the group of symmetries of an equilateral triangle (a group of order 6, the smallest nonabelian group).

4. i) Show that if G is a group and if $(ab)^n = 1$ for some elements $a, b \in G$ and some natural number n then $(ba)^n = 1$.
   ii) Show that if G is a group for which $a^2 = 1$ for any elements $a \in G$ then G is an abelian group.

5. Show that in an abelian group the mapping $a \longrightarrow a^n$ for a positive integer n is a group homomorphism. Find an example of a nonabelian group for which this mapping is not a group homomorphism.

6. In an ordered ring R the absolute value $|x|$ of an element $x \in R$ is defined by

$$|x| = \begin{cases} x & \text{if } x \geq 0, \\ -x & \text{if } x < 0. \end{cases}$$

Show that

i) $|x| \geq 0$ for all $x \in R$ and $|x| = 0$ if and only if $x = 0$;
ii) $|xy| = |x||y|$ for any $x, y \in R$;
iii) $|x + y| \leq |x| + |y|$ for any $x, y \in R$.

Problems, Group II

7. The addition and multiplication of cardinal numbers are defined by
    $\#(A) + \#(B) = \#(A \sqcup B)$ and $\#(A) \cdot \#(B) = \#(A \times B)$ for any sets A, B, finite or not.

    i) Show that $n + \aleph_0 = \aleph_0 + \aleph_0 = \aleph_0$ for any $n \in \mathbb{N}$.
    ii) Show that $\aleph_0 \cdot \aleph_0 = \aleph_0$.

8. i) Show that the ring $\mathbb{Z}/4\mathbb{Z}$ with 4 elements is not an integral domain.
    ii) Construct a field consisting of precisely 4 elements. [Suggestion: Begin with the field $\mathbb{F}_2$ of two elements, and considering the set of pairs (a_1, a_2) of elements $a_1, a_2 \in \mathbb{F}_2$ with the algebraic operations

    $$(a_1, a_2) + (b_1, b_2) = (a_1 + b_1, a_2 + b_2),$$
    $$(a_1, a_2) \cdot (b_1, b_2) = (a_1 b_1 + a_2 b_2, a_2 b_1 + a_1 b_2 + a_2 b_2).$$

    Remember that $-a = a$ in the field $\mathbb{F}_2$.] (An alternative way of describing the preceding algebraic operations is to write the pairs (a_1, a_2) as $a_1 + \epsilon a_2$ for some entity ϵ and to add and multiply these expressions as polynomials in ϵ where it is assumed that $\epsilon^2 + \epsilon + 1 = 0$; this may remind you of the standard definition of complex numbers.)

9. Show that a finite integral domain is a field.

10. Suppose that $\phi : F \longrightarrow G$ is a mapping from a field F to a field G that preserves the algebraic operations, in the sense that $\phi(a + b) = \phi(a) + \phi(b)$ and $\phi(ab) = \phi(a)\phi(b)$ for all $a, b \in F$.

i) Show that $\phi(a - b) = \phi(a) - \phi(b)$ for all $a, b \in F$.

ii) Show that $\phi(0_F) = 0_G$ where 0_F is the zero element of the field F and 0_G is the zero element in the field G.

iii) Show that either $\phi(a) = 0_G$ for all $a \in F$ or $\phi(a) \neq \phi(b)$ whenever $a, b \in F$ and $a \neq b$.

iv) Show that if $\phi(a) \neq 0$ for some element $a \in F$ then the image of ϕ is a subfield of G.

<h2>1.3   Vector Spaces</h2>

A **vector space** over a field F is defined to be a set V on which there are two operations, *addition*, which associates to $\mathbf{v}_1, \mathbf{v}_2 \in V$ an element $\mathbf{v}_1 + \mathbf{v}_2 \in V$, and *scalar multiplication*, which associates to $a \in F$ and $\mathbf{v} \in V$ an element $a\mathbf{v} \in V$, such that

(i) V is an abelian group under addition;

(ii) scalar multiplication is associative:

$$(ab)\mathbf{v} = a(b\mathbf{v}) \text{ for any } a, b \in F \text{ and } \mathbf{v} \in V;$$

(iii) the multiplicative identity $1 \in F$ is also the identity for scalar multiplication:

$$1\mathbf{v} = \mathbf{v} \text{ for all } \mathbf{v} \in V;$$

(iv) the distributive law holds for addition and scalar multiplication:

$$a(\mathbf{v}_1 + \mathbf{v}_2) = a\mathbf{v}_1 + a\mathbf{v}_2 \text{ and } (a_1 + a_2)\mathbf{v}$$
$$= a_1\mathbf{v} + a_2\mathbf{v} \text{ for all } a, a_1, a_2 \in F \text{ and } \mathbf{v}, \mathbf{v}_1, \mathbf{v}_2 \in V.$$

Note that the definition of a vector space involves both the set V and a particular field F. The elements of V are called **vectors**, and usually will be denoted by bold-faced letters; the elements of the field F are called **scalars** in this context. If $0 \in F$ is the additive identity in the field and $\mathbf{v} \in V$ is any vector in the vector space then $0\mathbf{v} = (0 + 0)\mathbf{v} = 0\mathbf{v} + 0\mathbf{v}$, so since V is a group under addition it follows that $0\mathbf{v}$ is the identity element of that group; that element is denoted by $\mathbf{0} \in V$ and is called the **zero vector**. The simplest vector space over any field F is the vector space consisting of the zero vector $\mathbf{0}$ alone; it is sometimes called the **trivial vector space** or the **zero vector space** and usually it is denoted just by $\mathbf{0}$.

A standard example of a vector space over a field F is the set F^n consisting of n-tuples of elements in the field F. The common practice, which will be followed systematically here, is to view a vector $\mathbf{v} \in F^n$ as a column vector

$$\mathbf{v} = \begin{pmatrix} x_1 \\ \vdots \\ x_n \end{pmatrix} \tag{1.34}$$

where $x_i \in F$ are the **coordinates** of the vector $\mathbf{v}$. Associated to any vector $\mathbf{v} \in F^n$ though there is also its **transposed vector** ${}^t\mathbf{v}$, the row vector

$${}^t\mathbf{v} = (x_1, \ldots, x_n); \tag{1.35}$$

this alternative version of a vector does play a significant role other than just notational convenience, for instance in matrix multiplication. Sometimes for notational convenience a vector is described merely by listing its coordinates, so as

$$\mathbf{v} = \{x_j\} = \{x_1, \ldots, x_n\}, \tag{1.36}$$

although the vector still will be viewed as a column vector. The sum of two vectors is the vector obtained by adding the coordinates of the two vectors, and the scalar product of a scalar $a \in F$ and a vector $\mathbf{v} \in V$ is the vector obtained by multiplying all the coordinates of the vector by a; thus

$$\begin{pmatrix} x_1 \\ \vdots \\ x_n \end{pmatrix} + \begin{pmatrix} y_1 \\ \vdots \\ y_n \end{pmatrix} = \begin{pmatrix} x_1 + y_1 \\ \vdots \\ x_n + y_n \end{pmatrix} \quad \text{and} \quad a \begin{pmatrix} x_1 \\ \vdots \\ x_n \end{pmatrix} = \begin{pmatrix} ax_1 \\ \vdots \\ ax_n \end{pmatrix} \quad \text{if } a \in F. \tag{1.37}$$

It is clear that F^n is a vector space over the field F with these operations.

A nonempty subset $V_0 \subset V$ for which $\mathbf{v}_1 + \mathbf{v}_2 \in V_0$ whenever $\mathbf{v}_1, \mathbf{v}_2 \in V_0$ and $a\mathbf{v} \in V_0$ whenever $a \in F$ and $\mathbf{v} \in V_0$ is called a **linear subspace** of V, or sometimes a **vector subspace** of V; clearly a linear subspace is itself a vector space over F. To a linear subspace $V_0 \subset V$ there can be associated an equivalence relation in the vector space V by setting $\mathbf{v}_1 \asymp \mathbf{v}_2 \pmod{V_0}$ whenever $\mathbf{v}_1 - \mathbf{v}_2 \in V_0$. This is clearly an equivalence relation: it is reflexive since $\mathbf{v}_1 - \mathbf{v}_1 = \mathbf{0} \in V_0$; it is symmetric since if $\mathbf{v}_1 - \mathbf{v}_2 \in V_0$ then $(\mathbf{v}_2 - \mathbf{v}_1) = -(\mathbf{v}_1 - \mathbf{v}_2) \in V_0$; and it is transitive since if $\mathbf{v}_1 - \mathbf{v}_2 \in V_0$ and $\mathbf{v}_2 - \mathbf{v}_3 \in V_0$ then $\mathbf{v}_1 - \mathbf{v}_3 = (\mathbf{v}_1 - \mathbf{v}_2) + (\mathbf{v}_2 - \mathbf{v}_3) \in V_0$. The equivalence class of a vector $\mathbf{v} \in V$ is the subset $\mathbf{v} + V_0 = \{\mathbf{v} + \mathbf{w} | \mathbf{w} \in V_0\} \subset V$. If $\mathbf{v}_1 \asymp \mathbf{v}_1'$ and $\mathbf{v}_2 \asymp \mathbf{v}_2'$ then $(\mathbf{v}_1 + \mathbf{v}_2) - (\mathbf{v}_1' + \mathbf{v}_2') = (\mathbf{v}_1 - \mathbf{v}_1') + (\mathbf{v}_2 - \mathbf{v}_2') \in V_0$ since $\mathbf{v}_1 - \mathbf{v}_1' \in V_0$ and $\mathbf{v}_2 - \mathbf{v}_2' \in V_0$, hence $\mathbf{v}_1 + \mathbf{v}_2 \asymp \mathbf{v}_1' + \mathbf{v}_2'$; and similarly $(a\mathbf{v}_1 - a\mathbf{v}_1') = a(\mathbf{v}_1 - \mathbf{v}_1') \in V_0$ so $a\mathbf{v}_1 \asymp a\mathbf{v}_1'$. Thus the algebraic operations on the vector space V extend to the set of equivalence classes, which then also has the structure of a vector space over the field F. This vector space is denoted by V/V_0 and is called the **quotient space** of V by the subspace V_0, or alternatively the vector space V **modulo** V_0. For example the subset of F^n consisting of vectors for which the first m coordinates are all zero where $0 < m < n$ is a vector subspace $V_0 \subset F^n$. Two vectors in F^n are equivalent modulo V_0 precisely when their first m coordinates are the same, so the equivalence class is described by the first m coordinates; hence the quotient space F^n/V_0 can be identified with the vector space F^m.

If $V_1, V_2 \subset V$ are two linear subspaces of a vector space V over a field F it is clear that their intersection $V_1 \cap V_2$ is a linear subspace of V. If the intersection is the zero subspace, consisting only of the zero vector $\mathbf{0}$, the two linear subspaces are said to be **linearly independent**. The **sum** of any two subspaces is the subset

$$V_1 + V_2 = \left\{ \mathbf{v}_1 + \mathbf{v}_2 \middle| \mathbf{v}_1 \in V_1, \quad \mathbf{v}_2 \in V_2 \right\} \subset V, \tag{1.38}$$

and it is easy to see that it also is a linear subspace of V. The sum of two linear subspaces is said to be a **direct sum** if the two subspaces are linearly independent, and in that case the sum is denoted by $V_1 \oplus V_2$. Any vector $\mathbf{v} \in V_1 + V_2$ can be written as the sum $\mathbf{v} = \mathbf{v}_1 + \mathbf{v}_2$ of vectors $\mathbf{v}_1 \in V_1$ and $\mathbf{v}_2 \in V_2$, by the definition of the sum of the subspaces. Any vector $\mathbf{v} \in V_1 \oplus V_2$ also can be written as the sum $\mathbf{v} = \mathbf{v}_1 + \mathbf{v}_2$ of vectors $\mathbf{v}_1 \in V_1$ and $\mathbf{v}_2 \in V$, since the direct sum is a special case of the sum of two linear subspaces; but in the case of a direct sum that decomposition is unique. Indeed if $\mathbf{v}_1 + \mathbf{v}_2 = \mathbf{w}_1 + \mathbf{w}_2$ for vectors $\mathbf{v}_1, \mathbf{w}_1 \in V_1$ and $\mathbf{v}_2, \mathbf{w}_2 \in V_2$ where $V_1 \cap V_2 = \mathbf{0}$ then $\mathbf{v}_1 - \mathbf{w}_1 = \mathbf{w}_2 - \mathbf{v}_2 \in V_1 \cap V_2 = \mathbf{0}$ hence $\mathbf{v}_1 = \mathbf{w}_1$ and $\mathbf{v}_2 = \mathbf{w}_2$. The uniqueness of this representation of a vector $\mathbf{v} \in V_1 \oplus V_2$ makes the direct sum of two linear subspaces a very useful construction.

On the other hand, to any two vector spaces V_1, V_2 over a field F there can be associated another vector space V over F that is the direct sum of subspaces V_i' isomorphic to V_i; indeed let V be the set of pairs

$$V = \left\{ (\mathbf{v}_1, \mathbf{v}_2) \middle| \mathbf{v}_1 \in V_1, \quad \mathbf{v}_2 \in V_2 \right\} \tag{1.39}$$

with the structure of a vector space over F where the sum is defined by

$$(\mathbf{v}_1', \mathbf{v}_2') + (\mathbf{v}_1'', \mathbf{v}_2'') = (\mathbf{v}_1' + \mathbf{v}_1'', \quad \mathbf{v}_2' + \mathbf{v}_2'')$$

and the scalar product is defined by

$$a(\mathbf{v}_1, \mathbf{v}_2) = (a\mathbf{v}_1, a\mathbf{v}_2).$$

It is a straightforward matter to verify that $V = V_1' \oplus V_2'$ where

$$V_1' = \left\{ (\mathbf{v}_1', \mathbf{0}) \mid \mathbf{v}_1' \in V_1 \right\} \quad \text{and} \quad V_2' = \left\{ (\mathbf{0}, \mathbf{v}_2') \mid \mathbf{v}_2' \in V_2 \right\},$$

and that V_i' is isomorphic to V_i for $i = 1, 2$. The vector space V so constructed also is called the **direct sum** of the two vector spaces V_1 and V_2, or sometimes their **exterior direct sum**, and also is indicated by writing $V = V_1 \oplus V_2$ as an abbreviation. Sometimes the direct sum of two vector spaces is also called the **product** of these vector spaces and is then denoted by $V_1 \times V_2$. It is of course possible to consider the direct sum, or product, of more than two vector spaces.

The **span** of a finite set $\mathbf{v}_1, \mathbf{v}_2, \ldots, \mathbf{v}_n$ of vectors in a vector space V over a field F is the set of vectors defined by

$$\text{span}(\mathbf{v}_1, \mathbf{v}_2, \ldots, \mathbf{v}_n) = \left\{ \sum_{j=1}^{n} a_j \mathbf{v}_j \mid a_j \in F \right\}; \tag{1.40}$$

clearly the span of any finite set of vectors in V is a linear subspace of V. By convention the span of the empty set of vectors is the trivial vector space $\mathbf{0}$ consisting of the zero vector alone. Vectors $\mathbf{v}_1, \mathbf{v}_2, \ldots, \mathbf{v}_n$ are said to be **linearly dependent** if $\sum_{j=1}^{n} a_j \mathbf{v}_j = \mathbf{0}$ for some $a_j \in F$ where $a_j \neq 0$ for at least one of the scalars a_j; and these vectors are said to be **linearly independent** if they are not linearly dependent, so if $\sum_{j=1}^{n} a_j \mathbf{v}_j = \mathbf{0}$ implies that $a_j = 0$ for all indices $1 \leq j \leq n$. It is easy to see that vectors $\mathbf{v}_1, \ldots, \mathbf{v}_n$ are linearly dependent if and only if $\text{span}(\mathbf{v}_1, \ldots, \mathbf{v}_n) = \text{span}(\mathbf{v}_1, \ldots, \mathbf{v}_{j-1}, \mathbf{v}_{j+1}, \ldots, \mathbf{v}_n)$ for some index $1 \leq j \leq n$. Indeed if $\text{span}(\mathbf{v}_1, \ldots, \mathbf{v}_n) = \text{span}(\mathbf{v}_1, \ldots, \mathbf{v}_{j-1}, \mathbf{v}_{j+1}, \ldots, \mathbf{v}_n)$ then $\mathbf{v}_j \in \text{span}(\mathbf{v}_1, \ldots, \mathbf{v}_n) = \text{span}(\mathbf{v}_1, \ldots, \mathbf{v}_{j-1}, \mathbf{v}_{j+1}, \ldots, \mathbf{v}_n)$ so $\mathbf{v}_j = \sum_{1 \leq i \leq n, i \neq j} a_i \mathbf{v}_i$, and that shows that the vectors $\mathbf{v}_1, \ldots, \mathbf{v}_n$ are linearly dependent; conversely if the vectors $\mathbf{v}_1, \ldots, \mathbf{v}_n$ are linearly dependent then $\sum_{i=1}^{n} a_i \mathbf{v}_i = \mathbf{0}$ for some scalars a_i, not all of which are zero; so if $a_j \neq 0$ then $\mathbf{v}_j = -\sum_{1 \leq i \leq n, i \neq j} a_i a_j^{-1} \mathbf{v}_i$ so $\text{span}(\mathbf{v}_1, \ldots, \mathbf{v}_n) = \text{span}(\mathbf{v}_1, \ldots, \mathbf{v}_{j-1}, \mathbf{v}_{j+1}, \ldots, \mathbf{v}_n)$.

Theorem 1.7 (Basis Theorem). *If m vectors $\mathbf{w}_1, \ldots, \mathbf{w}_m$ in a vector space V are contained in the span of n vectors $\mathbf{v}_1, \ldots, \mathbf{v}_n$ in V where $m > n$ then the vectors $\mathbf{w}_1, \ldots, \mathbf{w}_m$ are linearly dependent.*

Proof: Of course if $\mathbf{w}_i = \mathbf{0}$ for some index i then the vectors $\mathbf{w}_1, \ldots, \mathbf{w}_m$ are linearly dependent; so it can be assumed that $\mathbf{w}_i \neq \mathbf{0}$ for all indices i. Since $\mathbf{w}_1 \in \text{span}(\mathbf{v}_1, \ldots, \mathbf{v}_n)$ it follows that $\mathbf{w}_1 = \sum_{j=1}^{n} a_j \mathbf{v}_j$ for some scalars a_j; not all of the scalars a_j can vanish, so by relabeling the vectors $\mathbf{v}_j$ it can be assumed that $a_1 \neq 0$ hence that $\mathbf{w}_1 = a_1 \mathbf{v}_1 + \sum_{j=2}^{n} a_j \mathbf{v}_j$ or equivalently that $\mathbf{v}_1 = a_1^{-1} \mathbf{w}_1 - \sum_{j=2}^{n} a_1^{-1} a_j \mathbf{v}_j$, and it is clear from this that $\text{span}(\mathbf{v}_1, \ldots, \mathbf{v}_n) = \text{span}(\mathbf{w}_1, \mathbf{v}_2, \ldots, \mathbf{v}_n)$. Then since $\mathbf{w}_2 \in \text{span}(\mathbf{v}_1, \ldots, \mathbf{v}_n) = \text{span}(\mathbf{w}_1, \mathbf{v}_2, \ldots, \mathbf{v}_n)$ it follows that $\mathbf{w}_2 = b_1 \mathbf{w}_1 + \sum_{j=2}^{n} b_j \mathbf{v}_j$ for some scalars b_j, not all of which vanish. If $b_1 \neq 0$ but $b_j = 0$ for $2 \leq j \leq n$ this equation already shows that the vectors $\mathbf{w}_1, \mathbf{w}_2$ are linearly dependent, and hence of course so are the vectors $\mathbf{w}_1, \cdots, \mathbf{w}_n$. Otherwise $b_j \neq 0$ for some $j > 1$, and by relabeling the vectors $\mathbf{v}_j$ it can be assumed that $b_2 \neq 0$; and by the same argument as before it follows that $\operatorname{span}(\mathbf{v}_1, \ldots, \mathbf{v}_n) = \operatorname{span}(\mathbf{w}_1, \mathbf{w}_2, \mathbf{v}_3, \ldots, \mathbf{v}_n)$. The argument can be continued, so eventually either the set of vectors $(\mathbf{w}_1, \ldots, \mathbf{w}_n)$ is linearly dependent or $\operatorname{span}(\mathbf{w}_1, \ldots, \mathbf{w}_n) = \operatorname{span}(\mathbf{v}_1, \ldots, \mathbf{v}_n)$. Since $\mathbf{w}_{n+1} \in \operatorname{span}(\mathbf{v}_1, \ldots, \mathbf{v}_n) = \operatorname{span}(\mathbf{w}_1, \ldots, \mathbf{w}_n)$ it follows that $\mathbf{w}_{n+1} = \sum_{j=1}^{n} c_j \mathbf{w}_j$ and consequently the vectors $\mathbf{w}_1, \ldots, \mathbf{w}_{n+1}$ are linearly dependent and therefore so are the vectors $\mathbf{w}_1, \ldots, \mathbf{w}_m$ and that suffices for the proof.

A vector space V is **finite dimensional** if it is the span of finitely many vectors, that is, if $V = \operatorname{span}(\mathbf{v}_1, \ldots, \mathbf{v}_n)$ for some vectors $\mathbf{v}_1, \ldots, \mathbf{v}_n \in V$. If these vectors are not linearly independent then as already observed the span is unchanged by deleting at least one of these vectors; and that argument can be repeated until all the remaining vectors are linearly independent. Therefore any finite dimensional vector space can be written as the span of a finite set of linearly independent vectors, called a **basis** for the vector space. It is an immediate consequence of the Basis Theorem that the number of vectors in a basis is the same for all bases. Indeed if $\mathbf{w}_1, \ldots, \mathbf{w}_m$ and $\mathbf{v}_1, \ldots, \mathbf{v}_n$ are two bases for a vector space V then since $\mathbf{w}_1, \ldots, \mathbf{w}_m \in \operatorname{span}(\mathbf{v}_1, \ldots, \mathbf{v}_n)$ and since the vectors $\mathbf{w}_1, \ldots, \mathbf{w}_m$ are linearly independent the Basis Theorem shows that $m \leq n$; reversing the roles of these two bases and applying the same argument shows that $n \leq m$. The number of vectors in a basis for a vector space V is called the **dimension** of that vector space; that V has dimension n is indicated by writing $\dim V = n$.

Theorem 1.8. *If V is a finite dimensional vector space over a field F and $W \subset V$ is a subspace then W and the quotient space V/W are both finite dimensional vector spaces and*

$$\dim V = \dim W + \dim V/W. \tag{1.41}$$

Proof: If $W \subset V$ and if $\mathbf{w}_1, \ldots, \mathbf{w}_m$ is a basis for W then there is a basis for V of the form $\mathbf{w}_1, \ldots, \mathbf{w}_m, \mathbf{v}_1, \ldots, \mathbf{v}_n$ for some vectors $\mathbf{v}_1, \ldots, \mathbf{v}_n \in V$. Indeed if $W \subset V$ but $W \neq V$ then there is at least one vector $\mathbf{v}_1 \in V \sim W$, and the vectors $\mathbf{w}_1, \ldots, \mathbf{w}_m, \mathbf{v}_1$ are linearly independent; if these vectors do not span V then the argument can be repeated, providing a vector $\mathbf{v}_2$ such that the vectors $\mathbf{w}_1, \ldots, \mathbf{w}_m, \mathbf{v}_1, \mathbf{v}_2$ are linearly independent, and the argument can be continued. The Basis Theorem shows that this process finally stops; for since V is finite dimensional it has a basis of $m + n$ vectors for some n so there can be no more than $m + n$ linearly independent vectors in V. This argument starting from the empty set also gives a basis for any subspace of V. Vectors in the quotient space V/W are equivalence classes $\mathbf{v} + W$ for vectors $\mathbf{v} \in V$, and since $\mathbf{w}_i + W = W$ it follows that the equivalence classes $\mathbf{v}_i + W$ for $1 \leq i \leq n$ span V/W. If $\sum_{i=1}^n a_i(\mathbf{v}_i + W) = W$ for some $a_i \in F$ then $\sum_{i=1}^n a_i\mathbf{v}_i \in W$ so that $\sum_{i=1}^n a_i\mathbf{v}_i = \sum_{j=1}^m b_j\mathbf{w}_j$, which can only be the case if $a_i = b_j = 0$ for all i, j, since the vectors $\mathbf{w}_j, \mathbf{v}_i$ are linearly independent; that means that the equivalence classes $\mathbf{v}_i + W$ for $1 \leq i \leq n$ are linearly independent vectors in V/W. These vectors thus are a basis for the vector space so $\dim V/W = n = \dim V - \dim W$ since $\dim V = m + n$ and $\dim W = m$; and that demonstrates (1.41). It follows immediately from this that the vector spaces W and V/W are finite dimensional, which concludes the proof.

For example, the vector space F^n has a canonical basis that can be described in terms of the **Kronecker symbol**

$$\delta_j^i = \begin{cases} 1 & \text{if } i = j, \\ 0 & \text{if } i \neq j, \end{cases} \tag{1.42}$$

and the associated **Kronecker vector**

$$\delta_j = \begin{pmatrix} \delta_j^1 \\ \delta_j^2 \\ \vdots \\ \delta_j^n \end{pmatrix} \in F^n, \tag{1.43}$$

the column vector for which all coordinates are 0 except for the j-th coordinate which is 1. It is clear that the n vectors $\delta_1, \delta_2, \ldots, \delta_n$ are linearly independent and span the vector space F^n hence are a basis for that vector space F^n; indeed any vector (1.34) can be written uniquely in terms of this basis as

$$\mathbf{v} = \sum_{j=1}^n x_j\delta_j. \tag{1.44}$$

A **homomorphism** or **linear transformation** between vector spaces V and W over the same field F is defined to be a mapping $T : V \longrightarrow W$ such that

   (i) $T(\mathbf{v}_1 + \mathbf{v}_2) = T(\mathbf{v}_1) + T(\mathbf{v}_2)$ for all $\mathbf{v}_1, \mathbf{v}_2 \in V$, and
  (ii) $T(a\mathbf{v}) = aT(\mathbf{v})$ for all $a \in F$ and $\mathbf{v} \in V$.

A linear transformation $T : V \longrightarrow W$ is called an **injective linear transformation** if it is an injective mapping when viewed as a mapping between these two sets; and it is called a **surjective linear transformation** if it is a surjective mapping when viewed as a mapping between these two sets. A linear transformation $T : V \longrightarrow W$ that is both injective and surjective is called an **isomorphism** between the two vector spaces. A particularly simple example of an isomorphism is the **identity** linear transformation $I : V \longrightarrow V$ from a vector space to itself, the mapping defined by $I(\mathbf{v}) = \mathbf{v}$ for every vector $\mathbf{v} \in V$. That two vector spaces V and W are isomorphic is denoted by $V \cong W$, which indicates that there is some isomorphism between the two vector spaces but does not specify the isomorphism. It is clear that this is an equivalence relation between vector spaces over the field F.

The **kernel** or **null space** of a linear transformation $T : V \longrightarrow W$ is the subset

$$\ker(T) = \left\{ \mathbf{v} \in V \middle| T(\mathbf{v}) = \mathbf{0} \right\} \subset V; \tag{1.45}$$

it is clear that $\ker(T) \subset V$ is a linear subspace of V, for if $\mathbf{v}_1, \mathbf{v}_2 \in \ker(T)$ then $T(\mathbf{v}_1 + \mathbf{v}_2) = T(\mathbf{v}_1) + T(\mathbf{v}_2) = \mathbf{0} + \mathbf{0} = \mathbf{0}$ so $\mathbf{v}_1 + \mathbf{v}_2 \in \ker(T)$ and $T(a\mathbf{v}_1) = aT(\mathbf{v}_1) = a\mathbf{0} = \mathbf{0}$ so $a\mathbf{v}_1 \in \ker(T)$.

Lemma 1.9. *A linear transformation $T : V \longrightarrow W$ is injective if and only if* $\ker(T) = \mathbf{0}$.

Proof: If $\ker(T) = \mathbf{0}$ and if $T(\mathbf{v}_1) = T(\mathbf{v}_2)$ for two vectors $\mathbf{v}_1, \mathbf{v}_2 \in V$ then $T(\mathbf{v}_1 - \mathbf{v}_2) = \mathbf{0}$ so $(\mathbf{v}_1 - \mathbf{v}_2) \in \ker(T) = \mathbf{0}$ hence $\mathbf{v}_1 = \mathbf{v}_2$, so T is an injective mapping. Conversely if a linear transformation $T : V \longrightarrow W$ is an injective mapping then $T^{-1}(\mathbf{0})$ is a single point of V hence $\ker(T) = \mathbf{0}$.

The **image** of a linear transformation $T : V \longrightarrow W$ is the subset

$$T(V) = \left\{ T(\mathbf{v}) \middle| \mathbf{v} \in V \right\} \subset W, \tag{1.46}$$

the image of the mapping T in the usual sense. It is clear that $T(V) \subset W$ is a linear subspace of W; for if $\mathbf{w}_1, \mathbf{w}_2 \in T(V)$ then $\mathbf{w}_1 = T(\mathbf{v}_1)$ and $\mathbf{w}_2 = T(\mathbf{v}_2)$ for some vectors $\mathbf{v}_1, \mathbf{v}_2 \in V$ so $\mathbf{w}_1 + \mathbf{w}_2 = T(\mathbf{v}_1) + T(\mathbf{v}_2) = T(\mathbf{v}_1 + \mathbf{v}_2) \in T(V)$ and $a\mathbf{w}_1 = aT(\mathbf{v}_1) = T(a\mathbf{v}_1) \in T(V)$ for any $a \in F$. Clearly a linear transformation $T : V \longrightarrow W$ is surjective precisely when its image is the full vector space W.

Theorem 1.10. *If $T : V \longrightarrow W$ is a linear transformation between two finite dimensional vector spaces over a field F then T induces an isomorphism*

$$T^* : V/\ker(T) \xrightarrow{\cong} T(V) \quad \text{where} \quad T(V) \subset W, \tag{1.47}$$

so that

$$\dim V = \dim \ker(T) + \dim T(V). \tag{1.48}$$

In particular T itself is an isomorphism if and only if $\ker(T) = \mathbf{0}$ *and* $T(V) = W$.

Proof: If $V_0 = \ker(T)$ then V_0 is a linear subspace of V. The elements of the quotient vector space V/V_0 are the subsets $\mathbf{v} + V_0 \subset V$; and since $T(V_0) = \mathbf{0}$ the linear transformation T maps each of these subsets to the same element $T(\mathbf{v} + V_0) = T(\mathbf{v}) \in W$ and in that way T determines a mapping $T^* : V/V_0 \longrightarrow W$ which is easily seen to be a linear transformation having as its image the linear subspace $T(V) \subset W$. An element $\mathbf{v} + V_0 \in V/V_0$ is in the kernel of the linear transformation T^* if and only if $T(\mathbf{v} + V_0) = T(\mathbf{v}) = \mathbf{0}$, which is just the condition that $\mathbf{v} \in V_0$ hence that $\mathbf{v} + V_0 = V_0$ is the zero vector in V/V_0. That shows that the linear transformation (1.47) is an isomorphism, and (1.48) follows in view of Theorem 1.8 while it is clear that T itself is an isomorphism if and only if $\ker(T) = \mathbf{0}$ and $T(V) = W$, which suffices for the proof.

The explicit descriptions of linear transformations and the determination of their properties for the special case of mappings between finite dimensional vector spaces rest on their effect on bases of the vector spaces, as in the following observations.

Theorem 1.11 (Mapping Theorem). *Let $\mathbf{v}_1, \ldots, \mathbf{v}_n$ be a basis for a vector space V and $\mathbf{w}_1, \ldots, \mathbf{w}_m$ be a basis for a vector space W over a field F.*
(i) Any linear transformation $T : V \longrightarrow W$ is determined fully by the images $T(\mathbf{v}_i)$ of the basis vectors $\mathbf{v}_i$; and a linear transformation $T : V \longrightarrow W$ can be defined by specifying as the images $T(\mathbf{v}_j)$ any arbitrary n vectors in W.
(ii) The linear transformation defined by the values $T(\mathbf{v}_i) \in W$ is injective if and only if the vectors $T(\mathbf{v}_i)$ are linearly independent.
(iii) The linear transformation defined by the values $T(\mathbf{v}_i) \in W$ is surjective if and only if the vectors $T(\mathbf{v}_i)$ span W.
(iv) Two vector spaces over a field F are isomorphic if and only if they have the same dimension.

Proof: (i) If $T : V \longrightarrow W$ is a linear transformation then

$$T\left(\sum_{j=1}^{n} a_j \mathbf{v}_j\right) = \sum_{j=1}^{n} a_j T(\mathbf{v}_j), \tag{1.49}$$

hence the linear transformation T is determined fully by the images $T(\mathbf{v}_i)$; and for any specified vectors $T(\mathbf{v}_i) \in W$ the linear transformation defined by (1.49) is a linear transformation taking these values.
(ii) The linear transformation T determined by the images $T(\mathbf{v}_i)$ fails to be injective if and only if there are scalars $a_j \in F$, not all of which are 0, such that

$$\mathbf{0} = T\left(\sum_{j=1}^{n} a_j \mathbf{v}_j\right) = \sum_{j=1}^{n} a_j T(\mathbf{v}_j);$$

and that is precisely the condition that the vectors $T(\mathbf{v}_j)$ are linearly dependent.

(iii) Since the linear transformation T determined by the images $T(\mathbf{v}_i)$ is given by (1.49) it is evident that the linear transformation is surjective if and only if the vectors $T(\mathbf{v}_j)$ span W.

(iv) If $T : V \longrightarrow W$ is an isomorphism it must be both injective and surjective, and it then follows from the preceding parts of this theorem that the vectors $T(\mathbf{v}_i)$ are a basis for W and consequently dim W = dim V. On the other hand if dim V = dim $W = n$ and $\mathbf{v}_1, \ldots, \mathbf{v}_n$ is a basis for V while $\mathbf{w}_1, \ldots, \mathbf{w}_n$ is a basis for W then by the preceding part of this theorem there is a linear transformation $T : V \longrightarrow W$ sending each $\mathbf{v}_i$ to $\mathbf{w}_i$, and it is both injective and surjective so is an isomorphism of vector spaces. That suffices for the proof.

The set of all linear transformations from a vector space V to a vector space W is denoted by $\mathcal{L}(V, W)$; of course this is defined only for two vector spaces over the same field F. The sum of two linear transformations $S, T \in \mathcal{L}(V, W)$ is the mapping defined by $(S + T)(\mathbf{x}) = S(\mathbf{x}) + T(\mathbf{x})$ for any vector $\mathbf{x} \in V$, and is clearly itself a linear transformation $(S + T) \in \mathcal{L}(V, W)$; and the scalar product of an element $c \in F$ and a linear transformation $T \in \mathcal{L}(V, W)$ is the mapping defined by $(cT)(\mathbf{x}) = cT(\mathbf{x})$ for any vector $\mathbf{x} \in V$, and is clearly also a linear transformation. The distributive law obviously holds for addition and scalar multiplications, so the set $\mathcal{L}(V, W)$ is another vector space over the field F. Linear transformations are mappings from one space to another, so as in the case of any mappings the **composition** ST of two mappings S and T is the mapping defined by $(ST)(\mathbf{v}) = S(T(\mathbf{v}))$; the composition also is a linear transformation since $(ST)(\mathbf{v}_1 + \mathbf{v}_2) = S(T(\mathbf{v}_1 + \mathbf{v}_2)) = S(T(\mathbf{v}_1) + T(\mathbf{v}_2)) = S(T(\mathbf{v}_1)) + S(T(\mathbf{v}_2)) = ST(\mathbf{v}_1) + ST(\mathbf{v}_2)$ for any vectors $\mathbf{v}_1, \mathbf{v}_2 \in V$ and $(ST)(a\mathbf{v}) = S(T(a\mathbf{v})) = S(aT(\mathbf{v})) = aS(T(\mathbf{v})) = aST(\mathbf{v})$ for any vectors $\mathbf{v} \in V$ and any scalar $a \in F$.

Theorem 1.12. *A linear transformation $T : V \longrightarrow W$ between vector spaces V and W over a field F is an isomorphism if and only if there is a linear transformation $S : W \longrightarrow V$ such that $ST : V \longrightarrow V$ and $TS : W \longrightarrow W$ are the identity linear transformations, $ST = TS = I$.*

Proof: If $T : V \longrightarrow W$ is an isomorphism of vector spaces then T is surjective so for any vector $\mathbf{w} \in W$ there will be a vector $\mathbf{v} \in V$ such that $T(\mathbf{v}) = \mathbf{w}$; and T is injective so the vector $\mathbf{v}$ is uniquely determined and consequently can be viewed as a function $\mathbf{v} = S(\mathbf{w})$ of the vector $\mathbf{w} \in W$, thus defining a mapping $S : W \longrightarrow V$ for which $TS(\mathbf{w}) = T(\mathbf{v}) = \mathbf{w}$ so that $TS = I$. For any vector $\mathbf{v} \in V$ since $\mathbf{w} = T(\mathbf{v}) \in W$ it follows that $T(\mathbf{v}) = \mathbf{w} = TS(\mathbf{w}) = TST(\mathbf{v})$, so since T is injective necessarily $\mathbf{v} = ST(\mathbf{v})$ and consequently $ST = I$. The mapping S is a linear transformation; for if $\mathbf{w}_1 = T(\mathbf{v}_1)$ and $\mathbf{w}_2 = T(\mathbf{v}_2)$ then by linearity

$\mathbf{w}_1 + \mathbf{w}_2 = T(\mathbf{v}_1 + \mathbf{v}_2)$ hence $S(\mathbf{w}_1 + \mathbf{w}_2) = \mathbf{v}_1 + \mathbf{v}_2 = S(\mathbf{w}_1) + S(\mathbf{w}_2)$ and if $T(\mathbf{v}) = \mathbf{w}$ then also by linearity $T(a\mathbf{v}) = aT(\mathbf{v}) = a\mathbf{w}$ so $S(a\mathbf{w}) = aS(\mathbf{w})$ for any scalar $a \in F$.

Conversely suppose that there is a linear transformation S such that $ST = I$ and $TS = I$. For any vector $\mathbf{w} \in W$ it follows that $\mathbf{w} = TS(\mathbf{w})$ so that $\mathbf{w}$ is in the image of T hence T is surjective. If $T(\mathbf{v}) = \mathbf{0}$ for a vector $\mathbf{v} \in V$ then $\mathbf{v} = ST(\mathbf{v}) = S(\mathbf{0}) = \mathbf{0}$ so that the kernel of T is the zero vector hence T is injective, which suffices for the proof.

Theorem 1.13. *Any n-dimensional vector space V over the field F is isomorphic to the vector space F^n.*

Proof: If $\mathbf{v}_1, \dots, \mathbf{v}_n$ is a basis for V then by the Mapping Theorem there is a linear transformation $T : F^n \longrightarrow V$ defined by $T(\delta_j) = \mathbf{v}_j$. By the Mapping Theorem again the linear transformation T is injective, since the image vectors $T(\delta_j) = \mathbf{v}_j$ are linearly independent, and surjective, since the image vectors $T(\delta_j) = \mathbf{v}_j$ span the vector space V; hence the linear transformation is an isomorphism between the vector spaces F^n and V.

Thus the vector space F^n for a field F is not only a simple example of an n-dimensional vector space over the field F but is also a standard model for any such vector space. The choice of an isomorphism $T : V \longrightarrow F^n$ for a vector space V over a field F can be viewed as a **coordinate system** in V, for it permits any vector $\mathbf{v} \in V$ to be described by its image $T(\mathbf{v}) = \mathbf{x} = \{x_i\} \in F^n$, so to be described by the coordinates $x_i \in F$. A coordinate system for a vector space V is determined by the choice of a basis for V; and since there are a great many bases there are also a great many different coordinate systems for V. It is clear though that if $S, T : V \longrightarrow F^n$ are any two coordinate systems for a vector space V then $T = US$ where $U = TS^{-1} : F^n \longrightarrow F^n$ is an isomorphism of the vector space F^n; and conversely if $S : V \longrightarrow F^n$ is a coordinate system for the vector space V and $U : F^n \longrightarrow F^n$ is an isomorphism of the vector space F^n, then $T = US : V \longrightarrow F^n$ also is a coordinate system for the vector space V. Actually for many purposes it is sufficient just to consider the vector spaces F^n rather than abstract vector spaces over the field F, since some general properties of finite dimensional vector spaces can be proved by a simple calculation for the vector spaces F^n.

A linear transformation $T : F^m \longrightarrow F^n$ can be described quite explicitly in a way that is very convenient for calculation. The image of a basis vector $\delta_j \in F^m$ is a linear combination of the basis vectors $\delta_i \in F^n$, so

$$T(\delta_j) = \sum_{i=1}^{n} a_{ij}\delta_i \qquad (1.50)$$

for some scalars $a_{ij} \in F$; hence by linearity the image of an arbitrary vector $\mathbf{v} = \sum_{j=1}^{m} x_j \delta_j \in F^m$ with the coordinates $\{x_j\}$ is the vector

$$\mathbf{w} = T(\mathbf{v}) = \sum_{j=1}^{m} x_j T(\delta_j) = \sum_{j=1}^{m} \sum_{i=1}^{n} x_j a_{ij} \delta_i = \sum_{i=1}^{n} y_i \delta_i \tag{1.51}$$

with the coordinates $\{y_i\}$ given by

$$y_i = \sum_{j=1}^{m} a_{ij} x_j. \tag{1.52}$$

Thus the linear transformation T is described completely by the set of scalars $a_{ij} \in F$. It is customary and convenient to list these scalars in an array of the form

$$A = \begin{pmatrix} a_{11} & a_{12} & a_{13} & \cdots & a_{1m} \\ a_{21} & a_{22} & a_{23} & \cdots & a_{2m} \\ a_{31} & a_{32} & a_{33} & \cdots & a_{3m} \\ & & \cdots & & \\ a_{n1} & a_{n2} & a_{n3} & \cdots & a_{nm} \end{pmatrix} \tag{1.53}$$

called an $n \times m$ **matrix** over the field F; an $n \times m$ matrix A with entries in the field F thus describes a linear trasformation $T : F^m \longrightarrow F^n$, and that convention should be kept firmly in mind. The horizontal lines are called the **rows** of the matrix and the vertical lines are called the **columns** of the matrix, so an $n \times m$ matrix can be viewed as a collection of n row vectors or alternatively as a collection of m column vectors. The scalars a_{ij} are called the **entries** of the matrix, so $A = \{a_{ij}\}$ where a_{ij} is the entry in row i and column j. The linear relation between the coordinates $\{x_j\}$ of a vector in F^m and the coordinates $\{y_i\}$ of the image vector in F^n is (1.52). The preceding conventions and terminology are rather rigidly followed so should be kept in mind and used carefully and systematically.

Associating to a linear transformation $T : F^m \longrightarrow F^n$ the $n \times m$ matrix describing that linear transformation identifies the vector space $\mathcal{L}(F^m, F^n)$ of all linear transformations with the vector space of all $n \times m$ matrices over the field F, since that association clearly preserves addition of vectors and scalar multiplication; thus there is the natural isomorphism $\mathcal{L}(F^m, F^n) \cong F^{nm}$, since an $n \times m$ matrix is merely a collection of nm scalars, and consequently

$$\dim \mathcal{L}(F^m, F^n) = nm. \tag{1.54}$$

When the vector space F^{nm} is identified in this way with the vector space $\mathcal{L}(F^m, F^n)$ it is often denoted by $F^{n \times m}$; so this stands for the vector space of matrices over the field F having n rows and m columns. The zero vector in $F^{n \times m}$ is described by the **zero matrix 0**, the matrix having all entries 0; this represents the **trivial** linear transformation that maps all vectors in F^m to the zero vector in F^n. The matrix $I \in F^{n \times n}$ describing the identity linear transformation $F^n \longrightarrow F^n$, the linear transformation that takes the basis vectors δ_i to themselves, clearly has the entries $I = \{\delta_j^i\}$, the Kronecker symbols; it is called the **identity matrix**. It is often convenient to specify the size of an identity matrix, so to denote the $n \times n$ identity matrix by I_n; thus for instance

$$
I_4 = \begin{pmatrix} 1 & 0 & 0 & 0 \\ 0 & 1 & 0 & 0 \\ 0 & 0 & 1 & 0 \\ 0 & 0 & 0 & 1 \end{pmatrix}.
\tag{1.55}
$$

The equation (1.52) describing the relations between the coordinate of vectors $\mathbf{v} = \{x_j\}$ and $\mathbf{w} = \{y_i\}$ related by the linear transformation $\mathbf{w} = T\mathbf{v}$ described by a matrix A can be viewed as a collection of linear equations relating the variables x_i and y_j. This system of equations often is viewed as a matrix product, expressing the column vector or $n \times 1$ matrix $\mathbf{y} = \{y_i\}$ as the product of the $n \times m$ matrix A and the column vector or $m \times 1$ matrix $\mathbf{x} = \{x_j\}$ in the form

$$
\begin{pmatrix} y_1 \\ y_2 \\ y_3 \\ \cdots \\ y_n \end{pmatrix} = \begin{pmatrix} a_{11} & a_{12} & a_{13} & \cdots & a_{1m} \\ a_{21} & a_{22} & a_{23} & \cdots & a_{2m} \\ a_{31} & a_{32} & a_{33} & \cdots & a_{3m} \\ & & \cdots & & \\ a_{n1} & a_{n2} & a_{n3} & \cdots & a_{nm} \end{pmatrix} \begin{pmatrix} x_1 \\ x_2 \\ x_3 \\ \cdots \\ x_m \end{pmatrix}
\tag{1.56}
$$

where the entry y_i in row i of the product is the sum over the index j of the products of the entries a_{ij} in row i of the matrix A and the successive entries x_j of the column vector $\mathbf{x}$, so

$$
y_i = a_{i1}x_1 + a_{i2}x_2 + a_{i3}x_3 + \cdots + a_{im}x_m.
$$

More generally the **product** of an $n \times m$ matrix $A = \{a_{ij}\}$ and an $m \times l$ matrix $B = \{b_{ij}\}$ is defined to be the $n \times l$ matrix $AB = C = \{c_{ij}\}$ with entries

$$
c_{ij} = \sum_{k=1}^{m} a_{ik}b_{kj} \quad \text{for } 1 \le i \le n, \ 1 \le j \le l;
\tag{1.57}
$$

so column j of the product matrix $AB = C$ is the column vector $\mathbf{c}_j = A\,\mathbf{b}_j$ where $\mathbf{b}_j$ is column j of the matrix B, and row j of the product matrix $AB = C$ is the row vector ${}^t\mathbf{c}_j = {}^t\mathbf{a}_j B$ where ${}^t\mathbf{a}_j$ is row j of the matrix A. Matrices A and B can be multiplied to yield a product only when the number of columns of the matrix A is equal to the number of rows of the matrix B; and the number of rows of the product AB is the number of rows of A while the number of columns of the product AB is the number of columns of the matrix B. For example in the product

$$\begin{pmatrix} a_{11} & a_{12} & a_{13} \\ a_{21} & a_{22} & a_{23} \end{pmatrix} \begin{pmatrix} b_{11} & b_{12} \\ b_{21} & b_{22} \\ b_{31} & b_{32} \end{pmatrix} = \begin{pmatrix} c_{11} & c_{12} \\ c_{21} & c_{22} \end{pmatrix}$$

the first column of the product $AB = C$ is the product of the matrix A and the first column of the matrix B; explicitly

$$c_{11} = a_{11}b_{11} + a_{12}b_{21} + a_{13}b_{31}$$

$$c_{21} = a_{21}b_{11} + a_{22}b_{21} + a_{23}b_{31}.$$

It is easy to see from the definition that the matrix product is associative, in the sense that $A(BC) = (AB)C$ and consequently that multiple products such as ABC are well defined, and that it is distributive, in the sense that $A(B + C) = AB + BC$ and $(A + B)C = AC + BC$. It is worth calculating a few matrix products just to become familiar with the technique.

Theorem 1.14. *If a linear transformation $T : F^n \longrightarrow F^m$ is described by an $m \times n$ matrix B and a linear transformation $S : F^m \longrightarrow F^l$ is described by $l \times m$ matrix A then the composition $S \circ T : F^n \longrightarrow F^l$ is described by the product matrix AB.*

Proof: For any index $1 \le i \le n$

$$(S \circ T)(\delta_i) = S(T(\delta_i)) = S\left(\sum_{k=1}^{m} b_{ki}\delta_k\right) = \sum_{k=1}^{m} b_{ki}S(\delta_k)$$

$$= \sum_{k=1}^{m} b_{ki} \sum_{j=1}^{l} a_{jk}\delta_j = \sum_{j=1}^{l} \sum_{k=1}^{m} a_{jk}b_{ki}\delta_j = \sum_{j=1}^{l} c_{ji}\delta_j;$$

thus the composition $S \circ T$ is described by the matrix $C = \{c_{ji}\}$ where $c_{ji} = \sum_{k=1}^{m} a_{jk}b_{ki}$ so that $C = AB$, which suffices for the proof.

In view of the preceding theorem, the linear transformation described by a matrix A is usually also denoted by A, and AB denotes both the matrix product and the composition $A \circ B$ of the linear transformations described by these matrices. This convention will be followed henceforth. It follows from Theorem 1.12 that the linear transformation described by a matrix A is an isomorphism if and only if that linear transformation has an inverse linear transformation, or in matrix terms, if and only if there is a matrix A^{-1} such that $AA^{-1} = A^{-1}A = I$, the identity matrix; such a matrix A is called an **invertible**

matrix or a **nonsingular matrix**. A matrix that is not nonsingular of course is called a **singular** matrix. Not all matrices in $F^{n \times n}$ have inverses; the zero matrix of course has no inverse, nor does the matrix $A = \begin{pmatrix} 0 & 1 \\ 0 & 0 \end{pmatrix}$ since $AA = 0$, so these are examples of **singular** matrices. On the other hand the identity matrix is nonsingular, since it is its own inverse, while if A is nonsingular then so is A^{-1} with $(A^{-1})^{-1} = A$; and if A and B are nonsingular matrices then so is their product since clearly $(AB)(B^{-1}A^{-1}) = I$ hence $B^{-1}A^{-1} = (AB)^{-1}$. Thus the set of nonsingular $n \times n$ matrices is a group under multiplication called the **general linear group** over the field F and denoted by $\mathrm{Gl}(n, F)$.

Some other notational conventions involving matrices are widely used. In extension of the definition (1.35) of the transpose of a vector, the **transpose** of an $n \times m$ matrix $A = \{a_{ij}\}$ is defined to be the $m \times n$ matrix

$$^tA = B = \{b_{ij}\} \quad \text{where} \quad b_{ij} = a_{ji}; \tag{1.58}$$

thus if $\mathbf{a}_i$ is the i-th column vector of the matrix A then $^t\mathbf{a}_i$ is the i-th row vector of the matrix ^tA. If the matrix A describes a linear transformation $A : F^n \longrightarrow F^m$ then the transposed matrix ^tA describes a linear transformation $^tA : F^m \longrightarrow F^n$; but note that ^tA does not describe the inverse linear transformation to that described by A but rather something quite different altogether, since for instance the inverse of a linear transformation between two vector spaces of different dimensions is not even well defined. If $C = AB$ the entries of these matrices are related by $c_{ij} = \sum_k a_{ik} b_{kj}$, and that equation can be interpreted alternatively as $^tC = {}^tB {}^tA$, so transposition reverses the order of matrix multiplication. Another notational convention is that a matrix can be decomposed into **matrix blocks** by splitting the rectangular array of its entries into subarrays. Thus for example an $n \times m$ matrix $A = \{a_{ij}\}$ can be decomposed into 4 matrix blocks

$$A = \begin{pmatrix} A_{11} & A_{12} \\ A_{21} & A_{22} \end{pmatrix}$$

where A_{11} is a $k \times k$ matrix, A_{12} is a $k \times (m - k)$ matrix, A_{21} is an $(n - k) \times k$ matrix and A_{22} is an $(n - k) \times (m - k)$ matrix. A decomposition of this form in which A_{12} and A_{21} are both zero matrices is called a **direct sum** decomposition, denoted by $A = A_{11} \oplus A_{22}$. An $n \times n$ matrix A which has the direct sum decomposition $A = A_{11} \oplus A_{22} \oplus \cdots \oplus A_{nn}$ in which each of the component matrices A_{ii} is a 1×1 matrix, just a scalar, is called a **diagonal matrix** since all of its terms vanish aside from those terms along the **main diagonal**, as in

$$\begin{pmatrix} a_1 & 0 & 0 & 0 \\ 0 & a_2 & 0 & 0 \\ 0 & 0 & a_3 & 0 \\ 0 & 0 & 0 & a_4 \end{pmatrix}. \tag{1.59}$$

Another special form of a matrix is a **lower triangular** matrix, a matrix $A = \{a_{ij}\}$ such that $a_{ij} = 0$ whenever $i < j$, as for example

$$\begin{pmatrix} a_{11} & 0 & 0 & 0 \\ a_{21} & a_{22} & 0 & 0 \\ a_{31} & a_{32} & a_{33} & 0 \\ a_{41} & a_{42} & a_{43} & a_{44} \end{pmatrix}; \qquad (1.60)$$

an **upper triangular** matrix of course is defined correspondingly as a matrix $A = \{a_{ij}\}$ such that $a_{ij} = 0$ whenever $i > j$.

The direct sum $F^m \oplus F^n$ of vector spaces F^m and F^n was defined in (1.39) to be the set of pairs $(\mathbf{x}, \mathbf{y})$ of vectors $\mathbf{x} \in F^m$ and $\mathbf{y} \in F^n$; it is customary though in this context to write this pair as $\binom{\mathbf{x}}{\mathbf{y}}$, thereby naturally identifying the direct sum $F^m \oplus F^n$ with the vector space F^{m+n}. An $r \times m$ matrix A describes a linear transformation $A : F^m \longrightarrow F^r$ that takes the vector $\mathbf{x} \in F^m$ to the vector $A\mathbf{x} \in F^r$; and correspondingly an $s \times n$ matrix B describes a linear transformation $B : F^n \longrightarrow F^s$ that takes the vector $\mathbf{y} \in F^n$ to the vector $B\mathbf{y} \in F^s$. These two linear transformations can be combined into the direct sum $A \oplus B$, an $(r + s) \times (m + n)$ matrix, which acts on the direct sum $F^m \oplus F^n$ through the matrix product

$$\begin{pmatrix} A & \mathbf{0} \\ \mathbf{0} & B \end{pmatrix} \begin{pmatrix} \mathbf{x} \\ \mathbf{y} \end{pmatrix} = \begin{pmatrix} A\mathbf{x} \\ B\mathbf{y} \end{pmatrix}; \qquad (1.61)$$

thus direct sums of matrices act naturally on direct sums of vector spaces indicating the compatibility of the two notions of direct sum.

Two $n \times m$ matrices A and B over a field F are said to be **equivalent matrices** if there are a nonsingular $n \times n$ matrix S and a nonsingular $m \times m$ matrix T such that $B = SAT$. This is a basic equivalence relation among matrices; but there are other equivalence relations of equal importance and even greater interest that will be discussed in Section 4.2. It is clear that this is an equivalence relation among $n \times m$ matrices; indeed reflexivity and symmetry are quite obvious, and if B is equivalent to A so that $B = SAT$ for nonsingular matrices S and T and if C is equivalent to B so that $C = UBV$ for nonsingular matrices U and V then $C = (US)A(TV)$ where US and TV are nonsingular matrices. If a matrix A is viewed as a mapping $A : F^m \longrightarrow F^n$ that takes a vector $\mathbf{x} \in F^m$ to the vector $\mathbf{y} = A\mathbf{x} \in F^n$, it is natural to ask what matrix represents this linear transformation after changes of coordinate in F^m and F^n. Thus an isomorphism $T : F^m \longrightarrow F^m$ takes the vector $\mathbf{x}'$ to the vector $\mathbf{x} = T\mathbf{x}'$, and an isomorphism $S : F^n \longrightarrow F^n$ takes the vector $\mathbf{y}$ to the vector $\mathbf{y}' = S\mathbf{y}$; and making these substitutions in the equation $\mathbf{y} = A\mathbf{x}$ yields the equation $\mathbf{y}' = S\mathbf{y} = SA\mathbf{x} = SAT\mathbf{x}' = B\mathbf{x}'$ where $B = SAT \cong A$. In this sense equivalent matrices describe the same linear transformation but expressed in different coordinates. Alternatively, the equivalence of matrices A and B can be expressed in terms of the linear transformations defined by these matrices as the assertion that the following diagram is commutative:

$$
\begin{array}{ccc}
F^m & \xrightarrow{\;\;A\;\;} & F^n \\[2pt]
T\big\uparrow \text{iso} & & S\big\downarrow \text{iso} \\[2pt]
F^m & \xrightarrow{\;\;B\;\;} & F^n.
\end{array}
\tag{1.62}
$$

The arrows in the diagram indicate mappings, in this case linear transformations, labeled by the letters A, B, S, T; and the notation "iso" indicates that the two linear transformations S and T are isomorphisms. A diagram such as this is said to be **commutative** if any two sequences of mappings in the diagram that begin and end at the same point are equal; in this case that can only mean that the mapping in the path B is equal to the result of first applying the mapping T, then the mapping A, and finally the mapping S, which is just the assertion that $B = SAT$. Commutativity of diagrams is a very commonly used notion that can simplify considerably statements of the equality of various combinations of mappings. However the notion of equivalence of matrices is viewed, though, it leads to a natural classification problem: does an equivalence class of matrices have a simple standard representative, and are there simple invariants that determine when two matrices are equivalent? That is the motivation for the next segment of the discussion.

Theorem 1.15. *If $A \in F^{n \times m}$ is a matrix describing a linear transformation $A : F^m \longrightarrow F^n$ then there are a nonsingular $n \times n$ matrix S and a nonsingular $m \times m$ matrix T such that the matrix $SAT = B = \{b_{ij}\}$ has the entries*

$$
b_{ij} = \begin{cases} \delta^i_j & \text{for } 1 \leq i, j \leq k, \\ 0 & \text{otherwise}, \end{cases}
\tag{1.63}
$$

where $m - k$ is the dimension of the kernel of A; thus B is the $n \times m$ matrix

$$
B = \begin{pmatrix} I_k & \mathbf{0} \\ \mathbf{0} & \mathbf{0} \end{pmatrix}.
\tag{1.64}
$$

Proof: If the kernel of the linear transformation has dimension $m - k$ choose vectors $\mathbf{v}_{k+1}, \ldots, \mathbf{v}_m$ in F^m that form a basis for the kernel, and extend this set of vectors to a basis $\mathbf{v}_1, \ldots, \mathbf{v}_k, \mathbf{v}_{k+1}, \ldots, \mathbf{v}_m$ for the full vector space F^m. The image vectors $\mathbf{w}_i = A\mathbf{v}_i \in F^n$ for $1 \leq i \leq k$ are linearly independent; for if $\mathbf{0} = \sum_{i=1}^k c_i A\mathbf{v}_i = A(\sum_{i=1}^k c_i \mathbf{v}_i)$ where not all the scalars c_i vanish then $\sum_{i=1}^k c_i \mathbf{v}_i$ is a nontrivial vector in $\ker(A)$, but that is impossible since this vector is not in the span of the basis $\mathbf{v}_{k+1}, \ldots, \mathbf{v}_m$ for $\ker(A)$. Extend the vectors $\mathbf{w}_1, \ldots, \mathbf{w}_k$ to a basis $\mathbf{w}_1, \ldots, \mathbf{w}_n$ for the vector space F^n. Introduce the isomorphisms $T : F^m \longrightarrow F^m$

for which $T\delta_i = \mathbf{v}_i$ and $S : F^n \longrightarrow F^n$ for which $S\mathbf{w}_i = \delta_i$, and let $B = SAT : F^m \longrightarrow F^n$. If $1 \le i \le k$ then $B\delta_i = SAT\delta_i = SA\mathbf{v}_i = S\mathbf{w}_i = \delta_i$, while on the other hand if $k + 1 \le i \le m$ then $B\delta_i = SAT\delta_i = SA\mathbf{v}_i = \mathbf{0}$; that means that the entries of the matrix B are $b_{ji} = \delta_i^j$ for $1 \le i \le k$ and $b_{ji} = 0$ otherwise, so that the matrix B has the form (1.64), and that suffices for the proof.

There remain the questions whether the normal form (1.64) is uniquely determined, and if so whether it can be calculated more directly. To discuss these matters it is useful to introduce a further standard convention in linear algebra. When an $n \times m$ matrix A is viewed as a collection of m column vectors $A = (\mathbf{a}_1 \ \ \mathbf{a}_2 \ \ \cdots \ \ \mathbf{a}_m)$, where $\mathbf{a}_j \in F^n$, the linear subspace of F^n spanned by these column vectors $\mathbf{a}_j$ is called the **column space** of the matrix A and the dimension of this subspace is called the **column rank** of the matrix A, denoted by crank(A). The column space of course is just the image $AF^m \subset F^n$, since $A\mathbf{x} = \sum_{i=1}^{m} \mathbf{a}_j x_j$ for any vector $\mathbf{x} \in F^m$, so the column rank of A is just the dimension of the image AF^m. The **row space** of the matrix A and its **row rank** rrank(A) are defined correspondingly; or alternatively since the rows of a matrix A are the columns of the transposed matrix ${}^t A$ the row space of A is the transpose of the column space of ${}^t A$ so that rrank(A) = crank(${}^t A$).

Theorem 1.16. *Let A be an $n \times m$ matrix over a field F.*
(i) crank(SA) = crank(A) *and* rrank(SA) = rrank(A) *for any nonsingular $n \times n$ matrix S.*
(ii) crank(AT) = crank(A) *and* rrank(AT) = rrank(A) *for any nonsingular $m \times m$ matrix T.*
(iii) rrank(A) = crank(A) *for any matrix A.*

Proof: (i) If S is a nonsingular $n \times n$ matrix the mapping $S : F^n \longrightarrow F^n$ that takes a vector $\mathbf{v}$ to $S\mathbf{v}$ is an isomorphism of vector spaces. Therefore the column space of the matrix A, the subspace of F^n spanned by the column vectors $\mathbf{a}_j$ of the matrix A, is isomorphic to the column space of the matrix SA, the subspace of F^n spanned by the column vectors $S\mathbf{a}_j$ of the matrix SA, or equivalently crank(SA) = crank(A). On the other hand since the entries of the product matrix $B = SA$ are $b_{ij} = \sum_{k=1}^{n} s_{ik} a_{kj}$ it follows that the row vectors $\mathbf{b}_i = \{ b_{ij} \mid 1 \le j \le m \}$ of the matrix B are the linear combinations $\mathbf{b}_i = \sum_{k=1}^{n} s_{ik} \mathbf{a}_k$ of the row vectors $\mathbf{a}_k = \{ a_{kj} \mid 1 \le j \le m\}$ of the matrix A and therefore rrank$B \le$ rrank$A$. Conversely since S is nonsingular and $A = S^{-1}B$ it is also the case that rrank$B \le$ rrank$A$, so altogether rrank(B) = rrank(A).
(ii) If $B = AT$ then ${}^t B = {}^t T \, {}^t A$ and rrank${}^t B =$ crank$B$ and correspondingly for the matrix A; it therefore follows from (i) that crank(AT) = crank(A) and rrank(AT) = rrank(A) for any nonsingular $m \times m$ matrix T.
(iii) By Theorem 1.15, there are invertible matrices S, T so that SAT has the form (1.64); and it is obvious from that form that

crank(SAT) = rrank(SAT) = k. Since it was just observed that crank(SAT) = crank(A) and correspondingly for the row ranks it follows immediately that crank(A) = rrank(A), which suffices for the proof.

Since the row rank and the column rank of any matrix are always the same, the distinction between them is ordinarily ignored and the common value is just called the **rank** of the matrix and denoted by rank(A).

Corollary 1.17 (Equivalence Theorem). *Two $n \times m$ matrices over a field are equivalent if and only if they have the same rank; hence any $n \times m$ matrix A is equivalent to a unique matrix of the form $\begin{pmatrix} I_r & 0 \\ 0 & 0 \end{pmatrix}$ where $r = $ rank(A).*

Proof: Theorem 1.15 shows that any $n \times m$ matrix is equivalent to one of the form (1.64), which clearly is of rank k; and it follows from the preceding Theorem 1.16 (i) that equivalent matrices have the same rank; consequently the normal form (1.64) is uniquely determined, by the rank, which suffices for the proof.

Corollary 1.18. *An $n \times n$ matrix over a field F is invertible if and only if rank(A) = n.*

Proof: It is clear that a matrix of the form $\begin{pmatrix} I_r & 0 \\ 0 & 0 \end{pmatrix}$ is invertible if and only if $r = n$; any matrix is equivalent to a matrix of this form, by the Equivalence Theorem, and since any two equivalent matrices have the same rank by Theorem 1.16 and since any matrix equivalent to an invertible matrix is invertible that suffices for the proof.

An auxiliary tool that is required for the further discussion deals with rearrangements or permutations of a set of variables, where for example x_4, x_1, x_3, x_2 is a rearrangement or **permutation** of the set of variables x_1, x_2, x_3, x_4; thus a permutation of such a set of variables is just an element in the symmetric group $S(x_1, x_2, x_3, x_4)$. To the usual order of the variables there can be associated the polynomial $P(x) = \prod_{i<j}(x_i - x_j)$, a polynomial of degree $n(n-1)/2$ consisting of the product of the differences of any two distinct variables where the index of the first variable is less than the index of the second variable; the square $P(x)^2$ of this polynomial can be rewritten as the product of all the differences $x_i - x_j$ of distinct variables so is independent of the order of the variables. After a permutation π of the variables the resulting polynomial $\pi^*P(x)$ still consists of the products of the differences of any two distinct variables, but possibly with a reversal of the signs of the differences; for example if $P(x) = (x_1 - x_2)(x_1 - x_3)(x_2 - x_3)$ then if π is the permutation that interchanges the variables x_2 and x_3 the result of applying this permutation to the polynomial is the polynomial $\pi^*P(x) = (x_1 - x_3)(x_1 - x_2)(x_3 - x_2) = -P(x)$.

In general $\pi^*P(x) = \pm P(x)$ for some sign, which is called the **sign** of the permutation π and is denoted by $\operatorname{sgn}(\pi)$ so that $\pi^*P(x) = \operatorname{sgn}(\pi) \cdot P(x)$. A permutation π is **even** if $\operatorname{sgn}(\pi) = +1$ and is **odd** if $\operatorname{sgn}(\pi) = -1$.

Theorem 1.19. (i) $\operatorname{sgn}(\pi) = -1$ for the permutation $\pi \in S(x_1, \dots, x_n)$ that interchanges two variables x_i and x_j.
(ii) Any permutation can be written as a composition of permutations that interchange two variables.
(iii) $\operatorname{sgn}(\pi_1 \pi_2) = \operatorname{sgn}(\pi_1) \cdot \operatorname{sgn}(\pi_2)$ for any two permutations $\pi_1, \pi_2 \in S(x_1, \dots, x_n)$.

Proof: (i) A permutation π that interchanges x_i and x_j does not change the sign of any factor of the polynomial $P(x)$ that does not involve the variable x_i or the variable x_j. If $k < i < j$ or $i < j < k$ the product $(x_i - x_k)(x_j - x_k)$ is unchanged by the permutation; and if $i < k < j$ the product $(x_i - x_k)(x_k - x_j)$ is replaced by $(x_j - x_k)(x_k - x_i)$ so the sign of the product is unchanged. Thus the only effect of the permutation π on the sign of the polynomial $P(x)$ arises from the change in the sign of the factor $(x_i - x_j)$ hence $\pi^*P(x) = -P(x)$ so $\operatorname{sgn}(\pi) = -1$.
(ii) Suppose a permutation π replaces the variables $x_1, x_2, \dots, x_n$ by a rearranged set $x_{j_1}, x_{j_2}, \dots, x_{j_n}$ of variables. Composing the permutation π with the permutation π_{1,j_1} that interchanges the variables x_1 and x_{j_1} replaces the variables $x_{j_1}, x_{j_2}, \dots, x_{j_n}$ by $x_1, x_{k_2}, \dots, x_{k_n}$; composing this permutation with the permutation π_{2,k_2} replaces the variables $x_1, x_{k_2}, \dots, x_{k_n}$ by $x_1, x_2, x_{l_3}, \dots, x_{l_n}$, and so on. Thus the composition of the permutation π with the composition of these permutations which interchange two variables leads to the identity permutation.
(iii) For any two permutations π_1, π_2 of the variables $\pi_2^*P(x) = \operatorname{sgn}(\pi_2)P(x)$ by definition; applying the permutation π_1 to both sides of this equality yields $\pi_1^*(\pi_2^*P(x)) = \operatorname{sgn}(\pi_2) \cdot \pi_1^*P(x) = \operatorname{sgn}(\pi_2)\operatorname{sgn}(\pi_1)P(x)$, by definition. On the other hand $\pi_1^*(\pi_2^*P(x))$ is the result of applying first the permutation π_2 and then the permutation π_1 so is the result of applying the composite permutation $\pi_1 \circ \pi_2$, and consequently $\pi_1^*(\pi_2^*P(x)) = \operatorname{sgn}(\pi_1 \circ \pi_2)P(x)$, and that suffices for the proof.

The preceding theorem does provide a method that can be useful for determining whether a permutation is odd or even. By that theorem any permutation can be written as a composition of **transpositions**, permutations that interchange two variables, while each transposition has sign -1 and the sign of the permutation is the product of the signs of these transpositions. It follows that a permutation is even if it can be written as a product of an even number of transpositions and is odd if it can be written as a product of an odd number of transpositions. A permutation can be written as a product of transpositions in a number of different ways; but the theorem ensures that the parity of the total number of transpositions is an invariant of the permutation.

A linear transformation $T : V \longrightarrow F$ from a vector space V over a field F to the field F itself is often called a **linear function**. Also of interest of course are linear functions of several variables, or **multilinear functions**, mappings $T : V^k \longrightarrow F$ from Cartesian products V^k of a vector space V over the field F to the field F that are linear transformations in each separate variable; thus a multilinear function $T : V^k \longrightarrow F$ associates to any k vectors $\mathbf{v}_i \in V$ a value $T(\mathbf{v}_1, \mathbf{v}_2, \dots, \mathbf{v}_k) \in F$, where $T(\mathbf{v}_1, \mathbf{v}_2, \dots, \mathbf{v}_k)$ is a linear function of the vector $\mathbf{v}_j \in V$ whenever the remaining vectors $\mathbf{v}_1, \dots, \mathbf{v}_{j-1}, \mathbf{v}_{j+1}, \dots, \mathbf{v}_k$ are held fixed. A multilinear function $T : V^k \longrightarrow F$ is said to be a **symmetric** multilinear function if

$$T(\mathbf{v}_{j_1}, \mathbf{v}_{j_2}, \dots, \mathbf{v}_{j_k}) = T(\mathbf{v}_1, \mathbf{v}_2, \dots, \mathbf{v}_k) \tag{1.65}$$

for any permutation $j_1, j_2, \dots, j_k$ of the indices $1, 2, \cdots, k$; and it is said to be an **alternating** multilinear function if

$$T(\mathbf{v}_{j_1}, \mathbf{v}_{j_2}, \dots, \mathbf{v}_{j_k}) = \text{sgn} \begin{pmatrix} j_1 & j_2 & \cdots & j_k \\ 1 & 2 & \cdots & k \end{pmatrix} T(\mathbf{v}_1, \mathbf{v}_2, \dots, \mathbf{v}_k) \tag{1.66}$$

for any permutation $j_1, j_2, \dots j_k$ of the indices $1, 2, \cdots, k$, where the factor $\text{sgn} \begin{pmatrix} j_1 & j_2 & \cdots & j_k \\ 1 & 2 & \cdots & k \end{pmatrix}$ is the sign of the permutation $j_1, j_2, \cdots, j_k$ of the indices $1, 2, \cdots, k$, as defined on the preceding page. It is convenient to extend the definition by setting $\text{sgn} \begin{pmatrix} j_1 & j_2 & \cdots & j_k \\ 1 & 2 & \cdots & k \end{pmatrix} = 0$ if the indices $j_1, j_2, \cdots, j_k$ are not a permutation of the indices $1, 2, \cdots, k$, so if for example there are some repeated indices among $j_1, j_2, \cdots, j_k$. Both symmetric and alternating multilinear functions are of considerable interest; but actually for the purposes of the discussion here the alternating multilinear functions are of the greatest interest. The following special case is particularly important; other cases of great interest will be considered in the later discussion of differential forms.

Theorem 1.20. *For any field F and any n there is an alternating multilinear function*

$$\Phi : \underbrace{F^n \times \cdots \times F^n}_{n} \longrightarrow F \tag{1.67}$$

from n copies of the vector space F^n to F, and it is uniquely determined up to a constant factor.

Proof: If Φ is an alternating multilinear function as in (1.67) and the vectors $\mathbf{v}_i \in F^n$ are written in terms of the basis vectors δ_j as $\mathbf{v}_i = \sum_{j=1}^{n} v_{ji} \delta_j$ then since

Φ is multilinear it follows that

$$\Phi(\mathbf{v}_1,\cdots,\mathbf{v}_n) = \Phi\left(\sum_{j_1=1}^{n} v_{j_1 1}\delta_{j_1},\ldots,\sum_{j_n=1}^{n} v_{j_n n}\delta_{j_n}\right)$$

$$= \sum_{j_1,\ldots,j_n=1}^{n} v_{j_1 1}\cdots v_{j_n n}\Phi(\delta_{j_1},\ldots,\delta_{j_n})$$

where $\Phi(\delta_{j_1},\ldots,\delta_{j_n}) \in F$; and since Φ also is assumed to be alternating

$$\Phi(\delta_{j_1},\ldots,\delta_{j_n}) = \mathrm{sgn}\begin{pmatrix} j_1 & \cdots & j_n \\ 1 & \cdots & n \end{pmatrix}\Phi(\delta_1,\ldots,\delta_n).$$

Thus if there is an alternating multilinear function (1.67) then it must have the form

$$\Phi(\mathbf{v}_1,\ldots,\mathbf{v}_n) = C\sum_{j_1,\ldots,j_n=1}^{n} \mathrm{sgn}\begin{pmatrix} j_1 & \cdots & j_n \\ 1 & \cdots & n \end{pmatrix} v_{j_1 1}\cdots v_{j_n n} \qquad (1.68)$$

where $C = \Phi(\delta_1,\ldots,\delta_n)$. Conversely the mapping

$$\Phi : F^n \times \cdots \times F^n \longrightarrow F$$

defined by (1.68) is an alternating multilinear function, since each term $v_{j_1 1},\ldots,v_{j_n n}$ has exactly one factor from each column and

$$\Phi(\mathbf{v}_{k_1},\ldots,\mathbf{v}_{k_n}) = C\sum_{j_1,\ldots,j_n=1}^{n} \mathrm{sgn}\begin{pmatrix} j_1 & \cdots & j_n \\ 1 & \cdots & n \end{pmatrix} v_{j_1 k_1}\cdots v_{j_n k_n}$$

$$= C\sum_{j_1,\ldots,j_n=1}^{n} \mathrm{sgn}\begin{pmatrix} k_1 & \cdots & k_n \\ 1 & \cdots & n \end{pmatrix}\begin{pmatrix} j_1 & \cdots & j_n \\ k_1 & \cdots & k_n \end{pmatrix} v_{j_1 k_1}\cdots v_{j_n k_n}$$

$$= \mathrm{sgn}\begin{pmatrix} k_1 & \cdots & k_n \\ 1 & \cdots & n \end{pmatrix}\Phi(\mathbf{v}_1,\ldots,\mathbf{v}_n).$$

That suffices for the proof.

The preceding theorem shows that any alternating multilinear function has the form (1.68) for some constant C; for any $n \times n$ matrix A the value of the multilinear function (1.68) where $\mathbf{v}_i = \mathbf{a}_i$ are the columns of the matrix A and

$C = 1$ is called the **determinant** of the $n \times n$ matrix $A = (\mathbf{a}_1 \quad \mathbf{a}_2 \quad \cdots \quad \mathbf{a}_n)$ and is denoted by $\det A$, so that

$$\det A = \sum_{j_1,\ldots,j_n=1}^{n} \operatorname{sgn} \begin{pmatrix} j_1 & \cdots & j_n \\ 1 & \cdots & n \end{pmatrix} a_{j_1 1} \cdots a_{j_n n}; \qquad (1.69)$$

the summation is formally extended over all choices of the integers $j_1, \ldots, j_n$, with the understanding that $\operatorname{sgn} \begin{pmatrix} j_1 & \cdots & j_n \\ 1 & \cdots & n \end{pmatrix} = 0$ unless the indices $j_1, \ldots, j_n$ are a permutation of the indices $1, \ldots, n$. The basic properties of the determinant follow quite directly from its definition.

Theorem 1.21. *The determinant mapping over a field F satisfies*
(i) $\det I = 1$ *for the identity matrix I.*
(ii) $\det {}^t A = \det A$.
(iii) $\det(AB) = \det A \cdot \det B$ *for any $n \times n$ matrices A and B, and*
(iv) $\det A \neq 0$ *if and only if A is a nonsingular matrix.*

Proof: (i) By definition the determinant mapping is normalized so that $\det I = \Phi(\delta_1, \ldots, \delta_n) = 1$.
(ii) The products $a_{1j_1} \cdots a_{nj_n}$ and $a_{j_1 1} \cdots a_{j_n n}$ run over the same set of values as the indices $j_1, \ldots, j_n$ run through all permutations of the integers $1, \ldots, n$ so the basic formula (1.69) has the same value when the terms a_{ij} are replaced by a_{ji}.
(iii) If $A = \{a_{ij}\}$ and $B = \{b_{ij}\}$ are $n \times n$ matrices, the j-th column $\mathbf{v}_j$ of their product $AB = \{\sum_{k=1}^{n} a_{ik} b_{kj}\}$ is the linear combination $\mathbf{v}_j = \sum_{k=1}^{n} \mathbf{a}_k b_{kj}$ of the column vectors $\mathbf{a}_k$ of the matrix A; so for any linear function $\phi : F^n \longrightarrow F$ it follows that

$$\phi(\mathbf{v}_j) = \phi \left(\sum_{k=1}^{n} \mathbf{a}_k b_{kj} \right) = \sum_{k=1}^{n} b_{kj} \phi(\mathbf{a}_k).$$

The determinant is defined by (1.69), and since the function Φ is multilinear and alternating

$$\det AB = \Phi(\mathbf{v}_1, \ldots, \mathbf{v}_n) = \sum_{k_1,\ldots,k_n=1}^{n} b_{k_1 1} \ldots b_{k_n n} \Phi(\mathbf{a}_{k_1}, \ldots, \mathbf{a}_{k_n})$$

$$= \sum_{k_1,\ldots,k_n=1}^{n} b_{k_1 1} \cdots b_{k_n n} \operatorname{sgn} \begin{pmatrix} k_1 & \cdots & k_n \\ 1 & \cdots & n \end{pmatrix} \Phi(\mathbf{a}_1, \ldots, \mathbf{a}_n)$$

$$= \det B \cdot \det A.$$

(iv) If an $n \times n$ matrix A is nonsingular then it has a well-defined inverse matrix A^{-1} for which $AA^{-1} = I$, and it then follows from parts (i) and (iii) of this theorem that $\det A \det A^{-1} = \det(A \cdot A^{-1}) = \det I = 1$, and consequently $\det A \neq 0$. Conversely if A is an $n \times n$ matrix for which $\det A \neq 0$ then by the Equivalence Theorem, Theorem 1.15, there are nonsingular matrices S, T such that $SAT = B$ where $B = \begin{pmatrix} I_k & 0 \\ 0 & 0 \end{pmatrix}$ for some number k with $0 \leq k \leq n$; and since the matrices S and T are nonsingular $\det S \neq 0$ and $\det T \neq 0$ by what has just been proved, while $\det A \neq 0$ by assumption so $\det B \neq 0$ in view of (iii). That can be the case only if $k = n$, since if one of the columns of B is zero then $\det B = 0$ as a consequence of the defining formula (1.69); and if $k = n$ then $B = I$ is a nonsingular matrix, and therefore A is a product of nonsingular matrices so it is also nonsingular. That suffices for the proof.

Some further properties of determinants are useful in calculating their actual values.

Corollary 1.22. *The determinant mapping over a field F satisfies the following.*
(i) *$\det A$ changes signs if any two columns or rows are permuted.*
(ii) *$\det A = 0$ if the columns or rows are linearly dependent.*
(iii) *If a matrix A' arises from a matrix A by multiplying any row or column by $a \in F$ then $\det A' = a \det A$.*
(iv) *If a matrix A' arises from a matrix A by adding to one column a constant multiple of another column, or the corresponding operation on rows, then $\det A' = \det A$.*

Proof: (i) The determinant is defined as $\det A = \Phi(\mathbf{a}_1, \dots, \mathbf{a}_n)$ in terms of the columns of the matrix A; and since the function Φ is alternating and the sign of a transposition is -1 it follows that $\det A$ changes sign whenever two columns of A are interchanged. Since $\det {}^t A = \det A$ by Theorem 1.21 (ii) it follows from what has just been demonstrated that $\det A$ also changes sign whenever any two rows are interchanged.
(ii) If the columns of A are linearly dependent then a multiple of one of the columns is in the span of the remaining columns; if it is assumed for simplicity of notation that $\mathbf{a}_1 = \sum_{j=2}^{n} c_j \mathbf{a}_j$ then by linearity

$$\Phi(\mathbf{a}_1, \dots, \mathbf{a}_n) = \sum_{j=2}^{n} c_j \Phi(\mathbf{a}_j, \mathbf{a}_2, \dots, \mathbf{a}_n).$$

By (i) the function F changes sign when columns 1 and j are interchanged; but if the two columns are equal it also must be the case that the function does not change signs, and that can be the case only when the function F is zero.

(iii) Since the function Φ in (1.67) is multilinear it is multiplied by a if any column is multiplied by a; and since $\det {}^tA = \det A$ by Theorem 1.21 (ii) that also is the case when any row is multiplied by a.

(iv) Since the function Φ is multilinear

$$\Phi(\mathbf{a}_1 + c\mathbf{a}_2, \mathbf{a}_2, \ldots, \mathbf{a}_n) = \Phi(\mathbf{a}_1, \mathbf{a}_2, \ldots, \mathbf{a}_n) + c\Phi(\mathbf{a}_2, \mathbf{a}_2, \ldots, \mathbf{a}_n)$$

and $\Phi(\mathbf{a}_2, \mathbf{a}_2, \ldots, \mathbf{a}_n) = 0$ by (ii).

One rather straightforward way to calculate the explicit value of a determinant is to reduce the calculation to that of the determinants of smaller matrices. The basic formula (1.69) can be written

$$\det A = \sum_{j_1=1}^{n} \sum_{j_2,\ldots,j_n=1}^{n} \mathrm{sgn} \begin{pmatrix} j_1 & j_2 & \cdots & j_n \\ 1 & 2 & \cdots & n \end{pmatrix} a_{j_1 1} a_{j_2 2} \cdots a_{j_n n}$$

$$= a_{11} \sum_{j_2,\ldots,j_n=1}^{n} \mathrm{sgn} \begin{pmatrix} 1 & j_2 & \cdots & j_n \\ 1 & 2 & \cdots & n \end{pmatrix} a_{j_2 2} \cdots a_{j_n n}$$

$$+ a_{21} \sum_{j_2,\ldots,j_n=1}^{n} \mathrm{sgn} \begin{pmatrix} 2 & j_2 & \cdots & j_n \\ 1 & 2 & \cdots & n \end{pmatrix} a_{j_2 2} \cdots a_{j_n n}$$

$$+ \cdots +$$

$$+ a_{n1} \sum_{j_2,\ldots,j_n=1}^{n} \mathrm{sgn} \begin{pmatrix} n & j_2 & \cdots & j_n \\ 1 & 2 & \cdots & n \end{pmatrix} a_{j_2 2} \cdots a_{j_n n}$$

so

$$\det A = a_{11} \det A_{11} - a_{21} \det A_{21} + \cdots + (-1)^{n+1} a_{n1} \det A_{n1} \tag{1.70}$$

where A_{ij} is the matrix arising from the matrix A by deleting both row i and column j. Thus for instance

$$\det \begin{pmatrix} a & b \\ c & d \end{pmatrix} = a \det d - c \det b = ad - cb \tag{1.71}$$

and

$$\det \begin{pmatrix} 1 & 2 & 3 \\ 6 & 5 & 4 \\ 7 & 0 & 8 \end{pmatrix} = 1 \det \begin{pmatrix} 5 & 4 \\ 0 & 8 \end{pmatrix} - 6 \det \begin{pmatrix} 2 & 3 \\ 0 & 8 \end{pmatrix} + 7 \det \begin{pmatrix} 2 & 3 \\ 5 & 4 \end{pmatrix} = -105.$$

The columns of the matrix A can be permuted, with a change of sign, and the same formula can be applied to the rows, which provides a good deal of flexibility in this approach to calculating determinants. Note that if the entries a_{j1} in (1.70) are replaced by a_{ji} for $i \neq 1$ the result would be the determinant of the matrix obtained from A by replacing column 1 by column i, so the result would be 0, the determinant of a matrix with two identical columns. Consequently if $A^\dagger$ is the matrix with the entries $A^\dagger_{ij} = (-1)^{i+j} \det A_{ji}$, a matrix called the **adjugate** of the matrix A, it follows from (1.70) that

$$A \cdot A^\dagger = A^\dagger \cdot A = (\det A) \cdot I. \tag{1.72}$$

For example

$$\begin{pmatrix} a & b \\ c & d \end{pmatrix}^\dagger = \begin{pmatrix} d & -b \\ -c & a \end{pmatrix} \quad \text{where} \quad \det \begin{pmatrix} a & b \\ c & d \end{pmatrix} = ad - bc. \tag{1.73}$$

This can be applied to solve a system of linear equations $A\mathbf{x} = \mathbf{a}$ for a square matrix A explicitly as $\mathbf{x} = \frac{1}{\det A} A^\dagger \mathbf{a}$, an application customarily called **Cramer's rule**.

Matrices and vector spaces arise very commonly in other algebraic structures. A permutation of a set of variables can be described by a matrix operation; for example

$$\begin{pmatrix} 0 & 1 \\ 1 & 0 \end{pmatrix} \cdot \begin{pmatrix} x_1 \\ x_2 \end{pmatrix} = \begin{pmatrix} x_2 \\ x_1 \end{pmatrix} \quad \text{and} \quad \begin{pmatrix} 0 & 1 & 0 \\ 0 & 0 & 1 \\ 1 & 0 & 0 \end{pmatrix} \cdot \begin{pmatrix} x_1 \\ x_2 \\ x_3 \end{pmatrix} = \begin{pmatrix} x_2 \\ x_3 \\ x_1 \end{pmatrix}. \tag{1.74}$$

The matrices describing permutations in this way, called **permutation matrices**, are those matrices such that each row and each column contain a single entry of 1 but all other entries are 0. Multiplication in a group has the effect of permuting the elements of the group, so any finite group can be viewed as a subgroup of the set of all permutations of the elements of the group, and in that way any group is isomorphic to a group of matrices. This is actually an immensely useful tool in the investigation of abstract groups.

The algebraic discussion here will be concluded by examining yet another general algebraic structure. An **algebra** over field F is defined to be a vector space $\mathcal{A}$ over the field F with an additional operation that associates to any two vectors $\mathbf{v}_1, \mathbf{v}_2 \in \mathcal{A}$ their product $\mathbf{v}_1 \cdot \mathbf{v}_2 \in \mathcal{A}$ such that

(i) multiplication of vectors is associative and has an identity element $I \in \mathcal{A}$;

(ii) the distributive laws $\mathbf{v} \cdot (\mathbf{v}_1 + \mathbf{v}_2) = \mathbf{v} \cdot \mathbf{v}_1 + \mathbf{v} \cdot \mathbf{v}_2$ and
$(\mathbf{v}_1 + \mathbf{v}_2) \cdot \mathbf{v} = \mathbf{v}_1 \cdot \mathbf{v} + \mathbf{v}_2 \cdot \mathbf{v}$ hold for any vectors $\mathbf{v}, \mathbf{v}_1, \mathbf{v}_2 \in \mathcal{A}$;

(iii) scalar multiplication and the multiplication of vectors are related by
the laws $c(\mathbf{v}_1 \cdot \mathbf{v}_2) = (c\mathbf{v}_1) \cdot \mathbf{v}_2 = \mathbf{v}_1 \cdot (c\mathbf{v}_2)$ for all $c \in F$ and $\mathbf{v}_1, \mathbf{v}_2 \in \mathcal{A}$.

Note particularly that it is not assumed that the multiplication of vectors is commutative; if the multiplication of vectors is commutative the algebra is called a **commutative algebra**. The **dimension** of an algebra $\mathcal{A}$ is defined to be its dimension as a vector space. Just as in the case of rings, the zero element $\mathbf{0}$ plays a special role in multiplication in an algebra, so that $\mathbf{0} \cdot \mathbf{v} = \mathbf{v} \cdot \mathbf{0} = \mathbf{0}$ for any vector $\mathbf{v} \in \mathcal{A}$, with the same proof as in the preceding examination of rings. The matrix space $F^{n \times n}$ under multiplication of matrices is an example of an algebra, which of course is not necessarily commutative. The field F itself is a commutative one-dimensional algebra over F, indeed can readily be seen to be the unique one-dimensional algebra over F. The set $F[X]$ of all polynomials with coefficients in F is another commutative algebra, although not a finite dimensional algebra; and other infinite dimensional algebras will arise in the subsequent discussion.

It is not difficult to describe all two-dimensional commutative real algebras; indeed that is a good exercise in the study of algebras. If $\mathcal{A}$ is a commutative real algebra and $\dim \mathcal{A} = 2$, choose a basis $\mathbf{v}_1, \mathbf{v}_2$ for the vector space $\mathcal{A}$ so that $\mathbf{v}_1$ is the identity for multiplication in the algebra; the multiplication table in terms of this basis then must be of the form

$$\mathcal{A}_{x_1, x_2} : \quad \mathbf{v}_1 \cdot \mathbf{v}_1 = \mathbf{v}_1, \quad \mathbf{v}_1 \cdot \mathbf{v}_2 = \mathbf{v}_2 \cdot \mathbf{v}_1 = \mathbf{v}_2, \quad \mathbf{v}_2 \cdot \mathbf{v}_2 = x_1 \mathbf{v}_1 + x_2 \mathbf{v}_2$$

for some $x_1, x_2 \in \mathbb{R}$. This multiplication table can be simplified by replacing $\mathbf{v}_2$ by $\mathbf{v}_2' = \mathbf{v}_2 - t\mathbf{v}_1$ for a suitable real number $t \in \mathbb{R}$; for any choice of $t \in \mathbb{R}$ the vectors $\mathbf{v}_1, \mathbf{v}_2'$ are another basis for V, and by the distributive law

$$\begin{aligned}
\mathbf{v}_2' \cdot \mathbf{v}_2' &= (\mathbf{v}_2 - t\mathbf{v}_1) \cdot (\mathbf{v}_2 - t\mathbf{v}_1) = \mathbf{v}_2 \cdot \mathbf{v}_2 - 2t\mathbf{v}_1 \cdot \mathbf{v}_2 + t^2 \mathbf{v}_1 \cdot \mathbf{v}_1 \\
&= (x_1\mathbf{v}_1 + x_2\mathbf{v}_2) - 2t\mathbf{v}_2 + t^2\mathbf{v}_1 = (x_1 + t^2)\mathbf{v}_1 + (x_2 - 2t)\mathbf{v}_2 \\
&= (x_1 + t^2)\mathbf{v}_1 + (x_2 - 2t)(\mathbf{v}_2' + t\mathbf{v}_1) = (x_1 + tx_2 - t^2)\mathbf{v}_1 + (x_2 - 2t)\mathbf{v}_2'
\end{aligned}$$

so if $2t = x_2$ it follows that

$$\mathbf{v}_2' \cdot \mathbf{v}_2' = y\mathbf{v}_1 \quad \text{for some } y \in \mathbb{R}.$$

For a further simplification replace the vector $\mathbf{v}_2'$ by $\mathbf{v}_2'' = s\mathbf{v}_2'$ for another real number $s \in \mathbb{R}$, so that

$$\mathbf{v}_2'' \cdot \mathbf{v}_2'' = s^2 y \mathbf{v}_1.$$

The real number y appearing in this formula can be either 0 or positive or negative. Recalling that any positive real number can be written as the square of a positive real number, which was demonstrated at the end of Section 1.2, it is possible to choose s so that $s^2 y = 1$ if $y > 0$ and $s^2 y = -1$ if $y < 0$. Therefore after relabeling the vectors $\mathbf{v}_1$ and $\mathbf{v}_2$ the multiplication table for the algebra takes the simpler form

$$\mathcal{A}_\epsilon: \quad \mathbf{v}_1 \cdot \mathbf{v}_1 = \mathbf{v}_1, \quad \mathbf{v}_1 \cdot \mathbf{v}_2 = \mathbf{v}_2 \cdot \mathbf{v}_1 = \mathbf{v}_2, \quad \mathbf{v}_2 \cdot \mathbf{v}_2 = \epsilon \mathbf{v}_1 \tag{1.75}$$

where ϵ is either 0 or 1 or -1. For the case $\epsilon = 0$ the 2×2 real matrices

$$\mathbf{v}_1 = \begin{pmatrix} 1 & 0 \\ 0 & 1 \end{pmatrix} \quad \mathbf{v}_2 = \begin{pmatrix} 0 & 1 \\ 0 & 0 \end{pmatrix} \tag{1.76}$$

obviously satisfy (1.75); the vector space spanned by these two matrices is

$$\mathcal{A}_0 = \left\{ \begin{pmatrix} a & b \\ 0 & a \end{pmatrix} \,\middle|\, a, b \in \mathbb{R} \right\}, \tag{1.77}$$

and a straightforward calculation shows that this set of matrices is closed under multiplication and form a commutative real algebra. For the case $\epsilon = 1$ the 2×2 real matrices

$$\mathbf{v}_1 = \begin{pmatrix} 1 & 0 \\ 0 & 1 \end{pmatrix} \quad \mathbf{v}_2 = \begin{pmatrix} 0 & 1 \\ 1 & 0 \end{pmatrix} \tag{1.78}$$

clearly satisfy (1.75); the vector space spanned by these two matrices is

$$\mathcal{A}_1 = \left\{ \begin{pmatrix} a & b \\ b & a \end{pmatrix} \,\middle|\, a, b \in \mathbb{R} \right\}, \tag{1.79}$$

and a straightforward calculation shows that this set of matrices is closed under multiplication and form a commutative real algebra. For the case $\epsilon = -1$ the 2×2 real matrices

$$\mathbf{v}_1 = \begin{pmatrix} 1 & 0 \\ 0 & 1 \end{pmatrix} \quad \mathbf{v}_2 = \begin{pmatrix} 0 & 1 \\ -1 & 0 \end{pmatrix} \tag{1.80}$$

obviously satisfy (1.75); the vector space spanned by these two matrices is

$$\mathcal{A}_{-1} = \left\{ \begin{pmatrix} a & b \\ -b & a \end{pmatrix} \,\middle|\, a, b \in \mathbb{R} \right\}, \tag{1.81}$$

and a straightforward calculation shows that this set of matrices also is closed under multiplication and form a commutative real algebra. It is also easy to see that these three real commutative algebras are distinct algebras, in the sense that no two are isomorphic. Indeed there are elements in $\mathcal{A}_0$ the squares of which are zero, as for instance $\mathbf{v}_2 \cdot \mathbf{v}_2 = \mathbf{0}$; in $\mathcal{A}_1$ however $(x_1\mathbf{v}_1 + x_2\mathbf{v}_2) \cdot (x_1\mathbf{v}_1 + x_2\mathbf{v}_2) = (x_1^2 + x_2^2)\mathbf{v}_1 + 2x_1x_2\mathbf{v}_2$ for any real x_1, x_2, so nonzero elements always have nonzero squares, but $(\mathbf{v}_1 + \mathbf{v}_2) \cdot (\mathbf{v}_1 - \mathbf{v}_2) = \mathbf{0}$ so there are nonzero elements having products $\mathbf{0}$; finally if $(x_1\mathbf{v}_1 + x_2\mathbf{v}_2) \in \mathcal{A}_{-1}$ then $(x_1\mathbf{v}_1 + x_2\mathbf{v}_2) \cdot (x_1\mathbf{v}_1 - x_2\mathbf{v}_2) = (x_1^2 + x_2^2)\mathbf{v}_1$ for any real x_1, x_2, so any nonzero element $(x_1\mathbf{v}_1 + x_2\mathbf{v}_2) \in \mathcal{A}_{-1}$ has the multiplicative inverse $(x_1\mathbf{v}_1 + x_2\mathbf{v}_2)^{-1} = (x_1^2 + x_2^2)^{-1}(x_1\mathbf{v}_1 - x_2\mathbf{v}_2)$, showing that $\mathcal{A}_{-1}$ actually is a field.

The field $\mathcal{A}_{-1}$ is called the field of **complex numbers** and is denoted by $\mathbb{C}$. The basis vector $\mathbf{v}_1$ is customarily identified with the real identity element 1, and a product $x\mathbf{v}_1 \in \mathbb{C}$ is identified with the real number x viewed as being imbedded in $\mathbb{C}$; since $(x\mathbf{v}_1) \cdot (y\mathbf{v}_1) = (xy)\mathbf{v}_1$ the imbedding $\mathbb{R} \subset \mathbb{C}$ is compatible with the algebraic operations in the fields $\mathbb{R}$ and $\mathbb{C}$. The basis vector $\mathbf{v}_2$ is usually denoted by i, and the products $y\mathbf{v}_2 = yi$ for $y \in \mathbb{R}$ are known for historical reasons as **imaginary numbers**. The elements of the field $\mathbb{C}$, the complex numbers, thus are written $z = x + iy$, where $x = \Re(z) \in \mathbb{R}$ is called the **real part** of the complex number z and $y = \Im(z) \in \mathbb{R}$ is called the **imaginary part** of the complex number z. In (1.80) there is a choice whether to use $\mathbf{v}_2$ or $-\mathbf{v}_2$ as the second basis vector; whichever is chosen clearly leads to the same algebra (1.81) and the same algebraic operations. Thus it is possible to associate to any complex number $z = x + iy \in \mathbb{C}$ another complex number $\bar{z} = x - iy$ called the **complex conjugate** of z; and the mapping $z \longrightarrow \bar{z}$ is a field isomorphism since it is readily verified that $\overline{(z_1 + z_2)} = \overline{z_1} + \overline{z_2}$ and $\overline{(z_1 \cdot z_2)} = \overline{z_1} \cdot \overline{z_2}$. In terms of the matrices (1.81) complex conjugation corresponds to taking the transpose of the matrix. It is worth noting that obviously there is no nontrivial isomorphism of the field $\mathbb{R}$ with itself, exhibiting an interesting difference between the two fields $\mathbb{R}$ and $\mathbb{C}$. It is an interesting exercise to go through the preceding classification of two-dimensional algebras for the case of algebras over the complex numbers. The argument goes through as before; but any complex number is the square of another complex number, so there are only the two distinct algebras (1.75) over the complex numbers corresponding to the values $\epsilon = 0, 1$ and neither is a field. The world is in some ways not as interesting as it might be; alas there are no other finite dimensional real algebras that are fields.[8]

---

[8]See the *Princeton Companion to Mathematics*.

Problems, Group I

1. If W_1, W_2 are linear subspaces of a vector space V over a field F, which of the following subsets of V are also linear subspaces and why:
   (i) $W_1 \cap W_2$ (ii) $W_1 \cup W_2$ (iii) $W_1 \sim W_2$ (iv) $W_1 + W_2$.

2. Find a basis for the subspace of $\mathbb{R}^4$ spanned by the vectors $(1, 2, -1, 0)$, $(4, 8, -4, -3)$, $(0, 1, 3, 4)$, and $(2, 5, 1, 4)$.

3. Find an example of a vector space V and subspaces $M, N_1, N_2 \subset V$ such that $M \oplus N_1 = M \oplus N_2 = V$ but $N_1 \neq N_2$.

4. If A is an $m \times n$ matrix over a field F show that the vector space V consisting of those vectors $\mathbf{x} \in F^n$ such that $A\mathbf{x} = \mathbf{0}$ has dimension at least $n - m$. Find an example in which $n > m$ and the dimension of V is strictly greater than $n - m$.

5. If A is an $m \times n$ matrix over a field F with $n < m$ show that there do not exist any $n \times m$ matrices B over that field for which $AB = I$.

6. Find 2×2 matrices A, B for which $AB = 0$ but $BA \neq 0$.

7. Show that $\operatorname{rank} AB \leq \min(\operatorname{rank} A, \ \operatorname{rank} B)$ for any two matrices A, B for which the product is defined; and find an example for which this is an equality and another example for which this is a strict inequality.

8. Consider the set of $n \times n$ matrices

$$I_{i_1, i_2, \ldots, i_n} = (\delta_{i_1}, \delta_{i_2}, \ldots, \delta_{i_n})$$

   where $\delta_i \in \mathbb{R}^n$ is the Kronecker vector having the entry 1 in row i and entries 0 in the other rows and $i_1, i_2, \ldots, i_n$ are any n integers in the range $[1, n]$, not necessarily distinct integers.

   i) What is the rank of the matrix $I_{i_1, i_2, \ldots, i_n}$?
   ii) What is $\det I_{i_1, i_2, \ldots, i_n}$?
   iii) What is the vector $I_{i_1, i_2, \ldots, i_n} \mathbf{x}$ for a column vector $\mathbf{x} = \left(\begin{smallmatrix} x_1 \\ \vdots \\ x_n \end{smallmatrix} \right)$ in $\mathbb{R}^n$?

9. Let X be the real matrix

$$\begin{pmatrix} x_1 & x_2 & x_3 \\ x_2 & x_3 & x_1 \\ x_3 & x_1 & x_2 \end{pmatrix}.$$

     i) For what values x_i is rank$X = 2$?

     ii) For what values x_i is rank$X = 1$?

10.   A reasonably effective technique for calculating the rank and determinant of an $m \times n$ matrix A over a field F and examining explicitly the system of linear equations $A\mathbf{x} = \mathbf{y}$ is through the **elementary row operations** on A:

     i) add c times row i to row j for some scalar $c \in F$;

     ii) interchange rows i and j;

     iii) multiply row i by a nonzero scalar $c \in F$.

Show that these operations do not change the rank of a matrix, and that each can be realized by multiplying the matrix A on the left by a suitable $m \times m$ matrix (an **elementary matrix**). These operations can be applied to simplify the form of a matrix A by reducing it to a matrix A' that is as close to an identity matrix as possible, so that it is easy to calculate its rank and determinant. When the same product E of elementary matrices is applied to both sides the equation $A\mathbf{x} = \mathbf{y}$ becomes $A'\mathbf{x} = EA\mathbf{x} = E\mathbf{y}$, which has the same solution $\mathbf{x}$; so finding the solution $\mathbf{x}$ and determining the conditions on $\mathbf{y}$ under which there exists a solution $\mathbf{x}$ (or equivalently finding the image of the mapping defined by the matrix A) are simplified as well. To ensure that the same elementary operations are applied to the matrix A as to the vector $\mathbf{y}$ it is customary to apply the operations to the extended matrix $(A \; \mathbf{y})$ where the vector $\mathbf{y}$ is added as an additional last column of the matrix A. Apply this procedure to show that A can be reduced to A' in the following example:

$$A = \begin{pmatrix} 1 & 2 & 3 & 4 & 5 & y_1 \\ 1 & 0 & 1 & 0 & 1 & y_2 \\ 1 & 2 & 2 & 2 & 1 & y_3 \end{pmatrix} \quad A' = \begin{pmatrix} 1 & 0 & 0 & -2 & -3 & y_1' \\ 0 & 1 & 0 & 0 & -2 & y_2' \\ 0 & 0 & 1 & 2 & 4 & y_3' \end{pmatrix}.$$

Determine the vector $\mathbf{y}'$ explicitly in terms of the vector $\mathbf{y}$. Determine the rank of the matrix A. Show that the equation $A\mathbf{x} = \mathbf{y}$ can be solved for $\mathbf{x}$ whenever $\mathbf{y}$ is specified arbitrarily; and in particular find a solution $\mathbf{x}$ when $\mathbf{y} = \{1, 2, 3\}$. Is your solution unique?

Problems, Group II

11.   If V is an n-dimensional real vector space show that the kernel of any nonzero linear transformation $T : V \longrightarrow \mathbb{R}$ is a linear subspace of dimension $n - 1$ and conversely that any such linear subspace is the kernel of some linear transformation $T : V \longrightarrow \mathbb{R}$.

12.   If A is an $m \times n$ and B is an $n \times m$ matrix show that $I - AB$ is nonsingular if and only if $I - BA$ is nonsingular.

13.   Show that an $m \times n$ matrix A over a field F has rank 1 if and only if $A = \mathbf{b}\,^t\mathbf{c}$ for some nonzero (column) vectors $\mathbf{b} \in F^m$ and $\mathbf{c} \in F^n$.

14.   If A, B, C, D are square real matrices where $\det A \neq 0$ and $AC = CA$ show that $\det \begin{pmatrix} A & B \end{pmatrix} = \det(AD - CB)$. Find an example to show that this is not necessarily the case if the hypotheses are not assumed.

15.   An $n \times n$ matrix A is **symmetric** if $A = {}^tA$ and is **skew symmetric** if $A = -{}^tA$.

   i) Show that the set of symmetric matrices is a linear subspace of the vector space of all $n \times n$ matrices, as is the set of skew-symmetric matrices, and find the dimensions of these linear subspaces.

   ii) Show that the vector space of $n \times n$ matrices is the direct sum of the subspace of symmetric and the subspace of skew symmetric matrices, provided that $2 \neq 0$ in the field F. What happens for the field $\mathbb{F}_2 = \mathbb{Z}/2\mathbb{Z}$ of two elements in which $2 = 0$?

16.   The **trace** $\operatorname{tr}(A)$ of an $n \times n$ matrix is defined to be the sum of its diagonal terms.

   i) Show that $\operatorname{tr}(A + B) = \operatorname{tr}(A) + \operatorname{tr}(B)$, that $\operatorname{tr}(AB) = \operatorname{tr}(BA)$ and that $\operatorname{tr}(ABA^{-1}) = \operatorname{tr}(B)$ if the matrix A is invertible.

   ii) Is $\operatorname{tr}(AB) = \operatorname{tr}(A)\operatorname{tr}(B)$? Why?

   iii) Show that the equation $AB - BA = I$ has no solutions in $n \times n$ matrices A and B over a field F provided that $n \neq 0$ in F.

17.   If $T : F^n \longrightarrow F^n$ is a linear transformation such that the kernel of T is equal to the image of T show that n must be an even number. Find an example of such a mapping.

18.   If $T : V \longrightarrow V$ is a linear transformation on a finite dimensional vector space show that there is some number n for which $T^n(V) \cap \ker T^n = \mathbf{0}$.

19.   i) For any elements $a_1, \ldots, a_n$ in a field F show that

$$\det \begin{pmatrix} 1 & a_1 & \cdots & a_1^{n-1} \\ 1 & a_2 & \cdots & a_2^{n-1} \\ \cdots & \cdots & & \cdots \\ 1 & a_n & \cdots & a_n^{n-1} \end{pmatrix} = \prod_{1 \leq i < j \leq n} (a_i - a_j).$$

This determinant is called the **Vandermonde** determinant.

ii) Using the Vandermonde determinant show that for any n pairs

$$(a_1, b_1), \ldots, (a_n, b_n)$$

of elements in the field F, where a_i are distinct elements, there is a polynomial $p(x)$ of degree at most $n - 1$ such that $p(a_i) = b_i$ for $1 \leq i \leq n$.

20. A diagram of vector spaces V_i and linear transformations T_i of the form

$$V_1 \xrightarrow{T_1} V_2 \xrightarrow{T_2} V_3 \xrightarrow{T_3} V_4 \xrightarrow{T_4} \ldots$$

is said to be an **exact sequence** of vector spaces if in any subsegment of the form $V_{i-1} \xrightarrow{T_{i-1}} V_i \xrightarrow{T_i} V_{i+1}$ the image of the linear transformation T_{i-1} is precisely the kernel of the linear transformation T_i. For example, the assertion that the sequence $\mathbf{0} \xrightarrow{T_0} V_1 \xrightarrow{T_1} V_2$ is exact is equivalent to the assertion that the linear transformation T_1 is injective while on the other hand the assertion that the sequence $V_1 \xrightarrow{T_1} V_2 \xrightarrow{T_2} \mathbf{0}$ is exact is equivalent to the assertion that the linear transformation T_1 is surjective.

i) What is the interpretation of the exactness of the sequence

$$\mathbf{0} \xrightarrow{T_0} V_1 \xrightarrow{T_1} V_2 \xrightarrow{T_2} \mathbf{0}?$$

ii) What is the interpretation of the exactness of the sequence

$$\mathbf{0} \xrightarrow{T_0} V_1 \xrightarrow{T_1} V_2 \xrightarrow{T_2} V_3 \xrightarrow{T_3} \mathbf{0},$$

traditionally called a **short exact sequence**?

iii) Show that $\sum_{i=1}^{n} (-1)^i \dim V_i = 0$ for any exact sequence of the form

$$\mathbf{0} \xrightarrow{T_0} V_1 \xrightarrow{T_1} V_2 \xrightarrow{T_2} \ldots \xrightarrow{T_{n-1}} V_n \xrightarrow{T_n} \mathbf{0},$$

a sequence beginning and ending with the trivial vector space $\mathbf{0}$.

2

Topological Fundamentals

2.1 Normed Spaces

The geometric as distinct from the algebraic properties of vector spaces involve the notion of the length or size of a vector. Formally a norm on a real vector space V is a mapping $V \longrightarrow \mathbb{R}$ that associates to any vector $\mathbf{x} \in V$ a real number $\|\mathbf{x}\| \in \mathbb{R}$, the **norm** of the vector $\mathbf{x}$, with the following properties:

$$\text{(i) } positivity : \|\mathbf{x}\| \geq 0 \text{ and } \|\mathbf{x}\| = 0 \text{ if and only if } \mathbf{x} = \mathbf{0}; \qquad (2.1)$$

$$\text{(ii) } homogeneity : \|c\mathbf{x}\| = |c| \|\mathbf{x}\| \text{ for any } c \in \mathbb{R};$$

$$\text{(iii) } the\ triangle\ inequality : \|\mathbf{x} + \mathbf{y}\| \leq \|\mathbf{x}\| + \|\mathbf{y}\|.$$

For the one-dimensional real vector space $\mathbb{R}$ the norm of a number $x \in \mathbb{R}$, usually denoted by $|x|$ and called the **absolute value** of x, is defined by

$$|x| = \begin{cases} x & \text{if} \quad x \geq 0, \\ -x & \text{if} \quad x \leq 0. \end{cases} \qquad (2.2)$$

That the absolute value satisfies the condition required in a norm is quite apparent; and it is clear that aside from a constant factor the absolute value is the only norm on the vector space $\mathbb{R}^1$, since for any norm $\|x\|$ on $\mathbb{R}^1$ it follows from homogeneity that $\|x\| = \|x \cdot 1\| = |x| \|1\|$. However for general vector spaces there are a variety of norms in quite common use; in particular there are the following three norms in the vector space $\mathbb{R}^n$ of dimension $n > 1$:

(i) the ℓ_1 or **mean norm** of a vector $\mathbf{x} = \{x_j\}$:
   $\|\mathbf{x}\|_1 = \sum_{j=1}^{n} |x_j|$;

(ii) the ℓ_2 or **Cartesian norm** or **Euclidean norm** of a vector $\mathbf{x} = \{x_j\}$:
   $\|\mathbf{x}\|_2 = \sqrt{\sum_{j=1}^{n} x_j^2} = \sqrt{x_1^2 + \cdots + x_n^2}$ with the nonnegative square root;

(iii) the ℓ_∞ or **supremum norm** or **sup norm** of a vector $\mathbf{x} = \{x_j\}$:
   $\|\mathbf{x}\|_\infty = \max_{1 \leq j \leq n} |x_j|$.

That the ℓ_1 norm satisfies the properties of a norm is clear since the absolute value of a real number is a norm. That the ℓ_∞ norm satisfies the properties of a norm also is clear, except perhaps for the triangle inequality; to verify that inequality, if $\mathbf{x}, \mathbf{y} \in \mathbb{R}^n$ then since $|x_j| \leq \|\mathbf{x}\|_\infty$ and $|y_j| \leq \|\mathbf{y}\|_\infty$ for $1 \leq j \leq n$ it follows that $|x_j + y_j| \leq |x_j| + |y_j| \leq \|\mathbf{x}\|_\infty + \|\mathbf{y}\|_\infty$ for $1 \leq j \leq n$ and consequently that $\|\mathbf{x} + \mathbf{y}\|_\infty = \max_{1 \leq j \leq n} |x_j + y_j| \leq \|\mathbf{x}\|_\infty + \|\mathbf{y}\|_\infty$. That the ℓ_2 norm satisfies these three properties also is obvious except for the triangle inequality. That can be demonstrated quite conveniently by introducing another linear function, the **inner product** of two vectors $\mathbf{x}, \mathbf{y} \in \mathbb{R}^n$, defined by

$$(\mathbf{x}, \mathbf{y}) = \sum_{j=1}^{n} x_j y_j \quad \text{for vectors } \mathbf{x} = \{x_j\} \quad \text{and} \quad \mathbf{y} = \{y_j\} \in \mathbb{R}^n; \tag{2.3}$$

often the inner product is written $(\mathbf{x}, \mathbf{y}) = \mathbf{x} \cdot \mathbf{y}$ and is called the dot product of the two vectors $\mathbf{x}$ and $\mathbf{y}$. The ℓ_2 norm is obviously defined in terms of the inner product by

$$\|\mathbf{x}\|_2 = \sqrt{(\mathbf{x}, \mathbf{x})} \quad \text{(with the nonnegative square root).} \tag{2.4}$$

Actually inner products on vector spaces arise more generally and play a significant role in both algebra and analysis, so it is advantageous to consider them rather more generally. A **real inner product space** is defined as a real vector space V with a mapping that assigns to any vectors $\mathbf{x}, \mathbf{y} \in V$ a real number $(\mathbf{x}, \mathbf{y}) \in \mathbb{R}$ with the following properties:

(i) *linearity* : $(c_1 \mathbf{x}_1 + c_2 \mathbf{x}_2, \mathbf{y}) = c_1 (\mathbf{x}_1, \mathbf{y}) + c_2 (\mathbf{x}_2, \mathbf{y})$ for any $c \in \mathbb{R}$; $\tag{2.5}$

(ii) *symmetry* : $(\mathbf{x}, \mathbf{y}) = (\mathbf{y}, \mathbf{x})$;

(iii) *positivity* : $(\mathbf{x}, \mathbf{x}) \geq 0$ and $(\mathbf{x}, \mathbf{x}) = 0$ if and only if $\mathbf{x} = \mathbf{0}$.

It is evident that the inner product (2.3) does satisfy these conditions, and that for any inner product on V the function

$$\|\mathbf{x}\| = \sqrt{(\mathbf{x}, \mathbf{x})} \geq 0 \quad \text{for any } \mathbf{x} \in V \tag{2.6}$$

satisfies the first two conditions in the definition of a norm; but the triangle inequality is demonstrated most easily by using the inner product, and will be demonstrated using just the basic properties (2.5) of an inner product on a real vector space V. It is apparent from symmetry (ii) and linearity (i) that an inner product $(\mathbf{x}, \mathbf{y})$ on a vector space V is also a linear function of the vector $\mathbf{y}$. Not so obvious perhaps is that an inner product is determined by the norm (2.6) it defines, since

$$(\mathbf{x}, \mathbf{y}) = \tfrac{1}{4} \|\mathbf{x} + \mathbf{y}\|^2 - \tfrac{1}{4} \|\mathbf{x} - \mathbf{y}\|^2 \tag{2.7}$$

which is demonstrated by noting that

$$\|\mathbf{x}+\mathbf{y}\|^2 - \|\mathbf{x}-\mathbf{y}\|^2 = (\mathbf{x}+\mathbf{y}, \mathbf{x}+\mathbf{y}) - (\mathbf{x}-\mathbf{y}, \mathbf{x}-\mathbf{y})$$

$$= ((\mathbf{x}, \mathbf{x}) + 2(\mathbf{x}, \mathbf{y}) + (\mathbf{y}, \mathbf{y})) - ((\mathbf{x}, \mathbf{x}) - 2(\mathbf{x}, \mathbf{y}) + (\mathbf{y}, \mathbf{y}))$$

$$= 4(\mathbf{x}, \mathbf{y}).$$

Equation (2.7) is called the **polarization identity**.

Theorem 2.1. *For any vectors* $\mathbf{x}, \mathbf{y}$ *in a real inner product space* V
(i) $|(\mathbf{x}, \mathbf{y})| \leq \|\mathbf{x}\|\|\mathbf{y}\|$, *and this is an equality if and only if the two vectors are linearly dependent;*
(ii) $\|\mathbf{x}+\mathbf{y}\| \leq \|\mathbf{x}\| + \|\mathbf{y}\|$, *and this is an equality if and only if one of the two vectors is a nonnegative multiple of the other.*

Proof: (i) The inequality (i) is trivial if either $\mathbf{x} = \mathbf{0}$ or $\mathbf{y} = \mathbf{0}$, so assume that $\mathbf{x}, \mathbf{y} \in V$ are nonzero vectors and set

$$\mathbf{z} = \mathbf{y} - \frac{(\mathbf{y}, \mathbf{x})}{\|\mathbf{x}\|^2}\, \mathbf{x}, \quad \text{so that} \quad (\mathbf{z}, \mathbf{x}) = (\mathbf{y}, \mathbf{x}) - \frac{(\mathbf{y}, \mathbf{x})}{\|\mathbf{x}\|^2}(\mathbf{x}, \mathbf{x}) = 0.$$

Then

$$0 \leq (\mathbf{z}, \mathbf{z}) = (\mathbf{z}, \mathbf{y}) - \frac{(\mathbf{y}, \mathbf{x})}{\|\mathbf{x}\|^2}(\mathbf{z}, \mathbf{x})$$

$$= (\mathbf{z}, \mathbf{y}) = \|\mathbf{y}\|^2 - \frac{(\mathbf{y}, \mathbf{x})}{\|\mathbf{x}\|^2}(\mathbf{x}, \mathbf{y}),$$

so that $\|\mathbf{x}\|^2\|\mathbf{y}\|^2 - (\mathbf{y}, \mathbf{x})^2 \geq 0$, which implies the inequality (i); and this inequality is an equality if and only if $\mathbf{z} = \mathbf{0}$ so the vectors $\mathbf{x}$ and $\mathbf{y}$ are linearly dependent.
(ii) Assuming the inequality (i)

$$\|\mathbf{x}+\mathbf{y}\|^2 = (\mathbf{x}+\mathbf{y}, \mathbf{x}+\mathbf{y}) = \|\mathbf{x}\|^2 + 2(\mathbf{x}, \mathbf{y}) + \|\mathbf{y}\|^2$$

$$\leq \|\mathbf{x}\|^2 + 2|(\mathbf{x}, \mathbf{y})| + \|\mathbf{y}\|^2$$

$$\leq \|\mathbf{x}\|^2 + 2\|\mathbf{x}\|\|\mathbf{y}\| + \|\mathbf{y}\|^2 = (\|\mathbf{x}\| + \|\mathbf{y}\|)^2, \tag{2.8}$$

which implies the inequality (ii). If $\|\mathbf{x}+\mathbf{y}\| = \|\mathbf{x}\| + \|\mathbf{y}\|$ the inequalities in (2.8) are all equalities so $2|(\mathbf{x}, \mathbf{y})| = 2\|\mathbf{x}\|\|\mathbf{y}\|$ and then by (i) the vectors $\mathbf{x}$ and $\mathbf{y}$ must be linearly dependent; if $\mathbf{x} \neq \mathbf{0}$ then $\mathbf{y} = c\mathbf{x}$ for some $c \in \mathbb{R}$, and substituting this into the equality $\|\mathbf{x}+\mathbf{y}\| = \|\mathbf{x}\| + \|\mathbf{y}\|$ it follows that $|1 + c|\|\mathbf{x}\| = \|(1 + c)\mathbf{x}\| = \|\mathbf{x}+\mathbf{y}\| = \|\mathbf{x}\| + \|\mathbf{y}\| = \|\mathbf{x}\| + \|c\mathbf{x}\| = (1 + |c|)\|\mathbf{x}\|$ so $|1 + c| = 1 + |c|$ and consequently $c \geq 0$, which suffices for the proof.

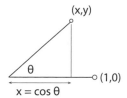

Figure 2.1. Geometrical interpretation of the Cauchy-Schwarz inequality in $\mathbb{R}^2$

The very useful inequality (i) in the preceding theorem, called the **Cauchy-Schwarz inequality**, can be written as the assertion that for any two nonzero vectors $\mathbf{x}, \mathbf{y} \in V$ the quotient

$$Q(\mathbf{x}, \mathbf{y}) = \frac{(\mathbf{x}, \mathbf{y})}{\|\mathbf{x}\|\|\mathbf{y}\|} \quad \text{satisfies} \quad -1 \leq Q(\mathbf{x}, \mathbf{y}) \leq 1. \tag{2.9}$$

Although the cosine function only will be defined later in Section 3.3, the elementary geometric properties of that function are probably sufficiently familiar to the readers that they can be assumed for the purposes of the next definitions. Thus it follows from (2.9) that there is a unique angle $\theta(\mathbf{x}, \mathbf{y})$ such that

$$0 \leq \theta(\mathbf{x}, \mathbf{y}) \leq \pi \quad \text{and} \quad \cos \theta(\mathbf{x}, \mathbf{y}) = Q(\mathbf{x}, \mathbf{y}). \tag{2.10}$$

It is customary to call this angle $\theta(\mathbf{x}, \mathbf{y})$ the **angle between the vectors x and y**. For the usual ℓ_2 norm and its inner product, as in the accompanying Figure 2.1, if $(x, y) \in \mathbb{R}^2$ is a vector of length $\|(x, y)\|_2 = 1$ at the origin in the plane of the variables x, y making an angle θ with the real axis then in the right triangle in the figure $x = \cos \theta$ and $((1, 0), (x, y)) = x$ and since $\|(1, 0)\|_2 = \|(x, y)\|_2 = 1$ then

$$Q((1, 0), (x, y)) = \frac{((1, 0), (x, y))}{\|(1, 0)\|_2 \|(x, y)\|_2} = x.$$

Thus by definition (2.10) the angle θ between the vectors $(1, 0)$ and (x, y) is defined by $\cos \theta = x$, showing that at least in this case the definition (2.10) of the angle between these two vectors agrees with the usual geometric sense of the angle between them. Two nonzero vectors $\mathbf{x}, \mathbf{y} \in V$ in a general real inner product space are said to be **parallel** and in the same direction if $Q(\mathbf{x}, \mathbf{y}) = 1$, so the angle between them is $\theta(\mathbf{x}, \mathbf{y}) = 0$; and the vectors are said to be parallel and in the opposite direction if $Q(\mathbf{x}, \mathbf{y}) = -1$, so the angle between them is $\theta(\mathbf{x}, \mathbf{y}) = \pi$. Equivalently two nonzero vectors are parallel and in the same direction if one of them is a positive multiple of the other, and they are parallel and in the opposite direction if one of them is a negative multiple of the other; that is evident from the proof of Theorem 2.1. The two vectors are said to be **orthogonal** or **perpendicular** if $Q(\mathbf{x}, \mathbf{y}) = 0$, so the angle between them is $\theta(\mathbf{x}, \mathbf{y}) = \pi/2$ or $3\pi/2$. With this interpretation the inner product can be described alternatively as

$$(\mathbf{x}, \mathbf{y}) = \|\mathbf{x}\| \|\mathbf{y}\| \cos\theta(\mathbf{x}, \mathbf{y}), \tag{2.11}$$

in terms of the angle $\theta(\mathbf{x}, \mathbf{y})$ between the vectors $\mathbf{x}$ and $\mathbf{y}$. This geometrical interpretation of the Cauchy-Schwarz inequality makes the classical ℓ_2 norm particularly useful in many applications.

One special case deserves some additional attention. The field of complex numbers $\mathbb{C}$ was defined at the end of Section 1.3 as one of the two-dimensional real algebras, and as such it has the natural structure of a two-dimensional real vector space. A complex number $z = x + iy$ can be identified with the vector $\{x, y\} \in \mathbb{R}^2$, so that $\|z\|_2^2 = x^2 + y^2$. On the other hand $z\bar{z} = (x + iy)(x - iy) = x^2 + y^2$, so the ℓ_2 norm also can be defined by $\|z\|_2^2 = z\bar{z}$. The advantage of this alternative expression is that for any two complex numbers z_1, z_2 the norm of the product satisfies $\|z_1 z_2\|_2^2 = (z_1 z_2)\overline{(z_1 z_2)} = (z_1\bar{z_1}) \cdot (z_2\bar{z_2}) = \|z_1\|_2^2 \|z_2\|_2^2$, and consequently $\|z_1 z_2\|_2 = \|z_1\|_2 \|z_2\|_2$; so the ℓ_2 norm respects the multiplication of the complex numbers, much the same way that the absolute value norm respects the multiplication of the real numbers. For that reason this norm is usually denoted by $|z|$ rather than $\|z\|_2$, and it is called the **absolute value** or in the older literature the **modulus** of the complex number z.

There are a great number of inequalities such as the Cauchy-Schwarz inequality that play significant roles throughout mathematics. Another such inequality is related to the l_1 norm, which sometimes is called the **mean norm** since $\frac{1}{n}\|\mathbf{x}\|_1$ is the mean or average value of the n coordinates x_j of any vector $\mathbf{x} = \{x_j\} \in \mathbb{R}^n$; the mean or average often also is called the **arithmetic mean** of the n real numbers x_j. The multiplicative analogue of the arithmetic mean is the **geometric mean**, the value $(|x_1||x_2| \cdots |x_n|)^{1/n}$.

Theorem 2.2 (Inequality of the Arithmetic and Geometric Means).
For any real numbers $x_j \in \mathbb{R}$

$$|x_1 x_2 \cdots x_n| \leq \mu^n \tag{2.12}$$

where

$$\mu = \frac{1}{n}(|x_1| + |x_2| + \cdots + |x_n|); \tag{2.13}$$

and this is an equality if and only if $|x_1| = \cdots = |x_n| = \mu$.

Proof: It is clearly sufficient to prove the result under the additional hypothesis that $x_j > 0$ for all j, since the result holds trivially if any variable is 0 and the absolute values of the variables are all nonnegative. The proof will be by induction on n. The case $n = 1$ holds trivially, so assume that the result has been demonstrated for some $n \geq 1$ and let

$$\mu = \frac{1}{n+1}(x_1 + \cdots + x_n + x_{n+1})$$

for real numbers $x_j > 0$. If $x_j = \mu$ for all indices $1 \leq j \leq n+1$ the desired result holds trivially; so it can be assumed that $x_n > \mu$ and $\mu > x_{n+1}$, by reordering the x_j suitably. In that case

$$y = x_n + x_{n+1} - \mu \geq x_n - \mu > 0$$

and the induction assumption can be applied to the n real numbers $x_1, \ldots, x_{n-1}, y$, for which $x_1 + \cdots + x_{n-1} + y = x_1 + \cdots + x_{n-1} + x_n + x_{n+1} - \mu = n\mu$ so that

$$x_1 \cdots x_{n-1} \cdot y \leq \mu^n$$

and consequently

$$x_1 \cdots x_{n-1} \cdot y \cdot \mu \leq \mu^{n+1}. \tag{2.14}$$

Now

$$y \cdot \mu - x_n \cdot x_{n+1} = (x_n + x_{n+1} - \mu)\mu - x_n \cdot x_{n+1}$$
$$= (x_n - \mu)(\mu - x_{n+1}) > 0$$

so $x_n \cdot x_{n+1} < y \cdot \mu$ and substituting this into (2.14) shows that

$$x_1 \cdots x_{n+1} < x_1 \cdots x_{n-1} \cdot y \cdot \mu \leq \mu^{n+1},$$

which suffices for the proof.

In any vector space there is a useful notion of the **equivalence** of two norms $\|\mathbf{x}\|_a$ and $\|\mathbf{x}\|_b$, defined by

$$\|\mathbf{x}\|_a \asymp \|\mathbf{x}\|_b \quad \text{if there are real numbers } c_a > 0, c_b > 0 \text{ such that} \tag{2.15}$$
$$\|\mathbf{x}\|_a \leq c_a \|\mathbf{x}\|_b \quad \text{and} \quad \|\mathbf{x}\|_b \leq c_b \|\mathbf{x}\|_a \quad \text{for all } \mathbf{x} \in V.$$

That this is actually an equivalence relation is quite straightforward. Indeed reflexivity and symmetry are clear; as for transitivity, if $\|\mathbf{x}\|_a \asymp \|\mathbf{x}\|_b$ there are positive real numbers c_a, c_b such that $\|\mathbf{x}\|_a \leq c_a\|\mathbf{x}\|_b$ and $\|\mathbf{x}\|_b \leq c_b\|\mathbf{x}\|_a$ while if $\|\mathbf{x}\|_b \asymp \|\mathbf{x}\|_c$ there are positive real numbers k_b, k_c such that $\|\mathbf{x}\|_b \leq k_b\|\mathbf{x}\|_c$ and

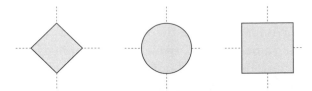

Figure 2.2. The unit balls $B(\mathbf{0}, 1)$ for the ℓ_1, ℓ_2, and ℓ_∞ norms on $\mathbb{R}^2$

$\|\mathbf{x}\|_c \leq k_c\|\mathbf{x}\|_b$, hence $\|\mathbf{x}\|_a \leq c_a\|\mathbf{x}\|_b \leq c_a k_b\|\mathbf{x}\|_c$ and $\|\mathbf{x}\|_c \leq k_c\|\mathbf{x}\|_b \leq k_c c_b\|\mathbf{x}\|_a$ so that $\|\mathbf{x}\|_a \asymp \|\mathbf{x}\|_c$.

A **ball** of radius $r \geq 0$ centered at a point $\mathbf{a}$ in a vector space V with the norm $\|\mathbf{x}\|$ is defined to be the set

$$B(\mathbf{a}, r) = \left\{ \mathbf{x} \in V \mid \|\mathbf{x} - \mathbf{a}\| \leq r \right\}. \tag{2.16}$$

The norm in a normed vector space V is actually determined by the unit ball at the origin in that vector space, for it is apparent that

$$\|\mathbf{x}\| = \inf \left\{ r \geq 0 \mid \mathbf{x} \in B(\mathbf{0}, r) \right\}. \tag{2.17}$$

Consequently two norms $\|\mathbf{x}\|_a$ and $\|\mathbf{x}\|_b$ on a vector space V are equivalent if and only if there are positive real numbers d_a and d_b such that

$$B_a(\mathbf{0}, 1) \subset d_a B_b(\mathbf{0}, 1) \quad \text{and} \quad B_b(\mathbf{0}, 1) \subset d_b B_a(\mathbf{0}, 1) \tag{2.18}$$

where $B_a(\mathbf{a}, r)$ is the ball defined in terms of the norm $\|\mathbf{x}\|_a$ while $B_b(\mathbf{a}, r)$ is the ball defined in terms of the norm $\|\mathbf{x}\|_b$ and $d_a B_b(\mathbf{0}, 1)$ denotes the set of all points $d_a \mathbf{x} \in V$ for which $\mathbf{x} \in B_b(\mathbf{0}, 1)$. The three norms defined previously are described by unit balls of rather different shapes, as indicated in Figure 2.2; it is rather clear geometrically from that figure that these three norms are equivalent, and it is easy to prove it explicitly.

Theorem 2.3. *For any vector* $\mathbf{x} \in \mathbb{R}^n$

$$\|\mathbf{x}\|_\infty \leq \|\mathbf{x}\|_1 \leq n\|\mathbf{x}\|_\infty \quad \text{and} \quad \|\mathbf{x}\|_\infty \leq \|\mathbf{x}\|_2 \leq \sqrt{n}\|\mathbf{x}\|_\infty; \tag{2.19}$$

consequently the ℓ_1, ℓ_2, *and* ℓ_∞ *norms are equivalent to one another.*

Proof: First $\|\mathbf{x}\|_1 = \sum_{j=1}^n |x_j| \geq |x_{j_0}|$ for any particular index j_0 so $\|\mathbf{x}_1\| \geq \|\mathbf{x}\|_\infty$, and similarly $\|\mathbf{x}\|_2^2 = \sum_{j=1}^n |x_j|^2 \geq |x_{j_0}|^2$ for any particular index j_0 so $\|\mathbf{x}_2\| \geq \|\mathbf{x}\|_\infty$. On the other hand $\|\mathbf{x}\|_1 = \sum_{j=1}^n |x_j| \leq \sum_{j=1}^n \|\mathbf{x}\|_\infty = n\|\mathbf{x}\|_\infty$ and similarly $\|\mathbf{x}\|_2^2 = \sum_{j=1}^n x_j^2 \leq \sum_{j=1}^n \|\mathbf{x}\|_\infty^2 = n\|\mathbf{x}\|_\infty^2$ hence $\|\mathbf{x}\|_2 \leq \sqrt{n}\|\mathbf{x}\|_\infty$. That demonstrates (2.19), which shows that $\|\mathbf{x}\|_1 \asymp \|\mathbf{x}\|_\infty$ and $\|\mathbf{x}\|_2 \asymp \|\mathbf{x}\|_\infty$, and consequently by symmetry and transitivity $\|\mathbf{x}\|_1 \asymp \|\mathbf{x}\|_2$, which suffices for the proof.

Another sort of norm is defined on the vector space $\mathcal{L}(V, W)$ of linear transformations between two normed vector spaces V and W; but since this norm depends explicitly on the norms chosen in the vector spaces V and W it is useful to modify the notation to indicate the norms in V and W explicitly, writing $\mathcal{L}(V, \ell_V; \ W, \ell_W)$ for the space of linear transformations between V and W when these vector spaces have ℓ_V and ℓ_W norms. A linear transformation $T \in \mathcal{L}(V, \ell_V; \ W, \ell_W)$ for which the quotients $\|T\mathbf{x}\|_w / \|\mathbf{x}\|_v$ are bounded above for all $x \in V \sim \{\mathbf{0}\}$ is called a **bounded linear transformation**. The **operator norm** $\|T\|_o$ for a bounded linear transformation s defined by

$$\|T\|_o = \sup\left\{ \frac{\|T\mathbf{x}\|_w}{\|\mathbf{x}\|_v} \ \middle| \ \mathbf{x} \in V, \ \mathbf{x} \neq \mathbf{0} \right\} \tag{2.20}$$

for any linear transformation $T \in \mathcal{L}(V, \ell_V; \ W, \ell_W)$. It is apparent from this definition that the operator norm $\|T\|_o$ can be defined alternatively as the infimum of all real numbers $C \geq 0$ for which $\|T\mathbf{x}\|_W \leq C\|\mathbf{x}\|_V$ for all $\mathbf{x} \in V$. As will be seen in the next paragraph, any linear transformation between finite dimensional vector spaces is a bounded linear transformation; but that is not the case even for some quite naturally defined linear transformations between infinite dimensional vector spaces. That the operator norm actually is a norm for bounded linear transformations is clear; it is obviously positive and homogeneous, and if $S, T \in \mathcal{L}(V, \|\mathbf{x}\|_v; \ W, \|\mathbf{x}\|_w)$ are bounded linear transformations then

$$\frac{\|(S + T)(\mathbf{x})\|_w}{\|\mathbf{x}\|_v} \leq \frac{\|S(\mathbf{x})\|_w}{\|\mathbf{x}\|_v} + \frac{\|T(\mathbf{x})\|_w}{\|\mathbf{x}\|_v} \quad \text{for all } \mathbf{x} \in V$$

hence $\|S + T\|_o \leq \|S\|_o + \|T\|_o$ so the triangle inequality holds.

Now linear transformations $A \in \mathcal{L}(\mathbb{R}^m, \mathbb{R}^n)$ are described by matrices $A \in \mathbb{R}^{n \times m}$, and when these matrices are viewed as vectors in $\mathbb{R}^{nm}$ the ℓ_1, ℓ_2, and ℓ_∞ norms can be defined on these matrices just as for any vectors; for instance

$$\|A\|_\infty = \max_{\substack{1 \leq i \leq n \\ 1 \leq j \leq m}} \left| a_{ij} \right| \quad \text{for a matrix} \quad A = \{a_{ij}\} \in \mathbb{R}^{n \times m},$$

and correspondingly for the ℓ_1 and ℓ_2 norms. For any vector $\mathbf{x} \in \mathbb{R}^m$ and any index $1 \leq i \leq n$

$$\left| \sum_{j=1}^{m} a_{ij}x_j \right| \leq \sum_{j=1}^{m} |a_{ij}||x_j| \leq \sum_{j=1}^{m} \|A\|_\infty \|\mathbf{x}\|_\infty = m \|A\|_\infty \|\mathbf{x}\|_\infty$$

so

$$\|A\mathbf{x}\|_\infty = \max_{1 \leq i \leq n} \left| \sum_{j=1}^{m} a_{ij}x_j \right| \leq m\|A\|_\infty \|\mathbf{x}\|_\infty. \tag{2.21}$$

Thus a matrix A as a linear transformation $A \in \mathcal{L}(\mathbb{R}^m, \ell_\infty;\ \mathbb{R}^n, \ell_\infty)$ is a bounded linear transformation so its operator norm is well defined and has the bound

$$\|A\|_o \le m\|A\|_\infty \quad \text{for } A \in \mathcal{L}(\mathbb{R}^m, \ell_\infty;\ \mathbb{R}^n, \ell_\infty). \tag{2.22}$$

On the other hand for any $A \in \mathcal{L}(\mathbb{R}^m, \|\mathbf{x}\|_\infty;\ \mathbb{R}^n, \|\mathbf{x}\|_\infty)$ it follows from the definition of the operator norm that for any basis vector $\delta_j \in \mathbb{R}^m$

$$\|A\|_o \ge \frac{\|A\delta_j\|_\infty}{\|\delta_j\|_\infty} = \|A\delta_j\|_\infty = \max_{1 \le i \le n} |a_{ij}|$$

and since that is the case for all basis vectors δ_j it follows that conversely

$$\|A\|_o \ge \|A\|_\infty \quad \text{for } A \in \mathcal{L}(\mathbb{R}^m, \ell_\infty;\ \mathbb{R}^n, \ell_\infty). \tag{2.23}$$

Equations from (2.22) and (2.23) show that the operator norm on a linear transformation $A \in \mathcal{L}(\mathbb{R}^m, \ell_\infty;\ \mathbb{R}^n, \ell_\infty)$ is equivalent to the ℓ_∞ norm on A; and since the ℓ_∞ norm is equivalent to the ℓ_1 and ℓ_2 norms on any vector space it follows that the operator norm on matrices is also equivalent to the ℓ_1 and ℓ_2 norms on matrices.

In many situations the particular norm is not terribly relevant since any equivalent norm could be used; if that is the case and if it is not necessary to specify just which norm is involved, the notation $\|\mathbf{x}\|$ will be used, meaning any one of a set of equivalent norms where that set of norms is either specified or understood. In particular when considering the vector spaces $\mathbb{R}^m$ or the vector spaces $\mathcal{L}(\mathbb{R}^m, \mathbb{R}^n)$ the notation $\|\mathbf{x}\|$ stands for any of the various equivalent norms. Some care must be taken, of course, for in some calculations the precise relations between the norms is significant; in particular $\|\mathbf{x}\|$ should have the same meaning in any single equation, or usually throughout any single proof, for quite obvious reasons.

Problems, Group I

1. Show that in a normed vector space $\big|\|\mathbf{x}\| - \|\mathbf{y}\|\big| \le \|\mathbf{x} - \mathbf{y}\|$.

2. Show that if $\|\mathbf{x}\|_V$ is a norm on a vector space V and $\|\mathbf{y}\|_W$ is a norm on a vector space W then both $\|(\mathbf{x}, \mathbf{y})\| = \|\mathbf{x}\|_V + \|\mathbf{y}\|_W$ and $\|(\mathbf{x}, \mathbf{y})\|_\infty = \max\big(\|\mathbf{x}\|_V, \|\mathbf{y}\|_W\big)$ are norms on the direct sum vector space $V \oplus W$, and the two norms are equivalent.

3. Show that if V is a normed vector space then in the vector space $\mathcal{L}(V, V)$ with the operator norm $\|\ \|_o$

    i) $\|I\|_o = 1$ for the identity mapping $I \in \mathcal{L}(V, V)$ and

    ii) $\|ST\|_o \leq \|S\|_o\|T\|_o$ for any $S, T \in \mathcal{L}(V, V)$.

    iii) Find an example in which $\|ST\|_o < \|S\|_o\|T\|_o$.

4.   Show that the closed unit ball $B(\mathbf{0}, 1)$ in a normed vector space is a convex set, in the sense that if $\mathbf{a} \in B(\mathbf{0}, 1)$ and $\mathbf{b} \in B(\mathbf{0}, 1)$ then the entire line segment

$$\ell(\mathbf{a}, \mathbf{b}) = \left\{ (1 - t)\mathbf{a} + t\mathbf{b} \mid 0 \leq t \leq 1 \right\}$$

from the point $\mathbf{a}$ to the point $\mathbf{b}$ is contained in $B(\mathbf{0}, 1)$.

Problems, Group II

5.   i) Show that a norm $\|\mathbf{x}\|$ on vectors in $\mathbb{R}^n$ can be defined by an inner product $(\mathbf{x}_1, \mathbf{x}_2)$ on $\mathbb{R}^n$, in the sense that there is an inner product for which $\|\mathbf{x}\|^2 = (\mathbf{x}, \mathbf{x})$, if and only if the norm satisfies the *parallelogram law*:

$$\|\mathbf{x} + \mathbf{y}\|^2 + \|\mathbf{x} - \mathbf{y}\|^2 = 2\|\mathbf{x}\|^2 + 2\|\mathbf{y}\|^2.$$

    ii) As a consequence, show that the ℓ_1 and ℓ_∞ norms on $\mathbb{R}^n$ cannot be defined by inner products.

6.   The set of all real polynomials in a single variable x is naturally a real vector space $\mathcal{P}$; although it is not a vector space of finite dimension, nonetheless it is possible to define norms on this vector space.

    i) Show that the expressions $\|p(x)\|_\infty$ and $\|p(x)\|_c$ defined for a polynomial $p(x) = a_0 + a_1 x + \cdots + a_n x^n$ of any degree n by

$$\|p(x)\|_\infty = \sup_{-\frac{1}{2} \leq x \leq \frac{1}{2}} |p(x)| \quad \text{and} \quad \|p(x)\|_c = \max(|a_0|, |a_1|, \ldots, |a_n|)$$

    are norms on the vector space $\mathcal{P}$.

    ii) Show that the norms $\|p(x)\|_\infty$ and $\|p(x)\|_c$ on $\mathcal{P}$ are not equivalent norms.

7.   Show that if $\|\mathbf{x}\|$ is any norm on $\mathbb{R}^n$ there is a real constant $M > 0$ such that $\|\mathbf{x}\| \leq M\|\mathbf{x}\|_\infty$ for all $\mathbf{x} \in \mathbb{R}^n$.

8.   Show that $\|\mathbf{x}\|_{\frac{1}{2}} = (|x_1|^{\frac{1}{2}} + |x_2|^{\frac{1}{2}})^2$ does not define a norm in $\mathbb{R}^2$. [Suggestion: Consider the result of problem (4). Comment: Compare this definition with that in the next problem.]

9. The ℓ_p norm on $\mathbb{R}^n$ for any $p \geq 1$ is defined by $\|\mathbf{x}\|_p = \left(\sum_{j=1}^{n} |x_j|^p\right)^{\frac{1}{p}}$.

i) Assuming the inequality $xy \leq \frac{x^p}{p} + \frac{y^q}{q}$ for $x > 0, y > 0$ and for $p \geq 1, q \geq 1$ where $\frac{1}{p} + \frac{1}{q} = 1$, demonstrate Hölder's inequality

$$|(\mathbf{x}, \mathbf{y})| \leq \|\mathbf{x}\|_p \|\mathbf{y}\|_q \quad \text{if} \quad p, q \geq 1, \quad \frac{1}{p} + \frac{1}{q} = 1$$

where $(\mathbf{x}, \mathbf{y}) = \sum_{j=1}^{n} x_j y_j$. [Comment: This is just the Cauchy-Schwarz inequality in the special case that $p = q = 2$, so in a sense it is a generalization of that inequality to the more general case of the ℓ_p norm.]

ii) Show that $\|\mathbf{x}\|_p$ is actually a norm on $\mathbb{R}^n$, in particular that

$$\|c\mathbf{x}\|_p = |c| \, \|\mathbf{x}\|_p, \quad \text{and} \quad \|\mathbf{x} + \mathbf{y}\|_p \leq \|\mathbf{x}\|_p + \|\mathbf{y}\|_p$$

for any $\mathbf{x}, \mathbf{y} \in \mathbb{R}^n$ and $c \in \mathbb{R}$. [Suggestion: $|x_j|^p = |x_j||x_j|^{p-1}$ to which the Hölder inequality can be applied.]

iii) Show that $\|\mathbf{x}\|_\infty = \lim_{p \to \infty} \|\mathbf{x}\|_p$ for any fixed point $\mathbf{x} \in \mathbb{R}^n$, thus in a sense explaining the notation for the supremum norm.

iv)* As a purely optional problem, in case you are interested in trying a challenge, prove the inequality $xy \leq \frac{x^p}{p} + \frac{y^q}{q}$ for $x, y > 0, p, q > 1$, and $\frac{1}{p} + \frac{1}{q} = 1$.

2.2 Metric Spaces

In a normed vector space the norm $\|\mathbf{x} - \mathbf{y}\|$ can be interpreted as the distance between the two vectors $\mathbf{x}$ and $\mathbf{y}$; and the basic properties of norms can be translated into basic properties for the notion of the distance between two vectors. The notion of distance between points is important in more general sets than just normed vector spaces, and it is convenient to axiomatize that notion and to derive the general properties that are applicable in a wide range of particular cases. For this purpose a **metric** on a set S is traditionally defined to be a mapping $\rho : S \times S \longrightarrow \mathbb{R}$ with the following properties:

(i) *positivity*, $\rho(x, y) \geq 0$ and $\rho(x, y) = 0$ if and only if $x = y$;
(ii) *symmetry*, $\rho(x, y) = \rho(y, x)$; and
(iii) *the triangle inequality*, $\rho(x, y) \leq \rho(x, z) + \rho(z, y)$.

A **metric space** is a set S together with a metric ρ defined on it; a metric space does not involve just a set but also a specified metric on that set. In particular a normed vector space V with a norm $\|\mathbf{x}\|$ is a metric space with the **norm**

metric defined by

$$\rho(\mathbf{x}, \mathbf{y}) = \|\mathbf{x} - \mathbf{y}\|; \tag{2.24}$$

the positivity and symmetry of the mapping ρ follow immediately from the positivity and homogeneity of the norm, and the triangle inequality of the mapping ρ follows equally immediately from that of the norm. However metrics can be defined on a wide range of sets other than vector spaces, and not even all metrics on a vector space are derived from norms on that vector space. For instance the **discrete metric** is defined on an arbitrary set S by

$$\rho_d(x, y) = \begin{cases} 0 & \text{if} \quad x = y, \\ 1 & \text{if} \quad x \neq y; \end{cases} \tag{2.25}$$

it is clear that this satisfies the conditions necessary to define a metric on S, but that if $S = \mathbb{R}^n$ this metric does not arise from any norm on $\mathbb{R}^n$, since for a norm metric $\rho(c\mathbf{x}, c\mathbf{y}) = |c|\rho(\mathbf{x}, \mathbf{y})$ for any scalar $c \in \mathbb{R}$. Any subset $R \subset S$ of a metric space S clearly also is a metric space under the restriction of the metric on S. The metric subspace R is said to be **dense** in S if for any point $s \in S$ and any $\epsilon > 0$ there is a point $r \in R$ such that $\rho(r, s) < \epsilon$. For example, the set $\mathbb{R}$ of real numbers is a metric space with the metric defined by the absolute value norm; the subset $\mathbb{Q} \subset \mathbb{R}$ of rational numbers is dense in $\mathbb{R}$ with this metric. It is occasionally the case that a metric subspace is easier to describe or to work with than the full metric space; and if the metric subspace is dense it in turn can be used to describe the full metric space.

It is possible to define a metric on a set S more generally as a mapping $\rho : S \times S \longrightarrow F$ for any ordered field F, such as the field $\mathbb{Q}$ of rational numbers, with the three basic properties of positivity, symmetry, and the triangle inequality; the order in F defines positivity and therefore what is meant by inequalities. The results to be discussed in this section hold for metrics defined as mappings to arbitrary ordered fields, unless specifically restricted to real metric spaces, as in Theorem 2.7 for instance; but since the principal interest here is analytical, the discussion will continue to be expressed in terms of real metrics, with fairly little note of the more general situation.

Some spaces are naturally equipped with a mapping $\rho : S \times S \longrightarrow \mathbb{R}$ that satisfies all of the conditions for a metric except possibly that $\rho(x, y) = 0$ for some elements $x, y \in S$ for which $x \neq y$; such a mapping is called a **pseudometric**, and a set equipped with such a mapping is called a **pseudometric space**. On any such space it is possible to introduce an equivalence relation by setting $x \asymp y$ whenever $\rho(x, y) = 0$. It is clear that this relation satisfies the reflexivity and symmetry conditions for an equivalence relation. If $x \asymp y$ and $y \asymp z$ so that $\rho(x, y) = \rho(y, z) = 0$ then by the triangle inequality $\rho(x, z) \leq \rho(x, y) + \rho(y, z) = 0$

hence $\rho(x, z) = 0$ so $x \asymp z$; thus this relation also satisfies the transitivity condition, hence it is an equivalence relation. If $x \asymp x'$ and $y \asymp y'$ then by the triangle inequality $\rho(x', y') \le \rho(x', x) + \rho(x, y) + \rho(y, y') = \rho(x, y)$; reversing the roles of x, y and x', y' yields the reverse inequality, so that $\rho(x', y') = \rho(x, y)$. Consequently it is possible to define a mapping $\rho^* : S^* \times S^* \longrightarrow \mathbb{R}$ by setting $\rho^*(x^*, y^*) = \rho(x, y)$ for any $x \in S$ representing the equivalence class x^* and any $y \in S$ representing the equivalence class y^*. It is easy to see that this defines a metric on the space S^*; and in this way there is a metric space S^* that is naturally associated to any pseudometric space S, and it often can be used in place of the pseudometric space S.

A space of course can have several metrics, just as normed vector spaces can have several different norms. In parallel to the discussion of norms, there is a notion of **equivalence** of two metrics ρ_a and ρ_b on a set S, defined by

$$\rho_a \asymp \rho_b \ \text{ if there are constants } c_a > 0, c_b > 0 \text{ such that} \tag{2.26}$$

$$\rho_a(x, y) \le c_a \, \rho_b(x, y) \ \text{ and } \ \rho_b(x, y) \le c_b \, \rho_a(x, y) \ \text{ for all } \ x, y \in S.$$

It is obvious from this definition and (2.15) that equivalent norms on a vector space V determine equivalent metrics on the set V; and it is fairly obvious that the discrete metric is not equivalent to the metric defined by a norm on a normed vector space. On the other hand different spaces can have essentially the same metric. A mapping $\phi : S \longrightarrow T$ from a metric space S with the metric ρ_S to a metric space T with the metric ρ_T is said to be an **isometry** if $\rho_T(\phi(x), \phi(y)) = \rho_S(x, y)$ for all points $x, y \in S$; clearly any isometry is an injective mapping. Two metric spaces S and T are said to be **isometric** if there is a bijective isometry $\phi : S \longrightarrow T$; and in that case clearly the inverse mapping is a bijective isometry $\phi^{-1} : T \longrightarrow S$. As far as the structure of a metric space is concerned, isometric metric spaces are essentially the same metric space.

A **sequence** in a metric space S is a set of points of S indexed by the natural numbers; thus a sequence in S is an indexed set such as $\{a_1, a_2, a_3, \dots\}$ where $a_\nu \in S$, often written as a set $\{a_\nu\}$ of points $a_\nu \in S$ for all $\nu \in \mathbb{N}$. A sequence $\{a_\nu\}$ does not necessarily consist of distinct points; for instance setting $a_\nu = a$ for all ν yields a perfectly acceptable sequence. A sequence $\{a_\nu\}$ **converges** to a or has the **limit** a, indicated by writing $\lim_{\nu \to \infty} a_\nu = a$, if for any $\epsilon > 0$ there is a number $N = N(\epsilon)$ depending on ϵ such that $\rho(a_\nu, a) < \epsilon$ whenever $\nu > N(\epsilon)$ where ρ is the metric in S. It is easy to see that if a sequence a_ν converges to a in terms of a metric ρ then it also converges to a in terms of any metric equivalent to ρ. A sequence does not necessarily converge or have a limit; for example in the metric space $\mathbb{R}$ with the metric defined by the norm $|x|$ the sequence of real numbers given by $a_\nu = \nu$ does not converge, nor does the sequence of real numbers given by $a_\nu = (-1)^\nu$. However if a sequence has a limit that limit is unique; for if a' and a'' are both limits of a sequence a_ν

then for any $\epsilon > 0$ there is a number N such that $\rho(a_\nu, a') < \epsilon$ and $\rho(a_\nu, a'') < \epsilon$ whenever $\nu > N$, and for any such ν it follows from the triangle inequality that $\rho(a', a'') \leq \rho(a', a_\nu) + \rho(a_\nu, a'') < 2\epsilon$, which holds for all $\epsilon > 0$ so $\rho(a', a'') = 0$ hence $a' = a''$. The sequence of the form $a_\nu = a$ clearly has the limit a and the sequence of real numbers $a_\nu = 1/\nu$ clearly has the limit 0. It is easy to see that if $\{a_\nu\}$ and $\{b_\nu\}$ are two convergent sequences then the sequence $\{a_\nu + b_\nu\}$ is also convergent and $\lim_{\nu\to\infty}(a_\nu + b_\nu) = \lim_{\nu\to\infty} a_\nu + \lim_{\nu\to\infty} b_\nu$; and if $a_\nu \leq b_\nu$ for all ν then $\lim_{\nu\to\infty} a_\nu \leq \lim_{\nu\to\infty} b_\nu$.

As might be expected, the question whether a sequence converges or not may be quite difficult to answer for some sequences.[1] For that reason it is important to have tests to see whether sequences converge or not that do not involve knowing or conjecturing what the limit of the sequence is. A sequence a_ν is a **Cauchy sequence** if for any $\epsilon > 0$ there is a number $N(\epsilon)$ depending on ϵ such that $\rho(a_\mu, a_\nu) < \epsilon$ whenever $\mu, \nu > N(\epsilon)$. It is easy to see that if a sequence a_ν is a Cauchy sequence in terms of a metric ρ then it is also a Cauchy sequence in terms of any metric equivalent to ρ. It is clear that any convergent sequence is a Cauchy sequence; for if $\lim_{n\to\infty} a_\nu = a$ then for any $\epsilon > 0$ there is a number N such that $\rho(a_\nu, a) < \epsilon/2$ whenever $\nu > N$, and therefore by the triangle inequality $\rho(a_\mu, a_\nu) \leq \rho(a_\mu, a) + \rho(a_\nu, a) < \epsilon/2 + \epsilon/2 = \epsilon$ whenever $\mu, \nu > N$. It is not necessarily the case though that a Cauchy sequence is convergent. For example the set S of real numbers x for which $0 < x \leq 1$ is a metric space with the metric defined by the absolute value norm; the sequence $1, 1/2, 1/3, 1/4, \ldots$ is clearly a Cauchy sequence, but it converges to the real number 0 which is not a point of S so the sequence does not converge in S. For a more interesting example, the set $\mathbb{Q}$ of rational numbers with the metric defined by the absolute value norm is also a metric space; it was demonstrated earlier that $\sqrt{2}$ can be expressed as the supremum of a collection of rational numbers, so it is possible to choose a sequence of rational numbers a_ν converging to $\sqrt{2}$; but since $\sqrt{2}$ is not a rational number the sequence a_ν does not converge in $\mathbb{Q}$.

Two Cauchy sequences $\{a_\nu\}$ and $\{b_\nu\}$ are said to be **equivalent** Cauchy sequences, indicated by writing $\{a_\nu\} \asymp \{b_\nu\}$, if for any $\epsilon > 0$ there is a number $N(\epsilon)$ depending on ϵ such that $\rho(a_\nu, b_\mu) < \epsilon$ whenever $\nu, \mu > N(\epsilon)$. The equivalence

---

[1] A well-known sequence begins by setting $a_1 = n$ for any natural number n and defining the further terms inductively by setting $a_{\nu+1} = a_\nu/2$ if a_ν is even and $a_{\nu+1} = 3a_\nu + 1$ if a_ν is odd; the conjecture, first made by Luther Collatz in 1937, is that no matter what natural number n is chosen the sequence eventually reaches the value 1, and then the sequence is periodic: $1, 4, 2, 1, 4, 2, 1, 4, 2, 1, \ldots$. That has not yet been proved, at the time this is being written, although it has been verified for all values $n \leq 5 \times 10^{18}$ and $1, 4, 2, 1, 4, \ldots$ is the only known cycle. More information can be found in the *On-Line Encyclopedia of Integer Sequences* founded by N.J.A. Sloane, with the web address https://oeis.org. There are extensions of the Collatz construction for which it is not actually provable that the sequence continues forever; see the discussion by John Conway, "On Unsettleable Arithmetical Problems" in *American Mathematical Monthly* 120 (2013): 192–198.

of Cauchy sequences can be characterized alternatively, and equivalent Cauchy sequences have the same basic convergence properties.

Theorem 2.4. *Let S be a metric space with the metric ρ.*
(i) *Two Cauchy sequences $\{a_\nu\}$ and $\{b_\mu\}$ in S are equivalent if and only if*

$$\lim_{\nu \to \infty} \rho(a_\nu, b_\nu) = 0. \tag{2.27}$$

(ii) *If $\{a_\nu\}$ and $\{b_\mu\}$ are equivalent Cauchy sequences in S then $\{a_\nu\}$ converges if and only if $\{b_\mu\}$ converges and the two sequences then have the same limit.*

Proof: (i) If two Cauchy sequences $\{a_\nu\}$ and $\{b_\mu\}$ are equivalent then for any $\epsilon > 0$ there is a number $N(\epsilon)$ depending on ϵ such that $\rho(a_\nu, b_\mu) < \epsilon$ whenever $\nu, \mu > N(\epsilon)$; for $\nu = \mu$ in particular $\rho(a_\nu, b_\nu) < \epsilon$ whenever $\nu > N(\epsilon)$, hence (2.27) holds. Conversely if two Cauchy sequences $\{a_\nu\}$ and $\{b_\mu\}$ satisfy (2.27) then for any $\epsilon > 0$ there is a number $N_1(\epsilon)$ such that $\rho(a_\nu, b_\nu) < \epsilon/2$ for all $\nu > N_1(\epsilon)$; and since the sequence a_ν is Cauchy there is also a number $N_2(\epsilon)$ so that $\rho(a_\nu, a_\mu) < \epsilon/2$ whenever $\nu, \mu > N_2(\epsilon)$. Consequently if $N(\epsilon) = \max\left(N_1(\epsilon), N_2(\epsilon)\right)$ then for any $\mu, \nu > N(\epsilon)$ by the triangle inequality $\rho(a_\nu, b_\mu) \leq \rho(a_\nu, a_\mu) + \rho(a_\mu, b_\mu) < \epsilon$ showing that $\{a_\nu\} \asymp \{b_\mu\}$.
(ii) If two Cauchy sequences $\{a_\nu\}$ and $\{b_\mu\}$ are equivalent then (2.27) holds; and if $\{a_\nu\}$ converges to a then $\lim_{\nu\to\infty} \rho(a_\nu, a) = 0$, so by the triangle inequality $\lim_{\nu\to\infty} \rho(b_\nu, a) \leq \lim_{\nu\to\infty} \left(\rho(b_\nu, a_\nu) + \rho(a_\nu, a)\right) \leq \lim_{\nu\to\infty} \rho(b_\nu, a_\nu) + \lim_{\nu\to\infty} \rho(a_\nu, a) = 0$ hence the sequence b_ν also converges to a, and that suffices for the proof.

The ordered field of real numbers is a metric space with the absolute value norm, and provides nice examples of Cauchy sequences and of equivalent Cauchy sequences. For any positive real number $r > 0$ let $n_0 \in \mathbb{N} \cup \{0\}$ be the largest nonnegative integer less than or equal to r, so that $r = n_0 + r_0$ where $0 \leq r_0 < 1$. If the interval from 0 to 1 is split into 10 equal intervals of length $1/10$ by points $n/10$ for $n = 0, 1, \ldots, 9$, the real number r_0 will be contained in at least one of these subintervals; if $n_1/10 \leq r_0 < (n_1 + 1)/10$ then $r = n_0 + n_1/10 + r_1$ where $0 \leq r_1 < 1/10$. Then if the interval from 0 to $1/10$ is split into 10 equal intervals of length $1/10^2$ by points $n/10^2$ for $n = 0, 1, \ldots, 9$, the real number r_1 will be contained in at least one of these subintervals; if $n_2/10^2 \leq r_1 < (n_2 + 1)/10^2$ then $r = n_0 + n_1/10 + n_2/10^2 + r_2$ where $0 \leq r_2 < 1/10^2$. The process can be continued, so the real number r can be written as the sum $r = a_\nu + r_\nu$ of the rational number $a_\nu = n_0 + n_1/10 + n_2/10^2 + \cdots + n_\nu/10^\nu$, customarily abbreviated as $a_\nu = n_0.n_1n_2\cdots n_\nu$, plus a real number r_ν in the interval $0 \leq r_\nu < 1/10^\nu$. The rational numbers a_ν thus defined are an increasing sequence of rational numbers bounded above, so by the completeness property of the real numbers the sequence $\{a_\nu\}$ converges to a real limit, which of course is just r itself. This convergent sequence is customarily abbreviated as

$$r = n_0.n_1n_2n_3\cdots \quad \text{where} \quad n_i = 0, 1, 2, \ldots, 8, 9 \quad \text{for} \quad i > 0 \qquad (2.28)$$

and is called the **decimal expansion** of the real number r, There is an alternative construction in which the subintervals in which the remainders are constrained have the form $n_\nu/10^\nu < r_{\nu-1} \le (n_\nu + 1)/10^\nu$, providing a different Cauchy sequence converging to the same real number r, hence an equivalent Cauchy sequence. The two constructions for the real number 1.234 lead to the two Cauchy sequences $1.2340000000\ldots$ and $1.2339999999\ldots$ describing the same real number, so to two distinct but equivalent Cauchy sequences of rational numbers. The corresponding construction can be used for decompositions of the interval into n equal segments for values of n other than 10, leading to an **expansion in the base** n of any real number. For example $\sqrt{2}$ can be expressed as a Cauchy sequence of the form

$$\sqrt{2} = 1.0110101000001001111\ldots$$

in an expansion in the base 2.

A metric space S is said to be **complete** if every Cauchy sequence in S converges. It was already noted that the space $\mathbb{Q}$ of rational numbers with the absolute value metric is not a complete metric space. On the other hand the space $\mathbb{R}$ of real numbers with the absolute value metric is a complete metric space, as will be demonstrated. Before doing so, perhaps it should be recalled that for ordered fields such as $\mathbb{Q}$ and $\mathbb{R}$ two distinct meanings have been associated to the adjective "complete": on the one hand a complete ordered field was defined as an ordered field in which any set of field elements bounded above has a least upper bound, while on the other hand an ordered field is a metric space with the absolute value metric and is a complete metric space if every Cauchy sequence converges. The two notions are virtually equivalent, so the use of the term "complete" is somewhat justified. The proof though requires that the fields involved satisfy the archimedean property, that there are no positive numbers in the field that are greater than any integer or less than any quotient $1/n$ for positive integers n, as discussed in section 1.2 as in that discussion, the field of real numbers is an archimedean field.

Theorem 2.5. *An ordered archimedean[2] field is a complete ordered field if and only if it is a complete metric space in the absolute value metric.*

---

[2]There is a slightly subtle point here. It was demonstrated in Theorem 1.6 that an ordered field that is complete, in the sense that there exist least upper bounds, is archimedean: there are no positive elements r of the field such that $r > n$ or $r < 1/n$ for all integers $n > 0$; but there are nonarchimedean fields that are complete metric spaces in terms of the metric defined by the absolute value. However for archimedean ordered fields the two notions of completeness are equivalent.

Proof: Suppose first that F is a complete ordered field and that a_ν is a Cauchy sequence in F for the absolute value norm. If there are only finitely many distinct points among the points a_ν it is clear that the sequence converges; so suppose that there are infinitely many distinct points. Since a_ν is a Cauchy sequence then there is a number N such that $|a_\nu - a_\mu| < 1$ for all $\mu, \nu > N$; so all these points, together with the finitely many points a_ν for $\nu \leq N$, are contained in some finite interval $[x_1, y_1] \subset F$. Split this interval by its midpoint into two subintervals. At least one of these halves will contain infinitely many points a_ν, so label that interval $[x_2, y_2]$, noting that $x_1 \leq x_2 \leq y_2 \leq y_1$. Continuing this process leads to a sequence of intervals $[x_i, y_i]$ each containing infinitely many points a_ν, where $x_i \leq x_{i+1} \leq y_{i+1} \leq y_1$. The points x_i are all bounded above by any y_j, so since F is a complete ordered field this set of points has a least upper bound $x_\infty = \sup x_i \leq y_j$; and similarly the points y_i are bounded below by x_j so they have a greatest lower bound $y_\infty = \inf y_i \geq x_j$, and $x_i \leq x_\infty \leq y_\infty \leq y_i$ for any i. By construction $|y_i - x_i| \leq |y_1 - x_1|/2^{i-1}$. The archimedean property shows that the quotients $1/2^{i-1}$ tend to 0 in the field, and consequently $x_\infty = y_\infty = a$. Each interval $[x_i, y_i]$ contains points a_ν for arbitrarily large indices ν, since the interval contains infinitely many points of the sequence; and since $|a_\mu - a_\nu| < y_i - x_i$ for all μ, ν sufficiently large it is evident that the sequence a_ν converges to the value $x_\infty = y_\infty = a$, thus demonstrating that a complete ordered field is a complete metric space with the metric defined by the absolute value.

Conversely suppose that F is an ordered field that is a complete metric space with the absolute value metric. If a subset $S \subset F$ is bounded above by M_1, so that $a \leq M_1$ for any $a \in S$, choose a point $a_1 \in S$. Then either there is a point $a_2 \in S$ such that $a_1 + \frac{1}{2}(M_1 - a_1) < a_2 \leq M_1$ or $a \leq a_1 + \frac{1}{2}(M_1 - a_1)$ for all $a \in S$. In the first case take the point a_2 and set $M_2 = M_1$, while in the second case set $a_2 = a_1$ and $M_2 = a_1 + \frac{1}{2}(M_1 - a_1)$; thus in either case $a_1 \leq a_2 \leq M_2 \leq M_1$ and $(M_2 - a_2) \leq \frac{1}{2}(M_1 - a_1)$. Repeating this construction leads to sequences a_ν and M_ν which are Cauchy sequences with the same limit point, again using the archimedean property; thus $A = \lim_{\nu \to \infty} a_\nu = \lim_{\nu \to \infty} M_\nu$, and it is clear that $A = \sup S$.

Corollary 2.6. *For any $n \in \mathbb{N}$ the vector space $\mathbb{R}^n$ is a complete metric space in terms of any of the equivalent standard norms $\ell_1, \ell_2, \ell_\infty$.*

Proof: For the special case $n = 1$ this follows from the preceding theorem, where the field $\mathbb{R}$ is defined as a complete ordered field. More generally for the vector space $\mathbb{R}^n$, if $\{A_\nu\}$ is a Cauchy sequence in the ℓ_∞ norm, where $A_\nu = \{a_{1\nu}, a_{2\nu}, \ldots, a_{n\nu}\} \in \mathbb{R}^n$, then the sequence $\{a_{i\nu}\}$ for each fixed i is a Cauchy sequence in $\mathbb{R}$ since $|a_{i\mu} - a_{i\nu}| \leq \|A_\mu - A_\nu\|_\infty$, so the sequence $\{a_{i\nu}\}$ converges to a real number a_i by the first part of the proof; and then the Cauchy sequence $\{A_\nu\}$ clearly converges to the vector $A = \{a_1, a_2, \ldots, a_n\}$, which suffices for the proof.

Theorem 2.7. *For any real metric space S there exists a canonical[3] isometry ϕ : $S \longrightarrow S^{**}$ into a complete real metric space S^{**}; and the image $\phi(S) \subset S^{**}$ is a dense subspace of S^{**}.*

Proof: To any real metric space S associate the set S^* of all Cauchy sequences in S. If $A = \{a_\nu\}, B = \{b_\nu\} \in S^*$ then

$$\rho(a_\mu, b_\mu) \leq \rho(a_\mu, a_\nu) + \rho(a_\nu, b_\nu) + \rho(b_\nu, b_\mu)$$

and consequently

$$\rho(a_\mu, b_\mu) - \rho(a_\nu, b_\nu) \leq \rho(a_\mu, a_\nu) + \rho(b_\nu, b_\mu);$$

interchanging the roles of μ and ν leads to the corresponding inequality with the left-hand side reversed, hence

$$|\rho(a_\mu, b_\mu) - \rho(a_\nu, b_\nu)| \leq \rho(a_\mu, a_\nu) + \rho(b_\nu, b_\mu). \tag{2.29}$$

Since A and B are Cauchy sequences then for any $\epsilon > 0$ there is a number $N(\epsilon)$ such that $\rho(a_\mu, a_\nu) < \epsilon/2$ and $\rho(b_\mu, b_\nu) < \epsilon/2$ whenever $\mu, \nu > N(\epsilon)$, hence from (2.29) that $|\rho(a_\mu, b_\mu) - \rho(a_\nu, b_\nu)| < \epsilon$ whenever $\mu, \nu > N(\epsilon)$; thus $\{\rho(a_\nu, b_\nu)\}$ is a Cauchy sequence of real numbers, and since the real field is complete this Cauchy sequence has a limit

$$\rho^*(A, B) = \lim_{\nu \to \infty} \rho(a_\nu, b_\nu). \tag{2.30}$$

The function $\rho^*(A, B)$ on $S^* \times S^*$ clearly satisfies all the conditions of a metric on S^* except that possibly $\rho^*(A, B) = 0$ for some sequences $A \neq B$ in S^*; thus S^* is at least a pseudometric space, and as noted earlier the space S^{**} of equivalence classes of elements of S^*, where two elements A, B are equivalent if and only if $\rho^*(A, B) = 0$, is a metric space with metric naturally induced by ρ^*. The mapping $\phi : S \longrightarrow S^{**}$ that associates to any $a \in S$ the equivalence class of the constant Cauchy sequence $S(a)$ for which $a_\nu = a$ for all ν is an isometry since $\rho^*\big(S(a), S(b)\big) = \rho(a, b)$ as an immediate consequence of (2.30). The metric space S^{**} and the isometry $\phi : S \longrightarrow S^{**}$ are defined canonically and uniquely in terms of the metric space S.

To show that $\phi(S)$ is a dense subspace of S^{**}, consider an equivalence class $\overline{A} \in S^{**}$ of Cauchy sequences $A = \{a_\nu\} \in S^*$. For any Cauchy sequence $A \in \overline{A}$ and any $\epsilon > 0$ there is a number $N(\epsilon) > 0$ such that $\rho(a_\mu, a_\nu) < \epsilon$ whenever $\mu, \nu > N(\epsilon)$. For any fixed value $\mu_0 > N(\epsilon)$ the constant Cauchy sequence $S(a_{\mu_0})$ satisfies $\rho^*(S(a_{\mu_0}), A) = \lim_{\nu \to \infty} \rho(a_{\mu_0}, a_\nu) < \epsilon$. It follows that $\rho^*(\phi(a_{\mu_0}), \overline{A}) < \epsilon$,

---

[3]That the isometry is canonical means that it is explicitly and uniquely defined in terms of the metric on the space S.

showing that $\phi(S)$ is dense in S^{**}. Note that the proof also shows that for any $\overline{A} \in S^{**}$ and any Cauchy sequence $A = \{a_\nu\} \in \overline{A}$ the sequence $\{\phi(a_\nu)\} \in S^{**}$ converges to $\overline{A}$.

Finally the preceding observations can be combined to show that S^{**} is a complete metric space. If $\overline{A}_n \in S^{**}$ are elements of S^{**} that form a Cauchy sequence $\{\overline{A}_n\}$ then for any $\epsilon > 0$ there is a number $N(\epsilon)$ such that $\rho^*(\overline{A}_m, \overline{A}_n) < \epsilon$ whenever $\mu, \nu > N(\epsilon)$. Since $\phi(S)$ is dense in S^{**} there are $b_n \in S$ such that $\rho^*(\overline{A}_n, \phi(b_n)) < \frac{1}{n}$. It follows from the triangle inequality that $\rho(b_m, b_n) = \rho^*(\phi(b_m), \phi(b_n)) \leq (\rho^*(\phi(b_m), \overline{A}_m) + \rho^*(\overline{A}_m, \overline{A}_n) + \phi^*(\overline{A}_n, b_n)) \leq \epsilon + \frac{1}{m} + \frac{1}{n}$ so $\{b_n\}$ is a Cauchy sequence in S; and as noted in the preceding paragraph the image sequence $\phi(b_n)$ converges to some $\overline{B} \in S^{**}$. Since $\rho^*(\overline{A}_n, \overline{B}) \leq \rho^*(\overline{A}_n, \phi(b_n)) + \rho(\phi(b_n), \overline{B})$ it follows that $\lim_{n\to\infty} \overline{A}_n = \overline{B}$, and that suffices for the proof.

It was discovered during the development of topology that many of the basic results did not really require the use of a norm or metric but only the use of the notion of nearness provided by a norm or metric. That notion is based on the concept of an ϵ-**neighborhood** $\mathcal{N}_a(\epsilon)$ of a point a in a metric space S with a metric $\rho(x, y)$, a subset of S defined by

$$\mathcal{N}_a(\epsilon) = \left\{ x \in S \mid \rho(a, x) < \epsilon \right\} \quad \text{where } \epsilon > 0. \tag{2.31}$$

An **open set** in the metric space S is defined in terms of these ϵ-neighborhoods as a subset $U \subset S$ such that for each point $a \in U$ there is an $\epsilon > 0$ such that $\mathcal{N}_a(\epsilon) \subset U$; intuitively an open set is one that contains with any point a all points that are sufficiently near a. For example an open ϵ-neighborhood $\mathcal{N}_a(\epsilon)$ of a point $a \in S$ is an open set; indeed if $b \in \mathcal{N}_a(\epsilon)$ then $\rho(a, b) < \epsilon$, and if $x \in \mathcal{N}_b(\epsilon')$ where $\epsilon' < \epsilon - \rho(a, b)$ then by the triangle inequality $\rho(a, x) \leq \rho(a, b) + \rho(b, x) < \rho(a, b) + \epsilon' < \epsilon$ so that $\mathcal{N}_b(\epsilon') \subset \mathcal{N}_a(\epsilon)$.

Theorem 2.8. *If $\mathcal{T}$ is the collection of all open subsets in a metric space S then:*
(i) *$\emptyset \in \mathcal{T}$ and $S \in \mathcal{T}$;*
(ii) *if $U_\alpha \in \mathcal{T}$ then $\bigcup_\alpha U_\alpha \in \mathcal{T}$;*
(iii) *if $U_i \in \mathcal{T}$ for $1 \leq i \leq N$ then $\bigcap_{i=1}^N U_i \in \mathcal{T}$.*

Proof: (i) The entire set S of course is open; and the condition that for each $a \in \emptyset$ there is an $\epsilon > 0$ such that $\mathcal{N}_a(\epsilon_i) \subset \emptyset$ is fulfilled vacuously since there are no points $a \in \emptyset$.
(ii) If U_α are open sets and $a \in U = \bigcup_\alpha U_\alpha$ then $a \in U_{\alpha_0}$ for some set U_{α_0}; and since U_{α_0} is open there is an $\epsilon > 0$ so that $a \in \mathcal{N}_a(\epsilon) \subset U_\alpha \subset U$ hence U is open.
(iii) If U_i are open sets and $a \in U = \bigcap_{i=1}^N U_i$ then for each i there is an $\epsilon_i > 0$ such that $\mathcal{N}_a(\epsilon_i) \subset U_i \subset U$; so if $\epsilon = \min \epsilon_i$ then $\epsilon > 0$ and $\mathcal{N}_a(\epsilon) \subset \mathcal{N}_a(\epsilon_i) \subset U_i$ for all U_i hence $\mathcal{N}_a(\epsilon) \subset U$ so U is open. That argument fails if there are infinitely

many such sets U_i since in that case there does not necessarily exist an $\epsilon > 0$ such that $\epsilon \leq \epsilon_i$ for all i.

The properties listed in the preceding theorem can be summarized as the assertion that the empty set and the entire space are open and that the union and finite intersection of open sets are open. It is easy to see that equivalent metrics on a metric space S determine the same collection of open subsets of S. Indeed if ρ_1 and ρ_2 are equivalent metrics on a set S then by definition there are constants $c_1 > 0$ and $c_2 > 0$ such that $\rho_1(a,b) \leq c_1 \rho_2(a,b)$ and $\rho_2(a,b) \leq c_2 \rho_1(a,b)$ for all points $a, b \in S$. If $U \subset S$ is an open subset in the topology defined by the metric ρ_1 then for any point $a \in U$ there is an $\epsilon > 0$ such that $\mathcal{N}_a(\epsilon) = \{ x \in S \mid \rho_1(a,x) < \epsilon \} \subset U$. Then $\mathcal{N}_a(\epsilon/c_1) = \{ x \in S \mid \rho_2(a,x) < \epsilon/c_1 \} \subset U$ so that U is also an open set in terms of the metric ρ_2. The converse results follow upon reversing the roles of the two metrics ρ_1 and ρ_2, so the two metrics determine the same open sets.

The normed vector spaces considered in Section 2.1 provide illustrative examples of metric spaces. The ϵ-neighborhoods of the origin $\mathbf{0} \in \mathbb{R}^n$ in the ℓ_1, ℓ_2, and ℓ_∞ norms are the subsets sketched in Figure 2.2, and are open sets in the metric topology defined by any of these norms. It is clear from the sketch that any one of these neighborhoods of the origin can be fit into another one by choosing a sufficiently small ϵ, and that is the geometric content of the assertion that these metrics are equivalent. An r-neighborhood of a point $\mathbf{a} \in \mathbb{R}^n$ in the ℓ_2 norm is also called an **open ball** of radius r centered at the point $\mathbf{a} \in \mathbb{R}^n$ and is commonly denoted by $B_r(\mathbf{a})$; here r is not expected to be particularly small, as distinct from the general expectation for ϵ-neighborhoods. The corresponding sets in an arbitrary metric space S are the sets

$$B_r(a) = \left\{ x \in S \mid \rho(a,x) < r \right\} \tag{2.32}$$

defined in terms of the metric ρ on S. In place of the ϵ-neighborhoods in the ℓ_∞ norm in $\mathbb{R}^n$ though it is more common to consider slightly more generally the **open cells** in $\mathbb{R}^n$, subsets of the form

$$\Delta = \left\{ \mathbf{x} = \{x_j\} \in \mathbb{R}^n \mid a_j < x_j < b_j \text{ for } 1 \leq j \leq n \right\} \tag{2.33}$$

for arbitrary real numbers $a_j < b_j$. In particular in $\mathbb{R}^1$ a cell is just the set of points $x \in \mathbb{R}$ such that $a < x < b$, also called an **open interval** and denoted by (a,b). In this notation (a,∞) stands for the set of points $x \in \mathbb{R}$ for which $x > a$ while $(-\infty, b)$ stands for the set of points $x \in \mathbb{R}$ for which $x < b$. It is also traditional to use notation such as this for intervals other than open intervals; thus $[a,b)$ denotes the set of points $x \in \mathbb{R}$ for which $a \leq x < b$ while $(a,b]$ denotes the set of points $x \in \mathbb{R}$ for which $a < x \leq b$ and $[a,b]$ denotes the set of points $x \in \mathbb{R}$ for which $a \leq x \leq b$. It is clear that these latter three sets are not open sets. In general open sets can be quite complicated sets with no simple general description as cells or balls; the exception though is the special case of the vector space $\mathbb{R}^1$, when all the open sets can be described quite simply.

Theorem 2.9. *An open subset $U \subset \mathbb{R}^1$ is a disjoint union of a countable number of open intervals.*

Proof: If $U \subset \mathbb{R}^1$ is an open set and $a \in U$ let $E = \{ x \in \mathbb{R} \mid (a, x) \subset U \}$. Clearly $E \neq \emptyset$, as a consequence of the assumption that U is open. If the set E is not bounded above then $(a, \infty) \subset U$. If the set E is bounded above and $a_+ = \sup(E)$, where the supremum exists as a consequence of the completeness of the real number system, then clearly $(a, a_+) \subset U$. If $a_+ \in U$ then $(a_+, a_+ + \epsilon) \subset U$ since U is open, so there are points $x \in E$ for which $x > a_+$, which is impossible since $a_+ = \sup(E)$; hence $a_+ \notin U$. The corresponding argument shows that there is a value a_- so that $(a_-, a_+) \subset U$ but $a_- \notin U$. Thus the open interval (a_-, a_+) is contained in U and is disjoint from the complement $U \sim (a_-, a_+)$. The same argument can be applied to any other point in U outside the interval (a_-, a_+), so U is the union of a disjoint collection of open intervals. Choosing a rational number in each of these intervals maps the set of intervals injectively to a subset of $\mathbb{Q}$, so the set of intervals is countable, and that suffices for the proof.

APPENDIX: CONSTRUCTION OF THE REAL NUMBERS

The field of real numbers was defined axiomatically as a complete ordered field, which sufficed to examine its basic properties. There remained the questions whether there actually exists such a field, and whether it is unique up to isomorphism. Since the completeness of an archimedean ordered field is equivalent to its completeness as a metric space with the absolute value metric, by Theorem 2.5, it might be expected that the process of completing a metric space could be applied to defining the real numbers; but the proof of Theorem 2.7 only handled the completion of a real metric space, assuming the existence of the real numbers. A modification of that proof does yield the existence of the reals and is included as the following appendix.

To proceed, then, let $\mathcal{R}$ denote the set of rational Cauchy sequences, Cauchy sequences $A = \{a_\nu\}$ where $a_\nu \in \mathbb{Q}$; and let $\mathcal{R}^*$ denote the set of equivalence classes of rational Cauchy sequences, where the equivalence $A \asymp B$ of Cauchy sequences is as previously discussed. The equivalence class of a Cauchy sequence $A \in \mathcal{R}$ will be denoted by $A^* \in \mathcal{R}^*$. The map $\phi : \mathbb{Q} \longrightarrow \mathcal{R}$ that associates to a rational $a \in \mathbb{Q}$ the Cauchy sequence $\phi(a) \in \mathcal{R}$ for which $a_\nu = a$ for all $\nu \in \mathbb{N}$ clearly is an injective mapping; and it is easily seen that the sequences $\phi(a)$ and $\phi(b)$ are equivalent if and only if $a = b$, so the mapping ϕ induces an injective mapping $\phi^* : \mathbb{Q} \longrightarrow \mathcal{R}^*$. The algebraic operations in $\mathbb{Q}$ readily extend to the corresponding algebraic operations in $\mathcal{R}$, upon defining for any Cauchy sequences $A = \{a_\nu\}$ and $B = \{b_\nu\}$ their sum and product to be the sequences $A + B = \{a_\nu + b_\nu\}$ and $AB = \{a_\nu b_\nu\}$; it is easy to see that these are indeed Cauchy sequences, which is quite apparent for the sum while for the product

$$|a_\mu b_\mu - a_\nu b_\nu| = |a_\mu(b_\mu - b_\nu) + (a_\mu - a_\nu)b_\nu| \leq |a_\mu| \cdot |b_\mu - b_\nu| + |a_\mu - a_\nu| \cdot |b_\nu|$$
(2.34)

and the terms $|a_\mu|$ and $|b_\nu|$ are bounded for any two Cauchy sequences. It is also easy to see that if $A \asymp A'$ and $B \asymp B'$ for any Cauchy sequences $A, A', B, B' \in \mathcal{R}$ then $(A + B) \asymp (A' + B')$ and $(AB) \asymp (A'B')$, using (2.34) in the case of the product; thus the operations of addition and multiplication are well defined in the set $\mathcal{R}^*$ of equivalence classes of rational Cauchy sequences. These operations are derived from the corresponding operations in $\mathbb{Q}$ so have the same properties as the operations in $\mathbb{Q}$, giving $\mathcal{R}^*$ the structure of a commutative ring where the additive identity $\phi^*(0)$ is the equivalence class represented by the Cauchy sequence $\phi(0)$ and the multiplicative identity $\phi^*(1)$ is the equivalence class represented by the Cauchy sequence $\phi(1)$; the mapping $\phi^* : \mathbb{Q} \longrightarrow \mathcal{R}^*$ is an injective ring homomorphism.

Lemma 2.10. *A Cauchy sequence $A = \{a_\nu\} \in \mathcal{R}$ is not equivalent to the Cauchy sequence $\phi(0)$ if and only if there are a natural number $N \in \mathbb{N}$ and a rational number $\epsilon > 0$ so that either*
(i) $a_\nu > \epsilon$ *for all indices $\nu > N$ or*
(ii) $-a_\nu > \epsilon$ *for all $\nu > N$.*

Proof: By definition a Cauchy sequence $A = \{a_\nu\} \in \mathcal{R}$ is equivalent to the Cauchy sequence $\phi(0)$ if and only if for any rational number $\epsilon > 0$ there is a natural number N such that $|a_\nu| < \epsilon$ for all $\nu > N$; hence the Cauchy sequence $A = \{a_\nu\}$ is not equivalent to the Cauchy sequence $\phi(0)$ if and only if there is a rational number $\epsilon > 0$ such that $|a_\nu| > \epsilon$ for infinitely many indices ν. Since A is a Cauchy sequence there is a natural number N such that $|a_\mu - a_\nu| < \epsilon/2$ for all $\mu, \nu > N$, so if $|a_{\nu_0}| > \epsilon$ for some particular index $\nu_0 > N$ then $|a_\nu| \geq |a_{\nu_0}| - |a_{\nu_0} - a_\nu| > \epsilon/2$ for all $\nu > N$. That is equivalent to condition (i) if $a_{\nu_0} > 0$ and condition (ii) if $a_{\nu_0} < 0$, and that suffices for the proof.

As a first application of the preceding lemma, if a Cauchy sequence $A = \{a_\nu\} \in \mathcal{R}$ is not equivalent to the Cauchy sequence $\phi(0)$, so that by the lemma there are a natural number $N \in \mathbb{N}$ and a rational number $\epsilon > 0$ so that $|a_\nu| > \epsilon$ if $\nu > N$, then the sequence B obtained by replacing a_ν by $b_\nu = 1$ if $\nu \leq N$ and by $b_\nu = 1/a_\nu$ for all $\nu > N$ is a Cauchy sequence, since

$$\left| \frac{1}{a_\mu} - \frac{1}{a_\nu} \right| = \left| \frac{a_\nu - a_\mu}{a_\mu a_\nu} \right| < \frac{1}{\epsilon^2} |a_\nu - a_\mu|;$$

and it is clear from the definition of the Cauchy sequence AB that $AB \asymp 1$, so it follows that the equivalence class $B^* \in \mathcal{R}^*$ is inverse to the equivalence class $A^* \in \mathcal{R}^*$, showing that $\mathcal{R}^*$ is actually a field. As a second application of the preceding lemma, if $\mathcal{P} \subset \mathcal{R}$ is the set of those Cauchy sequences $A = \{a_\nu\} \in \mathcal{R}$ for which there are a natural number $N \in \mathbb{N}$ and a rational number $\epsilon > 0$ such that $a_\nu > \epsilon$ for all $\nu > N$, then it follows from the lemma that for any Cauchy sequence $A \in \mathcal{R}$ either $A \in \mathcal{P}$ or $-A \in \mathcal{P}$ or $A \asymp \phi(0)$. Since it is clear that any Cauchy sequence that is equivalent to a sequence in the subset $\mathcal{P}$ also is contained in $\mathcal{P}$, it follows that if $\mathcal{P}^* \subset \mathcal{R}^*$ is the set of equivalence classes of Cauchy sequences that are contained in $\mathcal{P}$ then for any $A^* \in \mathcal{R}^*$ either $A^* \in \mathcal{P}^*$ or $-A^* \in \mathcal{P}^*$ or $A^* = \phi^*(0)$. Clearly $A^* + B^* \in \mathcal{P}^*$ and $A^*B^* \in \mathcal{P}^*$ whenever $A^* \in \mathcal{P}^*$ and $B^* \in \mathcal{P}^*$, so the field $\mathcal{R}^*$ is an ordered field if $\mathcal{P}^*$ is taken as the set of positive elements. The inequality $A^* \leq B^*$ for two equivalence classes $A^*, B^* \in \mathcal{R}^*$ then is defined as the condition that $B^* - A^* \in \mathcal{P}^*$ or $B^* - A^* = \phi^*(0)$, and the absolute value $|A^*|$ of an equivalence class $A^* \in \mathcal{R}^*$ is defined by

$$|A^*| = \begin{cases} A^* & \text{if} & A^* \in \mathcal{P}^* \quad \text{or} \quad A^* = \phi^*(0), \\ -A^* & \text{if} & -A^* \in \mathcal{P}^*. \end{cases} \tag{2.35}$$

The field $\mathcal{R}^*$ can be viewed as a vector space over the field of rationals, and the absolute value $|A^*|$ is a norm on this vector space; indeed the absolute value as defined in (2.35) is defined in the same way as the absolute value of the real numbers in (2.2), and the argument used to show the latter is a norm carries over to the norm (2.35).

Lemma 2.11. *For any $A^* \in \mathcal{R}^*$ and $E^* \in \mathcal{P}^*$ there is a rational number $q \in \mathbb{Q}$ such that $|A^* - \phi^*(q)| < E^*$.*

Proof: Choose Cauchy sequences $A = \{a_\nu\} \in \mathcal{R}$ and $E = \{e_\nu\} \in \mathcal{P}$ that represent the equivalence classes A^* and E^*. By the definition of the set $\mathcal{P}$ there are a rational number $\epsilon > 0$ and a natural number N such that $e_\nu > \epsilon$ for all $\nu > N$. Since A is a Cauchy sequence then if N is sufficiently large it is also the case that $|a_\mu - a_\nu| < \epsilon/2$ for all $\mu, \nu > N$. Let $q = a_{N+1} \in \mathbb{Q}$; then for all $\nu > N$ it follows that $|a_\nu - q| < \epsilon/2$ so $e_\nu - \pm(a_\nu - q) \geq e_\nu - |a_\nu - q| \geq \epsilon - \epsilon/2 \geq \epsilon/2 > 0$ showing that $E > |A - \phi(q)|$, which suffices for the proof.

To complete the construction it is only necessary to show that the ordered field $\mathcal{R}^*$ is a complete archimedean field and that it is uniquely determined. It follows from Lemma 2.11 that the field is archimedean, by taking $E^* = \phi^*(1)$ in the lemma and multiplying $\phi^*(q) + \phi^*(1)$ by a suitable positive integer. If $\{A_i^*\} \in \mathcal{R}^*$ is a Cauchy sequence, if $\{A_i\} \in \mathcal{R}$ is a representative Cauchy sequence, and if $\epsilon \in \mathbb{Q}$ then $|A_i - A_j| < \phi(\epsilon)$ whenever i, j are sufficiently large, or equivalently $|a_{i\nu} - a_{j\nu}| < \epsilon$ for all sufficiently large ν whenever i, j are sufficiently large. It follows from the preceding Lemma 2.11 that for any index i there is a rational $q_i \in \mathbb{Q}$ such that $|A_i - \phi(q_i)| < \phi(1/i)$ or equivalently such that $|a_{iv} - q_i| < 1/i$ whenever v is sufficiently large. Then whenever $i < j$ and i is sufficiently large

$$|q_i - q_j| \leq |q_i - a_{iv}| + |a_{iv} - a_{jv}| + |a_{jv} - q_j| \leq 1/i + |a_{iv} - a_{jv}| + 1/j;$$

and since $|a_{iv} - a_{jv}| < \epsilon$ whenever v is sufficiently large it follows that

$$|q_i - q_j| \leq 1/i + \epsilon + 1/j \quad \text{for any } \epsilon \in \mathbb{Q}$$

and consequently that

$$|q_i - q_j| \leq 2/i \quad \text{whenever } i \text{ is sufficiently large and } j > i.$$

That shows that $Q = \{q_v\}$ is a Cauchy sequence $Q \in \mathcal{R}$; and since for sufficiently large $v > i$

$$|a_{iv} - q_v| \leq |a_{iv} - q_i| + |q_i - q_v| < 3/i,$$

which means that $|A_i - Q| \leq 3/i$ for sufficiently large i, it follows that the Cauchy sequence $\{A_i^*\}$ converges to the Cauchy sequence Q^*. Therefore $\mathcal{R}^*$ is a complete ordered field. Since every complete ordered field contains the rational numbers $\mathbb{Q}$ as a subfield, and every element in the field is a limit of rationals, it further follows that there is indeed a unique complete ordered field, as desired.

That any real number can be represented as the limit of the sequence of sums of a decimal expansion was discussed earlier; and that conversely the sequence of sums of any decimal expansion converges to a real number follows from the result of Theorem 2.5, showing that a complete archimedean field is a complete metric space. Thus the field $\mathbb{R}$ of real numbers can be described alternatively as the set of equivalence classes of the Cauchy sequences formed by the sums of decimal expansions, which is perhaps the usual elementary way in which the real numbers are traditionally described. Of course the field $\mathbb{R}$ can be characterized as well in terms of binary or other expansions.

Problems, Group I

1. Show that if a sequence a_v in a metric space S converges in terms of a metric ρ then it converges in terms of any metric equivalent to ρ. Show that if the sequence is Cauchy for the metric ρ then it is Cauchy for any metric equivalent to ρ. Show that if the metric space S is complete in terms of a metric ρ then it is complete in terms of any metric equivalent to ρ.

2.   Show that a Cauchy sequence a_v in a metric space S is bounded, in the sense that for any point $a \in S$ there is a number N_a such that $\rho(a_v, a) < N_a$ for all a_v.

3.   Show that $\rho(x, y) = \left| \frac{1}{x} - \frac{1}{y} \right|$ is a metric on the set S of strictly positive real numbers. Is it equivalent to the usual metric $|x|$ on the set S? Why? Are there sequences of positive real numbers that are Cauchy for the metric ρ but not for the usual metric? If there are, give examples; if not, say why.

4.   Describe the Cauchy sequences in the discrete metric on a set S. Is S a complete metric space in this metric? Why?

5.   Show that on the vector space of $n \times n$ real matrices the function $\rho(A, B) = \text{rank}(A - B)$ is a metric that is equivalent to the discrete metric.

6.   Show that if ρ_1 and ρ_2 are two metrics on a set S then $\sigma(x, y) = \rho_1(x, y) + \rho_2(x, y)$ and $\tau(x, y) = \max(\rho_1(x, y), \rho_2(x, y))$ are also metrics on S. Are the metrics σ and τ equivalent? Why?

7.   i) Suppose that a_v is a sequence of real numbers such that $|a_{v+1} - a_v| \le r|a_v - a_{v-1}|$ for all $v \ge 2$. Is this a Cauchy sequence if $0 < r < 1$? Why? Is this a Cauchy sequence if $r = 1$? Why?
       ii) Suppose a_v is a sequence of real numbers such that for any $\epsilon > 0$ there is a natural number N for which $|a_{v+1} - a_v| < \epsilon$ whenever $v > N$. Is this sequence a Cauchy sequence? Why?
       iii) Suppose that a_v is a sequence of real numbers such that $0 \le a_v \le a_{v+1}$ and $a_v \le 10^{10}$ for all v. Is this sequence a Cauchy sequence? Why?

Problems, Group II

8.   Let S be the set consisting of all sequences of positive integers, and if $A = \{a_i\}$ and $B = \{b_i\}$ are such sequences set

$$\rho(A, B) = \begin{cases} 0 & \text{if} \quad a_i = b_i \quad \text{for} \quad i \ge 1, \\ \frac{1}{n} & \text{if} \quad a_1 = b_1, a_2 = b_2, \ldots, a_{n-1} = b_{n-1}, a_n \ne b_n \end{cases}$$

Show that this is a metric on S. What are the Cauchy sequences in this metric? Is S a complete space in terms of this metric?

9.   If ρ is a metric on a set S show that $\sigma(x, y) = \frac{\rho(x,y)}{1+\rho(x,y)}$ and $\tau(x, y) = \min\left(1, \rho(x, y)\right)$ also are metrics on S, and that both are bounded metrics in the sense that $\sigma(x, y) \le 1$ and $\tau(x, y) \le 1$ for all $x, y \in S$. Are the metrics σ and τ necessarily equivalent? Why?

10. Show that the set of bounded sequences (sequences of real numbers a_ν such that $\sup_\nu |a_\nu| < \infty$) is a metric space with the metric $\rho(\{a_\nu\}, \{b_\nu\}) = \sup_\nu |a_\nu - b_\nu|$. Is this a complete metric space? Why? Is the subspace consisting of sequences that are Cauchy sequences a complete metric space under the same metric? Why?

11. Define a sequence x_n of real numbers by choosing x_0 and x_1 arbitrarily and setting $x_n = \frac{1}{2}(x_{n-1} + x_{n-2})$ for all $n \geq 2$. What can you say about the convergence of this sequence?

12. Define a sequence x_n of real numbers by choosing $x_1 > 0$ arbitrarily and setting $x_n = \sqrt{a + x_{n-1}}$ for $n \geq 2$, where $a > 0$. Show that the sequence x_n converges to a real number $\xi > 0$ for which $\xi^2 - \xi - a = 0$.

13. Let S be a metric space with the metric ρ.

   i) Show that the set of all bounded real-valued functions on S (mappings from S to $\mathbb{R}$, each of which is bounded) form a vector space V, where the sum $f + g$ of two functions f, g is the function $(f + g)(x) = f(x) + g(x)$ and the scalar product rf of the function f and a real number r is the function $(rf)(x) = rf(x)$ for all $x \in S$.
   ii) Show that vector space V is a normed vector space under the norm $\|f\|_\infty = \sup_{x \in S} |f(x)|$, where the condition that a function f be bounded is interpreted as meaning that $\|f\|_\infty$ is finite.
   iii) Fix an $s_0 \in S$ and for any point $a \in S$ let $f_a(x) = \rho(a, x) - \rho(s_0, x)$ for any $x \in S$. Show that $f_a(x)$ is bounded and $f_a(x) \in V$.
   iv) Show that $\rho(a, b) = \|f_a - f_b\|_\infty$ for any points $a, b \in S$.
   v) Show that the mapping $\phi : a \longrightarrow f_a$ is an isometric mapping from the metric space S into the metric space V, where the metric on V is that defined by the norm $\|f\|_\infty$ on V.

   [Comment: The preceding argument shows that any metric space can be realized as a subset of a normed vector space.]

2.3 Topological Spaces

A **topological space** is defined to be a set S together with a collection $\mathcal{T}$ of subsets of S such that

   (i) if $U_\alpha \in \mathcal{T}$ then $\bigcup_\alpha U_\alpha \in \mathcal{T}$;
   (ii) if $U_i \in \mathcal{T}$ for $1 \leq i \leq N$ then $\bigcap_{i=1}^N U_i \in \mathcal{T}$;
   (iii) $\emptyset \in \mathcal{T}$ and $S \in \mathcal{T}$.

The sets in $\mathcal{T}$ are called the **open sets** in the topology $\mathcal{T}$; thus in a topological space any union or finite intersection of open sets is open, and the empty set $\emptyset$ and the full set S are open. The open sets in a metric space S provide a topology on S, called the **metric topology**; but there are many other possible topologies on a metric space. Any set S can be given the **discrete** or **maximal** topology, for which $\mathcal{T}$ contains all the subsets of S, or the **indiscrete** or **minimal** topology, for which $\mathcal{T}$ contains just the empty set $\emptyset$ and the set S itself. It is customary to say that one topology $\mathcal{T}_1$ is **larger** than another topology $\mathcal{T}_2$ if every open set in $\mathcal{T}_2$ is open in $\mathcal{T}_1$, that is, if $\mathcal{T}_2 \subset \mathcal{T}_1$; in that sense the discrete topology is the largest topology on any set. Correspondingly a topology $\mathcal{T}_1$ is **smaller** than another topology $\mathcal{T}_2$ if every open set in $\mathcal{T}_1$ is open in $\mathcal{T}_2$, that is, if $\mathcal{T}_1 \subset \mathcal{T}_2$; in that sense the indiscrete topology is the smallest, which explains the alternate terminology. It is possible to introduce topologies on any set of points. For instance on the set S consisting of two points 0 and 1 in addition to the discrete and indiscrete topologies there is an additional topology in which there are 3 open sets, the sets $\emptyset, \{0\}, \{0, 1\}$; this is the simplest case in which there is a topology other than the discrete and indiscrete topologies on a set. The set of two points with this topology is called the **Sierpinski space**. In general an open subset of a topological space S containing a point a is called an **open neighborhood**, or sometimes just a **neighborhood** of the point a, and often will be denoted by $\mathcal{N}_a$. An ϵ-neighborhood $\mathcal{N}_a(\epsilon)$ of a point a in a metric space is an open neighborhood of a; but there are considerably more general open neighborhoods of any point in a metric space. In the Sierpinski topology the point 0 is an open neighborhood of itself, while the only open neighborhood of the point 1 is the entire space $\{0, 1\}$. In a sense, open neighborhoods in a general topological space play the role of ϵ-neighborhoods in a metric space with the metric topology; however metric spaces really are a special category of topological spaces, with some quite special properties, so caution is required when attempting to extend properties of metric spaces to general topological spaces.

The complements of the open subsets in a topological space S are called the **closed sets** of S. It follows immediately from the defining properties of a topology that the collection $\mathcal{C}$ of closed sets in a topological space S has the following properties:

   (i) if $U_\alpha \in \mathcal{C}$ then $\bigcap_\alpha U_\alpha \in \mathcal{C}$;
  (ii) if $U_i \in \mathcal{C}$ for $1 \leq i \leq N$ then $\bigcup_{i=1}^{N} U_i \in \mathcal{C}$;
 (iii) $\emptyset \in \mathcal{C}$ and $S \in \mathcal{C}$.

Thus in a topological space an arbitrary intersection and a finite union of closed sets are closed, and the empty set $\emptyset$ and the full set S are closed; but an arbitrary union of closed sets is not necessarily closed. A topology on a space can be defined alternatively by specifying the collection of closed sets and defining the open sets to be the complements of the closed sets; however it is customary to define topologies in terms of the open sets.

For any subset E of a topological space S the **interior** of E is defined to be the set

$$E^o = \bigcup_{\substack{U \text{ open,} \\ U \subset E}} U, \tag{2.36}$$

the union of all open subsets of S contained in E; this is obviously an open subset of S such that $E^o \subset E$, indeed is the largest open subset of S contained in E since it is apparent from the definition that if U is any open set for which $U \subset E$ then $U \subset E^o$. It is clear that E is open if and only if $E = E^o$; indeed if $E = E^o$ then E is open since E^o is open, and conversely if E is open it can be taken as one of the sets U in the definition (2.36) hence $E = E^o$. Correspondingly the **closure** of E is defined to be the set

$$\bar{E} = \bigcap_{\substack{C \text{ closed,} \\ C \supset E}} C, \tag{2.37}$$

the intersection of all closed subsets of S containing E; this is obviously a closed subset of S such that $E \subset \bar{E}$, indeed is the smallest closed subset of S containing E since it is apparent from the definition that if C is any closed set for which $E \subset C$ then $\bar{E} \subset C$. It is clear that E is closed if and only if $E = \bar{E}$; indeed if $E = \bar{E}$ then E is closed since $\bar{E}$ is closed, and conversely if E is closed it can be taken as one of the sets C in the definition (2.37) hence $E = \bar{E}$. Thus any subset $E \subset S$ of a topological space is naturally bounded by an open and a closed set, in the sense that $E^o \subset E \subset \bar{E}$, and these are the best possible bounds. One extreme subset $E \subset S$ is that in which the closure of E is the entire space S, so that $\bar{E} = S$; such a subset E is said to be **dense**, or sometimes **dense in** S to be quite precise. For example the set $\mathbb{Q}^n$ of vectors consisting entirely of rational numbers is dense in the vector space $\mathbb{R}^n$ of real numbers with the metric topology. A topological space such as $\mathbb{R}^n$ that has a countable dense subset is called a **separable** topological space. The other extreme subset $E \subset S$ is that in which the interior of E is empty, so that $E^o = \emptyset$; such a subset E is said to be **sparse**. For example the set $\mathbb{Q}^n$ of rational vectors is sparse in the vector space $\mathbb{R}^n$ with the metric topology, even though it is dense in the space $\mathbb{R}^n$.

For some purposes, in particular for dealing with compositions of the operations of taking the complements, interiors, and closures of sets, it is clearer to use an alternative notation, setting

$$\text{int}\,(E) = E^o \quad \text{and} \quad \text{clo}\,(E) = \bar{E} \tag{2.38}$$

for the interior and closure of a set E. The complement of (2.36) is the equation

$$\sim E^o = \bigcap_{\substack{U \text{ open,} \\ U \subset E}} \sim U = \bigcap_{\substack{(\sim U) \text{ closed,} \\ (\sim U) \supset (\sim E)}} \sim U,$$

and comparing this and (2.37) shows that

$$\sim E^o = \overline{\sim E} \quad \text{or more clearly} \quad \sim \text{int}(E) = \text{clo}(\sim E), \tag{2.39}$$

since the alternative notation exhibits more clearly the order of the composition of the steps in the description of this set. This shows that the operations of interior and closure are dual through complementation, so that one of these operators can be written in terms of the other as

$$\text{int}(E) = \sim \text{clo}(\sim E) \quad \text{or equivalently} \quad \text{clo}(E) = \sim \text{int}(\sim E). \tag{2.40}$$

These two operations also can be described in terms of their effect on individual points. It is evident from the definition (2.36) that the set $E^o = \text{int}(E)$ can be described as

$$\text{int}(E) = \left\{ a \in S \,\middle|\, \mathcal{N}_a \subset E \quad \text{for some open neighborhood } \mathcal{N}_a \text{ of } a \right\}. \tag{2.41}$$

Therefore the complement $\sim \text{int}(E)$ is the set of points $a \in S$ such that no open neighborhood $\mathcal{N}_a$ of a is entirely contained in E, hence such that any open neighborhood $\mathcal{N}_a$ of a must contain some points in the complement of E; thus

$$\sim \text{int}(E) = \left\{ a \in S \,\middle|\, \mathcal{N}_a \cap (\sim E) \neq \emptyset \quad \text{for any open neighborhood } \mathcal{N}_a \text{ of } a \right\}. \tag{2.42}$$

In view of (2.40) the preceding equation applied to $\sim E$ yields the description

$$\text{clo}(E) = \left\{ a \in S \,\middle|\, \mathcal{N}_a \cap E \neq \emptyset \quad \text{for any open neighborhood } \mathcal{N}_a \text{ of } a \right\}. \tag{2.43}$$

The characterizations (2.41) and (2.43) are often the easiest ways of handling the interiors and closures of sets, and should be kept in mind as very useful tools.

The **boundary** of a subset E of a topological space S is the set defined by

$$\partial E = \overline{E} \cap \overline{\sim E} = \text{clo}(E) \cap \text{clo}(\sim E); \tag{2.44}$$

and it is evident from this and (2.40) that

$$\sim \partial E = (\sim E)^o \cup E^o = \text{int}(\sim E) \cup \text{int}(E). \tag{2.45}$$

It is clear from the definition (2.44) that the boundary of any set is a closed set and that

$$\partial E = \partial(\sim E). \tag{2.46}$$

It follows from (2.45) that

$$E \cap {\sim} \partial E = E \cap (\text{int}(E) \cup \text{int}({\sim} E))$$
$$= (E \cap \text{int}(E)) \cup (E \cap \text{int}({\sim} E))$$

and since $E \cap \text{int}(E) = \text{int}(E)$ and $E \cap \text{int}({\sim} E) \subset E \cap ({\sim} E) = \emptyset$ it follows that

$$\text{int}(E) = E \cap {\sim} \partial E. \tag{2.47}$$

From (2.40) and (2.47) it follows that

$$\text{clo}(E) = {\sim}\,\text{int}({\sim} E) = {\sim} ({\sim} E \cap {\sim} \partial {\sim} E) = E \cup (\partial {\sim} E),$$

so in view of (2.46)

$$\text{clo}(E) = E \cup \partial E. \tag{2.48}$$

From the characterization (2.43) of the closure of a set it follows immediately that the boundary ∂E can be described alternatively as

$$\partial E = \Big\{ a \in S \,\Big|\, \mathcal{N}_a \cap E \neq \emptyset \ \text{ and } \ \mathcal{N}_a \cap ({\sim} E) \neq \emptyset$$
$$\text{for any open neighborhood } \mathcal{N}_a \text{ of } a \Big\}. \tag{2.49}$$

This is perhaps the most natural and useful description of the boundary of a set. It is sometimes useful to note that

$$\partial E = \emptyset \ \text{ if and only if } E \text{ is both open and closed.} \tag{2.50}$$

Indeed if $\partial E = \emptyset$ then $\text{clo}(E) = E$ by (2.48) so E is closed, and since $\partial E = \partial ({\sim} E)$ the complement ${\sim} E$ is also closed so E is both open and closed; and conversely if E is both open and closed then both E and ${\sim} E$ are closed so $\partial E = \text{clo}(E) \cap \text{clo}({\sim} E) = E \cap {\sim} E = \emptyset$.

For some examples, in the metric space $\mathbb{R}^n$ the open ball $B_r(a)$ of radius r centered at a point $a \in \mathbb{R}^n$ as defined in (2.32) is an open set in the metric topology, and it is clear from (2.43) that if $r > 0$ its closure is just the set

$$\overline{B_r(a)} = \Big\{ x \in \mathbb{R}^n \,\Big|\, \rho(a, x) \leq r \Big\}. \tag{2.51}$$

The set defined by (2.51) of course is called the closed ball of radius r centered at the point $a \in \mathbb{R}^n$. The closed ball $\overline{B_0(a)}$ of radius 0 is just the point a itself, while the open ball $B_0(a)$ of radius 0 is the empty set, so in that special case the closure of the open ball of radius 0 is not the closed ball of radius 0. It is equally clear from (2.49) that the boundary of either the open or closed ball of radius $r > 0$ is

$$\partial B_r(a) = \partial \overline{B_r(a)} = \left\{ x \in S \,\middle|\, \rho(a, x) = r \right\}. \tag{2.52}$$

Similarly the open cell Δ defined in (2.33) is an open set, and its closure is the closed cell, formally defined by

$$\overline{\Delta} = \left\{ \mathbf{x} = \{x_j\} \in \mathbb{R}^n \,\middle|\, a_j \leq x_j \leq b_j \quad \text{for} \quad 1 \leq j \leq n \right\}. \tag{2.53}$$

For the special case $n = 1$ the open cell is the open interval (a, b) and the closed cell is the closed interval $[a, b]$, while $\partial(a, b) = \partial[a, b] = \{a, b\}$. Note that in the case of a closed cell it is always possible that $a = b$, in which case the set still is called a closed cell; but in that case its interior is the empty set.

Any subset $R \subset S$ of a topological space S can be given the **induced topology** or **relative topology** by defining the open sets in the topology of R to be the intersections $R \cap U$ of the subset $R \subset S$ with the open subsets U of the topological space S. It is quite obvious that the collection of these intersections does satisfy the conditions to define a topology on R, and that the closed sets in the relative topology of a subset $R \subset S$ can be described as the intersections of the subset R with the closed sets of S. In particular if S is a metric space with the metric ρ and the metric topology and if $R \subset S$ is any subset then R inherits a topology from the topology of S; it is easy to see that this topology is the metric topology on R in terms of the restriction to R of the metric on S. What can be a bit confusing is that although the open or closed sets in $R \subset S$ are subsets of S they are not necessarily open sets or closed sets in S. For example, if $S = \mathbb{R}^2$ and $R = \{ (x_1, x_2) \in \mathbb{R}^2 \mid x_2 = 0 \}$ then the relatively open subsets of R are just the open subsets of the x_1 axis viewed as the set $\mathbb{R}^1$; but these sets of course are not open subsets of $\mathbb{R}^2$. For another example, the subset $\mathbb{Q} \subset \mathbb{R}$ of rational numbers inherits from the metric topology on $\mathbb{R}$ a topology that is the metric topology for the metric associated to the norm $|x|$ on the rationals. The set of rationals in an interval (a, b) is an open subset of $\mathbb{Q}$ but of course is not an open subset of $\mathbb{R}$. The interval $(\sqrt{2}, \sqrt{3})$ is an open subset of the space $\mathbb{R}$ but not a closed subset. The same interval viewed as a subset of $\mathbb{Q}$ is an open subset of $\mathbb{Q}$; but it is also a closed subset of $\mathbb{Q}$, since its complement is the union of the open subset $(\sqrt{3}, \infty)$ and the open subset $(-\infty, \sqrt{2})$ of $\mathbb{Q}$. It must be kept in mind that the end points describing this interval really are not points of the space $\mathbb{Q}$ itself, although they describe an interval in $\mathbb{Q}$ quite satisfactorily.

A topological space S is said to be **connected** if the only subsets of S that are both open and closed are the empty set $\emptyset$ and the entire set S. A subset $E \subset S$ is said to be connected if it is a connected set in the relative topology. Any topological space with the indiscrete topology is connected; but a topological space with the discrete topology is not connected if it contains at least two points. Since the subset $(\sqrt{2}, \sqrt{3}) \subset \mathbb{Q}$ of the rationals was just observed to be both open and closed in the induced topology it follows that the set of rational numbers $\mathbb{Q}$ with the topology induced from the metric topology on $\mathbb{R}$ is not a connected topological space. To show that a subset E of a topological space S is not connected it suffices to exhibit a subset of E that is both open and closed in the induced topology; but it is often easier to express this condition just in terms of the topology of the ambient topological space S. For this purpose two subsets A, B of a topological space S are said to be **separated** if $A \cap \overline{B} = \overline{A} \cap B = \emptyset$; and this property can be used as follows.

Lemma 2.12. *If A and B are nonempty separated subsets of a topological space S then the union $A \cup B$ is not a connected subset of S.*

Proof: If $A \cap \overline{B} = \overline{A} \cap B = \emptyset$ then since $\overline{B}$ is a closed subset of S its complement $S \sim \overline{B}$ is an open subset of S hence the intersection $(S \sim \overline{B}) \cap (A \cup B) = A$ is an open subset of $A \cup B$ in the relative topology of that subset. The corresponding argument shows that B also is an open subset of $A \cup B$ in the relative topology of that subset. Therefore A and B are both open and closed subsets of $A \cup B$ in the relative topology of that subset, so $A \cup B$ is not connected. That suffices for the proof.

For instance in the real line $\mathbb{R}$ with the metric topology the sets $A = (0, 1)$ and $B = (1, 2)$ are separated as well as disjoint while the sets $A = (0, 1)$ and $\overline{B} = [1, 2]$ are disjoint but not separated. The union $E = A \cup B$ is not connected, by the preceding lemma; but the union $F = A \cup \overline{B}$ is the set $F = (0, 2]$ which will be shown in the following Theorem 2.13 to be connected. Showing that a space is connected is in many ways harder than showing a space is not connected.

Theorem 2.13. *Any nonempty interval (with or without the end points) of the set $\mathbb{R}$ of real numbers, with the topology induced from the usual metric topology on $\mathbb{R}$, is a connected set.*

Proof: If an interval $I \subset \mathbb{R}$ is not connected then it contains a subset $U \subset I$ other than the empty set $\emptyset$ or the entire interval I such that U is both open and closed in the induced topology on I; hence $I = U \cup V$ where $V = I \sim U$ also is both open and closed and $U \cap V = \emptyset$. The sets U and V are nonempty, so choose a point $a \in U$ and a point $b \in V$; it can be assumed that $a < b$, after relabeling the sets U and V if necessary. The subset $E = U \cap [a, b]$ then is a nonempty set of real numbers, since $a \in E$, and E is bounded above by b, since $E \subset [a, b]$; so by the basic property of the real number system the set E has a least upper bound $c = \sup E$. If $c \in U$ then $c < b$, for otherwise $c \in U \cap V = \emptyset$, and since U is open there is an $\epsilon > 0$ such that $I \cap (c - \epsilon, c + \epsilon) \subset U$; but then the points in $I \cap (c, c + \epsilon)$ are also in U, which is impossible since $c = \sup E$. On the other hand if $c \in V$ then $a < c$ similarly, and since V is open there is an $\epsilon > 0$ such that $I \cap (c - \epsilon, c + \epsilon) \subset V$; but then the points in $I \cap (c - \epsilon, c)$ are also in V hence are not in E and consequently $c - \epsilon$ is an upper bound for E, which is impossible since $c = \sup E$. Thus there can be no such point c, so the assumption that I is not connected rules out, and that suffices for the proof.

The preceding observation can be used to show that various other sets also are connected. For this purpose, a subset $E \subset \mathbb{R}^n$ is said to be **convex** if whenever $\mathbf{a}, \mathbf{b} \in E$ for two points $\mathbf{a}, \mathbf{b} \in \mathbb{R}^n$ then $\ell(\mathbf{a}, \mathbf{b}) \subset E$ where

$$\ell(\mathbf{a}, \mathbf{b}) = \left\{ \mathbf{x} = t\mathbf{a} + (1 - t)\mathbf{b} \in \mathbb{R}^n \,\middle|\, 0 \leq t \leq 1 \right\} \tag{2.54}$$

is the line segment joining these two points. For example the set of points $\mathbf{x} \in \mathbb{R}^n$ such that $\|\mathbf{x}\| < r$ for some real number $r > 0$ is a convex set for any norm on $\mathbb{R}^n$; indeed if $\|\mathbf{a}\| < r$ and $\|\mathbf{b}\| < r$ for any points $\mathbf{a}, \mathbf{b} \in \mathbb{R}^n$ then for any real number t in the interval $0 \leq t \leq 1$ it follows from the triangle inequality for norms that $\|t\mathbf{a} + (1 - t)\mathbf{b}\| \leq t\|\mathbf{a}\| + (1 - t)\|\mathbf{b}\| < tr + (1 - t)r = r$. In particular any ball in $\mathbb{R}^n$, open or closed, and any cell in $\mathbb{R}^n$, open or closed, is a convex subset, so by the following corollary each also is a connected subset of $\mathbb{R}^n$ in the usual topology.

Corollary 2.14. *Any nonempty convex subset of $\mathbb{R}^n$ is connected in the topology induced by the metric topology of $\mathbb{R}^n$.*

Proof: If E is a nonempty convex subset of $\mathbb{R}^n$ and E is not connected then there is a nonempty subset $U \subset E$ other than the entire set E that is both open and closed; and the complement $V = E \sim U$ then also is both open and closed. For any points $\mathbf{a} \in U$ and $\mathbf{b} \in V$ the line segment $\ell(\mathbf{a}, \mathbf{b})$ is also contained in E. This line segment can be identified with the closed interval $[0, 1] \subset \mathbb{R}$ through the parametrization (2.54); and the intersections $U \cap \ell(\mathbf{a}, \mathbf{b})$ and $V \cap \ell(\mathbf{a}, \mathbf{b})$ are then disjoint open subsets of $\ell(\mathbf{a}, \mathbf{b})$ whose union is $\ell(\mathbf{a}, \mathbf{b})$, which is impossible since by the preceding theorem the interval $\ell(\mathbf{a}, \mathbf{b})$ is connected. That suffices for the proof.

Special topological spaces, those with special regularity properties, occur frequently in analysis and topology. A topological space is called a **Hausdorff space**, or sometimes a T_2 space,[4] if for any two distinct points $a, b \in S$ there are disjoint open subsets U_a, U_b of S such that $a \in U_a$ and $b \in U_b$. The Sierpinski space and any topological space S with the indiscrete topology and with at least two points are examples of topological spaces that are not Hausdorff. Many

---

[4]Felix Hausdorff (1868–1942) was one of the founders of modern topoplogy; his book *Principles of Set Theory* surveyed the field and also contained a number of major contributions to modern set theory. A standard list of regularity properties were labeled T_1 through T_6 in honor of Andrey Tychonoff (1906–1993) who was also a major figure in the development of topology.

of the standard topological spaces are Hausdorff though, as will be apparent as the discussion continues. Indeed the Hausdorff condition is so natural and convenient that most of the subsequent discussion here will consider just Hausdorff spaces.

Theorem 2.15. *A metric space S with the metric topology is a Hausdorff space.*

Proof: If $a, b \in S$ are two distinct points in a metric space S with the metric ρ then $\rho(a, b) = 2\epsilon$ for some real number $\epsilon > 0$; and the ϵ neighborhoods $N_a(\epsilon)$ and $N_b(\epsilon)$ are disjoint open sets for which $a \in N_a(\epsilon)$ and $b \in N_b(\epsilon)$, which suffices for the proof.

Theorem 2.16. *Any point in a Hausdorff space is a closed set.*

Proof: If S is a Hausdorff space and $a \in S$ is a point in S then for any point $b \neq a$ in S there is an open subset $U_b \subset S$ such that $a \notin U_b$. The union $U = \bigcup_{b \in S \sim \{a\}} U_b$ is then an open subset of S, and since $\{a\} = S \sim U$ it follows that $\{a\}$ is a closed set. That suffices for the proof.

A point a in topological space S is contained in the closure of a subset $E \subset S$ if and only if $\mathcal{N}_a \cap E \neq \emptyset$ for any open neighborhood $\mathcal{N}_a$ of the point a, as in the characterization (2.43) of the closure of a set. A closely related condition is very useful in the study of properties of topological spaces. A point $a \in S$ is said to be a **limit point** of a subset $E \subset S$ if $(\mathcal{N}_a \sim \{a\}) \cap E \neq \emptyset$ for any open neighborhood $\mathcal{N}_a$ of the point a, or equivalently if any open neighborhood $\mathcal{N}_a$ of the point a contains a point of E *other than a*. The set of limit points of a subset $E \subset S$ is denoted by E' and is called the **derived set** of the set E. For example in the real line $\mathbb{R}$ with the metric topology defined by the absolute value norm the set of limit points of the open interval $E = (a, b)$ is the closed interval $E' = [a, b]$ so $E' = \overline{E}$; on the other hand the subset $E = \mathbb{Z} \subset \mathbb{R}$ is a closed subset of $\mathbb{R}$ but none of its points is a limit point so $E' = \emptyset$ although $\overline{E} = E$. Sets E for which $E' = E$ are called **perfect sets** and play an interesting role in topology. The real line and any closed interval in the real line $\mathbb{R}$ are examples of perfect sets, while the closed subset $\mathbb{Z} \subset \mathbb{R}$ is not a perfect set. In general the relations between the closure $\overline{E}$ and the set of limit points E' of a subset of a topological space can be summarized as follows.

Theorem 2.17. *For any subset E of a topological space S*

$$E' \subset \overline{E} \tag{2.55}$$

and

$$\overline{E} = E \cup E' \tag{2.56}$$

hence

$$E \quad \text{is closed if and only if} \quad E' \subset E. \tag{2.57}$$

Proof: It is evident from (2.43) and the definition of a limit point that $E' \subset \bar{E}$, which is (2.55). Since $E \subset \bar{E}$ and $E' \subset \bar{E}$ as just demonstrated it follows that $E \cup E' \subset \bar{E}$; so in order to prove (2.56) it is enough just to show that $\bar{E} \subset E \cup E'$. If $a \in \bar{E}$ then $\mathcal{N}_a \cap E \neq \emptyset$ for any open neighborhood $\mathcal{N}_a$ of the point a; and if $a \notin E$ then $a \notin (\mathcal{N}_a \cap E)$ so it must be the case that $(\mathcal{N}_a \sim \{a\}) \cap E \neq \emptyset$, which is just the condition that $a \in E'$. That shows that $\bar{E} \subset E \cup E'$ and thereby demonstrates (2.56). Since E is closed if and only if $E = \bar{E} = E \cup E'$ it is evident from (2.56) that E is closed if and only if $E' \subset E$, and that suffices for the proof of the theorem.

Any point a in a topological space S with the indiscrete topology is a limit point of any nonempty subset $E \neq \{a\}$ of S; and no point in a topological space with the discrete topology is a limit point of any subset. Limit points and the derived sets thus are not of much interest in these extreme topologies, but are of particular use in Hausdorff spaces.

Theorem 2.18. *A point a in a Hausdorff space S is a limit point of a subset $E \subset S$ if and only if every open neighborhood $\mathcal{N}_a$ of a contains infinitely many distinct points of E.*

Proof: Suppose that a is a limit point of E and U is an open neighborhood of a. It will be demonstrated by induction on n that the intersection $U \cap E$ contains at least n distinct points other than a for any integer $n > 0$. First since a is a limit point of E the intersection $U \cap E$ contains a point $a_1 \neq a$, which establishes the initial case $n = 1$. Assume then that there are n distinct points $a_1, a_2, \ldots, a_n$ other than a in $U \cap E$. Since S is Hausdorff each point is closed, hence so is any finite union of points; so $U_n = U \sim \{a_1, \ldots, a_n\}$ is an open neighborhood of a, and since a is a limit point of E it follows that $(U_n \sim \{a\}) \cap E \neq \emptyset$ so there is another point $a_{n+1} \in (U \cap E)$ distinct from the points $a, a_1, \ldots, a_n$, and that suffices for the proof.

It is evident that the proof of the preceding theorem really only required that each point in the space S be a closed set. That is the case for any Hausdorff space, by Theorem 2.16; but there are more general topological spaces, traditionally called T_1 spaces, for which the preceding theorem therefore also holds. A standard example of a topological space that is T_1 but not Hausdorff is an infinite set S with the topology in which the open subsets are the entire set, the empty set and any set E for which the complement $S \sim E$ is finite.

Problems, Group I

1. Find a subset of $\mathbb{R}$ with exactly 3 limit points, in the standard topology.

2. Prove that the following assertions are true for subsets E, F of any topological space:

i) $\text{int}(E \cup F) \supset \text{int}(E) \cup \text{int}(F)$ but this may be a proper inclusion;

ii) $\text{int}(E \cap F) = \text{int}(E) \cap \text{int}(F)$;

iii) $\text{clo}(E \cup F) = \text{clo}(E) \cup \text{clo}(F)$;

iv) $\text{clo}(E \cap F) \subset \text{clo}(E) \cap \text{clo}(F)$ but this may be a proper inclusion;

v) $\partial(E \cup F) \subset \partial E \cup \partial F$;

vi) $\partial(E \cap F) \subset \partial E \cup \partial F$.

3. Show that $\partial\partial E \subset \partial E$ in any topological space. Find an example in which $\partial\partial E \neq \partial E$.

4. If a set E in a topological space is connected is $\overline{E}$ necessarily connected? Is E^o necessarily connected?

5. i) Show that in a Hausdorff space the derived set E' of any set E is closed.
   ii) Is $(E')' = E'$ in any topological space?

6. Prove that equivalent metrics define the same topology. Two metrics ρ and σ on a set are said to be **weakly equivalent** if and only if for any ball $B_{\rho,r}(a)$ in terms of the metric ρ there is a ball $B_{\sigma,s}(a)$ in terms of the metric σ such that $B_{\sigma,s}(a) \subset B_{\rho,r}(a)$, and conversely for any ball $B_{\sigma,s}(a)$ there is a ball $B_{\rho,r}(a)$ such that $B_{\rho,r}(a) \subset B_{\sigma,s}(a)$. Show that two metrics on a set S determine the same topology if and only if they are weakly equivalent. Show that two metrics that are equivalent are weakly equivalent, but that the converse does not necessarily hold.

Problems, Group II

7. Show that if $A \subset \mathbb{R}$ is any closed subset in the standard topology there is a subset $E \subset \mathbb{R}$ such that $A = E'$.

8. i) Let $\mathcal{T}$ be the collection of subsets of $\mathbb{R}$ consisting of $\mathbb{R}$ itself, the empty set $\emptyset$, and all sets of the form $(-\infty, a)$ for any $a \in \mathbb{R}$. Show that this collection of sets defines a topology on $\mathbb{R}$.
   ii) Show that with this topology the set $\mathbb{R}$ is not a Hausdorff space.
   iii) Show that with this topology the derived set E' of the set E consisting just of the point 0 is not a closed set.

9. In a set S with a specified point s let $\mathcal{T}$ be the collection of subsets of S consisting of S itself and of all subsets $E \subset S$ that do not contain the point s. Show that this is a topology on the set S. Describe the closed sets in this topology. Is this topological space Hausdorff? Show that if S contains at least two points there are sets that are closed but not open.

10. Let T be a subset of a topological space S and give T the induced topology from that of S. The closure of a subset $X \subset T \subset S$ can mean either the closure $\mathrm{clo}_T(X)$ of X when that set is viewed as a subset of the topological space T, or alternatively the closure $\mathrm{clo}_S(X)$ of X when that set is viewed as a subset of the topological space S, and similarly for the interior of X. Show that $\mathrm{clo}_T(X) = T \cap \mathrm{clo}_S(X)$. Show that $\mathrm{int}_T(X) \supset T \cap \mathrm{int}_S(X)$ and give an example to show that the inclusion may be a proper inclusion.

11. The **cofinite topology** in an infinite set S is that for which a subset $E \subset S$ is open if and only if either $E = \emptyset$ or $\sim E$ is finite. Show that points are closed in this topology. Show that S is not Hausdorff in this topology. Show that the only closed sets are S or finite subsets of S.

12. Suppose that $\mathcal{T}_1$ and $\mathcal{T}_2$ are two topologies on a set S where $\mathcal{T}_1 \subset \mathcal{T}_2$, so that any open set in the topology $\mathcal{T}_1$ is also an open set in the topology $\mathcal{T}_2$.

   i) Show that $\mathrm{clo}_1(A) \supset \mathrm{clo}_2(A)$ for any subset $A \subset S$, where $\mathrm{clo}_1(A)$ denotes the closure of the subset A in the topology $\mathcal{T}_1$ and $\mathrm{clo}_2(A)$ denotes the closure of the subset A in the topology $\mathcal{T}_2$.
   ii) Show that $\mathrm{int}_1(A) \subset \mathrm{int}_2(A)$, with the corresponding notation.

13. Suppose that S_1, S_2 are topological spaces with the topologies $\mathcal{T}_1, \mathcal{T}_2$, respectively; and let $\mathcal{T}$ be the collection of all unions of products $U_1 \times U_2$ where $U_1 \in \mathcal{T}_1$ and $U_2 \in \mathcal{T}_2$.

   i) Show that $\mathcal{T}$ defines a topology on $S_1 \times S_2$. (This is called the **product topology** on the product $S_1 \times S_2$.)
      Show that for any subsets $E_1 \subset S_1$ and $E_2 \subset S_2$
   ii) $\mathrm{int}(E_1 \times E_2) = \mathrm{int}(E_1) \times \mathrm{int}(E_2)$,
   iii) $\mathrm{clo}(E_1 \times E_2) = \mathrm{clo}(E_1) \times \mathrm{clo}(E_2)$,
   iv) $\partial(E_1 \times E_2) = (\partial E_1 \times \mathrm{clo}(E_2)) \cup (\mathrm{clo}(E_1) \times \partial E_2)$.

2.4 Compact Sets

A particularly useful special class of subsets of a topological space has a somewhat subtle and indirect definition, and is rather harder to understand than the classes of subsets described thus far; but it is a very important class of subsets with a wide range of uses. For the purposes of this definition, an **open covering** of a subset $E \subset S$ of a topological space S is a collection of open subsets $U_\alpha \subset S$ such that $E \subset \bigcup_\alpha U_\alpha$. If some of the sets U_α are redundant they can be eliminated and the remaining sets are also an open covering of E, called a **subcovering** of the set E. A subset $E \subset S$ is said to be **compact** if every open covering of E has a finite subcovering, that is, if for any open covering $\{U_\alpha\}$ of E finitely many of the sets U_α actually cover all of E. To show that a set E is not compact it suffices to find a single open covering $\{U_\alpha\}$ of E such that no finite collection of the sets U_α can cover E; but to show that E is compact it is necessary to show that for *any* open covering $\{U_\alpha\}$ of E finitely many of the sets U_α already cover E. Examples of noncompact sets are quite easy: for instance $\mathbb{R}$ with its usual topology is not compact, since it is contained in the union of all the open intervals $(-n, n)$ but not in any finite set of these intervals. Even a finite interval $(-m, m)$ is not compact, since it is contained in the union of all the open intervals $(-m + \frac{1}{k}, m - \frac{1}{k})$ for all positive integers k but again it is not contained in any finite collection of these intervals. Of course any single point and any finite set of points are obviously compact sets. It is more difficult to demonstrate that other sets are compact, since it is necessary to show that *any* open covering has a finite subcovering. The basic nontrivial example of a compact set is a closed cell in $\mathbb{R}^n$; and the proof that a cell is compact rests on the completeness property of the real numbers. Just for the purposes of the proof the edgesize of the cell $\overline{\Delta} = \{ \mathbf{x} \in \mathbb{R}^n \mid a_i \leq x_i \leq b_i \}$ is defined to be the real number $e(\Delta) = \max_i (b_i - a_i)$; thus $e(\Delta) \geq 0$ and $e(\Delta) = 0$ if and only if the cell is just the single point $(a_1 = b_1, a_2 = b_2, \ldots, a_n = b_n)$.

Lemma 2.19. *If $\overline{\Delta}_\nu \subset \mathbb{R}^n$ are closed cells in $\mathbb{R}^n$ for $\nu = 1, 2, \ldots$ such that $\overline{\Delta}_{\nu+1} \subset \overline{\Delta}_\nu$ and $\lim_{\nu \to \infty} e(\overline{\Delta}_\nu) = 0$ then $\bigcap_\nu \overline{\Delta}_\nu$ is a single point of $\mathbb{R}^n$.*

Proof: If $\overline{\Delta}_\nu = \{ \mathbf{x} = \{x_j\} \mid a_j^\nu \leq x_j \leq b_j^\nu \}$ then for each j clearly $a_j^{\nu+1} \geq a_j^\nu$ and $b_j^{\nu+1} \leq b_j^\nu$; and since $\overline{\Delta}_\nu \subset \overline{\Delta}_1$ it is also the case that $a_j^\nu \leq b_j^1$ and $b_j^\nu \geq a_j^1$. The basic completeness property of the real number system implies that any increasing sequence of real numbers bounded from above and any decreasing sequence of real numbers bounded from below have limiting values; therefore $\lim_{\nu \to \infty} a_j^\nu = a_j$ and $\lim_{\nu \to \infty} b_j^\nu = b_j$ for some uniquely determined real numbers a_j, b_j. Since $a_j^\mu \leq b_j^\nu$ for any μ, ν it follows that $a_j \leq b_j^\nu$ for any ν, and from this in turn that $a_j \leq b_j$. Clearly the cell $\overline{\Delta} = \{ \mathbf{x} = \{x_j\} \mid a_j \leq x_j \leq b_j \}$ is contained in the intersection $\bigcap_\nu \overline{\Delta}_\nu$. On the other hand since $(b_j - a_j) \leq (b_j^\nu - a_j^\nu) \leq e(\overline{\Delta}_\nu)$ and $\lim_{\nu \to \infty} e(\overline{\Delta}_\nu) = 0$ then actually $b_j = a_j$ so the limiting cell $\overline{\Delta}$ is just a single point of $\mathbb{R}^n$, and that concludes the proof.

Theorem 2.20. *A closed cell in $\mathbb{R}^n$ is compact.*

Proof: If a closed cell $\overline{\Delta} = \{ \mathbf{x} = \{x_j\} \mid a_j \leq x_j \leq b_j \}$ is not compact there is an open covering $\{U_\alpha\}$ of $\overline{\Delta}$ that does not admit any finite subcovering. The cell $\overline{\Delta}$ can be written as the union of the closed cells arising from bisecting each of its sides. If finitely many of the sets $\{U_\alpha\}$ covered each of the subcells then finitely many would cover the entire set $\overline{\Delta}$, which is not the case; hence at least one of the subcells cannot be covered by finitely many of the sets $\{U_\alpha\}$. Then bisect each of the sides of that subcell, and repeat the process. The result is that there is a collection of closed cells $\overline{\Delta}_\nu$ which cannot be covered by finitely many of the open sets $\{U_\alpha\}$ and for which $\overline{\Delta}_{\nu+1} \subset \overline{\Delta}_\nu \subset \overline{\Delta}$, and $\lim_{\nu\to\infty} e(\overline{\Delta}_\nu) = 0$. It then follows from the preceding lemma that $\bigcap_\nu \overline{\Delta}_\nu = \{\mathbf{a}\}$, a single point in $\mathbb{R}^n$. This point must be contained within one of the sets $\{U_{\alpha_0}\}$, and if ν is sufficiently large then $\overline{\Delta}_\nu \subset U_{\alpha_0}$ as well; but that is a contradiction, since the cells $\overline{\Delta}_\nu$ were chosen so that none of them could be covered by finitely many of the sets $\{U_\alpha\}$. That contradiction shows that the cell $\overline{\Delta}$ must be compact, which concludes the proof.

Several basic and extremely useful properties of compact sets follow rather directly from the definition.

Theorem 2.21 (Compactness Theorem). *For any topological space S:*
(i) *A closed subset of a compact subset in S is compact.*
(ii) *If $K_n \subset S$ are nonempty closed subsets of S for $n \geq 0$, if at least one of them is compact and if $K_{n+1} \subset K_n$ for all n then $\bigcap_n K_n \neq \emptyset$.*
(iii) *If K_α are closed subsets of S, not necessarily just countably many sets, and if one of these sets is compact and any finite number of them have a nonempty intersection then $\bigcap_\alpha K_\alpha \neq \emptyset$.*
(iv) *If $R \subset S$ is a subset of S then a subset $K \subset R$ is compact when viewed as a subset of R in the induced topology if and only if K is compact when viewed as a subset of the topological space S.*

Proof: (i) If $F \subset K \subset S$ where F is closed and K is compact and if $U_\alpha \subset S$ are open subsets of S such that $F \subset \bigcup_\alpha U_\alpha$ then the sets U_α together with $U = S \sim F$ form an open covering of K, so finitely many of these sets serve to cover K and hence also serve to cover F; these finitely many sets, excluding the set U, are then finitely many of the sets U_α that cover F, hence F is compact.
(ii) It can be assumed that K_0 is compact, by ignoring the earlier sets K_n. If to the contrary $\bigcap_n K_n = \emptyset$ then the sets $U_n = S \sim K_n$ for $n \geq 0$ are open subsets of S such that

$$\bigcup U_n = \bigcup(S \sim K_n) = S \sim \bigcap K_n = S \sim \emptyset = S,$$

so the sets U_n are an open covering of K_0. By assumption K_0 is compact so finitely many of these open sets cover K_0, and since $U_n \subset U_{n+1}$ it must be the case that $K_0 \subset U_N = S \sim K_N$ for some N; but since $K_N \subset K_0$ then $K_N \subset (S \sim K_N)$ so that $K_N = \emptyset$, contradicting the hypothesis that these sets are nonempty. That contradiction shows that the intersection of all the sets must be nonempty.

(iii) Suppose to the contrary of the desired result that $\bigcap_\alpha K_\alpha = \emptyset$; then the sets $U_\alpha = S \sim K_\alpha$ are open subsets of S such that

$$\bigcup_\alpha U_\alpha = \bigcup_\alpha (S \sim K_\alpha) = S \sim \bigcap_\alpha K_\alpha = S \sim \emptyset = S,$$

so they form an open covering of S. If K_{α_0} is compact then since it is covered by the sets U_α it must be covered by finitely many of these sets, so that

$$K_{\alpha_0} \subset \bigcup_{i=1}^m U_{\alpha_i} = \bigcup_{i=1}^m (S \sim K_{\alpha_i}) = S \sim \bigcap_{i=1}^m K_{\alpha_i},$$

and consequently $K_{\alpha_0} \cap \bigcap_{i=1}^m K_{\alpha_i} = \emptyset$, contradicting the hypothesis that any intersection of finitely many of the sets K_i is nonempty. That contradiction serves to prove the initial assertion.

(iv) If $K \subset R$ is compact in the induced topology and if U_α are open sets in S that cover K then the intersections $U_\alpha \cap R$ are open sets in the relative topology of R that cover K so finitely many of these intersections serve to cover K; hence finitely many of the sets U_α also serve to cover K, so K is compact as a subset of S. Conversely if K is a compact subset of S and $K \subset R$, and if U_α are relatively open sets in R that cover K, there will be open subsets $V_\alpha \subset S$ for which $U_\alpha = V_\alpha \cap R$; the sets V_α of course cover K, so finitely many of them will cover K since K is compact in S and consequently finitely many of the sets U_α will cover K as well, so K is a compact subset of R in the relative topology. That concludes the proof of the theorem.

In the preceding theorem (ii) is really a special case of (iii) but it is listed separately since it is a very common form in which (iii) is applied. It was already noted that if R is a subset of a topological space S then open subsets of R in the induced topology need not be open subsets when viewed as subsets of S, and correspondingly for closed sets; thus the condition that a subset $E \subset R$ be open or closed can be quite different depending on whether that subset is viewed as a subset of S with the topology of S or as a subset of R with the induced topology of R. However (iv) of the preceding theorem shows that compactness is really an absolute property, in the sense that a subset $K \subset R$ is compact in the relative topology of R if and only if it is compact as a subset of S. Of course that does not imply that the restriction of a compact subset of S to R is necessarily a compact subset of R. As might be expected, for Hausdorff spaces such as $\mathbb{R}^n$ compact subsets have a number of further properties, as in the following theorem; but note though that these results are only demonstrated for Hausdorff spaces so cannot be assumed to hold in more general topological spaces.

Theorem 2.22 (Compactness Theorem for Hausdorff Spaces). *Compact sets in a Hausdorff topological space S have the following properties:*

(i) *Any compact subset of S is closed.*

(ii) *If $K \subset S$ is compact then any infinite subset $E \subset K$ has a limit point in K.*

(iii) *If $K_1, K_2 \subset S$ are disjoint compact subsets then there are disjoint open subsets $U_1, U_2 \subset S$ such that $K_1 \subset U_1$ and $K_2 \subset U_2$.*

Proof: (i) If $K \subset S$ is compact and $a \in (S \sim K)$ then since S is Hausdorff for any point $b \in K$ there are disjoint open neighborhoods U_b of b and $U_{b,a}$ of a. The collection of open sets U_b for all points $b \in K$ then cover the compact set K, so finitely many of these sets U_{b_i} already serve to cover K. The finite intersection $U = \bigcap_i U_{b_i,a}$ is then an open neighborhood of a that is disjoint from K and hence is an open neighborhood of a that is contained in $S \sim K$; and since that is true for any point a it follows that $S \sim K$ is open and consequently that K is closed.

(ii) Suppose $E \subset K$ is an infinite subset of the compact subset $K \subset S$ and E has no limit points in K. Then E has no limit points at all, since any limit point of E is a limit point of K and hence must be contained in K since K is closed by (i); so in particular E is a closed subset of S, and as a closed subset of the compact set K it follows that E also is compact. Every point $a \in E$ has an open neighborhood U_a that does not contain any of the other points of E. The collection of these sets U_a is an open covering of the compact set E, which must have a finite subcovering; but that is impossible, since no finite collection of the sets U_a can cover the entire infinite set E. Therefore E must have limit points in K.

(iii) Suppose first that $K_2 = \{a\}$ consists of a single point and is compact. Since S is Hausdorff then for any point $b \in K_1$ there are disjoint open subsets $U_b, U_{a,b} \subset S$ such that $b \in U_b$ and $a \in U_{a,b}$. The collection of open sets U_b for all $b \in K_1$ form an open covering of the compact set K_1, so finitely many of these sets U_{b_i} already serve to cover K_1. The union $U_1 = \bigcup_i U_{b_i}$ is an open set containing K_1 and the intersection $U_2 = \bigcap_i U_{a,b_i}$ is an open set containing a and $U_1 \cap U_2 = \emptyset$ as desired. Next if K_2 is any compact set it follows from the special case just demonstrated that for any point $a \in K_2$ there are disjoint open sets $U_{1,a}$ and U_a such that $K_1 \subset U_{1,a}$ and $a \subset U_a$. The sets U_a for all $a \in K_2$ form an open covering of K_2, and since K_2 is compact finitely many U_{a_i} of these sets already serve to cover K_2. Then as before the sets $U_1 = \bigcap_i U_{1,a_i}$ and $U_2 = \bigcup_i U_{a_i}$ are disjoint open subsets of S such that $K_1 \subset U_1$ and $K_2 \subset U_2$.

The following two basic theorems characterize compact subsets of $\mathbb{R}^n$; in the statement of the first of them a subset $E \subset \mathbb{R}^n$ is said to be **bounded** if it is contained in a cell $\Delta \subset \mathbb{R}^n$, or equivalently of course, if it is contained in a ball $B_r(a)$ centered at a point $a \in \mathbb{R}^n$.

Theorem 2.23 (Heine-Borel Theorem). *A subset $E \subset \mathbb{R}^n$ is compact if and only if it is closed and bounded.*

Proof: If E is a bounded closed subset of $\mathbb{R}^n$ then since it is bounded it is contained in some closed cell; since a closed cell is compact by Theorem 2.20 it follows from Theorem 2.21 (i) that the set E is compact. Conversely if E is a compact subset of $\mathbb{R}^n$ it follows from Theorem 2.22 (i) that E is closed since $\mathbb{R}^n$ is Hausdorff. If E is not bounded it can be covered by the open balls $B_r(\mathbf{0})$ of radius r for $r = 1, 2, 3, \ldots$ but not by any finite subset of these balls, so E is not compact. That suffices for the proof.

Theorem 2.24 (Bolzano-Weierstrass Theorem). *A subset $E \subset \mathbb{R}^n$ is compact if and only if every infinite subset of E has a limit point in E.*

Proof: If a subset $E \subset \mathbb{R}^n$ is compact then by Theorem 2.22 (ii) every sequence of distinct points in E has a limit point in E. To show conversely that if any infinite subset of E has a limit point in E then E is compact, it is equivalent to show that if $E \subset \mathbb{R}^n$ is not compact then it contains an infinite set of points with no limit point in E. If $E \subset \mathbb{R}^n$ is not compact then by the Heine-Borel Theorem either E is not bounded or E is not closed. If E is not bounded it must contain points $\mathbf{a}_\nu$ such that $\|\mathbf{a}_\nu\| \geq \nu$, and it is clear that this set of points has no limit point. If E is not closed then at least one of its limit points is not contained in E. That suffices to conclude the proof.

Compactness is useful in establishing the properties of some nontrivial and perhaps surprising examples of the topology of subsets of vector spaces $\mathbb{R}^n$; these illustrate some of the complications that are possible, so are useful to keep in mind in order not to leap to conclusions that may seem intuitively obvious but that are in fact false. The first example is the **rational neighborhood**. It was noted in Theorem 2.9 that open subsets of $\mathbb{R}^1$ consist of countably many disjoint open intervals (a, b), which sounds somewhat simpler than it may be in some cases. The rationals are countable, so list them as $a_1, a_2, a_3, \ldots$. Let S be the union $S = \bigcup_i I_i$ where I_i is the open interval centered at the point a_i and of length $\epsilon/2^i$. The set S thus is an open subset of $\mathbb{R}^1$ that contains all the rationals; but nonetheless it does not cover the entire real line. Indeed if S covered the closed interval $I = [0, 1]$ for instance then since that interval is compact finitely many N of the intervals I_i would serve to cover I; but the total length of these intervals is at most $\epsilon(\frac{1}{2} + (\frac{1}{2})^2 + (\frac{1}{2})^3 + \cdots + (\frac{1}{2})^N) < \epsilon$ so they cannot cover I. The complement $K = \mathbb{R} \sim S$ is thus a nontrivial closed subset of $\mathbb{R}$ that omits an open neighborhood of each rational number.

An even more interesting example is the **Cantor set**, one of the classical examples of a surprising set. Let I_0 be the closed unit interval $I_0 = [0, 1]$; let $I_1 \subset I_0$ be the set that arises from I_0 by removing the open middle third $(\frac{1}{3}, \frac{2}{3})$

of I_0; let I_2 be the set that arises from I_1 by removing the open middle third of each segment of I_1; and continue that process. The result is a collection of nonempty compact subsets $I_n \subset I_0$ such that $I_0 \supset I_1 \supset I_2 \supset \cdots$. The Cantor set is the intersection $C = \bigcap_n I_n$, which is a nonempty compact subset of the unit interval I_0 by Theorem 2.21 (i) and (ii). The end points of the segments in each set I_n belong to the Cantor set, since they are never removed during the course of the construction; any point in the Cantor set thus is a limit point of the set of these end points, so the Cantor set is an example of a perfect set. The Cantor set though contains no open or closed intervals, since the lengths of the intervals in each set I_n are equal to 3^{-n}. When the real numbers in the interval $[0, 1]$ are described by their ternary expansions, expansions to the base 3, it is clear that the Cantor set consists of those real numbers in $[0, 1]$ with a ternary expansion $a_1 a_2 a_3 a_4 \ldots$ where $a_i = 0$ or 2, so a ternary expansion for which $a_i \neq 1$.

The vector spaces $\mathbb{R}^n$ themselves are not compact; but by Theorem 2.23 any closed bounded subset of $\mathbb{R}^n$ is compact, so any point of $\mathbb{R}^n$ is contained in an open neighborhood which if small enough has a compact closure. That is the key to a number of properties of these vector spaces. More generally a topological space S is said to be **locally compact** if each point in S is an interior point of a compact subset of S. Locally compact Hausdorff spaces have a number of important special properties. Before discussing these properties though it is worth establishing a useful alternative characterization of locally compact Hausdorff spaces.

Lemma 2.25. *A Hausdorff space S is locally compact if and only if for any point $a \in S$ and any open neighborhood U of the point a there is an open set V in S such that $a \in V \subset \overline{V} \subset U$ and $\overline{V}$ is compact.*

Proof: If S satisfies the condition as in the statement of the lemma then S clearly is locally compact. Conversely suppose that S is a locally compact Hausdorff space, that $a \in S$ is a point in the interior K^o of a compact subset K of S, and that U is an open neighborhood of a. Then $a \in (U \cap K^o)$, and since the point a and the boundary $\partial(U \cap K^o)$ are disjoint closed subsets of the compact set K they are also compact sets and by Theorem 2.22 (iii) there are disjoint open subsets X, W in S for which $a \in X$ and $\partial(U \cap K^o) \subset W$. The intersection $V = X \cap (U \cap K^o)$ is an open neighborhood of a contained in $U \cap K^o$, and since $V \cap W = \emptyset$ it follows that $V \subset \overline{V} \subset (U \cap K^o) \subset U$; and as a closed subset of the compact set K the set $\overline{V}$ is compact. That suffices for the proof.

Lemma 2.26. *Any open or closed subset of a locally compact Hausdorff space is a locally compact Hausdorff space in the relative topology.*

Proof: If S is a locally compact Hausdorff space then any subset of S is clearly a Hausdorff space in the relative topology. It is clear from the preceding

Lemma 2.25 that an open subset of S is locally compact in the relative topology. On the other hand if $F \subset S$ is a closed subset of S then for each point $a \in F$ there is a compact set K in S such that $a \in K^o$; then $a \in K^o \cap F \subset K \cap F$, and $K \cap F$ is a closed subset of K hence is compact. That suffices for the proof.

Theorem 2.27 (Baire's Category Theorem). *The intersection of a countable number of dense open subsets of a locally compact Hausdorff space S is dense in S.*

Proof: Suppose that E_n are dense open subsets of the locally compact Hausdorff space S for all $n \geq 1$, so that $\overline{E}_n = S$; and let $E = \bigcap_{n=1}^{\infty} E_n$. What is to be proved is that for any open subset $U \subset S$ there is a point $a \in U \cap E$. For the proof it will be demonstrated by induction on n that for all $n \geq 1$ there are nonempty open subsets $U_n \subset S$ for which

$$U_n \subset \overline{U}_n \subset (U_{n-1} \cap E_n) \ \text{ and } \overline{U}_n \text{ is compact} \tag{2.58}$$

where $U_0 = U$. First $U \cap E_1 \neq \emptyset$ since E_1 is dense; and by Lemma 2.25 there is a nonempty open subset $U_1 \subset S$ such that $\overline{U}_1 \subset (U \cap E_1)$ and $\overline{U}_1$ is compact. For the inductive step, suppose that there are sets $U_1, \ldots, U_n$ satisfying (2.58). Since E_{n+1} is dense then $U_n \cap E_{n+1} \neq \emptyset$; and by Lemma 2.25 there is a nonempty open subset $U_{n+1} \subset S$ such that $\overline{U}_{n+1} \subset (U_n \cap E_{n+1})$ and $\overline{U}_{n+1}$ is compact, which is (2.58) for the case $n+1$. Then by Theorem 2.21 (ii) the intersection $K = \bigcap_{n=1}^{\infty} \overline{U}_n$ is a nonempty subset of U. If $a \in K$ then $a \in \overline{U}_n \subset (U_{n-1} \cap E_n)$ for all $n \geq 1$ so $a \in U \cap E$, and that suffices for the proof.

The preceding result has a wide variety of applications in analysis. Some of the first applications were phrased by René-Louis Baire in other terms that have since become quite standard; that may serve to explain why the preceding theorem is called Baire's Category Theorem. A sparse subset of a topological space was defined earlier as a subset $E \subset S$ for which $\text{int}(E) = \emptyset$. An even smaller subset, a **nowhere dense** subset $E \subset S$, is defined as a subset for which $\text{int}(\text{clo}(E)) = \emptyset$, so as a subset that has a sparse closure.

Lemma 2.28. *A subset E of a topological space is nowhere dense if and only if the complement of its closure is a dense open subset of S.*

Proof: Let $E \subset S$ be a subset of the topological space S and let $G = S \sim \overline{E}$ be the closure of its complement, so G is an open subset of S. If E is nowhere dense then by definition its closure $\overline{E}$ contains no open subsets of S; so any open neighborhood U of a point $a \in S$ must contain some point $b \in G$, hence $a \in \overline{G}$ so that G is dense in S. On the other hand if E is not nowhere dense then its closure $\overline{E}$ must contain an open neighborhood U of some point $a \in S$; this neighborhood then is not contained in G, so the point a cannot be a limit point of G showing that G is not dense, which suffices for the proof.

The rationals $\mathbb{Q}$ and the integers $\mathbb{Z}$ are both sparse subsets of the real line $\mathbb{R}$; the integers are nowhere dense but the rationals are not, although the rationals can be written as a countable union of the sets $\frac{1}{n}\mathbb{Z}$ for all $n \in \mathbb{N}$ thus as a countable union of nowhere dense subsets. A subset $E \subset S$ of a Hausdorff space S is said to be a set of the **first category** if it can be written as a countable union of nowhere dense subsets; the set of rational numbers thus is an example of a set of the first category. It is perhaps natural then to ask whether the set of real numbers $\mathbb{R}$ is a set of the first category; that is actually not the case, but not altogether easy to demonstrate. Subsets of a topological space that are not sets of the first category are called sets of the **second category**; and Baire's Category Theorem is the key to establishing the existence of such subsets.

Corollary 2.29. *A locally compact Hausdorff space S is a set of the second category.*

Proof: Suppose to the contrary that S is of the first category, so that $S = \bigcup_{\nu=1}^{\infty} \overline{E}_\nu$ for some nowhere dense subset $E_\nu \subset S$. Then $\bigcap_\nu (S \sim \overline{E}_\nu) = \emptyset$. However the preceding Lemma 2.28 shows that the complements $S \sim \overline{E}_\nu$ are dense open subsets of S, and Baire's Category Theorem shows that their intersection cannot be the empty set, indeed that their intersection is a dense subset of S. That contradiction serves to conclude the proof.

A simple application of the preceding corollary is the following possibly surprising result, the usefulness of which will be apparent later.

Corollary 2.30. *The set of rational numbers cannot be written as an intersection of countably many open subsets of $\mathbb{R}$.*

Proof: Suppose that $\mathbb{Q} = \bigcap U_i$ for some open subsets $U_i \subset \mathbb{R}$, so that set I of irrationals can be written as the union $I = \bigcup_i (\mathbb{R} \sim U_i)$ of the countably many closed sets $\mathbb{R} \sim U_i$; these sets must be nowhere dense, for if $\mathbb{R} \sim U_i$ has a non-empty interior the union I would also contain an open subset of $\mathbb{R}$. Then $\mathbb{R} = \mathbb{Q} \cup \bigcup_i (\mathbb{R} \sim U_i)$ is a representation of the real numbers $\mathbb{R}$ as a countable union of nowhere dense sets, so $\mathbb{R}$ would be a set of the first category in contradiction to the preceding corollary. That contradiction suffices to conclude the proof.

A traditional terminology is to say that a subset of a topological space is an F_σ set if it can be written as a countable union of closed sets, and that a subset is a G_δ set if it can be written as a countable intersection of open sets. Of course any Hausdorff topological space can be written as a union of closed sets, since it is the union of its points and points of a Hausdorff space are closed sets. The question whether a topological space or a subset of a topological space is an F_σ set or a G_δ when the space is uncountable can be both difficult and interesting.

The preceding corollary can be rephrased as the assertion that $\mathbb{Q}$ is not a G_δ subset of $\mathbb{R}$. Another consequence of Baire's Category Theorem can be used to establish a basic result about the real number system.

Theorem 2.31. *A nonempty perfect set in a locally compact Hausdorff space contains uncountably many points.*

Proof: Suppose that E is a nonempty perfect subset of a locally compact Hausdorff space S, a set such that $E' = E$. Since $E' \neq \emptyset$ the set E contains infinitely many points by Theorem 2.18. Suppose though in contradiction to the desired result that E contains only countably many distinct points $a_1, a_2, \ldots$. Each individual point is a closed subset $\{a_i\} \subset S$, indeed is a nowhere dense set since otherwise it would be both open and closed hence would be a component of S and therefore not a limit point of E, which is impossible since E is perfect. The set $E = \bigcup_{i=1}^{\infty} \{a_i\}$ thus is a countable union of nowhere dense sets so is a set of the first category in S. On the other hand E is a closed subset of S, so by Lemma 2.26 it is a locally compact Hausdorff space in the relative topology; but then it is a set of the second category by Corollary 2.29. That contradiction serves to conclude the proof.

Corollary 2.32. *The field $\mathbb{R}$ of real numbers is uncountable.*[5]

Proof: Since $\mathbb{R}$ is a perfect locally compact Hausdorff space this corollary is an immediate consequence of the preceding theorem.

Problems, Group I

1.   If $K \subset E_1 \cup E_2$, where K is compact and E_1, E_2 are disjoint open subsets of a topological space, is $K \cap E_1$ compact? Is that always the case if E_1, E_2 are not disjoint?

2.   Find an example of a metric space in which the Heine-Borel Theorem is not true.

3.   Find an example of a topological space in which there is a compact set that is not closed.

---

[5] An old question was whether there are subsets of the set $\mathbb{R}$ of real numbers that have cardinality strictly greater than that of the set $\mathbb{Q}$ of the rationals but strictly less than that of the entire set $\mathbb{R}$ of real numbers, or equivalently whether there are subsets S of the set of real numbers such that $\aleph_0 < \#(S) < \#(\mathbb{R})$; that is the continuum hypothesis, going back to the work of Georg Cantor. Kurt Gödel showed in 1940 that it is consistent with the rest of mathematics to assume that there are no such sets, and Paul Cohen showed in 1963 that it is consistent with the rest of mathematics to assume that there are such sets.

4.   If K is a compact subset of a topological space is clo(K) also compact?

5.   Suppose that $\mathcal{T}_1$ and $\mathcal{T}_2$ are two topologies on a set S where $\mathcal{T}_1 \subset \mathcal{T}_2$, so that any open set in the topology $\mathcal{T}_1$ is also an open set in the topology $\mathcal{T}_2$.

   i) If K is a subset of S that is compact in the topology $\mathcal{T}_1$ is it necessarily compact in the topology $\mathcal{T}_2$? Why?
   ii) If K is a subset of S that is compact in the topology $\mathcal{T}_2$ is it necessarily compact in the topology $\mathcal{T}_1$? Why?

6.   Show that if A and B are disjoint compact subsets of a metric space S with a metric ρ then there are points $a \in A$ and $b \in B$ such that $\rho(a,b) = \inf_{x \in A, y \in B} \rho(x,y)$. (This limit is usually denoted by $\rho(A,B)$.)

7.   What are the compact subsets of $\mathbb{Q}$ with the usual metric topology?

Problems, Group II

8.   The sum $A + B$ of any two subsets $A, B \subset \mathbb{R}$ of the real line is the subset defined by $A + B = \{a + b \mid a \in A,\ b \in B\}$.

   i) If A and B are open is $A + B$ open? Why?
   ii) If A and B are closed is $A + B$ closed? Why?
   iii) If A and B are compact is $A + B$ closed? Why?
   iv) If A and B are compact is $A + B$ compact? Why?

9.   Show that the empty set and the collection of subsets of $\mathbb{R}$ having compact complements form a topology on $\mathbb{R}$. Is $\mathbb{R}$ with this topology compact? Hausdorff? Connected?

10.   Show that if S is a locally compact Hausdorff space then for any point $a \in S$ and any closed subset $C \subset S$ not containing the point a there exist disjoint open subsets U, V in S such that $a \in U$ and $C \subset V$. (A topological space is said to be **regular** or T_3 if any point and closed set not containing that point are contained in disjoint open subsets; thus any locally compact Hausdorff space is regular.)

11.   If K_n are nonempty compact subsets of a Hausdorff space S such that $K_{n+1} \subset K_n$ and $\bigcap_n K_n \subset U$ for some open subset $U \subset S$ show that there is some number N such that $K_n \subset U$ for all $n > N$.

12. Show that in a complete metric space S the intersection of any sequence of dense open subsets is dense. [Suggestion: Examine the proof given earlier here of the corresponding result for locally compact Hausdorff spaces.]

13. Show that a nonempty perfect set in a complete metric space contains uncountably many points. [Suggestion: Consider the proof of Theorem 2.31.]

14. Show that a compact metric space is complete.

3

Mappings

3.1 Continuous Mappings

A **mapping** $f : S \longrightarrow T$ from a set S to a set T associates to each point $x \in S$ a point $f(x) = y \in T$. A good deal of analysis deals with mappings $f : S \longrightarrow \mathbb{R}$ from a set S to the real numbers; it is traditional to call such a mapping a **function** on the set S, or a real-valued function on the set S since complex-valued functions often also are considered. The mappings involved usually have some regularity properties, either continuity or differentiability for instance. If S and T are metric spaces, with metrics ρ_S and ρ_T, a mapping $f : S \longrightarrow T$ is defined to be **continuous at a point** $a \in S$ if for any $\epsilon > 0$ there is a $\delta > 0$ such that

$$\rho_T(f(x), f(a)) < \epsilon \quad \text{whenever} \quad \rho_S(x, a) < \delta; \tag{3.1}$$

and it is said to be **continuous in** S if it is continuous at each point $a \in S$. In particular, if S and T are normed vector spaces with the metrics associated to norms $\|\mathbf{x}\|_S$ and $\|\mathbf{x}\|_T$, a mapping $\mathbf{f} : S \longrightarrow T$ is continuous at a point $\mathbf{a} \in S$ if for any $\epsilon > 0$ there is a $\delta > 0$ such that

$$\|\mathbf{f}(\mathbf{x}) - \mathbf{f}(\mathbf{a})\|_T < \epsilon \quad \text{whenever} \quad \|\mathbf{x} - \mathbf{a}\|_S < \delta. \tag{3.2}$$

The simplest normed space is the real line $\mathbb{R}$ where the norm on $\mathbb{R}$ is the absolute value; a mapping $f : \mathbb{R} \longrightarrow \mathbb{R}$ is a real-valued function of a real variable, and it is continuous at a point $a \in \mathbb{R}$ if for any $\epsilon > 0$ there is a $\delta > 0$ such that

$$|f(x) - f(a)| < \epsilon \quad \text{whenever} \quad |x - a| < \delta, \tag{3.3}$$

undoubtedly the most familiar form of the definition (3.2).

For example, a constant mapping $f : \mathbb{R} \longrightarrow \mathbb{R}$ defined by $f(x) = b \in \mathbb{R}$ for all $x \in \mathbb{R}$ is continuous in $\mathbb{R}$ since for any point $a \in \mathbb{R}$ clearly $|f(x) - f(a)| = 0 < \epsilon$ whenever $|x - a| < \delta$ for any δ. The mapping $f(x) = x^n$ for any integer $n \geq 0$ is continuous in $\mathbb{R}$, since for any point $a \in \mathbb{R}$ it is easy to see that $|f(x) - f(a)| = |x^n - a^n| = |x - a| \cdot |p(x)|$ where $p(x)$ is a polynomial in x; if $|x - a| < 1$ then $|p(x)| < M$ for some constant $M > 0$ so $|f(x) - f(a)| < \epsilon$ whenever $|x - a| < \min(\epsilon/M, \ 1)$. The sum of two continuous mappings $f, g : \mathbb{R} \longrightarrow \mathbb{R}$ is continuous at any point $a \in \mathbb{R}$, since $|f(x) - f(a)| < \epsilon/2$ if $|x - a| < \delta_f$ for some δ_f since

f is continuous and $|g(x) - g(a)| < \epsilon/2$ if $|x - a| < \delta_g$ for some δ_g since g is continuous, so $|f(x) + g(x) - f(a) - g(a)| < \epsilon$ whenever $|x - a| < \min(\delta_f, \delta_g)$. A mapping $f : \mathbb{R}^1 \longrightarrow \mathbb{R}^1$ for which $f(x) = 1/x$ is continuous at any point $a \neq 0$; for if $|x - a| < \min(\frac{1}{2}|a|, \frac{1}{2}|a|^2\epsilon)$ then $|x| > \frac{1}{2}|a|$ and $|f(x) - f(a)| = |ax|^{-1}|x - a| < \epsilon$. On the other hand the mapping $f : \mathbb{R} \longrightarrow \mathbb{R}$ defined by

$$f(x) = \begin{cases} 0 & \text{if } x \le 0, \\ 1 & \text{if } x > 0, \end{cases} \tag{3.4}$$

is continuous at each point $x \in \mathbb{R}$ except for the origin, at which it is not continuous since

$$|f(x) - f(0)| = \begin{cases} 0 & \text{if } x \le 0, \\ 1 & \text{if } x > 0. \end{cases}$$

Slightly less obviously, the real-valued function

$$f(x) = \begin{cases} 0 & \text{if } x \text{ is rational,} \\ 1 & \text{if } x \text{ is irrational,} \end{cases} \tag{3.5}$$

is continuous at no point; and considerably less obviously, the real-valued function

$$g(x) = \begin{cases} 0 & \text{if } x \text{ is irrational,} \\ \frac{1}{q} & \text{if } x = \frac{p}{q} \text{ for coprime } p, q, \end{cases} \tag{3.6}$$

is continuous at all irrational numbers but is not continuous at any rational number.

If a mapping $\mathbf{f} : \mathbb{R}^m \longrightarrow \mathbb{R}^n$ is continuous in terms of some norms on the normed vector spaces $\mathbb{R}^m$ and $\mathbb{R}^n$ it is clearly continuous in terms of any equivalent norms on these vector spaces; so it is customary to assume without further mention that if $\mathbb{R}^m$ is viewed as a normed space it is with respect to any one of the standard equivalent norms discussed in Section 2.1, the ℓ_1 or ℓ_2 or ℓ_∞ norm. A mapping $\mathbf{f} : \mathbb{R}^m \longrightarrow \mathbb{R}^n$ is described by its coordinate functions $f_i(\mathbf{x})$, where $\mathbf{f}(\mathbf{x}) = \{f_i(\mathbf{x})\}$ for $1 \le i \le n$; and the mapping $\mathbf{f}$ is easily seen to be continuous at a point $\mathbf{a} \in S$ if and only if the coordinate functions $f_i(\mathbf{x})$ are continuous at that point, since for instance $\|\mathbf{f}(\mathbf{x}) - \mathbf{f}(\mathbf{a})\|_\infty = \max_i |f_i(\mathbf{x}) - f_i(\mathbf{a})|$. As a consequence it is often possible to reduce questions and proofs about mappings $\mathbf{f} : \mathbb{R}^m \longrightarrow \mathbb{R}^n$ to the special case of mappings $f : \mathbb{R}^m \longrightarrow \mathbb{R}$. Mappings $f : \mathbb{R}^n \longrightarrow \mathbb{R}$ can fail to be continuous for more complicated reasons than those discussed in the examples for real-valued functions of a single real variable. For instance the mapping $h : \mathbb{R}^2 \longrightarrow \mathbb{R}$

$$h(x_1, x_2) = \begin{cases} \dfrac{x_1 x_2}{x_1^2 + x_2^2} & \text{if } |x_1| + |x_2| \neq 0, \\ 0 & \text{if } x_1 = x_2 = 0, \end{cases} \tag{3.7}$$

is continuous at every point $(x_1, x_2) \in \mathbb{R}^2$ except at the origin; but the restriction $h(x_1, cx_1) = \frac{c}{1+c^2}$ of this function to a line through the origin described by a nonzero constant c also is a nonzero constant depending on the choice of c except at the origin, where by definition the function takes the value 0, so the function fails to be continuous at the origin. Of course the mappings naturally expected to be continuous are indeed continuous. A linear function $f(\mathbf{x}) = \sum_{i=1}^{m} c_i x_i$ is continuous at any point $\mathbf{a}$, since

$$|f(\mathbf{x}) - f(\mathbf{a})| = \left| \sum_{i=1}^{m} c_i (x_i - a_i) \right| \leq m \|\mathbf{c}\|_\infty \|\mathbf{x} - \mathbf{a}\|_\infty;$$

if $\|\mathbf{c}\| = 0$ the mapping is constant hence continuous while on the other hand if $\|\mathbf{c}\| \neq 0$ then $|f(\mathbf{x}) - f(\mathbf{a})| < \epsilon$ whenever $\|\mathbf{x} - \mathbf{a}\|_\infty < \epsilon / m \|\mathbf{c}\|_\infty$ so the mapping is also continuous. That holds at any point $\mathbf{a} \in \mathbb{R}^m$ so a linear transformation $T : \mathbb{R}^m \longrightarrow \mathbb{R}^n$ is continuous in $\mathbb{R}^m$. A mapping $f : \mathbb{R}^2 \longrightarrow \mathbb{R}^1$ for which $f(x_1, x_2) = x_1 x_2$ is continuous at any point (a_1, a_2), since if $\epsilon < 1$

$$|f(x_1, x_2) - f(a_1, a_2)| = |(x_1 - a_1)(x_2 - a_2) + a_1(x_2 - a_2) + a_2(x_1 - a_1)|$$

$$\leq |x_1 - a_1||x_2 - a_2| + |a_1||x_2 - a_2| + |a_2||x_1 - a_1| < \epsilon$$

whenever $|x_i - a_i| < \frac{1}{3} \epsilon \min(1, (1 + |a_1|)^{-1}, (1 + |a_2|)^{-1})$.

Theorem 3.1. (i) *If S is a normed vector space with the norm $\|x\|$ the function $f(x) = \|x\|$ is a continuous function on S.*
(ii) *If S is a metric space with the metric $\rho(x, y)$ then for any fixed point $b \in S$ the function $f(x) = \rho(x, b)$ is a continuous function on S.*

Proof: (i) For any norm $\|\mathbf{x}\|$ on S by the triangle inequality $\|\mathbf{x}\| \leq \|\mathbf{x} - \mathbf{a}\| + \|\mathbf{a}\|$ so $\|\mathbf{x}\| - \|\mathbf{a}\| \leq \|\mathbf{a} - \mathbf{x}\|$; the same result holds when the variables $\mathbf{a}$ and $\mathbf{x}$ are exchanged so actually

$$\left| \|\mathbf{x}\| - \|\mathbf{a}\| \right| \leq \|\mathbf{a} - \mathbf{x}\|. \tag{3.8}$$

Therefore the function $f(x) = \|x\|$ satisfies $|f(x) - f(a)| < \epsilon$ whenever $\|x - a\| < \delta = \epsilon$, hence $f(x)$ is continuous at any point $a \in S$.
(ii) For any metric ρ by the triangle inequality $\rho(x, b) \leq \rho(x, a) + \rho(a, b)$ hence $\rho(x, b) - \rho(a, b) \leq \rho(x, a)$; the same result holds when the variables a and b are

reversed, hence the function $f(x) = \rho(x, b)$ satisfies

$$|f(x) - f(a)| = |\rho(x, b) - \rho(a, b)| \le \rho(x, a). \tag{3.9}$$

Consequently $|f(x) - f(a)| < \epsilon$ whenever $\rho(x, a) < \delta = \epsilon$ showing that $f(x) = \rho(x, b)$ is continuous at any point $a \in S$. That suffices for the proof.

The definition (3.1) of continuity, usually called the ϵ, δ definition of continuity, is often the most convenient to use in explicit calculations, and is basic in discussing additional aspects of continuity such as uniform continuity. However simple continuity actually is a property of the underlying topological structure, a result of fundamental importance as well as of convenience for establishing some general properties of continuous functions.

Theorem 3.2. *A mapping $f : S \longrightarrow T$ between two metric spaces is continuous at a point $a \in S$ if and only if a is an interior point of the inverse image $f^{-1}(\mathcal{N}_b)$ of any open neighborhood $\mathcal{N}_b$ of the point $b = f(a) \in T$, that is, if and only if $a \in \mathrm{int}(f^{-1}(\mathcal{N}_{f(a)}))$ for any open neighborhood $\mathcal{N}_{f(a)}$ of the point $f(a) \in T$.*

Proof: The definition (3.1) of a continuous mapping at a point $a \in S$ can be rephrased in terms of ϵ-neighborhoods of the points in these metric spaces as the condition that for any $\epsilon > 0$ there is a $\delta > 0$ such that $f(\mathcal{N}_a(\delta)) \subset \mathcal{N}_b(\epsilon)$ where $b = f(a)$, or equivalently, such that $\mathcal{N}_a(\delta) \subset f^{-1}(\mathcal{N}_b(\epsilon))$. If $f : S \longrightarrow T$ is a continuous mapping at a point $a \in S$ and $\mathcal{N}_b$ is any open neighborood of the image $b = f(a)$ then $\mathcal{N}_b(\epsilon) \subset \mathcal{N}_b$ for a sufficiently small ϵ hence $\mathcal{N}_a(\delta) \subset f^{-1}(\mathcal{N}_b(\epsilon)) \subset f^{-1}(\mathcal{N}_b)$ so a is an interior point of $f^{-1}(\mathcal{N}_b)$. On the other hand if a is an interior point of the inverse image $f^{-1}(\mathcal{N}_b)$ of any open neighborhood $\mathcal{N}_b$ of the image $b = f(a)$ then in particular for the open neighborhood $\mathcal{N}_b = \mathcal{N}_b(\epsilon)$ the point a must be an interior point of $f^{-1}(\mathcal{N}_b(\epsilon))$ hence $\mathcal{N}_a(\delta) \subset f^{-1}(\mathcal{N}_b(\epsilon))$ for some sufficiently small δ so the mapping f is continuous at the point a. That suffices for the proof.

In view of Theorem 3.2 a mapping $f : S \longrightarrow T$ between two arbitrary topological spaces, not necessarily metric spaces, is defined to be **continuous at a point** $a \in S$ if

$$a \in \mathrm{int}\left(f^{-1}(\mathcal{N}_{f(a)})\right) \quad \text{for any open neighborhood } \mathcal{N}_{f(a)} \text{ of } f(a). \tag{3.10}$$

Theorem 3.2 amounts to the assertion that for metric spaces the criterion (3.10) for continuity is equivalent to the criterion (3.1). The mapping $f : S \longrightarrow T$ is defined to be **continuous in** S if it is continuous at each point $a \in S$; that condition can be rephrased equivalently as follows.

Theorem 3.3. *A mapping $f : S \longrightarrow T$ between two topological spaces is continuous in S if and only if $f^{-1}(E)$ is an open subset of S for any open subset E of T, or* *equivalently, if and only if $f^{-1}(E)$ is a closed subset of S for any closed subset E of T.*

Proof: If $f : S \longrightarrow T$ is a continuous mapping it follows immediately from the definition (3.10) that any point $a \in f^{-1}(E)$ is an interior point of $f^{-1}(E)$ for any open subset $E \subset T$ hence that $f^{-1}(E) \subset S$ is open whenever $E \subset T$ is open; and conversely if $f^{-1}(E)$ is an open subset of S for any open subset E of T then any point $a \in f^{-1}(E)$ is an interior point for any open subset $E \subset T$ so f is continuous at a. Since $f^{-1}(T \sim E) = S \sim f^{-1}(E)$ it follows that for any mapping $f : S \longrightarrow T$ and any subset $E \subset T$ the inverse image $f^{-1}(E)$ of any closed subset $E \subset T$ is closed if and only if the inverse image $f^{-1}(T \sim E)$ of any open subset $T \sim E$ is open, which is just the condition that $f : S \longrightarrow T$ is a continuous mapping. That suffices for the proof.

An illustration of the convenience of the characterization (3.10) of continuity of mappings is in the proof of the following standard property of continuous mappings.

Theorem 3.4. *If R, S, T are topological spaces, if a mapping $f : R \longrightarrow S$ is continuous at a point $a \in R$, and if a mapping $g : S \longrightarrow T$ is continuous at the point $b = f(a) \in S$, then the composition $h = g \circ f : R \longrightarrow T$ is continuous at the point $a \in R$; and if mappings $f : R \longrightarrow S$ and $g : S \longrightarrow T$ are continuous then their composition $g \circ f : R \longrightarrow T$ is continuous.*

Proof: Since g is continuous at the point $b = f(a) \in S$, if $\mathcal{N}_c$ is an open neighborhood of $c = g(b)$ then $b \in \text{int}\,(g^{-1}(\mathcal{N}_c))$ so there is an open subset $\mathcal{N}_b \subset S$ such that $b \in \mathcal{N}_b \subset g^{-1}(\mathcal{N}_c)$. Since f is continuous at the point $a \in R$ then $a \in \text{int}\,(f^{-1}(\mathcal{N}_b))$ so there is an open subset $\mathcal{N}_a \subset R$ such that $a \in \mathcal{N}_a \subset f^{-1}(\mathcal{N}_b) \subset h^{-1}(\mathcal{N}_c)$ hence that $a \in \text{int}\,(h^{-1}(\mathcal{N}_c))$, showing that h is continuous at the point a. The second statement of the theorem is an immediate consequence, in view of the preceding theorem, and that suffices for the proof.

As an application of the preceding theorem, if mappings $f, g : S \longrightarrow \mathbb{R}$ from a topological space S to the real numbers are continuous then their sum $f + g$ is continuous since it is the composition $f + g = s \circ t$ of the continuous mapping $t : S \longrightarrow \mathbb{R}^2$ for which $t(x) = (f(x), g(x))$ and the the continuous mapping $s : \mathbb{R}^2 \longrightarrow \mathbb{R}^1$ for which $s(x_1, x_2) = x_1 + x_2$; their product fg is continuous since it is the composition $fg = p \circ t$ of the continuous mapping $t : S \longrightarrow \mathbb{R}^2$ and the continuous mapping $p : \mathbb{R}^2 \longrightarrow \mathbb{R}$ for which $p(x_1, x_2) = x_1 x_2$; and the mapping $1/f$ is continuous at any point $a \in S$ at which $f(a) \neq 0$ since it is the composition of the continuous mapping $f : S \longrightarrow \mathbb{R}$ and the continuous mapping $i : \mathbb{R} \sim \{0\} \longrightarrow \mathbb{R}$ for which $i(x) = 1/x$. In particular any polynomial is continuous, since it is a composition of addition and multiplication of continuous functions beginning with the function x; and any quotient of polynomials is continuous, so long as the denominator is nonzero.

The continuity of a mapping $f : S \longrightarrow T$ between two topological spaces is characterized by the condition that the inverse image of an open set is open, or alternatively by the condition that the inverse image of a closed set is closed. However the image of an open set need not be open, and the image of a closed set need not be closed. For example if $S = [0, 1) \cup [2, 3]$ with the induced topology as a subset of $\mathbb{R}^1$ the mapping $f : S \longrightarrow [0, 3]$ defined by

$$f(x) = \begin{cases} 2x & \text{if} \quad 0 \le x < 1, \\ x & \text{if} \quad 2 \le x \le 3, \end{cases} \tag{3.11}$$

clearly is a continuous mapping. The subset $[2, 3] \subset S$ is open in S but its image $[2, 3]$ is closed in $[0, 3]$; and the subset $[0, 1) \subset S$ is closed in S but its image $[0, 2)$ is open in $[0, 3]$. However compactness is preserved under continuous mappings rather than under their inverses.

Theorem 3.5. *The image of a compact subset $K \subset S$ under a continuous mapping $f : S \longrightarrow T$ is compact.*

Proof: If $K \subset S$ is compact and $f(K)$ is contained in a union of open sets U_α then K is contained in the union of the open sets $f^{-1}(U_\alpha)$; and since K is compact it is contained in a union of finitely many of the sets $f^{-1}(U_\alpha)$, hence $f(K)$ is contained in the union of the images $f(f^{-1}(U_\alpha)) \subset U_\alpha$ of these finitely many open sets. That suffices for the proof.

However the inverse image of a compact set under a continuous mapping need not be compact; indeed for the continuous mapping (3.11) the inverse image $f^{-1}([0, 2]) = [0, 1) \cup \{2\}$ of the compact set $[0, 2]$ is not compact. This result has a number of applications, only two of which will be discussed here.

Corollary 3.6 (Maximum-Minimum Theorem). *If $f : S \longrightarrow \mathbb{R}$ is a continuous real-valued function on a compact topological space S then there are points $a, b \in S$ for which*

$$f(a) = \sup_{x \in S} f(x) \quad and \quad f(b) = \inf_{x \in S} f(x). \tag{3.12}$$

Proof: The image $f(S) \subset \mathbb{R}$ is compact by the preceding theorem, hence is a closed set; so if $\alpha = \sup_{x \in S} f(x)$ then since α is a limit point of the set $f(S)$ it must be contained in the set $f(S)$ hence $\alpha = f(a)$ for some point $a \in S$, and correspondingly for $\beta = \inf_{x \in S} f(x)$. That suffices for the proof.

Corollary 3.7. *All norms on a finite dimensional real vector space are equivalent.*

Proof: It was already observed in Section 2.1 that the three standard norms on the vector space $\mathbb{R}^n$, the ℓ_1, ℓ_2, and ℓ_∞ norms, are equivalent; so it is only necessary to show that any other norm $\|\mathbf{x}\|$ on $\mathbb{R}^n$ is equivalent to one of these norms, for instance to the ℓ_∞ norm. When $\mathbb{R}^n$ is viewed as a normed vector space with the ℓ_∞ norm the set

$$S = \left\{ \mathbf{x} \in \mathbb{R}^n \,\middle|\, \|\mathbf{x}\|_\infty = 1 \right\}$$

is a closed set, since $\|\mathbf{x}\|_\infty$ is a continuous function by Theorem 3.1 (i), and is a bounded subset of $\mathbb{R}^n$, from the definition of the norm; hence S is a compact subset, by the Heine-Borel Theorem. In terms of the canonical basis $\delta_1, \ldots, \delta_n$ for the vector space $\mathbb{R}^n$ any vector $\mathbf{x} \in \mathbb{R}^n$ can be written as the sum $\mathbf{x} = \sum_{i=1}^{n} x_i \delta_i$ and $\|\mathbf{x}\|_\infty = \max_{1 \le i \le n} |x_i|$. In addition if $M = \max_i \|\delta_i\|$ it follows from homogeneity and the triangle inequality that

$$\|\mathbf{x}\| \le \sum_{i=1}^{n} \|x_i \delta_i\| \le \sum_{i=1}^{n} |x_i| M \le Mn \max_{1 \le i \le n} |x_i| = Mn\|\mathbf{x}\|_\infty \; ;$$

from the preceding inequality and (3.8) it follows that

$$\Big| \|\mathbf{x}\| - \|\mathbf{a}\| \Big| \le \|\mathbf{x} - \mathbf{a}\| \le Mn\|\mathbf{x} - \mathbf{a}\|_\infty$$

hence the function $f(\mathbf{x}) = \|\mathbf{x}\|$ is continuous on $\mathbb{R}^n$ in terms of the ℓ_∞ norm. By Corollary 3.6 if $C = \sup_{\mathbf{x} \in S} \|\mathbf{x}\|$ and $c = \inf_{\mathbf{x} \in S} \|\mathbf{x}\|$ these values are taken at some points of S so are finite and

$$c \le f(\mathbf{x}) \le C \quad \text{for any point} \quad \mathbf{x} \in S.$$

If $\mathbf{x} \neq \mathbf{0}$ then $\mathbf{x}/\|\mathbf{x}\|_\infty \in S$ so the preceding equation shows that

$$c \le f\left(\frac{\mathbf{x}}{\|\mathbf{x}\|_\infty} \right) = \frac{f(\mathbf{x})}{\|\mathbf{x}\|_\infty} \le C \quad \text{for any point} \quad \mathbf{x} \in S,$$

or equivalently $c\|\mathbf{x}\|_\infty \le f(\mathbf{x}) \le C\|\mathbf{x}\|_\infty$, showing that the norms $\|\mathbf{x}\| = f(x)$ and $\|\mathbf{x}\|_\infty$ are equivalent and thereby concluding the proof.

It follows that all real normed vector spaces of a finite dimension n are isomorphic, so that there is basically only one such space. That is definitely not the case for vector spaces of infinite dimension though, and the study of various such spaces is an important topic in analysis. Connectivity as well as compactness is preserved under continuous mappings rather than under their inverses.

Theorem 3.8. *The image of a connected topological space S under a continuous mapping $f : S \longrightarrow T$ is connected.*

Proof: If the image $f(S)$ is not connected then it contains a nonempty proper subset $E \subset f(S)$ that is both open and closed; the inverse image $f^{-1}(E) \subset S$ then is a nonempty proper subset of S that is also both open and closed, which is impossible since S is assumed connected. That suffices for the proof.

The inverse image of a connected topological space under a continuous mapping need not be connected however; indeed the mapping (3.11) once again is a continuous bijective mapping from the set S that is not connected to the connected set $[0, 3]$, so the inverse image of the connected set $[0, 3]$ is not connected. The preceding theorem has a very commonly used corollary.

Corollary 3.9 (Intermediate Value Theorem). *If $f : [a, b] \longrightarrow \mathbb{R}$ is a continuous function in the interval $[a, b] \subset \mathbb{R}$ where $f(a) = A < f(b) = B$ then any $C \in [A, B]$ is the image $C = f(c)$ of some point $c \in [a, b]$.*

Proof: Suppose to the contrary that there is some value C such that $A < C < B$ but $f(x) \neq C$ for all $a \leq x \leq b$. The sets $E_- = f^{-1}(-\infty, C)$ and $E_+ = f^{-1}(C, \infty)$ are nonempty disjoint open subsets of the interval $[a, b]$ for which $[a, b] = E_- \cup E_+$, since there are no points $x \in [a, b]$ such that $f(x) = C$; but these sets are then closed as well as open, which is impossible since $[a, b]$ is connected. That contradiction suffices to conclude the proof.

A bijective continuous mapping $f : S \longrightarrow T$ between two topological spaces is called a **homeomorphism** if its inverse mapping $g = f^{-1} : T \longrightarrow S$ also is continuous; two topological spaces S and T are said to be **homeomorphic** if there is a homeomorphism $f : S \longrightarrow T$, or equivalently of course if there is a homeomorphiosm $g : T \longrightarrow S$. It is worth pointing out that the inverse of a bijective continuous mapping between two topological spaces is not necessarily a continuous mapping; for example the mapping (3.11) is a continuous bijective mapping but its inverse fails to be continuous, since the inverse mapping does not map the interval $[0, 3]$ to a connected set. Thus a critical part of the definition of homeomorphism is that the inverse mapping is required to be continuous. In some cases however the inverse of a continuous bijective mapping actually is automatically continuous.

Theorem 3.10. *A bijective continuous mapping $f : S \longrightarrow T$ from a compact Hausdorff space S onto a Hausdorff space T is a homeomorphism.*

Proof: If the mapping $f : S \longrightarrow T$ is bijective it has a well-defined inverse mapping $g : T \longrightarrow S$. To show that g is continuous it suffices to show that $g^{-1}(E)$ is closed for any closed subset $E \subset S$. If E is closed then by Theorem 2.21 (i) it is compact, since S is compact; and then $g^{-1}(E) = f(E)$ is compact by Theorem 3.5, hence is closed by Theorem 2.22 (i), and that suffices for the proof.

It is evident from the definition that a homeomorphism $f : S \longrightarrow T$ establishes a bijective correspondence between the open subsets of S and those of T: the inverse image of any open subset of T is an open subset of S and the image of any open subset of S is the inverse image of that set under the continuous mapping f^{-1} so is an open subset of T. The corresponding assertion holds for closed sets, or for compact sets or for connected sets; so the two spaces S and T really can be identified as far as their structure of a topological space is concerned. Homeomorphism is an equivalence relation among topological spaces. In particular any two isometric metric spaces are homeomorphic as topological spaces with the metric topology; but on the other hand homeomorphic metric spaces are not necessarily isometric, since for example the mapping $f : (0, \infty) \longrightarrow (0, \infty)$ defined by $f(x) = x^2$ is a homeomorphism, with the inverse mapping $g : (0, \infty) \longrightarrow (0, \infty)$ given by $g(x) = \sqrt{x}$ with the positive square root, but obviously it is not an isometry. Some homeomorphisms are perhaps not entirely obvious; for example if

$$f(\mathbf{x}) = \frac{1}{1 + \|\mathbf{x}\|} \, \mathbf{x} \quad \text{and} \quad g(\mathbf{x}) = \frac{1}{1 - \|\mathbf{x}\|} \, \mathbf{x} \tag{3.13}$$

in terms of a norm $\|\mathbf{x}\|$ on $\mathbb{R}^n$ then $f(\mathbf{x})$ is well defined for all $\mathbf{x} \in \mathbb{R}^n$ and $\|f(\mathbf{x})\| < 1$ so that f is a continuous mapping $f : \mathbb{R}^n \longrightarrow D$ where $D = \{ \mathbf{x} \in \mathbb{R}^n | \|\mathbf{x}\| < 1 \}$, and $g(\mathbf{x})$ is well defined so long as $\|\mathbf{x}\| < 1$ so is a continuous mapping $g : D \longrightarrow \mathbb{R}^n$. If $\mathbf{y} = f(\mathbf{x})$ then $(1 + \|\mathbf{x}\|)\mathbf{y} = \mathbf{x}$ so $(1 + \|\mathbf{x}\|)\|\mathbf{y}\| = \|\mathbf{x}\|$ and consequently $\|\mathbf{x}\| = (1 - \|\mathbf{y}\|)^{-1}\|\mathbf{y}\|$; therefore

$$\mathbf{x} = (1 + \|\mathbf{x}\|)\mathbf{y} = (1 - \|\mathbf{y}\|)^{-1}\mathbf{y} = g(\mathbf{y}),$$

showing that f and g are inverse to one another hence are homeomorphisms. Thus the topological spaces $\mathbb{R}^n$ and $D \subset \mathbb{R}^n$ are homeomorphic; in particular D is the unit ball in $\mathbb{R}^n$ for the ℓ_2 norm and a unit cell for the ℓ_∞ norms, and at first glance these two subsets may not seem to be homeomorphic to the full space $\mathbb{R}^n$.

There are convenient alternative approaches to continuity in the special case of metric spaces; although some results extend to more general topological spaces, for the later applications here it is sufficient just to consider metric spaces. A mapping $f : S \longrightarrow T$ between two metric spaces S and T with the metrics ρ_S and ρ_T is said to have the **limit** b at a point $a \in S'$, indicated by writing $\lim_{x \to a} f(x) = b$, if for any $\epsilon > 0$ there is a $\delta > 0$ so that

$$\rho_T \left(f(x), b \right) < \epsilon \quad \text{whenever} \quad 0 < P_s(x, a) < \delta. \tag{3.14}$$

It is important to keep the inequality $0 < P_s(x, a)$ clearly in mind in the preceding definition; the actual value $f(a)$ does not play any role in the definition. For points that are not limit points of the set S the limit is undefined. This definition of limit is reminiscent of the earlier definition (3.1) of continuity; indeed the two concepts are closely related.

Theorem 3.11. *A mapping $f : S \longrightarrow T$ between two metric spaces S is continuous at a point $a \in S'$ if and only if $\lim_{x \to a} f(x) = f(a)$.*

Proof: This result follows immediately upon comparing the definition (3.1) of continuity with the definition (3.14) of limit, so no further proof is needed.

Although the preceding theorem has a trivial proof, the result is sufficiently important and useful to merit being pointed out explicitly; and it is a standard alternative definition of continuity. It is perhaps worth noting that for an isolated point of S, a point $a \in S \sim S'$, the previous definitions imply that any function is automatically continuous at that point. A mapping $f : S \longrightarrow T$ between two metric spaces does not necessarily have a limit at all points of S'; for example the mapping $f : \mathbb{R} \longrightarrow \mathbb{R}$ defined by (3.5) does not have a limit at any point $a \in \mathbb{R}$. On the other hand the mapping $g : \mathbb{R} \longrightarrow \mathbb{R}$ defined by (3.6) has the limit $\lim_{x \to a} g(x) = 0$ at all points $a \in \mathbb{R}$; and as observed earlier $g(x)$ is continuous precisely at those points $a \in \mathbb{R}$ at which $g(a) = 0$, which are the irrational numbers. Another characterization of the limit of a mapping is the following, which is occasionally useful in providing an alternative characterization of continuity.

Theorem 3.12. *If $f : S \longrightarrow T$ is a mapping between two metric spaces S and T then $\lim_{x \to a} f(x) = b$ at a point $a \in S'$ if and only if $\lim_{n \to \infty} f(a_n) = b$ for any sequence of points $a_n \in S$ such that $a_n \neq a$ and $\lim_{n \to \infty} a_n = a$.*

Proof: If $\lim_{x \to a} f(x) = b$ then by definition for any $\epsilon > 0$ there is a $\delta > 0$ such that $\rho_T(f(x), b) < \epsilon$ whenever $x \neq a$ and $\rho_S(x, a) < \delta$. If $a_n \in S$ is a sequence of points for which $a_n \neq a$ and $\lim_{n \to \infty} a_n = a$ there is a number N such that $0 < \rho_S(a_n, a) < \delta$ for all $n > N$ and then $\rho_T(f(a_n), b) < \epsilon$; therefore $\lim_{n \to \infty} f(a_n) = b$. On the other hand if it is not the case that $\lim_{x \to a} f(x) = b$ then by definition there is an $\epsilon > 0$ such that for all $\delta > 0$ there are points $x \neq a$ for which $\rho_S(x, a) < \delta$; but $\rho_T(f(x), b) > \epsilon$. In particular then there are points $a_n \neq a$ for which $0 < \rho_S(a_n, a) < 1/n$ but $\rho_T(f(a_n), b) > \epsilon$, hence for which it is not the case that $\lim_{n \to \infty} f(a_n) = b$. That suffices for the proof.

Corollary 3.13. *If $f : S \longrightarrow T$ is a mapping between two metric spaces S and T then f is continuous at a point $a \in S'$ if and only if $\lim_{n \to \infty} f(a_n) = f(a)$ for any sequence of points $a_n \in S$ such that $a_n \neq a$ and $\lim_{n \to \infty} a_n = a$.*

Proof: By Theorem 3.11 the mapping f is continuous at $a \in S'$ if and only if $\lim_{x \to a} f(x) = f(a)$; and by Theorem 3.12 $\lim_{x \to a} f(x) = f(a)$ if and only if

$\lim_{n\to\infty} f(a_n) = f(a)$ for any sequence of points $a_n \in S$ such that $a_n \neq a$ and $\lim_{n\to\infty} a_n = a$. That suffices for the proof.

For the special case of mappings $f : S \longrightarrow \mathbb{R}$ from a metric space to the real numbers, that is, for real-valued functions on a metric space, there is still another useful characterization of continuity; it will play a significant role in the later discussion of integration. If a function $f : S \longrightarrow \mathbb{R}$ is bounded in S its **oscillation** in any subset $E \subset S$ is defined by

$$o_f(E) = \sup_{x \in E} f(x) - \inf_{x \in E} f(x). \tag{3.15}$$

It is evident that if $E_1 \subset E_2$ then $o_f(E_1) \leq o_f(E_2)$. In particular if a function f is defined and bounded in an open neighborhood $\mathcal{N}_a(r)$ of a point $a \in S$ then $o_f(\mathcal{N}_a(r_1)) \leq o_f(\mathcal{N}_a(r_2))$ whenever $r_1 \leq r_2 \leq r$ and it is then clear that the limit

$$o_f(a) = \lim_{r \to 0} o_f\big(\mathcal{N}_a(r)\big) \tag{3.16}$$

exists; it is called the **oscillation** of the function $f(x)$ at the point a.

Theorem 3.14. *A bounded function f in an open neighborhood of a point a in a metric space S is continuous at the point a if and only if $o_f(a) = 0$.*

Proof: If $o_f(a) = 0$ then for any $\epsilon > 0$ there is a $\delta > 0$ such that $o_f(\mathcal{N}_a(\delta)) < \epsilon$, so in particular $|f(x) - f(a)| < \epsilon$ whenever $x \in \mathcal{N}_a(\delta)$ and consequently f is continuous at the point a. Conversely if f is continuous at a then for any $\epsilon > 0$ there is a $\delta > 0$ such that $|f(x) - f(a)| < \frac{1}{2}\epsilon$ whenever $x \in \mathcal{N}_a(\delta)$; therefore $f(a) - \frac{1}{2}\epsilon < f(x) < f(a) + \frac{1}{2}\epsilon$ for any $x \in \mathcal{N}_a(\delta)$, hence $o_f(\mathcal{N}_a(\delta)) \leq \epsilon$ and consequently $o_f(a) = 0$. That suffices for the proof.

Lemma 3.15. *If f is a bounded function in a subset S of a metric space then*

$$E = \left\{ x \in S \,\middle|\, o_f(x) < \epsilon \right\}$$

is an open subset of S for any $\epsilon > 0$.

Proof: If $a \in E$ then $o_f(a) < \epsilon$ so $o_f(\mathcal{N}_a(\delta) \cap S) < \epsilon$ if δ is sufficiently small. If $b \in \mathcal{N}_a(\delta) \cap S$ and δ_0 is chosen sufficiently small that $\mathcal{N}_b(\delta_0) \subset \mathcal{N}_a(\delta)$ then $o_f(\mathcal{N}_b(\delta_0) \cap S) \leq o_f(\mathcal{N}_a(\delta) \cap S) < \epsilon$ therefore $b \in E$ and hence $(\mathcal{N}_a(\delta) \cap S) \subset E$. That shows that E is an open subset of S and thereby concludes the proof.

Theorem 3.16. *If f is a bounded function in a metric space S then the set of those points of S at which f is continuous is a G_δ set.*

Proof: By the preceding Lemma 3.15 the set

$$E_n = \left\{ x \in S \,\middle|\, o_f(x) < \frac{1}{n} \right\}$$

is an open subset of S for any natural number n. The intersection $E = \bigcap_{n \in \mathbb{N}} E_n$ is consequently a G_δ set, a countable intersection of open sets. Clearly $x \in E$ if and only if $o_f(x) = 0$, which is just the condition that f is continuous at the point $x \in S$ by Theorem 3.14, and that suffices for the proof.

Corollary 3.17. *There does not exist a function that is continuous at the rational numbers and discontinuous at the irrational numbers.*

Proof: It was demonstrated in Corollary 2.30 as an application of Baire's Category Theorem that the rational numbers do not form a G_δ subset of the real numbers; therefore it follows from the preceding theorem that there are no functions that are continuous precisely at the rationals, which suffices for the proof.

Some aspects of continuity on metric spaces are not purely topological properties but are really metric properties, in the sense that they cannot be stated just in terms of open and closed sets but essentially involve the metrics used to define the topology. A mapping $f : S \longrightarrow T$ between two metric spaces S and T with metrics ρ_S and ρ_T is continuous at a point $a \in S$ if and only if for every $\epsilon > 0$ there is a $\delta > 0$ such that $\rho_T(f(a), f(x)) < \epsilon$ whenever $\rho_S(x, a) < \delta$; and the mapping f is continuous in S if it is continuous at each point $a \in S$, that is, if for each point $a \in S$ and every $\epsilon > 0$ there is a $\delta_a > 0$, which may depend on the point a, such that $\rho_T(f(a), f(x)) < \epsilon$ whenever $\rho_S(x, a) < \delta_a$. The mapping f is said to be **uniformly continuous** in S if it is possible to find values $\delta_a > 0$ that are independent of the point $a \in S$, or equivalently if for any $\epsilon > 0$ there exists $\delta > 0$ such that $\rho_T(f(x), f(y)) < \epsilon$ for all points $x, y \in S$ for which $\rho_S(x, y) < \delta$. The metrics permit a detailed comparison of the properties of a mapping $f : S \longrightarrow T$ at different points of the spaces S and T, a uniformity of the topology in a sense, that is not possible on general topological spaces. A bounded linear mapping $\mathbf{f} : S \longrightarrow T$ between two normed vector spaces is uniformly continuous, since $\|\mathbf{f}(\mathbf{x}) - \mathbf{f}(\mathbf{y})\|_T = \|\mathbf{f}(\mathbf{x} - \mathbf{y})\|_T \leq \|\mathbf{f}\|_o \|\mathbf{x} - \mathbf{y}\|_S$ in terms of the operator norm $\|\mathbf{f}\|_o$ of the linear mapping $\mathbf{f}$, so $\|\mathbf{f}(\mathbf{x}) - \mathbf{f}(\mathbf{y})\|_T < \epsilon$ whenever $\|\mathbf{x} - \mathbf{y}\|_S < \epsilon/\|\mathbf{f}\|_o$. In general the function $f(x) = \|x\|$ on a normed vector space is uniformly continuous, by equation (3.8); and the function $f(x) = \rho(a, x)$ for a fixed point a in a metric space S with the metric ρ is uniformly continuous, by equation (3.9). Not all continuous mappings are uniformly continuous though; for example the mapping $f : \mathbb{R} \longrightarrow \mathbb{R}$ defined by $f(x) = x^2$ is easily seen not to be uniformly continuous.

Theorem 3.18. *Any continuous mapping $f : S \longrightarrow T$ from a compact metric space S to a metric space T is uniformly continuous.*

Proof: If $f : S \longrightarrow T$ is a continuous mapping and $\epsilon > 0$ then for any point $a \in S$ there is a $\delta_a > 0$ such that $\rho_T(f(a), f(x)) < \frac{1}{2}\epsilon$ whenever $x \in \mathcal{N}_a(\delta_a)$. The collection of open sets $\mathcal{N}_a(\delta_a/2)$ for all points $a \in S$ covers the compact space S, so finitely many of these neighborhoods $\mathcal{N}_i = \mathcal{N}_{a_i}(\delta_{a_i}/2)$ for some points $a_i \in S$ for $1 \leq i \leq N$ already suffice to cover S. If $\delta = \min_i \delta_{a_i}$ then $\delta > 0$; and if $x, y \in S$ are any two points such that $\rho_S(x, y) < \frac{1}{2}\delta$, then $x \in \mathcal{N}_i$ for one of these finitely many neighborhoods, and since $\rho_S(x, y) < \frac{1}{2}\delta$ then

$$\rho_S(a_i, y) \leq \rho_S(a_i, x) + \rho_S(x, y) < \delta$$

so $y \in \mathcal{N}_{a_i}(\delta_{a_i})$ as well. It follows that

$$\rho_T\big(f(x), f(y)\big) \leq \rho_T\big(f(x), f(a_i)\big) + \rho_T\big(f(a_i), f(y)\big) \leq \tfrac{1}{2}\epsilon + \tfrac{1}{2}\epsilon = \epsilon,$$

so f is uniformly continuous and that concludes the proof.

Uniformity properties are of particular importance when considering families of mappings between metric spaces. A sequence of mappings $f_n : S \longrightarrow T$ between two metric spaces is said to **converge at a point** $a \in S$ if the sequence of points $f_n(a) \in T$ converges to a point of T. The sequence is said to **converge pointwise** in S, or for short just to **converge** in S, if it converges at each point of S; and in that case the **limit** of the sequence is the function $f(x)$ for which $f(x) = \lim_{n \to \infty} f_n(x)$ for all points $x \in S$. The limit of a convergent sequence of continuous mappings is not necessarily continuous; for instance the functions $f_n(x) = x^n$ are continuous in the interval $[0, 1] \subset \mathbb{R}$ and converge to the limit function

$$f(x) = \lim_{n \to \infty} x^n = \begin{cases} 0 & \text{if } 0 \leq x < 1, \\ 1 & \text{if } x = 1, \end{cases}$$

which fails to be continuous at the point $x = 1$. More interesting, though, is the limit

$$f(x) = \lim_{m \to \infty} \lim_{n \to \infty} \big(\cos(m!\pi x)\big)^{2n}; \tag{3.17}$$

if x is rational then $m!\pi x$ is an integral multiple of π whenever m is sufficiently large, so $\cos(m!\pi x) = \pm 1$ and hence $f(x) = 1$, but if x is irrational then $|\cos(m!\pi x)| < 1$, so $f(x) = 0$, and consequently $f(x)$ is the function (3.5) which is continuous at no point. However there is a finer notion of convergence which ensures that limits of continuous functions are continuous. A sequence of mappings $f_n : S \longrightarrow T$ from a topological space S to a metric space T with the metric ρ is said to be **uniformly convergent** to a mapping $f : S \longrightarrow T$ if for any $\epsilon > 0$ there is a number N such that $\rho(f_n(x), f(x)) < \epsilon$ for all $x \in S$ whenever $n > N$. A sequence of points in a complete metric space converges if and only if it is a Cauchy sequence; it follows immediately that a sequence of mappings $f_n : S \longrightarrow T$ from a topological space S to a complete metric space T converges in S if and only if the sequence $\{f_n(x)\}$ is a Cauchy sequence at each point $x \in S$. A sequence of mappings f_n from a topological space S to a complete metric space T with the metric ρ is said to be a **uniformly Cauchy sequence** if for any $\epsilon > 0$ there is a number N such that $\rho\left(f_m(x), f_n(x)\right) < \epsilon$ for all $x \in S$ whenever $m, n > N$.

Theorem 3.19. *A sequence of mappings f_n from a topological space S to a complete metric space T is uniformly convergent if and only if it is uniformly Cauchy.*

Proof: If the sequence of mappings $f_n : S \longrightarrow T$ is uniformly convergent it is clearly uniformly Cauchy. Conversely if the sequence of mappings $f_n : S \longrightarrow T$ is a uniformly Cauchy sequence then for any $\epsilon > 0$ there is a number N such that $\rho(f_m(x), f_n(x)) < \epsilon$ for all $x \in S$ whenever $m, n > N$. In particular for any point $x \in S$ the sequence $\{f_n(x)\}$ is a Cauchy sequence, so it converges to some value $f(x)$; hence there is some index $M_x > N$ such that $\rho(f(x), f_{M_x}(x)) < \epsilon$. Since also $\rho(f_n(x), f_{M_x}(x)) < \epsilon$ for any $n > N$ it follows from the triangle inequality that

$$\rho\left(f(x), f_n(x)\right) \leq \rho\left(f(x), f_{M_x}(x)\right) + \rho\left(f_{M_x}(x), f_n(x)\right) < 2\epsilon$$

for any $n > N$, showing that the sequence $f_n(x)$ converges uniformly to $f(x)$, to conclude the proof.

Theorem 3.20 (Uniform Convergence Theorem). *The limit of a uniformly convergent sequence of continuous mappings from a topological space to a metric space is a continuous mapping.*

Proof: Suppose that the sequence of continuous mappings $f_n : S \longrightarrow T$ from a topological space S to a metric space T with the metric ρ converges uniformly to a mapping f; so for any $\epsilon > 0$ there is a number N such that $\rho(f_N(x), f(x)) < \epsilon$ for all points $x \in S$. For any point $a \in S$ the function f_N is continuous at a so there is an open neighborhood U of the point a in S such that $\rho(f_N(x), f_N(a)) < \epsilon$ for all points $x \in U$. Then for all $x \in U$ it follows from the triangle inequality that

$$\rho\left(f(x), f(a)\right) \leq \rho\left(f(x), f_N(x)\right) + \rho\left(f_N(x), f_N(a)\right) + \rho\left(f_N(a), f(a)\right)$$
$$< \epsilon + \epsilon + \epsilon = 3\epsilon,$$

which shows that the limit mapping is continuous at the point a. That holds for any $a \in S$ so the limit mapping is continuous in S, which suffices for the proof.

Questions involving the interchange of orders of limits in sequences of mappings, such as whether $\lim_{n\to\infty} \lim_{x\to a} f_n(x)$ is actually equal to the other order of the limits $\lim_{x\to a} \lim_{n\to\infty} f_n(x)$, arise frequently in analysis. In general some caution is advisable, since it is not always possible to interchange the orders in which the limits are taken. For example

$$\lim_{n\to\infty} \lim_{x\to 0} \frac{1}{1+nx} = 1 \quad \text{but} \quad \lim_{x\to 0} \lim_{n\to\infty} \frac{1}{1+nx} = 0.$$

However if the sequences of mappings converge uniformly then such interchanges of limits is possible; this result has a great many uses so should be kept firmly in mind, not only for its usefulness but also as a warning that limits cannot always be interchanged.

Theorem 3.21. *If a sequence of mappings $f_n : S \longrightarrow T$ from a topological space S to a complete metric space T converges uniformly to a mapping $f : S \longrightarrow T$, and if the limits $\lim_{x\to a} f_n(x) = b_n \in T$ exist, then the points b_n converge to a point $b \in T$ and $\lim_{x\to a} f(x) = b$, or equivalently*

$$\lim_{n\to\infty} \lim_{x\to a} f_n(x) = \lim_{x\to a} \lim_{n\to\infty} f_n(x). \tag{3.18}$$

Proof: Since the sequence of mappings $\{f_n\}$ is uniformly convergent it is uniformly Cauchy; so for any $\epsilon > 0$ there is an N such that

$$\rho\big(f_m(x), f_n(x)\big) < \epsilon \quad \text{for all } m, n \geq N, \; x \in S. \tag{3.19}$$

The function $\rho(x, y)$ is continuous in both variables, as an obvious consequence of Theorem 3.1 (ii); so since $b_n = \lim_{x\to a} f_n(x)$ it follows upon taking the limit as x tends to a in (3.19) that

$$\rho(b_m, b_n) \leq \epsilon \quad \text{for all } m, n \geq N. \tag{3.20}$$

Thus $\{b_n\}$ is a Cauchy sequence, so $\lim_{n\to\infty} b_n = b$ for a point $b \in T$; and it follows upon taking the limit as m tends to ∞ in (3.20) for the particular value $n = N$ that

$$\rho(b, b_N) \leq \epsilon. \tag{3.21}$$

Then returning to (3.19) again, since $\lim_{m\to\infty} f_m(x) = f(x)$ for all points $x \in S$ taking the limit in (3.19) as m tends to infinity for the fixed value $n = N$ it follows as well that

$$\rho\big(f(x), f_N(x)\big) \leq \epsilon \quad \text{for all} \quad x \in S. \tag{3.22}$$

Finally since $\lim_{x \to a} f_N(x) = b_N$ there is a $\delta > 0$ so that

$$\rho\big(f_N(x), b_N\big) \le \epsilon \quad \text{if} \quad 0 < \rho(x, a) < \delta. \tag{3.23}$$

Using the triangle inequality and combining the three inequalities (3.21), (3.22), and (3.23) show that

$$\rho(f(x), b) \le \rho(f(x), f_N(x)) + \rho(f_N(x), b_N) + \rho(b_N, b) \le 3\epsilon$$

whenever $0 < \rho(x, a) < \delta$. Therefore $\lim_{x \to a} f(x) = b$ and that suffices for the proof.

The uniform convergence of a sequence of mappings is actually a surprisingly strict condition on the mappings appearing in the sequence. A collection $\mathcal{F}$ of mappings $f : S \longrightarrow T$ between two metric spaces with the metrics ρ_S, ρ_T is said to be a **uniformly equicontinuous** family of mappings if for any $\epsilon > 0$ there is a $\delta > 0$ such that $\rho_T(f(x), f(y)) < \epsilon$ whenever $\rho_S(x, y) < \delta$ for any mapping $f \in \mathcal{F}$ and any points $x, y \in S$. Thus each function $f \in \mathcal{F}$ is uniformly continuous, so that for a given $\epsilon > 0$ there is a $\delta > 0$ that is independent of the choice of the points $x, y \in S$; but moreover δ is even independent of the particular function $f \in \mathcal{F}$.

Theorem 3.22. *If $f_n : S \longrightarrow T$ is a uniformly convergent sequence of continuous mappings from a compact metric space S to a metric space T then the set of mappings f_n form a uniformly equicontinuous family of mappings.*

Proof: By the Uniform Convergence Theorem, Theorem 3.20, the limit mapping $f = \lim_{n \to \infty} f_n$ is continuous in S; and since S is compact then by Theorem 3.18 the mapping f is also uniformly continuous in S. Thus for any $\epsilon > 0$ there is a $\delta_0 > 0$ such that $\rho_T(f(x), f(y)) < \frac{1}{3}\epsilon$ for any points $x, y \in S$ for which $\rho_S(x, y) < \delta_0$. Now since the mappings f_n converge uniformly there is a number N such that $\rho_T(f_n(x), f(x)) < \frac{1}{3}\epsilon$ for all $x \in S$ for all $n > N$. Combining the two preceding results shows that

$$\rho_T\big(f_n(x), f_n(y)\big) \le \rho_T\big(f_n(x), f(x)\big) + \rho_T\big(f(x), f(y)\big) + \rho_T\big(f(y), f_n(y)\big) < \epsilon \tag{3.24}$$

for any points $x, y \in S$ for which $\rho_S(x, y) < \delta_0$ and for all $n > N$. Each of the mappings $f_n : S \longrightarrow T$ is also uniformly continuous by Theorem 3.18; so for each n there is a $\delta_n > 0$ so that $\rho_T(f_n(x), f_n(y)) < \epsilon$ for any points $x, y \in S$ for which $\rho_S(x, y) < \delta_n$. If $\delta = \min(\delta_0, \delta_1, \dots, \delta_N)$ then $\delta > 0$ and $\rho_T(f_n(x), f_n(y)) < \epsilon$ for any points $x, y \in S$ for which $\rho_S(x, y) < \delta$ and for all n, and that suffices for the proof.

What is possibly even more interesting is that there is a rough converse to the preceding result. For the statement of this result, a **subsequence** of a sequence $\{f_n\}$ of mappings $f_n : S \longrightarrow T$ between two sets is a subset f_{n_i} of these mappings for indices $1 \le n_1 < n_2 < n_3 < \dots$.

Lemma 3.23. *A compact metric space S has a countable dense set of points.*

Proof: For any number n the set of open neighborhoods $\mathcal{N}_a(\frac{1}{n})$ of all points $a \in S$ covers S; so since S is compact there are finitely many points $a_{i,n} \in S$ so that the open neighborhoods $\mathcal{N}_{a_{i,n}}(\frac{1}{n})$ for these finitely many indices i also cover S. The collection of all these points $\{a_{i,n}\}$ is countable, since it is the union over the countably many indices n of finite sets, and is clearly a dense subset of S, which suffices for the proof.

Theorem 3.24 (Ascoli's Theorem). *If $\mathcal{F}$ is a uniformly equicontinuous family of mappings from a compact metric space S to a compact metric space T then any sequence of mappings $f_n \in \mathcal{F}$ has a uniformly convergent subsequence.*

Proof: Consider a sequence of mappings $f_n \in \mathcal{F}$. The preceding Lemma 3.23 shows that there is a countable dense set of points $a_i \in S$. If the images $f_n(a_1) \in T$ for all indices n are an infinite set then since T is compact this infinite set of points will have a limit point, hence a suitable subsequence $f_{n,1}(a_1) \in T$ will converge. If there are only finite many image points then a suitable subsequence $f_{n,1}(a_1) \in T$ will consist just of the same point repeated, so also a convergent subsequence of points. In either case there is a subsequence $f_{n,1}$ of the mappings f_n such that the sequence $f_{n,1}(a_1)$ converges. The argument can be repeated for the sequence $f_{n,1}(a_2)$; so there is a subsequence $f_{n,2}$ of the sequence $f_{n,1}$ so that the subsequences $f_{n,2}(a_1)$ and $f_{n,2}(a_2)$ both converge, and the process can be continued. Then as in Cantor's diagonalization construction the sequence consisting of the first mapping in $f_{n,1}$, the second mapping in $f_{n,2}$, the third mapping in $f_{n,3}$, and so on, will be a subsequence g_n of the mappings f_n so that for each point a_i the sequence $g_n(a_i)$ converges.

Since every compact metric space is complete, the theorem will be demonstrated by showing that the sequence of mappings g_n is uniformly Cauchy, hence is uniformly convergent, in S. For any $\epsilon > 0$ there will be a $\delta > 0$ so that $\rho_T(f(x), f(y)) < \frac{1}{3}\epsilon$ for any $f \in \mathcal{F}$ and any points $x, y \in S$ for which $\rho_S(x, y) < \delta$; in particular that is the case for any of the mappings g_n, since these mappings are contained in $\mathcal{F}$. The open neighborhoods $\mathcal{N}_{a_i}(\delta)$ of the dense set of points a_i are an open covering of S; so since S is compact finitely many of these neighborhoods, say the neighborhoods $\mathcal{N}_{a_i}(\delta)$ for indices $1 \leq i \leq M$, cover S. For each of these points a_i the subsequence $g_n(a_i)$ converges, so for some N_i suitably large $\rho(g_n(a_i), g_m(a_i)) < \frac{1}{3}\epsilon$ for all $m, n > N_i$; then set $N = \max_{1 \leq i \leq M} N_i$. Now any point $x \in S$ is contained in one of the neighborhoods $\mathcal{N}_{a_i}(\delta)$; and for any $m, n > N$ it follows that

$$\rho\big(g_n(x), g_m(x)\big) \leq \rho\big(g_n(x), g_n(a_i)\big) + \rho\big(g_n(a_i), g_m(a_i)\big) + \rho\big(g_m(a_i), g_m(x)\big)$$

$$\leq \frac{1}{3}\epsilon + \frac{1}{3}\epsilon + \frac{1}{3}\epsilon = \epsilon,$$

which shows that the sequence g_n is uniformly Cauchy in S, thereby concluding the proof.

APPENDIX: THE FUNDAMENTAL THEOREM OF ALGEBRA

One of the basic results of algebra, often called the **Fundamental Theorem of Algebra**, is the assertion that any nonconstant complex polynomial $P(z)$ has a root, a complex number a such that $P(a) = 0$. Actually this is not really an algebaic theorem since it involves the complex, hence real, number system and that number system is really a topological rather than purely algebraic construction; hence the proofs of the Fundamental Theorem of Algebra are necessarily topological rather than algebraic. The proof included here rests on compactness and simple geometric constructions, so can be viewed as another example of the usefulness of the notion of compactness. There are versions of this proof that appear much simpler, involving properties of analytic functions rather than real estimates.

Theorem 3.25 (Fundamental Theorem of Algebra). *Any nonconstant complex polynomial has a root.*

Proof: Suppose to the contrary that $P(z)$ is a complex polynomial of degree $n > 0$ that has no roots, so that $|P(z)| > 0$ for all points $z \in \mathbb{C}$. If the polynomal $P(z)$ is written out explicitly as

$$P(z) = a_0 + a_1 z + \cdots + a_n z^n = z^n \left(a_n + a_{n-1}\frac{1}{z} + \cdots + a_0 \frac{1}{z^n} \right)$$

where $a_n \neq 0$ then $\left| a_n + a_{n-1}\frac{1}{z} + \cdots + a_0\frac{1}{z^n} \right| > \frac{1}{2}|a_n|$ whenever $|z| \geq R$ for a sufficiently large value of R so $|P(z)| > \frac{1}{2}|a_n||z|^n$ whenever $|z| \geq R$; in particular if R is chosen sufficiently large so that also $\frac{1}{2}|a_n|R^n > |P(0)|$ then actually $|P(z)| > |P(0)|$ whenever $|z| \geq R$. The restriction of the continuous function $|P(z)|$ to the compact disc $|z| \leq R$ for such a choice of R thus does not take its minimal value on the boundary $|z| = R$, since $|P(0)|$ is smaller than its value on that boundary, so $|P(z)|$ must take its minimal value at some point z_0 for which $|z_0| < R$. The polynomial $P(z)$ can be rewritten

$$P(z) = b_0 + b_k(z - z_0)^k + \cdots + b_n(z - z_0)^n \quad \text{for some index } k$$

where $b_k \neq 0$ and $b_0 = P(z_0) \neq 0$; and if $Q(z) = b_0 + b_k(z - z_0)^k$ then for some positive constants M and r

$$\left| \frac{P(z) - Q(z)}{(z - z_0)^{k+1}} \right| = \left| b_{k+1} + \cdots + b_n(z - z_0)^{n-k-1} \right| \leq M \quad \text{whenever} \quad |z - z_0| \leq r$$

and consequently

$$|P(z)| \leq |Q(z)| + M |z - z_0|^{k+1} \quad \text{whenever} \quad |z - z_0| \leq r.$$

Now if t is real and $t > 0$, and if ζ is any k-th root of $-b_0/b_k$, then

$$|Q(z_0 + t\zeta)| = \left|b_0 + b_k(t\zeta)^k\right| = |b_0|\left|1 + \frac{b_k}{b_0}(t\zeta)^k\right| = |b_0|(1 - t^k);$$

hence if t is sufficiently small

$$|P(z_0 + t\zeta)| \leq |Q(z_0 + t\zeta)| + M|t\zeta|^{k+1}$$
$$\leq |b_0|(1 - t^k) + M|\zeta|^{k+1}t^{k+1} = |b_0| - |b_0|t^k(1 - M't)$$

where $M' = M|\zeta|^{k+1}/|b_0| > 0$. If t is sufficiently small that $(1 - M't) > 0$ then $|P(z_0 + t\zeta)| < |b_0| = |P(z_0)|$, which is a contradiction since $|P(z_0)|$ is the minimal value of $|P(z)|$; and that contradiction suffices to conclude the proof.

Problems, Group I

(Note: In these problems the topology on $\mathbb{R}$ is the usual metric topology unless explicitly stated otherwise.)

1. Is the mapping $f : \mathbb{Q} \longrightarrow \mathbb{Q}$, defined by $f(x) = 0$ if $x^2 < 2$ and $f(x) = 1$ if $x^2 \geq 2$, a continuous mapping, where the rationals have the induced topology as a subset of $\mathbb{R}$? Why?

2. Show that if $f : S \longrightarrow \mathbb{R}$ is a continuous real-valued function on a compact topological space S then either there is a point $a \in S$ for which $f(a) = 0$ or there is a positive number $\delta > 0$ such that $|f(x)| \geq \delta$ for all $x \in S$.

3. What are the continuous mappings $f : R \longrightarrow S$ between two topological spaces R and S in the four cases in which each space has either the discrete or indiscrete topology? (One case is that in which both have the discrete topology, and so on.)

4. i) If $f : R \longrightarrow S$ is a continuous surjective mapping between two topological spaces and $E \subset R$ is a dense subset of R show that $f(E)$ is a dense subset of S.
    ii) If $f, g : R \longrightarrow S$ are two continuous mappings between two topological spaces where S is Hausdorff, and if $g(a) = f(a)$ for all $a \in E$ where E is a dense subset of R, show that actually $g(x) = f(x)$ for all $x \in R$.

5. Show that if $f, g : X \longrightarrow Y$ are continuous mappings between two topological spaces where Y is Hausdorff then $\{x \in X | f(x) = g(x)\}$ is closed. Is that necessarily the case if Y is not Hausdorff?

Problems, Group II

6.   Show that if $f : [0, 1] \longrightarrow [0, 1]$ is a continuous surjective mapping there is at least one point $a \in [0, 1]$ for which $f(a) = a$.

7.   If $E \subset \mathbb{R}$ is a closed subset show that there is a continuous real-valued function on $\mathbb{R}$ such that $f(x) = 0$ if and only if $x \in E$.

8.   Show that if $f : R \longrightarrow S$ is a mapping between two topological spaces and $A \subset R$ it is not true that the restriction $f|A : A \longrightarrow S$ of the mapping f to A is continuous in the induced topology on A if and only if the mapping $f : R \longrightarrow S$ is continuous at each point $a \in A$. What is true (and relevant)?

9.   Show that if $f : \mathbb{R} \longrightarrow \mathbb{R}$ is a mapping that is continuous at the point $0 \in \mathbb{R}$ and satisfies $f(x + y) = f(x) + f(y)$ for all $x, y \in \mathbb{R}$ then there is a real constant c such that $f(x) = cx$ for all $x \in \mathbb{R}$.

10.   i) Show that $f : R \longrightarrow S$ is a continuous mapping between two topological spaces if and only if $f(\mathrm{clo}(E)) \subset \mathrm{clo}(f(E))$ for all subsets $E \subset R$.
            ii) Show that $f : R \longrightarrow S$ is a continuous mapping between two topological spaces if and only if $f^{-1}(\mathrm{clo}(E)) \supset \mathrm{clo}(f^{-1}(E))$ for all subsets $E \subset S$.
            iii) Show that $f : R \longrightarrow S$ is a continuous and closed mapping between two topological spaces if and only if $f(\mathrm{clo}(E)) = \mathrm{clo}(f(E))$ for all subsets $E \subset R$, where a mapping f is said to be **closed** if the image of any closed set is a closed set.

11.   Show that if a subset $A \subset \mathbb{R}$ has the property that any continuous real-valued function $f : A \longrightarrow \mathbb{R}$ is bounded then A is a compact set.

12.   Show that there is no continuous mapping $f : [0, 1] \longrightarrow [0, 1]$ such that for each point $x \in [0, 1]$ the inverse image $f^{-1}(x)$ is either the empty set or contains exactly two points of $[0, 1]$. Show however that there are such mappings if they are not required to be continuous.

13.   Show that if $f : [0, 1] \longrightarrow \mathbb{R}$ is a monotonically increasing function for which $f([0, 1]) = [a, b]$ for some interval $[a, b] \subset \mathbb{R}$ then f is a continuous function in $[0, 1]$.

3.2   Differentiable Mappings

A mapping $\mathbf{f} : U \longrightarrow \mathbb{R}^n$ from an open subset $U \subset \mathbb{R}^m$ into $\mathbb{R}^n$ is said to be **differentiable** at a point $\mathbf{a} \in U$ if there is a linear mapping $A : \mathbb{R}^m \longrightarrow \mathbb{R}^n$, described by an $n \times m$ matrix A, such that for all $\mathbf{h}$ in an open neighborhood of the origin in $\mathbb{R}^m$

$$\mathbf{f}(\mathbf{a} + \mathbf{h}) = \mathbf{f}(\mathbf{a}) + A\mathbf{h} + \epsilon(\mathbf{h}) \quad \text{where} \quad \lim_{\mathbf{h} \to 0} \frac{||\epsilon(\mathbf{h})||}{||\mathbf{h}||} = 0; \qquad (3.25)$$

here $\| \ \|$ is the ℓ_∞ norm or any equivalent norm on $\mathbb{R}^m$ and $\mathbb{R}^n$. Note that in the quotient $||\epsilon(\mathbf{h})||/||\mathbf{h}||$ the vectors and hence the norms involved are from two different vector spaces. It is easy to see that equivalent norms on the vector spaces $\mathbb{R}^m$ and $\mathbb{R}^n$ yield the same condition of differentiability. Differentiability can be viewed as asserting that the change in the value of the mapping $\mathbf{f}$ effected by a small change in the variable is approximately linear aside from the error $\epsilon(\Delta\mathbf{x})$; to emphasize that interpretation the definition (3.25) is sometimes written

$$\Delta\mathbf{f}(\mathbf{x}) = \mathbf{f}(\mathbf{x} + \Delta\mathbf{x}) - \mathbf{f}(\mathbf{x}) = A\Delta\mathbf{x} + \epsilon(\Delta\mathbf{x}), \qquad (3.26)$$

where $\Delta\mathbf{f}(\mathbf{x})$ indicates a change in the value of $\mathbf{f}(\mathbf{x})$ resulting from a change $\Delta\mathbf{x}$ in the variable $\mathbf{x}$ and the error function $\epsilon(\Delta\mathbf{x})$ satisfies

$$\lim_{\Delta\mathbf{x} \to 0} \frac{||\epsilon(\Delta\mathbf{x})||}{||\Delta\mathbf{x}||} = 0. \qquad (3.27)$$

Theorem 3.26. *If a mapping $\mathbf{f} : \mathbb{R}^m \longrightarrow \mathbb{R}^n$ is differentiable at a point $\mathbf{a} \in \mathbb{R}^m$ then it is continuous at that point.*

Proof: If (3.25) holds and $\|A\|_o$ is the operator norm of the linear transformation A then for any $\epsilon > 0$ it follows from the triangle inequality that

$$||\mathbf{f}(\mathbf{a} + \mathbf{h}) - \mathbf{f}(\mathbf{a})|| = ||A\mathbf{h} + \epsilon(\mathbf{h})|| \leq ||A\mathbf{h}|| + ||\epsilon(\mathbf{h})|| \leq ||A||_o||\mathbf{h}|| + ||\epsilon(\mathbf{h})|| < \epsilon$$

whenever $\|\mathbf{h}\|$ is sufficiently small, which suffices for the proof.

The simplest case is that of mappings $f : \mathbb{R}^1 \longrightarrow \mathbb{R}^1$, **functions** of a real variable; in that case the norm is just the absolute value of a real number and it is possible to divide equation (3.25) by the real number h so long as $h \neq 0$, so

$$\frac{|\epsilon(h)|}{|h|} = \left| \frac{f(a + h) - f(a)}{h} - A \right| \quad \text{for} \quad h \neq 0;$$

consequently the function $f(x)$ is differentiable at the point a precisely when

$$\lim_{\substack{h \to 0 \\ h \neq 0}} \frac{f(a + h) - f(a)}{h} = A \quad \text{for some } A \in \mathbb{R}. \qquad (3.28)$$

Since the limit is uniquely defined, it follows that the constant A is uniquely determined; it is called the **derivative** of the function f at the point a and is denoted by $f'(a)$. The case $m = 3$ and $n = 2$, that of a mapping $\mathbf{f} : \mathbb{R}^3 \to \mathbb{R}^2$, illustrates the general form of the criterion (3.25) for differentiability, which in this case takes the form

$$\begin{pmatrix} f_1(\mathbf{a} + \mathbf{h}) \\ f_2(\mathbf{a} + \mathbf{h}) \end{pmatrix} = \begin{pmatrix} f_1(\mathbf{a}) \\ f_2(\mathbf{a}) \end{pmatrix} + \begin{pmatrix} a_{11} & a_{12} & a_{13} \\ a_{21} & a_{22} & a_{23} \end{pmatrix} \begin{pmatrix} h_1 \\ h_2 \\ h_3 \end{pmatrix} + \begin{pmatrix} \epsilon_1(\mathbf{h}) \\ \epsilon_2(\mathbf{h}) \end{pmatrix}. \tag{3.29}$$

Some simple mappings are easily seen to be differentiable. For example a constant mapping $\mathbf{f} : \mathbb{R}^m \longrightarrow \mathbb{R}^n$ is differentiable at any point $\mathbf{a} \in \mathbb{R}^m$ since $\mathbf{f}(\mathbf{a} + \mathbf{h}) = \mathbf{f}(\mathbf{a})$, which is (3.25) for $A = 0$ and $\epsilon(\mathbf{h}) = \mathbf{0}$. A linear transformation $\mathbf{f}(\mathbf{x}) = A\mathbf{x}$ for a matrix $A \in \mathbb{R}^{n \times m}$ is differentiable at any point $\mathbf{a} \in \mathbb{R}^m$ since $\mathbf{f}(\mathbf{a} + \mathbf{h}) = A(\mathbf{a} + \mathbf{h}) = \mathbf{f}(\mathbf{a}) + A\mathbf{h}$, which is (3.25) for the matrix A with $\epsilon(\mathbf{h}) = \mathbf{0}$. If $f(x) = cx^n$ for some natural number n then $f(a + h) - f(a) = c(a + h)^n - ca^n = cna^{n-1}h + h^2 P(h)$ for some polynomial $P(h)$ in the variable h, and consequently $f(x)$ is differentiable at any point $a \in \mathbb{R}$ and its derivative is $f'(A) = cna^{n-1}$.

Theorem 3.27. *A mapping $\mathbf{f} : \mathbb{R}^m \longrightarrow \mathbb{R}^n$ is differentiable at a point $\mathbf{a}$ if and only if each of the coordinate functions f_i of the mapping $\mathbf{f}$ is differentiable at that point.*

Proof: If $\mathbf{f}$ is a differentiable mapping it follows from (3.25) that each coordinate function f_i satisfies

$$f_i(\mathbf{a} + \mathbf{h}) = f_i(\mathbf{a}) + \sum_{j=1}^{m} a_{ij} h_j + \epsilon_i(\mathbf{h}) \quad \text{where } \lim_{\mathbf{h} \to 0} \frac{|\epsilon_i(\mathbf{h})|}{\|\mathbf{h}\|} = 0, \tag{3.30}$$

since $\dfrac{|\epsilon_i(\mathbf{h})|}{\|\mathbf{h}\|_\infty} \leq \dfrac{\|\epsilon(\mathbf{h})\|_\infty}{\|\mathbf{h}\|_\infty}$; and this is just the condition that each of the coordinate functions f_i of the mapping $\mathbf{f}$ is differentiable at the point $\mathbf{a}$. Conversely if each of the coordinate mappings f_i is differentiable at the point $\mathbf{a}$ then (3.30) holds for $1 \leq i \leq n$. The collection of these n equations taken together is the equation (3.25) in which $\epsilon(\mathbf{h}) = \{\epsilon_i(\mathbf{h})\}$; and since $\dfrac{\|\epsilon(\mathbf{h})\|_\infty}{\|\mathbf{h}\|_\infty} = \max_{1 \leq i \leq n} \dfrac{|\epsilon_i(\mathbf{h})|}{\|\mathbf{h}\|_\infty}$ it follows that $\lim_{\mathbf{h} \to 0} \dfrac{\|\epsilon(\mathbf{h})\|_\infty}{\|\mathbf{h}\|_\infty} = 0$, so the mapping $\mathbf{f}$ is differentiable. That concludes the proof.

For the special case of a vector $\mathbf{h} = \{0, \ldots, 0, h_k, 0, \ldots, 0\}$ having all components zero except for the k-th component h_k equation (3.30) takes the form

$$f_i(a_1, \ldots, a_k + h_k, \ldots, a_m) = f_i(a_1, \ldots, a_k, \ldots, a_m) + a_{ik} h_k + \epsilon_i(h_k)$$

where $\lim_{h_k \to 0} \frac{|\epsilon_i(h_k)|}{|h_k|} = 0$; that is just the condition that $f_i(\mathbf{x})$, viewed as a function of the variable x_k alone for fixed values $x_j = a_j$ for the remaining variables for $j \neq k$, is a differentiable function of the variable x_k and that its derivative at the point $x_k = a_k$ is the real number a_{ik}. The constant a_{ik} is called the **partial derivative** of the function f_i with respect to the variable x_k at the point $\mathbf{a}$, and is denoted by $a_{ik} = \partial_k f_i(\mathbf{a})$. It follows that the entries in the matrix $A = \{a_{ik}\}$ are the uniquely determined partial derivatives of the coordinate functions of the mapping, so the matrix A is uniquely determined; it is called the **derivative** of the mapping $\mathbf{f}$ at the point $\mathbf{a}$ and is denoted by $\mathbf{f}'(\mathbf{a})$. The derivative thus is an $n \times m$ matrix with the entries

$$\mathbf{f}'(\mathbf{a}) = \left\{ a_{ik} = \partial_k f_i(\mathbf{a}) \middle| 1 \leq k \leq m, \ 1 \leq i \leq n \right\}. \tag{3.31}$$

It is evident from (3.25) that a sum $\mathbf{f}_1 + \mathbf{f}_2$ of two differentiable mappings $\mathbf{f}_1, \mathbf{f}_2$ is also differentiable and that $(\mathbf{f}_1 + \mathbf{f}_2)'(\mathbf{a}) = \mathbf{f}'_1(\mathbf{a}) + \mathbf{f}'_2(\mathbf{a})$; and it is also evident from (3.25) that if $\mathbf{f} : \mathbb{R}^l \longrightarrow \mathbb{R}^m$ is differentiable and $T : \mathbb{R}^m \longrightarrow \mathbb{R}^n$ is a linear transformation then $T\mathbf{f} : \mathbb{R}^l \longrightarrow \mathbb{R}^n$ is differentiable and $(T\mathbf{f})'(\mathbf{a}) = T\mathbf{f}'(\mathbf{a})$. In particular for the special case of functions $f : \mathbb{R}^m \longrightarrow \mathbb{R}^1$ any linear combination $c_1 f_1(\mathbf{x}) + c_2 f_2(\mathbf{x})$ of differentiable functions is differentiable and $(c_1 f_1 + c_2 f_2)'(\mathbf{a}) = c_1 f'_1(\mathbf{a}) + c_2 f'_2(\mathbf{a})$, so in that sense differentiation is a linear operator. There are various alternative notations for derivatives and partial derivatives of functions of several variables that are in common use. For instance $D\mathbf{f}(\mathbf{a})$ is often used in place of $\mathbf{f}'(\mathbf{a})$, and $D_k f(\mathbf{a})$ or $\frac{\partial f}{\partial x_k}(\mathbf{a})$ in place of $\partial_k f(\mathbf{a})$, when $f(\mathbf{x})$ is a real-valued function of the variable $\mathbf{x} \in \mathbb{R}^n$; and when the mapping $\mathbf{f} : \mathbb{R}^m \longrightarrow \mathbb{R}^n$ is viewed as giving the coordinates y_i of a point $\mathbf{y} \in \mathbb{R}^n$ as functions $y_i = f_i(\mathbf{x})$ of the coordinates x_j of points $\mathbf{x} \in \mathbb{R}^m$ the notation $\frac{\partial y_i}{\partial x_j}$ is quite commonly used in place of $\partial_j f_i$.

One of many useful properties of the derivative is its role in determining local maxima or minima of functions. A real-valued function $f : U \longrightarrow \mathbb{R}^1$ defined in an open subset $U \subset \mathbb{R}^m$ has a **local minimum** at a point $\mathbf{a} \in U$ if there is an open neighborhood $U_\mathbf{a} \subset U$ of the point $\mathbf{a}$ in U such that $f(\mathbf{x}) \geq f(\mathbf{a})$ for all points $\mathbf{x} \in U_\mathbf{a}$; the notion of a **local maximum** is defined correspondingly of course, and a **local extremum** is either a local minimum or a local maximum. A point $\mathbf{a} \in U$ is called a **critical point** of the function f if $f'(\mathbf{a}) = 0$.

Theorem 3.28. *If $f : U \longrightarrow \mathbb{R}^1$ is a function in an open subset $U \subset \mathbb{R}^m$, and if f has a local extremum at a point $\mathbf{a} \in U$ and is differentiable at that point, then $f'(\mathbf{a}) = 0$ so $\mathbf{a}$ is a critical point of the function f.*

Proof: First for the special case of a function of a single variable, if $f(x)$ has a local minimum at a point a then $(f(a+h) - f(a))/h \geq 0$ for $h > 0$ so the limit (3.28) for $h \geq 0$ is nonnegative; but $(f(a+h) - f(a))/h \leq 0$ for $h < 0$ so the limit (3.28) for $h \leq 0$ is nonpositive. Since the limits are the same whether $h > 0$

or $h < 0$ it follows that $f'(a) = 0$. The corresponding argument gives the same result at a local maximum. Next for the case of a function of several variables, when all the variables except the j-th are held fixed the function $f(\mathbf{x})$ as a function of the single variable x_j has an extremum at the point $x_j = a_j$ so by the result established in the special case it follows that $\partial_j f(\mathbf{a}) = 0$; and since that holds for each variable x_j it follows that all the partial derivatives are 0 at the point $\mathbf{a}$ so $f'(\mathbf{a}) = 0$, which suffices for the proof.

Of course a critical point of a function need not be a local maximum or minimum of the function; for instance the origin is a critical point for the function $f(x) = x^3$ on the real line, but it is neither a local maximum nor minimum of the function. However if a function $f : U \longrightarrow \mathbb{R}$ is differentiable in an open subset $U \subset \mathbb{R}^m$ then a search for local extrema normally begins by finding the critical points of that function in the subset U and examining each to see whether or not it is a local maximum or minimum. Sometimes it is quite clear that a particular critical point is a local maximum or minimum, while at other times it may take more careful examination; a simple general test that is sometimes useful involves second derivatives, which will be discussed in Section 3.3, and Sylvester's Criterion, Theorem 4.19, which will be discussed in Section 4.2. Another consequence of Theorem 3.28 is a rather theoretical result that is used mostly for functions of a single variable.

Theorem 3.29 (Mean Value Theorem). *If f and g are continuous functions in a closed interval $[a, b]$ and are differentiable in the open interval (a, b) then*

$$\big(f(b) - f(a)\big)\, g'(c) = \big(g(b) - g(a)\big) f'(c) \tag{3.32}$$

for some point $c \in (a, b)$. In particular

$$\big(f(b) - f(a)\big) = (b - a) f'(c) \tag{3.33}$$

for some point $c \in (a, b)$.

Proof: Introduce the function $h(x)$ defined in the interval $[a, b]$ by

$$h(x) = \big(f(b) - f(a)\big) g(x) - \big(g(b) - g(a)\big) f(x).$$

A direct calculation shows that $h(a) = h(b) = f(b)g(a) - g(b)f(a)$. The conclusion of the theorem is just the statement that $h'(x) = 0$ at a point $x \in (a, b)$. Since h is a continuous function on the compact set $[a, b]$ it takes a maximum value at some point $x_1 \in [a, b]$ and a minimum value at some point $x_2 \in [a, b]$. If both points x_1, x_2 are end points of the interval then $h(x)$ is constant, since $h(a) = h(b)$, and then $h'(x) = 0$ for all points $x \in (a, b)$. If at least one of the points x_1, x_2, say the point x_i, is an interior point of (a, b) then the function $h(x)$ is differentiable at that point x_i and $h'(x_i) = 0$ by the preceding theorem, which establishes (3.32). Since (3.33) is just the special case of (3.32) for which $g(x) = x$ that suffices for the proof.

It is perhaps worth considering the preceding theorem a bit more closely. The hypotheses are that the function is continuous in the closed interval $[a, b]$ but is differentiable only in the open subinterval (a, b), so not necessarily at the end points a, b; and the point c is contained in the open subinterval (a, b), so is not one of the end points. It is quite easy to remember equation (3.33), since that equation is rather like the defining equation (3.25) for the derivative, and is even more so if (3.33) is rewritten in the form

$$f(a + h) = f(a) + f'(c)h \quad \text{for a point } c \text{ between } a \text{ and } a + h. \tag{3.34}$$

The theorem is in some senses basically about functions of a single variable, for the statement really involves the behavior of a function in an interval $[a, b] \subset \mathbb{R}$; the direct extension of the theorem to functions of several variables in the subsequent Theorem 3.36 and its corollary really amounts to applying the theorem to the restriction of a function to intervals of line segments in the space of several variables. The Mean Value Theorem is surprisingly important, but is often ignored; it should be kept firmly in mind though as a possibly very useful tool in proofs, as illustrated in the following corollaries.

Corollary 3.30. *If $f : (a, b) \longrightarrow \mathbb{R}$ is a differentiable mapping in the interval $(a, b) \subset \mathbb{R}$ and if $f'(x) = 0$ at each point $x \in (a, b)$ then f is constant in the interval (a, b).*

Proof: If $a < x_1 < x_2 < b$ then it follows from the Mean Value Theorem that there is a point x_3 in the interval (x_1, x_2) for which $(f(x_2) - f(x_1)) = f'(x_3)(x_2 - x_1) = 0$, showing that $f(x_1) = f(x_2)$ and thereby concluding the proof.

Corollary 3.31. *If $f : (a, b) \longrightarrow \mathbb{R}$ is a differentiable mapping in the interval $(a, b) \subset \mathbb{R}$ and if $f'(x) > 0$ at each point $x \in (a, b)$ then f is strictly monotonically increasing in the interval (a, b), while if $f'(x) < 0$ at each point $x \in (a, b)$ then f is strictly monotonically decreasing in the interval (a, b) .*

Proof: If $f'(x) > 0$ then for any points $a < x_1 < x_2 < b$ it follows from the Mean Value Theorem that there is a point x_3 in the interval $x_1 < x_3 < x_2$ for which $(f(x_2) - f(x_1)) = f'(x_3)(x_2 - x_1) > 0$; and that is just the condition that f is strictly monotonically increasing. The same formula when $f'(x) < 0$ shows that f is monotonically decreasing, and that suffices for the proof.

Corollary 3.32 (L'Hôpital's Rule). *If f and g are continuous functions in an open neighborhood U of the origin $0 \in \mathbb{R}$, if they are differentiable in the complement $U \sim \{0\}$, if $f(0) = g(0) = 0$, if the limits $\lim_{x \to 0} f'(x)$ and $\lim_{x \to 0} g'(x)$ exist and*

$\lim_{x \to 0} g'(x) \neq 0$, *then*

$$\lim_{x \to 0} \frac{f(x)}{g(x)} = \lim_{x \to 0} \frac{f'(x)}{g'(x)} = \frac{\lim_{x \to 0} f'(x)}{\lim_{x \to 0} g'(x)}. \tag{3.35}$$

Proof: For any sufficiently small $\delta > 0$ the open interval $(-\delta, \delta)$ will be contained in the neighborhood U. The functions f and g then are continuous in $(-\delta, \delta)$ and are differentiable there except possibly at the origin 0. An application of the Mean Value Theorem to the appropriate interval $(-\delta, 0)$ or $(0, \delta)$ shows that if $0 < |x| < \delta$ then

$$\frac{f(x)}{g(x)} = \frac{f(x) - f(0)}{g(x) - g(0)} = \frac{f'(x_0)}{g'(x_0)} \tag{3.36}$$

for some x_0 for which $0 < |x_0| < |x| < \delta$. By hypothesis the limits $F = \lim_{x \to 0} f'(x)$ and $G = \lim_{x \to 0} g'(x)$ exist, so the functions $f'(x)$ and $g'(x)$ can be extended by continuity to the origin 0; since G is nonzero and a quotient of two continuous functions is continuous if the denominator is nonzero it follows that

$$\lim_{x_0 \to 0} \frac{f'(x_0)}{g'(x_0)} = \frac{\lim_{x_0 \to 0} f'(x_0)}{\lim_{x_0 \to 0} g'(x_0)} = \frac{F}{G},$$

which denonstrates (3.35) and thereby concludes the proof.

If a mapping $\mathbf{f} : \mathbb{R}^m \longrightarrow \mathbb{R}^n$ is differentiable at a point $\mathbf{a} \in \mathbb{R}^m$ then it has partial derivatives $\partial_j f_i(\mathbf{a})$ with respect to each variable x_j at that point; but it is not true that conversely if the coordinate functions of a mapping $\mathbf{f}$ have partial derivatives at a point $\mathbf{a}$ with respect to each variable x_j then $\mathbf{f}$ is a differentiable mapping. For example the mapping $f : \mathbb{R}^2 \longrightarrow \mathbb{R}$ defined by

$$f(\mathbf{x}) = \begin{cases} \dfrac{x_1 x_2}{x_1^2 + x_2^2} & \text{if } \mathbf{x} \neq \mathbf{0}, \\ 0 & \text{if } \mathbf{x} = \mathbf{0}, \end{cases} \tag{3.37}$$

vanishes identically in the variable x_2 if $x_1 = 0$, so $\partial_2 f(0,0) = 0$, and similarly $\partial_1 f(0,0) = 0$. This function is not even continuous at the origin, since for instance it takes the value $\frac{1}{2}$ whenever $x_1 = x_2$ except at the origin where it takes the value 0; hence it is not differentiable at the origin. However if the partial derivatives of a mapping not only exist but also are continuous then the mapping is differentiable, by another application of the Mean Value Theorem.

Theorem 3.33. *If the partial derivatives of a mapping $\mathbf{f} : \mathbb{R}^m \to \mathbb{R}^n$ exist in an open neighborhood of a point $\mathbf{a} \in \mathbb{R}^m$ and are continuous at that point then the mapping $\mathbf{f}$ is differentiable at $\mathbf{a}$.*

Proof: In view of Theorem 3.27 it is enough to prove this for the special case that $n = 1$ so that $f : \mathbb{R}^m \to \mathbb{R}$ is just a real-valued function; and for convenience only the case $m = 2$ will be demonstrated in detail, since it is easier to follow the proof in that simple case and all the essential ideas are present. Assume that the partial derivatives $\partial_j f(\mathbf{x})$ exist at all points in an open ball $B_r(\mathbf{a})$ of radius r centered at the point $\mathbf{a}$, and consider a vector $\mathbf{h} = \{h_j\} \in \mathbb{R}^2$ for which $\|\mathbf{h}\|_2 < r$. The function $f(x_1, a_2)$ as a function of the variable x_1 alone is differentiable in the variable x_1 in the interval from a_1 to $a_1 + h_1$, since that segment is contained in the ball $B_r(\mathbf{a})$, so the Mean Value Theorem can be applied to this function, showing that

$$f(a_1 + h_1, a_2) - f(a_1, a_2) = h_1 \, \partial_1 f(\alpha_1, a_2)$$

for some value α_1 between a_1 and $a_1 + h_1$; and similarly

$$f(a_1 + h_1, a_2 + h_2) - f(a_1 + h_1, a_2) = h_2 \, \partial_2 f(a_1 + h_1, \alpha_2)$$

for some value α_2 between a_2 and $a_2 + h_2$. From these two results it follows that

$$
\begin{aligned}
f(\mathbf{a} + \mathbf{h}) - f(\mathbf{a}) &= f(a_1 + h_1, a_2 + h_2) - f(a_1, a_2) \\
&= (f(a_1 + h_1, a_2 + h_2) - f(a_1 + h_1, a_2)) \\
&\quad + (f(a_1 + h_1, a_2) - f(a_1, a_2)) \\
&= h_2 \, \partial_2 f(a_1 + h_1, \alpha_2) + h_1 \, \partial_1 f(\alpha_1, a_2)
\end{aligned}
$$

hence that

$$f(\mathbf{a} + \mathbf{h}) - f(\mathbf{a}) = h_2 \, \partial_2 f(a_1, a_2) + h_1 \, \partial_1 f(a_1, a_2) + \epsilon(\mathbf{h}) \tag{3.38}$$

where

$$\epsilon(\mathbf{h}) = h_2 \, (\partial_2 f(a_1 + h_1, \alpha_2) - \partial_2 f(a_1, a_2)) + h_1 \, (\partial_1 f(\alpha_1, a_2) - \partial_1 f(a_1, a_2)).$$

By the triangle inequality

$$\frac{|\epsilon(\mathbf{h})|}{\|\mathbf{h}\|_2} \le |\partial_2 f(a_1 + h_1, \alpha_2) - \partial_2 f(a_1, a_2)| + |\partial_1 f(\alpha_1, a_2) - \partial_1 f(a_1, a_2)|$$

since $\frac{|h_1|}{\|\mathbf{h}\|_2} \le 1$ and $\frac{|h_2|}{\|\mathbf{h}\|_2} \le 1$; by assumption the partial derivatives are continuous at the point $\mathbf{a}$, so for any given $\epsilon > 0$ there is a radius r sufficiently small that the points $(a_1 + h_1, \alpha_2)$ and (α_1, a_2) are near enough to (a_1, a_2) that $\epsilon(\mathbf{h})/\|\mathbf{h}\|_2 < \epsilon$, showing that

$$\lim_{\mathbf{h} \to 0} \frac{|\epsilon(\mathbf{h})|}{\|\mathbf{h}\|_2} = 0. \tag{3.39}$$

The combination of equations (3.38) and (3.39) shows that the function $f(\mathbf{x})$ is differentiable at the point $\mathbf{a}$, which concludes the proof.

The differentiability of a function implies that its partial derivatives exist, but does not imply that they are continuous, as will be demonstrated by examples later in the discussion; and the existence of the partial derivatives does not imply differentiability unless it is also assumed that the partial derivatives are continuous, as illustrated by the example (3.37). There can be some confusion about this point, so care should be taken to distinguish carefully between the differentiability of a mapping, as defined in (3.25), and the existence of the partial derivatives of the mapping. A mapping $\mathbf{f}: U \longrightarrow \mathbb{R}^n$ in an open subset $U \subset \mathbb{R}^m$ is said to be **continuously differentiable** or of **class** $\mathcal{C}^1$ in an open subset $U \subset \mathbb{R}^m$ if the partial derivatives of its coordinate functions exist and are continuous throughout the set U. If a mapping is continuously differentiable then it is differentiable in the sense of (3.25); but if a mapping is differentiable in the sense of (3.25) it is not necessarily continuously differentiable.

There is a useful standard notation dealing with differentiable functions of several variables. Points in $\mathbb{R}^n$ are customarily viewed as column vectors; but the derivative $f'(\mathbf{x})$ of a function is naturally presented as a row vector, so in order to treat a derivative $f'(\mathbf{x})$ as a standard vector in $\mathbb{R}^n$ it is necessary to consider instead its transpose ${}^t f'(\mathbf{x})$. To clarify the situation that transpose is often denoted by $\nabla f(\mathbf{x})$, or alternatively by $\mathbf{grad} f(\mathbf{x})$, and is called the **gradient** of the function $f(\mathbf{x})$; thus by definition

$$\nabla f(\mathbf{x}) = \mathbf{grad} f(\mathbf{x}) = \begin{pmatrix} \partial_1 f(\mathbf{x}) \\ \partial_2 f(\mathbf{x}) \\ \cdots \\ \partial_n f(\mathbf{x}) \end{pmatrix} = {}^t f'(\mathbf{x}). \tag{3.40}$$

The gradient of a function in $\mathbb{R}^n$ thus is a vector of the same form as any other vector in $\mathbb{R}^n$, so it is possible to consider for example the vector $\mathbf{x} + \nabla f(\mathbf{x})$ or a linear transform $A \nabla f(\mathbf{x})$ described by a matrix $A \in \mathbb{R}^{n \times n}$. Although this is really a rather trivial point the gradient notation is frequently used and frequently useful.

Another approach to the derivative $f'(\mathbf{x})$ of a function of several variables is to avoid the question whether it is a row or column vector by writing it as an explicit linear combination of basis vectors in the vector space $\mathbb{R}^m$. If $f(\mathbf{x}) = x_j$ then

$$f'(\mathbf{x}) = (0, \quad \cdots, \quad 0, \quad 1, \quad 0, \quad \cdots, \quad 0),$$

where the entry 1 is in column j; thus $f'(\mathbf{x})$ is independent of $\mathbf{x}$ and actually is one of the standard basis vectors for the vector space $\mathbb{R}^m$. It is tempting to avoid introducing a separate notation and to denote this function just by x_j; the derivative of this function then normally would be denoted by x'_j, which is somewhat confusing since x'_j commonly is used to denote another set of variables in $\mathbb{R}^n$. It is clearer and rather more customary to denote the derivative of the function x_j by dx_j, so that

$$dx_j = (0, \quad \cdots, \quad 0, \quad 1, \quad 0, \quad \cdots, \quad 0), \tag{3.41}$$

where the entry in column j is 1 and all other entries are 0. With this notation the vector $f'(\mathbf{x})$ for an arbitrary differentiable function f is denoted correspondingly by $df(\mathbf{x})$ and is written in terms of the basis (3.41) as the linear combination

$$df(\mathbf{x}) = \sum_{j=1}^{m} \partial_j f(\mathbf{x})\, dx_j. \tag{3.42}$$

This expression for the derivative $f'(\mathbf{x})$ is called a **differential form**, or to be more explicit, a differential form of degree 1. More generally a mapping $\mathbf{f} : U \longrightarrow \mathbb{R}^m$ when viewed as associating to a point $\mathbf{x} \in \mathbb{R}^m$ a vector also of dimension m is called a **vector field** on the subset $U \subset \mathbb{R}^m$; this is familiar from physics, for the electric, magnetic, and gravitational fields. A vector field $\mathbf{f}$ in an open subset $U \subset \mathbb{R}^m$ with the coordinate functions $f_j(\mathbf{x})$ also can be written as a linear combination of the standard basis vectors in $\mathbb{R}^m$ and consequently can be viewed as the differential form of degree 1

$$\omega_{\mathbf{f}}(\mathbf{x}) = \sum_{j=1}^{m} f_j(\mathbf{x})\, dx_j. \tag{3.43}$$

There is an extensive use of this and higher order differential forms that will be discussed in Chapter 7.

The differentiation of functions that arise through the composition of several other functions can be reduced to the differentiation of the various factors in the composition; but the calculation can be somewhat complicated. The basic situation is that of the composition of a mapping $\mathbf{g} : U \longrightarrow \mathbb{R}^m$ defined in an open neighborhood U of a point $\mathbf{a} \in \mathbb{R}^l$ and a mapping $\mathbf{f} : V \longrightarrow \mathbb{R}^n$ defined in an open neighborhood V of the point $\mathbf{b} = \mathbf{g}(\mathbf{a}) \in \mathbb{R}^m$, where $\mathbf{g}(U) \subset V$; the composition $\phi = \mathbf{f} \circ \mathbf{g} : U \longrightarrow \mathbb{R}^n$ is the mapping defined by $\phi(\mathbf{x}) = \mathbf{f}(\mathbf{g}(\mathbf{x}))$ for any $\mathbf{x} \in U$, as in the following diagram.

$$
\begin{array}{ccccccc}
\mathbb{R}^l & & \mathbb{R}^m & & \mathbb{R}^n & & \\
\cup & & \cup & & \cup & & \\
U & \xrightarrow{\;\mathbf{g}\;} & V & \xrightarrow{\;\mathbf{f}\;} & W & \phi = \mathbf{f} \circ \mathbf{g} & \qquad (3.44)\\
\iota\uparrow & & \iota\uparrow & & \iota\uparrow & & \\
\mathbf{a} & \xrightarrow{\;\mathbf{g}\;} & \mathbf{b} = \mathbf{g}(\mathbf{a}) & \xrightarrow{\;\mathbf{f}\;} & \mathbf{f}(\mathbf{b}) = \phi(\mathbf{a}) & &
\end{array}
$$

where the mappings ι are simply inclusions of the indicated points in the corresponding subsets.

Theorem 3.34 (Chain Rule). *With the notation as in the preceding diagram, if the mapping* $\mathbf{g}$ *is differentiable at the point* $\mathbf{a}$ *and the mapping* $\mathbf{f}$ *is differentiable at the point* $\mathbf{b} = \mathbf{g}(\mathbf{a})$ *then the composite mapping* $\phi = \mathbf{f} \circ \mathbf{g}$ *is differentiable at the point* $\mathbf{a}$ *and its derivative is* $\phi'(\mathbf{a}) = \mathbf{f}'(\mathbf{g}(\mathbf{a})) \cdot \mathbf{g}'(\mathbf{a}).$

Proof: Since the mapping $\mathbf{f}$ is differentiable at the point $\mathbf{b}$

$$\mathbf{f}(\mathbf{b} + \mathbf{h}) = \mathbf{f}(\mathbf{b}) + \mathbf{f}'(\mathbf{b})\mathbf{h} + \epsilon_{\mathbf{f}}(\mathbf{h}) \text{ where } \lim_{\mathbf{h} \to 0} \frac{\|\epsilon_{\mathbf{f}}(\mathbf{h})\|_{\infty}}{\|\mathbf{h}\|_{\infty}} = 0, \qquad (3.45)$$

and since the mapping $\mathbf{g}$ is differentiable at the point $\mathbf{a}$

$$\mathbf{g}(\mathbf{a} + \mathbf{k}) = \underbrace{\mathbf{g}(\mathbf{a})}_{\mathbf{b}} + \underbrace{\mathbf{g}'(\mathbf{a})\mathbf{k} + \epsilon_{\mathbf{g}}(\mathbf{k})}_{\mathbf{h}} \text{ where } \lim_{\mathbf{k} \to 0} \frac{\|\epsilon_{\mathbf{g}}(\mathbf{k})\|_{\infty}}{\|\mathbf{k}\|_{\infty}} = 0. \qquad (3.46)$$

Substituting (3.46) into (3.45) where $\mathbf{b} = \mathbf{g}(\mathbf{a})$ and $\mathbf{h} = \mathbf{g}'(\mathbf{a})\mathbf{k} + \epsilon_{\mathbf{g}}(\mathbf{k})$ leads to the result that

$$\begin{aligned} \phi(\mathbf{a} + \mathbf{k}) = \mathbf{f}(\mathbf{g}(\mathbf{a} + \mathbf{k})) &= \mathbf{f}(\mathbf{b} + \mathbf{h}) \\ &= \mathbf{f}(\mathbf{b}) + \mathbf{f}'(\mathbf{b})\mathbf{h} + \epsilon_{\mathbf{f}}(\mathbf{h}) \\ &= \phi(\mathbf{a}) + \mathbf{f}'(\mathbf{b})(\mathbf{g}'(\mathbf{a})\mathbf{k} + \epsilon_{\mathbf{g}}(\mathbf{k})) + \epsilon_{\mathbf{f}}(\mathbf{h}) \end{aligned}$$

hence that

$$\phi(\mathbf{a} + \mathbf{k}) = \phi(\mathbf{a}) + \mathbf{f}'(\mathbf{b})\mathbf{g}'(\mathbf{a})\mathbf{k} + \epsilon(\mathbf{k}) \qquad (3.47)$$

where $\epsilon(\mathbf{k}) = \mathbf{f}'(\mathbf{b})\epsilon_{\mathbf{g}}(\mathbf{k}) + \epsilon_{\mathbf{f}}(\mathbf{h})$. From the triangle inequality it follows that for $\mathbf{k} \neq \mathbf{0}$, if $\mathbf{h} \neq \mathbf{0}$

$$\frac{\|\epsilon(\mathbf{k})\|_{\infty}}{\|\mathbf{k}\|_{\infty}} \leq \frac{\|\mathbf{f}'(\mathbf{b})\epsilon_{\mathbf{g}}(\mathbf{k})\|_{\infty}}{\|\mathbf{k}\|_{\infty}} + \frac{\|\epsilon_{\mathbf{f}}(\mathbf{h})\|_{\infty}}{\|\mathbf{h}\|_{\infty}} \cdot \frac{\|\mathbf{g}'(\mathbf{a})\mathbf{k} + \epsilon_{\mathbf{g}}(\mathbf{k})\|_{\infty}}{\|\mathbf{k}\|_{\infty}}$$

$$\leq \frac{\|\mathbf{f}'(\mathbf{b})\|_{o}\|\epsilon_{\mathbf{g}}(\mathbf{k})\|_{\infty}}{\|\mathbf{k}\|_{\infty}} + \frac{\|\epsilon_{\mathbf{f}}(\mathbf{h})\|_{\infty}}{\|\mathbf{h}\|_{\infty}} \cdot \left(\frac{\|\mathbf{g}'(\mathbf{a})\|_{o}\|\mathbf{k}\|_{\infty}}{\|\mathbf{k}\|_{\infty}} + \frac{\|\epsilon_{\mathbf{g}}(\mathbf{k})\|_{\infty}}{\|\mathbf{k}\|_{\infty}} \right)$$

in terms of the operator norms of the matrices, while if $\mathbf{h} = \mathbf{0}$

$$\frac{\|\epsilon(\mathbf{k})\|_{\infty}}{\|\mathbf{k}\|_{\infty}} = \frac{\|\mathbf{f}'(\mathbf{b})\epsilon_{\mathbf{g}}(\mathbf{k})\|_{\infty}}{\|\mathbf{k}\|_{\infty}} \leq \frac{\|\mathbf{f}'(\mathbf{b})\|_{o}\|\epsilon_{\mathbf{g}}(\mathbf{k})\|_{\infty}}{\|\mathbf{k}\|_{\infty}},$$

also in terms of the operator norms of the matrices. Since $\lim_{\mathbf{k} \to 0} \frac{\|\epsilon_{\mathbf{g}}(\mathbf{k})\|_{\infty}}{\|\mathbf{k}\|_{\infty}} = 0$ and $\lim_{\mathbf{h} \to 0} \frac{\|\epsilon_{\mathbf{f}}(\mathbf{h})\|_{\infty}}{\|\mathbf{h}\|_{\infty}} = 0$ while $\lim_{\mathbf{k} \to 0} \mathbf{h} = \mathbf{0}$ it follows from the preceding equation that $\lim_{\mathbf{k} \to \mathbf{0}} \frac{\|\epsilon(\mathbf{k})\|_\infty}{\|\mathbf{k}\|_\infty} = 0$, and it then follows from (3.47) that $\phi = \mathbf{f} \circ \mathbf{g}$ is differentiable at the point $\mathbf{a}$ and that $\phi'(\mathbf{a}) = \mathbf{f}'(\mathbf{g}(\mathbf{a})) \cdot \mathbf{g}'(\mathbf{a})$, which concludes the proof.

In particular if $l = m = n = 1$, so all the vector spaces involved are one-dimensional, then f and g are real-valued functions as is their composition $h(x) = f(g(x))$; and the conclusion of the preceding theorem takes the simpler form

$$h'(x) = f'(g(x)) \cdot g'(x) \quad \text{where } h(x) = f(g(x)). \tag{3.48}$$

For a slightly more complicated example, since the function $g(y_1, y_2) = y_1 y_2$ is differentiable at any point $(y_1, y_2) \in \mathbb{R}$ and $g'(y_1, y_2) = (y_2, y_1)$ it follows that for any differentiable mapping $\mathbf{f} = \{f_1, f_2\} : U \longrightarrow \mathbb{R}^2$ in an open subset $U \subset \mathbb{R}^m$ the composition $\phi = g \circ \mathbf{f} : U \longrightarrow \mathbb{R}$ is a differentiable mapping and

$$\phi'(\mathbf{x}) = g'\left(\mathbf{f}(\mathbf{x})\right)\mathbf{f}'(\mathbf{x}) = \left(f_2(\mathbf{x}), f_1(\mathbf{x})\right) \cdot \begin{pmatrix} f_1'(\mathbf{x}) \\ f_2'(\mathbf{x}) \end{pmatrix} = f_2(\mathbf{x})f_1'(\mathbf{x}) + f_1(\mathbf{x})f_2'(\mathbf{x});$$

since $\phi(\mathbf{x}) = f_1(\mathbf{x})f_2(\mathbf{x})$ the preceding amounts to a formula for the derivative of the product of two functions of several variables, which in terms of partial derivatives takes the form

$$\partial_k(f_1(\mathbf{x})f_2(\mathbf{x})) = f_2(\mathbf{x})\partial_k f_1(\mathbf{x}) + f_1(\mathbf{x})\partial_k f_2(\mathbf{x}). \tag{3.49}$$

The corresponding argument shows that the quotient $\psi(\mathbf{x}) = f_1(\mathbf{x})/f_2(\mathbf{x}) = f_1(\mathbf{x})f_2(\mathbf{x})^{-1}$ of two differentiable functions is differentiable at any point $\mathbf{x}$ at which $f_2(\mathbf{x}) \neq 0$ and that

$$\partial_k \left(\frac{f_1(\mathbf{x})}{f_2(\mathbf{x})} \right) = \frac{f_2(\mathbf{x})\partial_k f_1(\mathbf{x}) - f_1(\mathbf{x})\partial_k f_2(\mathbf{x})}{f_2(\mathbf{x})^2}. \tag{3.50}$$

Yet another application of the chain rule establishes the following observation, which will be considered further in Chapter 5.

Lemma 3.35. *If $\mathbf{f} : U \longrightarrow V$ is a bijective differentiable mapping between two open subsets $U, V \subset \mathbb{R}^m$ and if the inverse mapping $\mathbf{f}^{-1} = \mathbf{g} : V \longrightarrow U$ is also differentiable then $\mathbf{f}'(\mathbf{x})$ is an invertible matrix at each point $\mathbf{x} \in U$ and $\mathbf{g}'(\mathbf{f}(\mathbf{x})) = \mathbf{f}'(\mathbf{x})^{-1}$ at each point $\mathbf{x} \in U$.*

Proof: The composition $\mathbf{g} \circ \mathbf{f} : U \longrightarrow U$ is the identity mapping $(\mathbf{g} \circ \mathbf{f})(\mathbf{x}) = \mathbf{x}$ so $(\mathbf{g} \circ \mathbf{f})'(\mathbf{x}) = I$ where I is the $m \times m$ identity matrix; and by the chain rule $\mathbf{g}'(\mathbf{f}(\mathbf{x})) \cdot \mathbf{f}'(\mathbf{x}) = (\mathbf{g} \circ \mathbf{f})'(\mathbf{x}) = I$ so the matrix $\mathbf{g}'(\mathbf{f}(\mathbf{x}))$ is the inverse of the matrix $\mathbf{f}'(\mathbf{x})$ at each point $\mathbf{x} \in U$, and that suffices for the proof.

An alternative notation for the chain rule is suggestive and sometimes quite useful. When mappings $\mathbf{g} : \mathbb{R}^l \longrightarrow \mathbb{R}^m$ and $\mathbf{f} : \mathbb{R}^m \longrightarrow \mathbb{R}^n$ are described in terms of the coordinates $\mathbf{t} = \{t_1, \ldots, t_l\} \in \mathbb{R}^l$, $\mathbf{x} = \{x_1, \ldots, x_m\} \in \mathbb{R}^m$ and $\mathbf{y} = \{y_1, \ldots, y_n\} \in \mathbb{R}^n$, the coordinate functions of the mappings $\mathbf{f}$, $\mathbf{g}$ and $\boldsymbol{\phi} = \mathbf{f} \circ \mathbf{g}$ have the form $y_i = f_i(\mathbf{x}) = \phi_i(\mathbf{t})$ and $x_j = g_j(\mathbf{t})$. The partial derivatives are sometimes denoted by

$$(\phi')_{ik} = \partial_k \phi_i(\mathbf{x}) = \frac{\partial y_i}{\partial t_k}, \quad (f')_{ij} = \partial_j f_i(\mathbf{x}) = \frac{\partial y_i}{\partial x_j}, \quad (g')_{jk} = \partial_k g_j(\mathbf{t}) = \frac{\partial x_j}{\partial t_k}.$$

By the preceding theorem the derivative of the composite function $\boldsymbol{\phi} = \mathbf{f} \circ \mathbf{g}$ is the matrix product $\phi' = \mathbf{f}'\mathbf{g}'$, which in terms of the entries of these matrices is $(\phi')_{ik} = \sum_{j=1}^{m}(f')_{ij}(g')_{jk}$ or equivalently $\partial_k \phi_i = \sum_{j=1}^{m} \partial_j f_i \cdot \partial_k g_j$; and in the alternative notation this takes the form

$$\frac{\partial y_i}{\partial t_k} = \sum_{j=1}^{m} \frac{\partial y_i}{\partial x_j} \cdot \frac{\partial x_j}{\partial t_k}. \tag{3.51}$$

This is the extension to mappings in several variables of the traditional formulation of the chain rule for functions of a single variable as the identity $\frac{dy}{dt} = \frac{dy}{dx} \cdot \frac{dx}{dt}$; this form of the chain rule is in some ways easier to remember, and with some caution, easier to use, than the version of the chain rule in the preceding theorem. It is customary in this notation though to omit any explicit mention of the points at which the derivatives are taken; so some care must be taken to remember that the derivative $\frac{\partial y_i}{\partial x_j}$ is evaluated at the point $\mathbf{x}$ while the derivatives $\frac{\partial y_i}{\partial t_k}$ and $\frac{\partial x_j}{\partial t_k}$ are evaluated at the point $\mathbf{t}$. This lack of clarity means that some caution must be taken when this notation is used. In particular when considering an expression such as

$$\frac{\partial}{\partial x_1} f(x_1, x_2, x_3) \quad \text{where} \quad x_3 = g(x_1, x_2),$$

it is not clear whether this means the derivative of the function f with respect to its first variable or the derivative of the composite function of the two variables x_1, x_2 with respect to the variable x_1; the notation $\partial_1 f(x_1, x_2, x_3)$ where $x_3 = g(x_1, x_2)$ is less ambiguous.

With practice the chain rule for differentiation can be applied without going through the step of writing a function as an explicit composition of mappings. For instance if $f(y_1, y_2, y_3)$ is a differentiable function of three variables and $g_i(x_1, x_2)$ for $1 \leq i \leq 3$ are three differentiable functions of two variables the composition

$$h(x_1, x_2) = f\big(g_1(x_1, x_2), g_2(x_1, x_2), g_3(x_1, x_2)\big)$$

is a well-defined differentiable function of two variables. This function is the composition $h(\mathbf{x}) = f(G(\mathbf{x}))$ of the mapping $f : \mathbb{R}^3 \longrightarrow \mathbb{R}^1$ and the mapping $G : \mathbb{R}^2 \longrightarrow \mathbb{R}^3$ for which

$$G(\mathbf{x}) = \begin{pmatrix} g_1(\mathbf{x}) \\ g_2(\mathbf{x}) \\ g_3(\mathbf{x}) \end{pmatrix} \quad \text{where } \mathbf{x} = \begin{pmatrix} x_1 \\ x_2 \end{pmatrix};$$

so by the chain rule

$$h'(\mathbf{x}) = f'(G(\mathbf{x})) \cdot G'(\mathbf{x})$$

$$= \begin{pmatrix} \partial_1 f(G(\mathbf{x})) & \partial_2 f(G(\mathbf{x})) & \partial_3 f(G(\mathbf{x})) \end{pmatrix} \begin{pmatrix} \partial_1 g_1(\mathbf{x}) & \partial_2 g_1(\mathbf{x}) \\ \partial_1 g_2(\mathbf{x}) & \partial_2 g_2(\mathbf{x}) \\ \partial_1 g_3(\mathbf{x}) & \partial_2 g_3(\mathbf{x}) \end{pmatrix}$$

$$= \begin{pmatrix} \sum_{i=1}^{3} \partial_i f(G(\mathbf{x})) \, \partial_1 g_i(\mathbf{x}) & \sum_{i=1}^{3} \partial_i f(G(\mathbf{x})) \, \partial_2 g_i(\mathbf{x}) \end{pmatrix}.$$

Thus for instance

$$\partial_1 h(\mathbf{x}) = \partial_1 f(g_1(\mathbf{x}), g_2(\mathbf{x}), g_3(\mathbf{x})) \, \partial_1 g_1(\mathbf{x}) + \partial_2 f(g_1(\mathbf{x}), g_2(\mathbf{x}), g_3(\mathbf{x})) \, \partial_1 g_2(\mathbf{x})$$
$$+ \partial_3 f(g_1(\mathbf{x}), g_2(\mathbf{x}), g_3(\mathbf{x})) \, \partial_1 g_3(\mathbf{x});$$

this formula can be interpreted as asserting that the partial derivative with respect to the first variable of the composite function $h(\mathbf{x}) = f(g_1(\mathbf{x}), g_2(\mathbf{x}), g_3(\mathbf{x}))$ can be calculated by taking the partial derivative of the function f with respect to its i-th variable, evaluated at the point $G(\mathbf{x})$, multiplied by the partial derivative with respect to the variable x_1 of the function $g_i(\mathbf{x})$ that is substituted for the i-th variable, and adding the results for $i = 1, 2, 3$. With that in mind, it is possible to write the preceding formula down directly, without going through the explicit calculation of the derivatives of the individual mappings involved. As another example, if $\phi(x_1, x_2) = f(x_1, x_2, g(x_1, x_2))$ for a differentiable function $f(x_1, x_2, x_3)$ of three variables and a differentiable function $g(x_1, x_2)$ of two variables, the partial derivatives of the composite function $\phi(x_1, x_2)$ are

$$\partial_1 \phi(\mathbf{x}) = \partial_1 f(x_1, x_2, g(x_1, x_2)) + \partial_3 f(x_1, x_2, g(x_1, x_2)) \partial_1 g(\mathbf{x}),$$
$$\partial_2 \phi(\mathbf{x}) = \partial_2 f(x_1, x_2, g(x_1, x_2)) + \partial_3 f(x_1, x_2, g(x_1, x_2)) \partial_2 g(\mathbf{x}).$$

If there is any question about the application of this technique of calculation, it can always be checked by writing the composition as an explicit combination of mappings and applying the chain rule as stated in the preceding theorem. For instance the function $\phi(x_1, x_2) = f(x_1, x_2, g(x_1, x_2))$ is the composition

$\phi(\mathbf{x}) = f(G(\mathbf{x}))$ where $G : \mathbb{R}^2 \longrightarrow \mathbb{R}^3$ is the mapping $G(x_1, x_2) = (x_1, x_2, g(x_1, x_2))$, and the preceding formula can be derived from a straightforward application of the chain rule formula to this composition.

The chain rule also is useful in deriving information about the derivatives of functions that are defined only implicitly. For example if a function $f(x_1, x_2)$ satisfies the equation

$$f(x_1, x_2)^5 + x_1 f(x_1, x_2) + f(x_1, x_2) = 2x_1 + 3x_2$$

and the initial condition that $f(0,0) = 0$ then the values of that function are determined implicitly but not explicitly by the preceding equation. This equation is the condition that the composition of the mapping $F : \mathbb{R}^2 \longrightarrow \mathbb{R}^3$ defined by $F(x_1, x_2) = (x_1, x_2, f(x_1, x_2))$ and the mapping $G : \mathbb{R}^3 \longrightarrow \mathbb{R}$ defined by $G(x_1, x_2, y) = y^5 + x_1 y + y - 2x_1 - 3x_2$ is the trivial mapping $G \circ F(x_1, x_2) = 0$, so that $(G \circ F)'(0,0) = 0$; and if the function f is differentiable it follows from the chain rule that

$$\partial_1(G \circ F) = 5f(x_1, x_2)^4 \partial_1 f(x_1, x_2) + x_1 \partial_1 f(x_1, x_2) + f(x_1, x_2) + \partial_1 f(x_1, x_2) - 2,$$

so since $\partial_1(G \circ F) = 0$ and $f(0,0) = 0$ the preceding equation reduces to $\partial_1 f(0,0) = 2$. A similar calculation yields the value of $\partial_2 f(0,0)$. Thus although an implicit description of a function can be quite difficult to solve explicitly, it is often quite easy to obtain a good deal of information about the derivative of that function by an application of the chain rule; one might just say that differentiation linearizes functions.

For another application of the chain rule, a straight line through a point $\mathbf{a} \in \mathbb{R}^m$ in the direction of a unit vector $\mathbf{u}$ can be described parametrically as the set of points $\mathbf{x} = G(t)$ for $t \in \mathbb{R}$, where $G : \mathbb{R} \longrightarrow \mathbb{R}^m$ is the mapping $G(t) = \mathbf{a} + t\mathbf{u}$. If $f : U \longrightarrow \mathbb{R}$ is a differentiable function in an open set $U \subset \mathbb{R}^m$ containing the point $\mathbf{a}$ the restriction of f to this straight line can be viewed as a function $(f \circ G)(t) = f(\mathbf{a} + t\mathbf{u})$ of the parameter $t \in \mathbb{R}$ near the origin. The derivative at $t = 0$ of this restriction is called the **directional derivative** of the function f at the point $\mathbf{a}$ in the direction of the unit vector $\mathbf{u}$, and is denoted by $\partial_{\mathbf{u}} f(\mathbf{a})$. It follows from the chain rule that $\partial_{\mathbf{u}} f(\mathbf{a}) = f'(\mathbf{a}) G'(0)$; the coordinate functions of the mapping $G(t)$ are $g_j(t) = a_j + u_j t$ so the derivative $G'(0)$ is just the $m \times 1$ matrix or vector $G'(0) = \{u_j\}$, hence

$$\partial_{\mathbf{u}} f(\mathbf{a}) = \sum_{j=1}^{m} \partial_j f(\mathbf{a}) u_j. \tag{3.52}$$

The preceding equation describes the directional derivative as the sum of the products of the entries in the matrix $f'(\mathbf{x})$ and the vector $\mathbf{u}$; and when expressed in terms of the transposed matrix ${}^t f'(\mathbf{x}) = \nabla f(\mathbf{x})$ that equation can be interpreted in vector terms as the dot product of two vectors, so as

$$\partial_{\mathbf{u}} f(\mathbf{a}) = \big(\nabla f(\mathbf{a})\big) \cdot \mathbf{u}. \tag{3.53}$$

This is one way in which the gradient notation is useful; it expresses the directional derivative in terms of operations on ordinary vectors in the vector space $\mathbb{R}^m$. When the dot product is described in terms of the angle between two vectors as in (2.11) the preceding equation can be written alternatively as

$$\partial_{\mathbf{u}} f(\mathbf{a}) = \|\nabla f(\mathbf{a})\|_2 \, \cos\theta\big(\nabla f(\mathbf{a}), \mathbf{u}\big) \tag{3.54}$$

where $\theta(\nabla f(\mathbf{a}), \mathbf{u})$ is the angle between the vectors $\nabla f(\mathbf{a})$ and $\mathbf{u}$, since $\|\mathbf{u}\|_2 = 1$. Thus the largest value of the directional derivative of the function f is in the direction of the gradient vector $\nabla f(\mathbf{a})$, and the actual value of the directional derivative in that direction is $\|\nabla f(\mathbf{a})\|_2$; and the smallest value of the directional derivative is in the opposite direction, in which the actual value of the directional derivative is $-\|\nabla f(\mathbf{a})\|_2$. The directional derivative vanishes in the directions perpendicular to the gradient vector $\nabla f(\mathbf{a})$.

Theorem 3.36 (Mean Value Theorem). *If $f : U \longrightarrow \mathbb{R}$ is a differentiable function in an open set $U \subset \mathbb{R}^m$ and if two points $\mathbf{a}, \mathbf{b}$ and the line λ joining them lie in U then*

$$f(\mathbf{b}) - f(\mathbf{a}) = \nabla f(\mathbf{c}) \cdot (\mathbf{b} - \mathbf{a}) \tag{3.55}$$

for some point $\mathbf{c}$ between $\mathbf{a}$ and $\mathbf{b}$ on the line λ joining these two points.

Proof: The function $g(t) = f(\mathbf{a} + t(\mathbf{b} - \mathbf{a}))$ is a differentiable function of the variable t in an open neighborhood of the interval $[0, 1]$, and by the Mean Value Theorem for functions of one variable $g(1) - g(0) = g'(\tau)$ for a point $\tau \in (0, 1)$. Since $g(1) = f(\mathbf{b})$ and $g(0) = f(\mathbf{a})$, while by the chain rule

$$g'(t) = \sum_{j=1}^{m} \partial_j f(\mathbf{a} + t(\mathbf{b} - \mathbf{a}))(b_j - a_j),$$

it follows that

$$f(\mathbf{b}) - f(\mathbf{a}) = g(1) - g(0) = g'(\tau) = \sum_{j=1}^{m} \partial_j f(\mathbf{a} + \tau(\mathbf{b} - \mathbf{a}))(b_j - a_j)$$

$$= \sum_{j=1}^{m} \partial_j f(\mathbf{c})(b_j - a_j) = \nabla f(\mathbf{c}) \cdot (\mathbf{b} - \mathbf{a})$$

where $\mathbf{c} = \mathbf{a} + \tau(\mathbf{a} - \mathbf{b})$, and that suffices for the proof.

Although the preceding theorem can be applied to each coordinate function of the mapping $\mathbf{f} : U \longrightarrow \mathbb{R}^n$ for $n > 1$, the points at which the derivatives of the different coordinate functions are evaluated may be different. However for some purposes an estimate is useful enough and that problem can be avoided.

Corollary 3.37 (Mean Value Inequality). *If* $\mathbf{f} : U \longrightarrow \mathbb{R}^n$ *is a differentiable mapping in an open set* $U \subset \mathbb{R}^m$ *and if two points* $\mathbf{a}, \mathbf{b}$ *and the line* λ *joining them lie in U then*

$$\|\mathbf{f}(\mathbf{b}) - \mathbf{f}(\mathbf{a})\|_2 \le \|\mathbf{f}'(\mathbf{c})\|_o \|\mathbf{b} - \mathbf{a}\|_2 \tag{3.56}$$

for some point $\mathbf{c}$ *between* $\mathbf{a}$ *and* $\mathbf{b}$ *on the line* λ, *where* $\| \, \|_o$ *is the operator norm.*

Proof: For any vector $\mathbf{u} \in \mathbb{R}^n$ for which $\|\mathbf{u}\|_2 = 1$ the dot product $f_{\mathbf{u}}(\mathbf{x}) = \mathbf{u} \cdot \mathbf{f}(\mathbf{x})$ is a real-valued function to which the Mean Value Theorem can be applied, so

$$f_{\mathbf{u}}(\mathbf{b}) - f_{\mathbf{u}}(\mathbf{a}) = \nabla f_{\mathbf{u}}(\mathbf{c}) \cdot (\mathbf{b} - \mathbf{a}) = \sum_{j=1}^{m} \partial_j (\mathbf{u} \cdot \mathbf{f})(\mathbf{c}) \; (b_j - a_j)$$

$$= \sum_{i=1}^{n} \sum_{j=1}^{m} u_i \partial_j f_i(\mathbf{c})(b_j - a_j) = \sum_{i=1}^{n} \sum_{j=1}^{m} u_i \cdot f'(\mathbf{c})_{ij}(b_j - a_j)$$

$$= \mathbf{u} \cdot \big(\mathbf{f}'(\mathbf{c})(\mathbf{b} - \mathbf{a})\big)$$

for some point $\mathbf{c}$ between $\mathbf{a}$ and $\mathbf{b}$ on the line λ joining these two points. If the unit vector $\mathbf{u}$ is chosen to lie in the direction of the vector $\mathbf{f}(\mathbf{b}) - \mathbf{f}(\mathbf{a})$ then

$$\|\mathbf{f}(\mathbf{b}) - \mathbf{f}(\mathbf{a})\|_2 = |f_{\mathbf{u}}(\mathbf{b}) - f_{\mathbf{u}}(\mathbf{a})| = |\mathbf{u} \cdot \big(\mathbf{f}'(\mathbf{c})(\mathbf{b} - \mathbf{a})\big)|$$

$$\le \|\mathbf{f}'(\mathbf{c})(\mathbf{b} - \mathbf{a})\|_2 \le \|\mathbf{f}'(\mathbf{c})\|_o \|(\mathbf{b} - \mathbf{a})\|_2$$

by the Cauchy-Schwarz inequality, which suffices for the proof.

Of course there is an analogous result in terms of any norms equivalent to the ℓ_2 norm, for the appropriate operator norm and a suitable constant factor. The constant factor is irrelevant for many applications, such as the following.

Theorem 3.38. *If* $\mathbf{f} : \Delta \longrightarrow \mathbb{R}^n$ *is a differentiable mapping in a cell* $\Delta \subset \mathbb{R}^m$ *such that* $\|\mathbf{f}'(\mathbf{x})\|_o \le M$ *for all* $\mathbf{x} \in \Delta$ *then the mapping* $\mathbf{f}$ *is uniformly continuous in* Δ.

Proof: If $\|\mathbf{f}'(\mathbf{x})\|_o \le M$ for all $\mathbf{x} \in \Delta$ then for any two points $\mathbf{x}, \mathbf{y} \in \Delta$ the line joining them is contained in Δ so it follows from the Mean Value Inequality that

$$\|\mathbf{f}(\mathbf{x}) - \mathbf{f}(\mathbf{y})\|_\infty \le M' \|\mathbf{x} - \mathbf{y}\|_\infty,$$

for some constant M', hence the mapping $\mathbf{f}$ is uniformly continuous in Δ and thereby concludes the proof.

It was noted in the preceding section that while the limit of a sequence of continuous functions is not necessarily continuous, that is the case if the sequence converges uniformly; there is a corresponding result for sequences of differentiable functions.

Theorem 3.39. *If* $\{f_n\}$ *is a sequence of differentiable functions in the open interval* (a,b), *if the derivatives* f_n' *converge uniformly in* (a,b) *to a function* g, *and if the functions* f_n *converge at least at one point* $c \in (a,b)$, *then the sequence of functions* f_n *is uniformly convergent to a differentiable function* f *in* (a,b) *and* $f'(x) = g(x)$ *at all points* $x \in (a,b)$.

Proof: For any points $x, y \in (a,b)$ it follows from the Mean Value Theorem that

$$\big(f_m(y) - f_n(y)\big) - \big(f_m(x) - f_n(x)\big) = (y - x)\big(f_m'(t) - f_n'(t)\big)$$

for some point $t \in (a,b)$ between x and y, hence that

$$\left|\big(f_m(y) - f_m(x)\big) - \big(f_n(y) - f_n(x)\big)\right| \leq |y - x||f_m'(t) - f_n'(t)| \tag{3.57}$$

where $|y - x| \leq |b - a|$. The derivatives f_n' converge uniformly in (a,b), so for any $\epsilon > 0$ there is a number N such that $|f_m'(t) - f_n'(t)| < \epsilon$ for all $t \in (a,b)$ whenever $m, n > N$, hence such that

$$\left|\big(f_m(y) - f_m(x)\big) - \big(f_n(y) - f_n(x)\big)\right| \leq |b - a|\epsilon \quad \text{whenever } m, n > N. \tag{3.58}$$

Thus the sequence of functions $f_n(y) - f_n(x)$ is uniformly Cauchy in (a,b) so it converges uniformly in (a,b); and since the sequence $f_n(c)$ converges then for $y = c$ it follows that the sequence of functions $f_n(x)$ converges uniformly in (a,b) to a function $f(x)$, which by Theorem 3.20 is at least continuous in (a,b). Now for any fixed point $x \in (a,b)$ let δ be sufficiently small that $\mathcal{N}_x(\delta) \subset (a,b)$; it then follows from (3.57) for $x + h$ in place of y that for any $\epsilon > 0$ there is a number N such that

$$\left|\frac{f_m(x + h) - f_m(x)}{h} - \frac{f_n(x + h) - f_n(x)}{h}\right| \leq |f_m'(t) - f_n'(t)| < \epsilon$$

if $-\delta < h < \delta$, $h \neq 0$ and $m, n > N$. Consequently the sequence of functions $\frac{f_n(x+h)-f_n(x)}{h}$ of the variable $h \in (-\delta, \delta) \sim \{0\}$ is uniformly Cauchy and

$$\lim_{n \to \infty} \frac{f_n(x + h) - f_n(x)}{h} = \frac{f(x + h) - f(x)}{h}$$

uniformly in $h \in (-\delta, \delta) \sim \{0\}$. It then follows from Theorem 3.21 that

$$\lim_{n \to \infty} \lim_{h \to 0} \frac{f_n(x + h) - f_n(x)}{h} = \lim_{h \to 0} \lim_{n \to \infty} \frac{f_n(x + h) - f_n(x)}{h}$$

hence that

$$g(x) = \lim_{n \to \infty} f_n'(x) = \lim_{h \to 0} \frac{f(x+h) - f(x)}{h} = f'(x),$$

which concludes the proof.

Problems, Group I

1. Find the derivatives of the following mappings:

    i) $f : \mathbb{R}^4 \longrightarrow \mathbb{R}^1$ where $f(\mathbf{x}) = (x_1^2 + x_2^3 + x_3 x_4)$

    ii) $f : \mathbb{R}^3 \longrightarrow \mathbb{R}^3$ where $f(\mathbf{x}) = \begin{pmatrix} x_1 + x_2 \\ x_2^3 + x_3^2 \\ x_1 x_3 \end{pmatrix}$

2. Consider the function $f : \mathbb{R}^2 \longrightarrow \mathbb{R}$ defined by

    $$f(\mathbf{x}) = \frac{2x_1 - x_2}{1 + x_1^2 + x_2^2}.$$

    Find the points in $\mathbb{R}^2$ where f has a local maximum or minimum and the values of f at those points.

3. If f_1, f_2 are differentiable functions defined in an open neighborhood of the point $(2, 1)$ in $\mathbb{R}^2$ such that $f_1(2, 1) = 4$, that $f_2(2, 1) = 3$, and that

    $$f_1^2 + f_2^2 + 4 = 29 \quad \text{and} \quad \frac{f_1^2}{x_1^2} + \frac{f_2^2}{x_2^2} + 4 = 17,$$

    find $\partial_1 f_2(2, 1)$ and $\partial_2 f_1(2, 1)$.

4. If g is a $\mathcal{C}^1$ real-valued function in $\mathbb{R}^2$ and

    $$f(x_1, x_2, x_3) = g(x_1, g(x_1, g(x_2, x_3))),$$

    find the derivative of the function f in terms of the partial derivatives of the function g.

5. Find the directional derivative at the point $(1, -2)$ and in the direction $(\frac{3}{5}, \frac{4}{5})$ for the function $f(\mathbf{x}) = (x_1 + 2x_2 - 4)/(7x_1 + 3x_2)$.

Problems, Group II

6.   If $f : \mathbb{R} \to \mathbb{R}$ is a differentiable function such that $|f'(x)| < (x^2 + 1)^{-1}$ prove that $f(x)$ is bounded.

7.   i) Show that the function $f(\mathbf{x}) = \sqrt{|x_1 x_2|}$ (for the positive square root) is not differentiable at the origin.

      ii) Show that if $f(\mathbf{x})$ is a real-valued function defined in an open neighborhood of the origin in $\mathbb{R}^2$ and if $|f(\mathbf{x})| \leq \|\mathbf{x}\|_2^2$ then $f(\mathbf{x})$ is differentiable at the origin. What is its derivative at the origin?

      iii) If $f(\mathbf{x})$ is a real-valued function defined in an open neighborhood of the origin in $\mathbb{R}^2$ and if $|f(\mathbf{x})| \leq \|\mathbf{x}\|_2$ in that neighborhood is $f(x)$ necessarily differentiable at the origin? Why?

8.   Show that if $f(x)$ is a real-valued continuous function on $[0, \infty) \subset \mathbb{R}$ such that $f(0) = 0$, if $f(x)$ is differentiable on $(0, \infty)$, and if its derivative $f'(x)$ is monotonically increasing, then $g(x) = f(x)/x$ is a monotonically increasing function on $(0, \infty)$.

9.   Find all C^1 functions $f(x)$ on the real line for which $f(0) = 0$ and $|f'(x)| \leq |f(x)|$ for all $x \in \mathbb{R}$.

10.   Show that there is a unique real number c such that for any function f that is differentiable on $[0, 1]$ and that takes the values $f(0) = 0$ and $f(1) = 1$ there is some point $x \in (0, 1)$ (depending on the function f) such that $f'(x) = cx$.

11.   If $f(x)$ is a continuous real-valued function on $\mathbb{R}$, if f is differentiable at all points except the origin, and if $\lim_{x \to 0} f'(x) = 3$, is $f(x)$ differentiable at the origin? Why?

12.   Show that the function

$$f(\mathbf{x}) = \begin{cases} \dfrac{x_1 |x_2|}{\sqrt{x_1^2 + x_2^2}} & \text{if } (x_1, x_2) \neq (0, 0), \quad \text{positive square root} \\ 0 & \text{if } (x_1, x_2) = (0, 0), \end{cases}$$

has a well-defined directional derivative in each direction at the origin but is not differentiable at the origin.

13. The derivative of a mapping $\mathbf{f} : \mathbb{R}^2 \longrightarrow \mathbb{R}^2$ is a matrix $\mathbf{f}'(\mathbf{x}) = \{\partial_j f_i(\mathbf{x})\} \in \mathbb{R}^{2\times 2}$. For any real algebra $\mathcal{A} \subset \mathbb{R}^{2\times 2}$ (as defined in Section 1.3) let $\mathcal{S}_\mathcal{A}$ be the collection of differentiable mappings from open subsets of $\mathbb{R}^2$ into $\mathbb{R}^2$ such that $\mathbf{f}'(\mathbf{x}) \in \mathcal{A}$ at all points at which the mapping is defined. This thus is a set of mappings satisfying some linear partial differential equations.

  i) Show that $\mathcal{S}_\mathcal{A}$ is a real vector space, that is, that the sum and scalar product of any elements of $\mathcal{S}_\mathcal{A}$ belong to $\mathcal{S}_\mathcal{A}$.
  ii) Show that if $\mathbf{f}, \mathbf{g} \in \mathcal{S}_\mathcal{A}$ then $\mathbf{g} \circ \mathbf{f} \in \mathcal{S}_\mathcal{A}$. (Collections of mappings satisfying this condition are said to form a pseudogroup.)
  iii) Describe as explicitly as you can the pseudogroups $\mathcal{S}_{\mathcal{A}_0}$ and $\mathcal{S}_{\mathcal{A}_1}$ for the algebras $\mathcal{A}_0$ and $\mathcal{A}_1$.
  iv) When a mapping $\mathbf{f} : \mathbb{R}^2 \longrightarrow \mathbb{R}^2$ is identified with a mapping $\tilde{f} : \mathbb{C} \longrightarrow \mathbb{C}$, by associating to any vector $\binom{x}{y} \in \mathbb{R}^2$ the complex number $x + iy \in \mathbb{C}$, show that the pseudogroup $\mathcal{S}_{\mathcal{A}_{-1}}$ associated to the algebra $\mathcal{A}_{-1}$ can be identified with the set of those mappings $\tilde{f}$ that are differentiable mappings as functions of a complex variable, in the sense that $\tilde{f}(a + h) = \tilde{f}(a) + ch + \epsilon(h)$ where $a, c, h, \epsilon(h)$, and $\tilde{f}(a) \in \mathbb{C}$ and $\lim_{h\to 0} \frac{|\epsilon(h)|}{|h|} = 0$. (These are called holomorphic or complex analytic functions.)

3.3 Analytic Mappings

If $f : U \longrightarrow \mathbb{R}$ is a function defined in an open set $U \subset \mathbb{R}^2$, and if the partial derivative $\partial_{j_1} f(\mathbf{x})$ exists at all points $\mathbf{x} \in U$, the function $\partial_{j_1} f(\mathbf{x})$ may itself have partial derivatives, such as $\partial_{j_2}(\partial_{j_1} f(\mathbf{x}))$, which for convenience is shortened to $\partial_{j_2}\partial_{j_1} f(\mathbf{x})$; and the process can continue, leading to $\partial_{j_3}\partial_{j_2}\partial_{j_1} f(\mathbf{x})$ and so on. The order in which successive derivatives are taken may be significant; for example, a straightforward calculation shows that $\partial_1\partial_2 f(0,0) = 1$ but $\partial_2\partial_1 f(0,0) = -1$ for the function

$$f(x_1, x_2) = \begin{cases} \dfrac{x_1 x_2 (x_1^2 - x_2^2)}{x_1^2 + x_2^2} & \text{if } (x_1, x_2) \neq (0,0), \\ 0 & \text{if } (x_1, x_2) = (0,0). \end{cases}$$

However for continuously differentiable functions the order of differentiation is irrelevant.

Theorem 3.40. *If $f : U \longrightarrow \mathbb{R}$ is a function in open subset $U \subset \mathbb{R}^2$, if the partial derivatives $\partial_1 f(\mathbf{x}), \partial_2 f(\mathbf{x}), \partial_1\partial_2 f(\mathbf{x}), \partial_2\partial_1 f(\mathbf{x})$ exist at all points $\mathbf{x} \in U$, and if the mixed partial derivatives $\partial_1\partial_2 f(\mathbf{x}), \partial_2\partial_1 f(\mathbf{x})$ are continuous at a point $\mathbf{a} \in U$, then $\partial_1\partial_2 f(\mathbf{a}) = \partial_2\partial_1 f(\mathbf{a})$.*

Proof: For points $\mathbf{h} = (h_1, h_2) \in \mathbb{R}^2$ near the origin note that

$$\Delta = \big(f(a_1 + h_1, a_2 + h_2) - f(a_1 + h_1, a_2)\big) - \big(f(a_1, a_2 + h_2) - f(a_1, a_2)\big)$$
$$= \phi(a_1 + h_1) - \phi(a_1) \quad \text{where } \phi(x_1) = f(x_1, a_2 + h_2) - f(x_1, a_2)$$
$$= \phi'(a_1')h_1 \quad \text{where } a_1' \text{ is between } a_1 \text{ and } a_1 + h_1,$$

obtained by applying the Mean Value Theorem for functions of one variable to the function $\phi(x)$; and when $\phi'(a_1') = \partial_1 f(a_1', a_2 + h_2) - \partial_1 f(a_1', a_2)$ is viewed as a function of the variable a_2 then by applying the Mean Value Theorem for functions of one variable to this function it follows that

$$\phi'(a_1') = \partial_2 \partial_1 f(a_1', a_2')h_2 \quad \text{where } a_2' \text{ is between } a_2 \text{ and } a_2 + h_2,$$

so altogether

$$\Delta = \partial_2 \partial_1 f(a_1', a_2')h_2 h_1.$$

On the other hand it is possible to group the terms in the equation for Δ in another way, so that

$$\Delta = \big(f(a_1 + h_1, a_2 + h_2) - f(a_1, a_2 + h_2)\big) - \big(f(a_1 + h_1, a_2) - f(a_1, a_2)\big).$$

The same argument used for the first grouping when applied to this grouping amounts to interchanging the roles of the two variables, which leads to the result that

$$\Delta = \partial_1 \partial_2 f(a_1'', a_2'')h_1 h_2$$

for some points a_1'' between a_1 and $a_1 + h_1$ and a_2'' between a_2 and $a_2 + h_2$; the points a_1'' and a_2'' are not necessarily the same as the points a_1' and a_2' since they are derived by different applications of the Mean Value Theorem in one variable. Comparing the two expressions for Δ and dividing by $h_1 h_2$ shows that

$$\partial_2 \partial_1 f(a_1', a_2') = \partial_1 \partial_2 f(a_1'', a_2'').$$

Since the functions $\partial_1 \partial_2 f(x_1, x_2)$ and $\partial_2 \partial_1 f(x_1, x_2)$ are continuous at the point $\mathbf{a} = (a_1, a_2)$ and a_1' and a_1'' both approach a_1 as h_1 tends to 0 while a_2' and a_2'' both approach a_2 as h_2 tends to 0 it follows in the limit that $\partial_2 \partial_1 f(a_1, a_2) = \partial_1 \partial_2 f(a_1, a_2)$, which suffices to conclude the proof.

The preceding result of course applies to functions defined in open subsets of $\mathbb{R}^n$ for any n, since it just involves a change in the order of differentiation for a pair of variables; and it applies to higher derivatives of any orders, for the same reason. Consequently for functions that are continuously differentiable of the appropriate orders the notation can be simplified, for example by writing $\partial_1^2 \partial_2 f(\mathbf{x})$ in place of $\partial_1 \partial_2 \partial_1 f(\mathbf{x})$ or $\partial_2 \partial_1 \partial_1 f(\mathbf{x})$, or by writing $\partial_{j_1 j_2} f(\mathbf{x})$ in place of $\partial_{j_1} \partial_{j_2} f(\mathbf{x})$. It is sometimes convenient to use the **multi-index notation**, in which $\partial^I f(\mathbf{x})$ where $I = (3, 2, 1)$ stands for $\partial_1^3 \partial_2^2 \partial_3 f(\mathbf{x})$ for instance. Another notation frequently used is

$$\frac{\partial^3 f(\mathbf{x})}{\partial x_1^2 \partial x_2} = \partial_1^2 \partial_2 f(\mathbf{x}).$$

For a function $f(x)$ of a single variable the derivative $f'(x)$ is another function of a single variable; and its derivative, the second derivative of the initial function $f(x)$, is denoted by $f''(x)$ or $f^{(2)}(x)$; the derivative of this function in turn is denoted by $f'''(x)$ or $f^{(3)}(x)$. A traditional alternative notation, paralleling the notation for partial derivatives, is

$$\frac{df}{dx} = \frac{d}{dx} f(x) = f'(x) \quad \text{and} \quad \frac{d^2 f}{dx^2} = \frac{d^2}{dx^2} f(x) = f''(x).$$

The definition (3.25) of differentiability involved an approximation of the function $f(a + h)$ of the auxiliary variable h by a polynomial of degree 1 in that variable. The existence of derivatives of higher orders can be interpreted correspondingly as involving an approximation of the function $f(x + h)$ by polynomials in h of higher degree.

Theorem 3.41 (Taylor Series in One Variable). *If $f : U \longrightarrow \mathbb{R}^1$ is a function with derivatives up to order $k + 1$ in an open neighborhood $U \subset \mathbb{R}^1$ of a point $a \in \mathbb{R}^1$ then for any $h \in \mathbb{R}^1$ sufficiently small*

$$f(a + h) = f(a) + f'(a)h + \frac{1}{2!} f''(a)h^2 + \frac{1}{3!} f'''(a)h^3 + \dots$$

$$\dots + \frac{1}{k!} f^{(k)}(a)h^k + \frac{1}{(k + 1)!} f^{(k+1)}(\alpha)h^{k+1} \quad (3.59)$$

where α is between a and $a + h$.

Proof: If h is sufficiently small that $[a, a + h] \subset U$ let

$$R(x) = f(a + x) - f(a) - f'(a)x - \frac{1}{2!} f''(a)x^2 - \frac{1}{3!} f'''(a)x^3 - \dots$$

$$\dots - \frac{1}{k!} f^{(k)}(a)x^k - \frac{1}{(k + 1)!} cx^{k+1} \quad (3.60)$$

for any $x \in [0, h]$, where $c \in \mathbb{R}$ is chosen so that $R(h) = 0$. Clearly $R(0) = 0$ as well so by the Mean Value Theorem for functions of a single variable $R'(\alpha_1) = 0$ for some value α_1 between 0 and h. Note that

$$R'(x) = f'(a + x) - f'(a) - f''(a)x - \frac{1}{2!} f'''(a)x^2 -$$

$$\dots - \frac{1}{(k - 1)!} f^{(k-1)}(a)x^{k-1} - \frac{1}{k!} cx^k, \quad (3.61)$$

which is much the same as (3.60) but for the index $k - 1$ in place of the index k. In particular $R'(0) = 0$, and since $R'(\alpha_1) = 0$ as well it follows from the Mean Value Theorem for functions of a single variable that $R''(\alpha_2) = 0$ for some α_2 between 0 and α_1. The process continues, with further differentiation of the expression (3.60), yielding a sequence of points $\{\alpha_1, \alpha_2, \ldots, \alpha_{k+1}\}$ where α_{i+1} is between 0 and α_i, hence is between 0 and h, and $R^{(i)}(\alpha_i) = 0$. Finally for the case $i = k + 1$ it follows that

$$R^{(k+1)}(x) = f^{(k+1)}(a + x) - c,$$

and since $R^{(k+1)}(\alpha_{k+1}) = 0$ it follows that $c = f^{(k+1)}(a + \alpha_{k+1}) = f^{(k+1)}(\alpha)$ where $\alpha = a + \alpha_{k+1}$. That demonstrates the desired result for the case that $h > 0$; the corresponding argument holds for $h < 0$, so that suffices for the proof.

It may clarify the preceding proof to write out the special case $k = 2$ in detail. The corresponding result in several variables can be deduced from the result in a single variable by an application of the chain rule.

Theorem 3.42 (Taylor Series in Several Variables). *If $f : U \longrightarrow \mathbb{R}$ has continuous partial derivatives up to order $k + 1$ in an open neighborhood $U \subset \mathbb{R}^m$ of a point $\mathbf{a} \in \mathbb{R}^m$ then for any $\mathbf{h} = \{h_j\} \in \mathbb{R}^m$ sufficiently small*

$$f(\mathbf{a} + \mathbf{h}) = f(\mathbf{a}) + \sum_{j=1}^{m} \partial_j f(\mathbf{a}) h_j + \frac{1}{2!} \sum_{j_1, j_2 = 1}^{m} \partial_{j_1 j_2} f(\mathbf{a}) h_{j_1} h_{j_2} + \cdots$$

$$\cdots + \frac{1}{k!} \sum_{j_1, \ldots, j_k = 1}^{m} \partial_{j_1 \ldots j_k} f(\mathbf{a}) h_{j_1} \ldots h_{j_k}$$

$$+ \frac{1}{(k+1)!} \sum_{j_1, \ldots, j_{k+1} = 1}^{m} \partial_{j_1 \ldots j_{k+1}} f(\boldsymbol{\alpha}) h_{j_1} \ldots h_{j_{k+1}} \quad (3.62)$$

where $\boldsymbol{\alpha}$ is between $\mathbf{a}$ and $\mathbf{a} + \mathbf{h}$ on the line segment connecting them.

Proof: Let $\boldsymbol{\phi}(t) = \mathbf{a} + t\mathbf{h}$ for any $t \in \mathbb{R}$ and consider the function $g(t) = f(\boldsymbol{\phi}(t))$, for which $g(0) = f(\boldsymbol{\phi}(0)) = f(\mathbf{a})$ and $g(1) = f(\boldsymbol{\phi}(1)) = f(\mathbf{a} + \mathbf{h})$. The Taylor series (3.59) of the function $g(t)$ has the form

$$g(1) = g(0) + g'(0) + \frac{1}{2!} g''(0) + \ldots + \frac{1}{k!} g^{(k)}(0) + \frac{1}{(k+1)!} g^{(k+1)}(\tau) \quad (3.63)$$

for some $\tau \in (0,1)$. Since $\phi'(t) = \frac{d}{dt}(\mathbf{a} + t\mathbf{h}) = \mathbf{h} = {}^t(h_1, h_2, \ldots, h_m)$ repeated applications of the chain rule show that

$$g'(t) = \sum_{j_1=1}^{m} \partial_{j_1} f(\mathbf{a} + t\mathbf{h}) h_{j_1},$$

$$g''(t) = \sum_{j_1,j_2=1}^{m} \partial_{j_1 j_2} f(\mathbf{a} + t\mathbf{h}) h_{j_1} h_{j_2}$$

and in general

$$g^{(\nu)}(t) = \sum_{j_1,j_2,\ldots,j_\nu=1}^{m} \partial_{j_1 j_2 \cdots j_\nu} f(\mathbf{a} + t\mathbf{h}) h_{j_1} h_{j_2} \cdots h_{j_\nu};$$

substituting these values into (3.63) for $t = 0$ yields (3.62) for $\boldsymbol{\alpha} = \mathbf{a} + \tau\mathbf{h}$ and thereby concludes the proof.

If $f(x)$ is a C^∞ function at a point $a \in \mathbb{R}$, a function defined in an open neighborhood of a that has derivatives of all orders at the point a, then for any N the Taylor series (3.59) provides an approximation

$$f(a + x) = P_N(x) + \epsilon_N(x) \tag{3.64}$$

by the polynomial

$$P_N(x) = c_0 + c_1 x + c_2 x^2 + \cdots + c_N x^N, \tag{3.65}$$

where $c_n = \frac{1}{n!} f^{(n)}(a)$ for $0 \le n \le N$, with the error given by

$$\epsilon_N(x) = \frac{1}{(N+1)!} f^{(N+1)}\big(\alpha(x)\big) x^{N+1} \tag{3.66}$$

for some point $\alpha(x)$ between a and $a + x$ depending on x. There are a number of alternative expressions for the error in this polynomial approximation; but for present purposes the form used here will suffice. The collection of these polynomial approximations (3.65) for all N is the **Taylor series** of the function f at the point a, the formal infinite series

$$f(a + x) \sim \sum_{n=0}^{\infty} c_n x^n \quad \text{where} \quad c_n = \frac{1}{n!} f^{(n)}(a). \tag{3.67}$$

The **partial sums** of the Taylor series for functions of a single variable are the approximating polynomials

$$P_N(x) = \sum_{n=0}^{N} c_n x^n. \tag{3.68}$$

If $\lim_{N\to\infty} \epsilon_N(x) = 0$ for all points $x \in \mathcal{N}_0$ for some open neighborhood $\mathcal{N}_0$ of the origin then the function $f(a + x)$ is the limit $f(a + x) = \lim_{N\to\infty} P_N(x)$ of the partial sums $P_N(x)$ in the neighborhood $\mathcal{N}_0$; the Taylor series is said to **converge** to the function $f(a + x)$ or to have the **sum** $f(a + x)$, which is indicated by writing

$$f(a + x) = \sum_{n=0}^{\infty} c_n x^n \quad \text{for all } x \in \mathcal{N}_0. \tag{3.69}$$

In that case the function f is said to be **real analytic** at the point a; and if the function f is real analytic at each point in an open set $U \subset \mathbb{R}$ it is said to be real analytic in U. Every real analytic function of course is a $\mathcal{C}^\infty$ function, but it is not always the case that a $\mathcal{C}^\infty$ function is real analytic. Indeed for some $\mathcal{C}^\infty$ functions the Taylor series does not converge; and even if the Taylor series of a $\mathcal{C}^\infty$ function f converges it does not necessarily converge[1] to the function f. The further investigation of real analytic functions requires a more detailed examination of the convergence of series.

More generally, a **power series** in a single variable is a formal sum

$$\sum_{n=0}^{\infty} c_n x^n; \tag{3.70}$$

it is a real power series if $c_n \in \mathbb{R}$ for all n, and is a complex power series if $c_n \in \mathbb{C}$ for all n. The **partial sums** of the power series (3.70) are the polynomials

$$P_N(x) = \sum_{n=0}^{N} c_n x^n, \tag{3.71}$$

viewed as a polynomial in the real variable $x \in \mathbb{R}$ for a real power series and as a polynomial in the complex variable $x \in \mathbb{C}$ for a complex power series, although in the latter case the variable is traditionally denoted by z. The power series is said to **converge** to a function $f(x)$ in a subset U in $\mathbb{R}$ or $\mathbb{C}$, or to have the **sum** $f(x)$ in U, if $f(x) = \lim_{N\to\infty} P_N(x)$ for each point $x \in U$; that also is indicated by writing

$$f(x) = \sum_{n=0}^{\infty} c_n x^n \quad \text{for} \quad x \in U. \tag{3.72}$$

---

[1]A standard example of this sort of behavior is illustrated in problem 13 at the end of this section.

The series is said to **converge uniformly** in U if the partial sums $P_N(x)$ converge uniformly in the set $U \subset \mathbb{R}$. The corresponding definitions hold for power series in several variables.

Even more generally, any sequence of real or complex numbers c_n can be viewed as describing a formal series $\sum_{n=0}^{\infty} c_n$, with the **partial sums** $S_N = \sum_{n=0}^{N} c_n$. The series is said to **converge** to S, or to have the **sum** S, if $\lim_{N \to \infty} S_N = S$; that is indicated by writing $\sum_{n=0}^{\infty} c_n = S$. It is clear that whether a series converges or not is quite independent of any finite number of initial terms of the series; the sum of the series of course depends on all the terms of the series. The simplest series are real series $\sum_{n=0}^{\infty} c_n$ for which $c_n \geq 0$ for all n; they are usually called **positive series**, with the caution that in a positive series it is not necessarily the case that $c_n > 0$ for all n rather just that $c_n \geq 0$ for all n. For a positive series the sequence of partial suns S_N is monotonically increasing; so clearly the series converges if and only if the sequence of partial sums is bounded, and in that case the sum can be described alternatively as $S = \sup\{S_N\}$. Note that if $\sum_{n=0}^{\infty} c'_n$ and $\sum_{n=0}^{\infty} c''_n$ are positive series with the partial sums S'_N and S''_N, respectively, and if the first series converges while $c'_n \geq c''_n$ for all n, then the second series also converges, since $\sup\{S''_N\} \leq \sup\{S'_N\}$.

There are actually two rather more precise forms of the notion of the convergence of a series. Some series are inherently convergent: those are the easiest series to handle and they have the most convenient and useful properties. Some series are only accidentally convergent though: whether they converge or not, and the limit to which they converge, depend on the particular order of the terms in the series. To be more precise, a series $\sum_{n=0}^{\infty} c_n$ of real or complex terms is said to be **absolutely convergent** if the associated positive series $\sum_{n=0}^{\infty} |c_n|$ is convergent. Of course a positive series is absolutely convergent if and only if it is convergent. A series of complex terms can be viewed as a pair of series of real terms, the real and imaginary parts of the series. Since a complex series converges if and only if both the real and imaginary parts converge, and a complex series converges absolutely if and only if both the real and imaginary parts converge absolutely, it follows that for the most part it is sufficient to consider just real series in detail. Clearly if $\sum_{n=0}^{\infty} c'_n$ is an absolutely convergent real or complex series and if $|c''_n| \leq |c'_n|$ for all n then the series $\sum_{n=0}^{\infty} c''_n$ is also absolutely convergent.

Theorem 3.43. (i) *If a series of real or complex numbers is absolutely convergent then it is convergent.*

(ii) *If a series of real or complex numbers is absolutely convergent then all rearrangements[2] of the series are convergent and have the same sum.*

---

[2]The terms in a series $\sum_{n=0}^{\infty} c_n$ are naturally ordered by the indices n. A rearrangement of the series $\sum_{n=0}^{\infty} c_n$ is a series $\sum_{n=0}^{\infty} c'_n$ where $c'_n = c_{\phi(n)}$ for some bijective mapping $\phi : \mathbb{Z}^+ \longrightarrow \mathbb{Z}^+$ of

Proof: (i) That the series $\sum_{n=0}^{\infty} c_n$ is absolutely convergent means that the partial sums $S_N^* = \sum_{n=0}^{N} |c_n|$ of the positive series $\sum_{n=0}^{\infty} |c_n|$ converge. The series $\sum_{n=0}^{\infty}(c_n + |c_n|)$ is also a positive series, and since $|c_n + |c_n|| \leq 2|c_n|$ this series also converges so its partial sums S_N^{**} converge. Since $S_N = S_N^{**} - S_N^*$ where $S_N = \sum_{n=0}^{N} c_n$ it follows that the partial sums S_N converge, so the series $\sum_{n=0}^{\infty} c_n$ is convergent.

(ii) If $S = \sum_{n=0}^{\infty} c_n$ is a convergent positive series and if $S_M' = \sum_{m=0}^{M} c_m'$ are the partial sums of a rearrangement of that series then since all the entries c_m' are among the entries c_n clearly $S_M' \leq S$ for all M. On the other hand for any $\epsilon > 0$ there is a number N for which $S_N > S - \epsilon$; and for any M sufficiently large the entries c_m' of the partial sum $\sum_{m=0}^{M} c_m'$ include all the entries c_n of the partial sum $\sum_{n=0}^{N} c_n$ so $\sum_{m=0}^{M} c_m' \geq \sum_{n=0}^{N} c_n > S - \epsilon$. Thus $S - \epsilon < S_M' \leq S$ for sufficiently large M and consequently $\lim_{M \to \infty} S_M' = S$ so the rearranged series has the same sum as the original series.

Next if $S = \sum_{n=0}^{\infty} c_n$ is an absolutely convergent real series then $S^* = \sum_{n=0}^{\infty} |c_n|$ and $S^{**} = \sum_{n=0}^{\infty}(c_n + |c_n|)$ are convergent positive series, so by part (i) any rearrangements of these series converge to the same sum as the initial series; it follows that the same rearrangements of these two series yields a rearrangement of the series $S = \sum_{n=0}^{\infty} c_n = \sum_{n=0}^{\infty}(c_n + |c_n|) - \sum_{n=0}^{\infty} |c_n|$ which also converges to the same sum as the initital series. If $S = \sum_{n=0}^{\infty} c_n$ is an absolutely convergent complex series then the real and imaginary parts are absolutely convergent real series, which converge for any rearrangement to the sum of the initial series, and that suffices for the proof.

A series $\sum_{n=0}^{\infty} c_n$ where $c_n \in \mathbb{R}$ or $\mathbb{C}$ is convergent if it is absolutely convergent, by the preceding theorem; however it may be convergent but not absolutely convergent, in which case it is said to be **conditionally convergent**. Thus any convergent real or complex series is either absolutely convergent or conditionally convergent, and none are both. If a convergent real series $\sum_{n=0}^{\infty} c_n$ is absolutely convergent then the series formed from just the nonnegative terms c_n is also convergent, since the partial sums of this series are bounded by the partial sums of the series $\sum_{n=0}^{\infty} |c_n|$. Conversely if a real series $\sum_{n=0}^{\infty} c_n$ is convergent and the series formed from just the nonnegative terms c_n is convergent then the series formed from the negative terms c_n clearly also is convergent hence the series $\sum_{n=0}^{\infty} c_n$ is absolutely convergent. It follows that if a real series $\sum_{n=0}^{\infty} c_n$ is conditionally convergent then the series formed from the nonnegative terms c_n is divergent, and of course similarly the series formed from the negative terms c_n also is divergent. A classical example of a conditionally

---

the nonnegative integers $\mathbb{Z}^+$; thus the coefficients c_n' are a rearrangement of the coefficients c_n. In the older literature a rearrangement of a series is sometimes called a derangement of the series.

convergent series is the **alternating harmonic series**

$$1 - \frac{1}{2} + \frac{1}{3} - \frac{1}{4} + \frac{1}{5} - \frac{1}{6} + \cdots . \tag{3.73}$$

That this series is convergent is quite easy to see; indeed more generally the following is true.

Theorem 3.44. *If c_n is a monotonically decreasing sequence of nonnegative real numbers, so that $c_n \geq c_{n+1} \geq 0$, and if $\lim_{n \to \infty} c_n = 0$, then the series*

$$c_1 - c_2 + c_3 - c_4 + c_5 - c_6 + \cdots \tag{3.74}$$

is convergent.

Proof: If S_N is the partial sum of the series (3.74) then

$$S_{2N} = \underbrace{(c_1 - c_2)}_{\geq 0} + \underbrace{(c_3 - c_4)}_{\geq 0} + \cdots + \underbrace{(c_{2N-1} - c_{2N})}_{\geq 0},$$

so clearly S_{2N} is an increasing sequence of real numbers. On the other hand

$$S_{2N} = c_1 \underbrace{-(c_2 - c_3)}_{\leq 0} \underbrace{-(c_4 - c_5)}_{\leq 0} \cdots \underbrace{-(c_{2N-2} - c_{2N-1})}_{\leq 0} \underbrace{-c_{2N}}_{\leq 0},$$

so clearly $S_{2N} \leq c_1$. Therefore the sequence S_{2N} converges, since its partial sums are a monotonically increasing sequence of real numbers bounded above. Since $|S_{2N} - S_{2N-1}| = |c_{2N}|$ and by assumption $\lim_{N \to \infty} c_{2N} = 0$ it follows that the sequence S_{2N-1} also converges and has the same limit as the sequence S_{2N}; therefore the sequence S_N converges, which suffices for the proof.

In particular the alternating harmonic series converges; but it does not converge absolutely since the **harmonic series**, the series $\sum_{n=1}^{\infty} 1/n$, can be decomposed as

$$\sum_{n=1}^{\infty} \frac{1}{n} = 1 + \frac{1}{2} + \underbrace{\left(\frac{1}{3} + \frac{1}{4}\right)}_{\geq \frac{1}{2}} + \underbrace{\left(\frac{1}{5} + \frac{1}{6} + \frac{1}{7} + \frac{1}{8}\right)}_{\geq \frac{1}{2}} + \cdots \tag{3.75}$$

where

$$\frac{1}{2^n + 1} + \frac{1}{2^n + 2} + \cdots + \frac{1}{2^{n+1}} \geq 2^n \frac{1}{2^{n+1}} = \frac{1}{2}$$

for all n so it is a divergent series. Therefore the alternating harmonic series is conditionally convergent; the series consisting of just the odd terms diverges, and its partial sums increase without limit, while the series consisting of just the even terms also diverges, and its partial sums decrease without limit.

Riemann showed that by rearranging the order of the terms of any conditionally convergent series it can be made to diverge or to converge to any chosen limit. Indeed the subseries consisting just of the positive terms and that consisting just of the negative terms both diverge. Take enough positive terms so that the partial sum of these is greater than N, then take enough negative terms so that the whole partial sum is between $N/2$ and N, then take enough positive terms so that the whole partial sum is greater than $2N$, and so on; the rearranged series then diverges. On the other hand to rearrange the series to converge to $\sqrt{2}$, for example, take enough positive terms so that the partial sum is just greater than $\sqrt{2}$, then take enough negative terms to bring the sum to just less than $\sqrt{2}$, then take enough positive terms to bring the partial sum to just greater than $\sqrt{2}$, and so on; and since the terms c_n of the series satisfy $\lim_{n \to \infty} c_n = 0$ the resulting oscillation of the partial sums converges to $\sqrt{2}$. Thus a real series is absolutely convergent if and only if all rearrangements of the series converge to the same limit; but of course that is not a very effective test for absolute convergence.

The preceding discussion focused on Taylor series in a single variable or the corresponding series $\sum_{n=0}^{\infty} c_n$ for real or complex numbers c_n. The study of Taylor series in several variables or the corresponding series $\sum_{n_1,n_2,\ldots,n_r=0}^{\infty} c_{n_1 n_2 \ldots n_r}$ for real or complex numbers $c_{n_1 n_2 \ldots n_r}$ is important for many purposes, but will not be pursued extensively here; however some results about series in several variables are sufficiently useful even in considering functions of a single variable that a few words about them are perhaps in order here. For present purposes it is sufficient just to consider **double series**, those of the form $\sum_{n_1,n_2=0}^{\infty} c_{n_1 n_2}$ for some real or complex constants $c_{n_1 n_2}$. The partial sums

$$S_{N_1 N_2} = \sum_{\substack{0 \le n_1 \le N_1 \\ 0 \le n_2 \le N_2}} c_{n_1 n_2} \tag{3.76}$$

are simply sums of a finite collection of numbers, so the order in which the summation is carried out is irrelevant; but in an expression $\sum_{n_1,n_2=0}^{\infty} c_{n_1 n_2}$ it is not at all clear what is meant, since there are infinitely many numbers $c_{n_1 n_2}$ and the order in which the summation is carried out may be quite relevant indeed. However with the Taylor series in mind it is natural to say that a double series **converges**[3] to the sum $S = \sum_{n_1,n_2=0}^{\infty} c_{n_1 n_2}$ if for any $\epsilon > 0$ there is a number M such that the partial sums (3.76) satisfy $|S - S_{N_1 N_2}| < \epsilon$ for all $N_1 > M$ and $N_2 > M$. If that is not the case the series is said to **diverge**. A **positive double series** is defined to be a double series $\sum_{n_1,n_2=0}^{\infty} c_{n_1 n_2}$ for which $c_{n_1 n_2} \ge 0$ for all n_1, n_2. If a

---

[3]This definition, which goes back to Pringsheim in the 1870s, seems to be the most commonly used one; but there are some alternatives that are used from time to time, so when reading about double series some care should be taken to be sure which definition actually is being used.

positive double series converges then the partial sums $S_{N_1 N_2}$ are bounded, since by definition there is a number M such that $S_{N_1 N_2} < S + \epsilon$ whenever $N_1 > M$ and $N_2 > M$ while $S_{n_1 n_2} \leq S_{N_1 N_2}$ whenever $n_1 \leq N_1$ and $n_2 \leq N_2$ since the partial sums are monotonically increasing. Conversely if the partial sums of a positive double series are bounded and $S = \sup\{S_{N_1 N_2}\}$ then the series converges to the sum S since $S_{N_1 N_2} \leq S$ for all N_1, N_2 while for any $\epsilon > 0$ there is some partial sum $S_{M_1 M_2}$ for which $S_{M_1 M_2} > S - \epsilon$ and $S_{N_1 N_2} \geq S_{M_1 M_2} > S - \epsilon$ whenever $N_1 > M_1$ and $N_2 > M_2$. A double series $\sum_{n_1, n_2 = 0}^{\infty} c_{n_1 n_2}$ is said to **converge absolutely** if the positive double series $\sum_{n_1, n_2 = 0}^{\infty} |c_{n_1 n_2}|$ converges.

Theorem 3.45. *If a double series of real or complex numbers is absolutely convergent then it is convergent.*

Proof: That the double series $\sum_{n_1, n_2 = 0}^{\infty} c_{n_1 n_2}$ is absolutely convergent means that the positive double series $\sum_{n_1, n_2 = 0}^{\infty} |c_{n_1 n_2}|$ converges to a sum S^*; thus for any $\epsilon > 0$ there exists a number M^* so that the partial sums

$$S_{N_1 N_2}^* = \sum_{\substack{0 \leq n_1 \leq N_1 \\ 0 \leq n_2 \leq N_2}} |c_{n_1 n_2}|$$

satisfy $|S_{N_1 N_2}^* - S^*| < \epsilon$ whenever $N_1, N_2 > M^*$. The terms of the positive double series $\sum_{n_1, n_2 = 0}^{\infty} (c_{n_1 n_2} + |c_{n_1 n_2}|)$ satisfy $|c_{n_1 n_2} + |c_{n_1 n_2}|| \leq 2|c_{n_1 n_2}|$ so its partial sums $S_{N_1 N_2}^{**}$ are bounded and hence this series converges to a sum S^{**}; thus there exists a number M^{**} so that the partial sums

$$S_{N_1 N_2}^{**} = \sum_{\substack{0 \leq n_1 \leq N_1 \\ 0 \leq n_2 \leq N_2}} (c_{n_1 n_2} + |c_{n_1 n_2}|)$$

satisfy $|S_{N_1 N_2}^{**} - S^{**}| < \epsilon$ whenever $N_1, N_2 > M^{**}$. The partial sums

$$S_{N_1 N_2} = \sum_{\substack{0 \leq n_1 \leq N_1 \\ 0 \leq n_2 \leq N_2}} c_{n_1 n_2}$$

of the original series are $S_{N_1 N_2} = S_{N_1 N_2}^{**} - S_{N_1 N_2}^*$, so if $N_1, N_2 > \max(M^*, M^{**})$ then

$$|S_{N_1, N_2} - (S^{**} - S^*)| = |(S_{N_1, N_2}^{**} - S_{N_1, N_2}^*) - (S^{**} - S^*)|$$
$$\leq |S_{N_1, N_2}^{**} - S^{**}| + |S_{N_1, N_2}^* - S^*| < 2\epsilon$$

so the series $\sum_{n_1, n_2 = 0}^{\infty} c_{n_1 n_2}$ is convergent to the sum $S = S^{**} - S^*$. This applies to the real and imaginary double series associated to any complex double series, and that suffices for the proof.

One way to simplify the examination of double series is to rewrite the double series as a single series by ordering the terms of the series appropriately; for example the Cantor diagonalization of the double series $\sum_{n_1,n_2=0}^{\infty} c_{n_1 n_2}$ is the single series $\sum_{n=0}^{\infty} \tilde{c}_n$ where

$$\tilde{c}_0 = c_{00}, \ \tilde{c}_1 = c_{10}, \ \tilde{c}_2 = c_{01}, \ \tilde{c}_3 = c_{20}, \ \tilde{c}_4 = c_{11}, \ \tilde{c}_5 = c_{02}, \ \tilde{c}_6 = c_{30}, \ \ldots$$

as in Cantor's Diagonalization Theorem.

Theorem 3.46. *A real or complex double series is absolutely convergent if and only if its Cantor diagonalization is absolutely convergent; and the double series and its Cantor diagonalization have the same sum.*

Proof: If $\tilde{S}_N = \sum_{n=0}^{N} \tilde{c}_n$ is a partial sum of the Cantor diagonalization of a convergent positive double series with the sum $S = \sum_{n_1,n_2=0}^{\infty} c_{n_1 n_2}$ then since all the entries $\tilde{c}_n$ in $\tilde{S}_N$ occur among the entries $c_{n_1 n_2}$ in the double series it follows that $\tilde{S}_N \leq S$. That is true for all N so the Cantor diagonalization converges to a sum $\tilde{S} \leq S$. The corresponding argument in the converse direction completes the proof in this special case.

If $S_N = \sum_{n_1,n_2=0}^{N} c_{n_1 n_2}$ is a partial sum of real double series then $S_N^* = \sum_{n_1,n_2=0}^{N} |c_{n_1 n_2}|$ and $S_N^{**} = \sum_{n_1,n_2=0}^{\infty} (|c_{n_1 n_2}| + c_{n_1 n_2})$ are the partial sums of positive double series. By the first part of the proof the positive double series with partial sums S_N^* and S_N^{**} converge if and only if their Cantor diagonalizations converge, and the sums of the double series are the same as the sums of their Cantor diagonalizations. That must then also be the case for the real double series $S^{**} - S^* = \sum_{n_1 n_2=0}^{\infty} c_{n_1 n_2}$, which yields the proof in this case. Complex double series correspond to a pair of real double series, and that suffices for the proof.

Of course there are many other diagonalizations of double series in addition to the Cantor diagonalization. Another standard diagonalization arises by grouping the terms $c_{n_1 n_2}$ according to the sum $n_1 + n_2$ and then in each group ordering the terms by the first index n_1, the summation around the borders of squares of increasing sizes. Any diagonalization can be viewed as a rearrangement of the Cantor diagonalization, so it follows from Theorem 3.43 that any diagonalization of an absolutely convergent double series is convergent to the same sum as the double series. There is another approach to double series that is even more common; it involves summing over the two separate indices separately. Thus for any double series $\sum_{n_1,n_2=0}^{\infty} c_{n_1 n_2}$ with the partial sums $S_{N_1 N_2}$ it is traditional to introduce the summation

$$\sum_{n_1=0}^{\infty} \sum_{n_2=0}^{\infty} c_{n_1 n_2} = \lim_{N_1 \to \infty} \lim_{N_2 \to \infty} S_{N_1 N_2} \qquad (3.77)$$

or in the other order

$$\sum_{n_2=0}^{\infty} \sum_{n_1=0}^{\infty} c_{n_1 n_2} = \lim_{N_2 \to \infty} \lim_{N_1 \to \infty} S_{N_1 N_2}; \qquad (3.78)$$

if these limits exist they are called the **iterated sums** of the double series. Even if the limits exist, though, they may not be equal; for, as observed in the earlier discussion of limits, exchanges of the orders of double limits do not always preserve the values of the limits.

Theorem 3.47. (i) *The iterated sums of an absolutely convergent double series exist and are equal to the sum of the double series.*
(ii) *If one iterated sum of a double series is absolutely convergent then the double series itself is absolutely convergent.*

Proof: (i) First if a positive double series converges to a sum $S = \sum_{n_1 n_2 = 0}^{\infty} c_{n_1 n_2}$ then

$$\sum_{n_1=0}^{N_1} \sum_{n_2=0}^{N_2} c_{n_1 n_2} \leq S$$

for any N_1, N_2. For each index n_1 the preceding inequality holds for any N_2 hence

$$\sum_{n_1=0}^{N_1} \sum_{n_2=0}^{\infty} c_{n_1 n_2} \leq S;$$

and since that is the case for all N_1 it follows that

$$\sum_{n_1=0}^{\infty} \sum_{n_2=0}^{\infty} c_{n_1 n_2} \leq S,$$

so the iterated sums of the positive double series exist and are at most the sum S of the series. On the other hand for any $\epsilon > 0$ there are some values $N_1 N_2$ so that

$$\sum_{n_1=0}^{N_1} \sum_{n_2=0}^{N_2} c_{n_1 n_2} \geq S - \epsilon,$$

since for finite values N_1, N_2 the iterated sums are also the corresponding sums of the double series. This inequality continues to hold for any fixed N_1 as N_2 tends to infinity, and then as N_1 tends to infinity, so

$$\sum_{n_1=0}^{\infty} \sum_{n_2=0}^{\infty} c_{n_1 n_2} \geq S - \epsilon$$

and since that is the case for evey $\epsilon > 0$ it follows that

$$\sum_{n_1=0}^{\infty} \sum_{n_2=0}^{\infty} c_{n_1 n_2} \geq S,$$

which shows that the iterated sum is equal to the sum of the double series.

Next if $\sum_{n_1 n_2=0}^{\infty} c_{n_1 n_2}$ is an absolutely convergent double series with partial sums $S_{N_1 N_1}$ and sum S then the two positive double series $\sum_{n_1,n_2=0}^{\infty} |c_{n_1 n_2}|$ and $\sum_{n_1,n_2=0}^{\infty} (c_{n_1 n_2} + |c_{n_1 n_2}|)$ are convergent, with the partial sums $S_{N_1 N_2}^*$ and $S_{N_1 N_2}^{**}$ and the sums S^* and S^{**}, respectively. The results already established apply to these series, so the iterated sums of both series exist and have the same value as the sums of the series. Since $S_{N_1 N_2} = S_{N_1 N_2}^{**} - S_{N_1 N_2}^*$ and the iterated sums of the series $S_{N_1 N_2}^*$ and $S_{N_1 N_2}^{**}$ converge it follows that the iterated limits of these two series exist and are equal to the sum of the series. That is the case for the real and imaginary parts of complex series.

(ii) If the iterated sum $\sum_{n_1=0}^{\infty} \sum_{n_2=0}^{\infty} |c_{n_1 n_2}| = S$ exists then the partial sums satisfy $S_{N_1 N_2}^* = \sum_{n_1=0}^{N_1} \sum_{n_2=0}^{N_2} |c_{n_1 n_2}| \leq S$ for all N_1, N_2, so the double series is absolutely convergent, and the rest of the theorem then follows from (i) to conclude the proof.

A possibly surprising amount of the study of the convergence of series rests on the properties of just one special series, the **geometric series** $\sum_{n=0}^{\infty} z^n$ where z is either a real or a complex number. Since

$$(1 - z)(1 + z + z^2 + \cdots + z^N) = 1 - z^{N+1} \tag{3.79}$$

it follows that

$$1 + z + z^2 + \cdots + z^N = \frac{1}{1 - z} - \frac{z^{N+1}}{1 - z};$$

if $|z| \leq r < 1$ then $\left| \frac{z^{N+1}}{1-z} \right| \leq \frac{r^{N+1}}{1-r}$ and consequently $\lim_{N \to \infty} \frac{z^{N+1}}{1-z} = 0$ so

$$\frac{1}{1 - z} = \sum_{n=0}^{\infty} z^n \quad \text{for } z \in \mathbb{C}, |z| < 1, \tag{3.80}$$

and this series is absolutely convergent, since the same formula holds for $|z|$ in place of z. The series is not convergent for $|z| > 1$ since the individual terms increase in absolute value so the partial sums do not converge. For real values of the variable z this is the Taylor series of the function $\frac{1}{1-x}$, as can be verified by calculating the value of the derivative $\frac{d}{dx}(1 - x)^{-1}$ at the origin $x = 0$; thus that particular Taylor series converges only in the interval $|x| < 1$, although the function $1/(1 - x)$ is well defined for values $x > 1$. The following criteria for convergence of general series rest on the convergence of the geometric series.

Theorem 3.48 (Convergence Tests). *Consider an infinite series*

$$\sum_{n=0}^{\infty} c_n \quad \text{where} \quad c_n \in \mathbb{R} \ \text{or} \ \mathbb{C}. \tag{3.81}$$

(i) *A necessary but not sufficient condition for the series (3.81) to converge is that* $\lim_{n \to \infty} |c_n| = 0$.

(ii) **Comparison Test** *If* $|c_n| \leq C_n$ *for all large* n, *where* $C_n \geq 0$ *and* $\sum_{n=0}^{\infty} C_n$ *converges, then the series (3.81) is absolutely convergent.*

(iii) **Root Test** *If* $|c_n| \leq Mr^n$ *for all large* n, *where* $M \geq 0$ *and* $r < 1$, *then the series (3.81) is absolutely convergent.*

(iv) **Ratio Test** *If* $|c_{n+1}/c_n| \leq r$ *for all large* n, *where* $0 \leq r < 1$, *then the series (3.81) is absolutely convergent.*

Proof: (i) If the series (3.81) converges then the partial sums P_N form a Cauchy sequence, so $\lim_{N \to \infty} |P_N - P_{N-1}| = \lim_{N \to \infty} |c_N| = 0$; but that is not sufficient to guarantee that the series converges, since for instance the terms of the harmonic series (3.75) tend to zero but the series does not converge.

(ii) If Q_N are the partial sums of the series $\sum_{n=0}^{\infty} |c_n|$ then for any $\epsilon > 0$

$$|Q_N - Q_M| \leq \sum_{n=M+1}^{N} |c_n| \leq \sum_{n=M+1}^{N} C_n < \epsilon$$

whenever M, N are sufficiently large, since the series $\sum_{n=0}^{\infty} C_n$ converges; thus the partial sums Q_N of the series $\sum_{n=0}^{\infty} |c_n|$ are a Cauchy sequence so that series converges.

(iii) The series $M \sum_{n=0}^{\infty} r^n$ converges by (3.80), hence the series $\sum_{n=0}^{\infty} |c_n|$ converges by (ii), the comparison test.

(iv) If $|c_{n+1}/c_n| \leq r < 1$ for all $n \geq N$ it follows immediately by induction on n that $|c_{N+n}| \leq |c_N| r^n$, hence the series converges by (iii), the root test.

The tests (iii) and (iv) in the preceding theorem are the most commonly used simple tests for the convergence of series. It should be noted though that they only test for the absolute convergence of series; a series may well converge even if it does not converge absolutely. It should also be noted that these tests are sufficient but are not necessary for absolute convergence; so if a series is absolutely convergent it does not necessarily satisfy the conditions either of the root or the ratio test. When the entries c_n in a series are sufficiently simple functions of the index n the root and ratio tests can be rephrased and extended as follows.

Corollary 3.49. *Consider an infinite series $\sum_{n=0}^{\infty} c_n$ where $c_n \in \mathbb{R}$ or $\mathbb{C}$.*
(i) **Root Test** *If the limit $\rho = \lim_{n\to\infty} \sqrt[n]{|c_n|}$ exists the series is absolutely convergent if $\rho < 1$ and divergent if $\rho > 1$.*
(ii) **Ratio Test** *If the limit $\rho = \lim_{n\to\infty} \left|\frac{c_{n+1}}{c_n}\right|$ exists the series is absolutely convergent if $\rho < 1$ and divergent if $\rho > 1$.*

Proof: (i) If $\lim_{n\to\infty} \sqrt[n]{|c_n|} < 1$ then $\sqrt[n]{|c_n|} \leq r < 1$ for all large n so $|c_n| < r^n$ and the series converges by Theorem 3.48 (iii), while if $\lim_{n\to\infty} \sqrt[n]{|c_n|} > 1$ then $\sqrt[n]{|c_n|} \geq r > 1$ for all large n so the terms c_n do not tend to zero and the series diverges by Theorem 3.48 (i).
(ii) If $\lim_{n\to\infty} |c_{n+1}/c_n| < 1$ then $|c_{n+1}/c_n| \leq r < 1$ for all large n and the series converges by Theorem 3.48 (iv), while if $\lim_{n\to\infty} \sqrt[n]{|c_n|} > 1$ then $\left|\frac{c_{n+1}}{c_n}\right| \geq r > 1$ for all large n so the terms c_n do not tend to zero and the series diverges by Theorem 3.48 (i). That suffices for the proof.

As in the case of the preceding Theorem 3.48 the convergence tests in Corollary 3.49 only test for absolute convergence, and they are sufficient but not necessary tests for convergence. In particular if $\rho = 1$ the tests are not applicable so the series may or may not converge. For example the series $\sum_{n=1}^{\infty} c_n$ diverges if $c_n = n^{-1}$ as already observed but converges if $c_n = n^{-2}$, as will be discussed next; but $\lim_{n\to\infty} \sqrt[n]{|c_n|} = \lim_{n\to\infty} \left|\frac{c_{n+1}}{c_n}\right| = 1$ for both series. The ratio test is perhaps the easiest to use, since it does not require taking the n-th roots of the coefficients; however it is a weaker test than the root test, since for example it does not demonstrate that the series $\frac{1}{2} + \frac{1}{2} + \frac{1}{4} + \frac{1}{4} + \frac{1}{8} + \frac{1}{8} + \cdots$ converges while the root test does. There are more sophisticated tests that sometimes provide information in cases in which the root and ratio tests fail.

Theorem 3.50 (Raabe's Test). *For any series $\sum_{n=0}^{\infty} c_n$ where $c_n \in \mathbb{R}$ or $\mathbb{C}$ and $c_n \neq 0$ for all n let*

$$R_n = n\left(1 - \frac{|c_{n+1}|}{|c_n|}\right). \tag{3.82}$$

If there is a number N_0 such that $R_n \geq R > 1$ for all $n \geq N_0$ then the series is absolutely convergent.

Proof: To simplify the notation replace $|c_n|$ by c_n, so it can be assumed that $c_n > 0$ for all n. From the definition (3.82) it follows that

$$R_n - 1 = n - 1 - n\frac{c_{n+1}}{c_n} = \frac{1}{c_n}((n-1)c_n - nc_{n+1}). \tag{3.83}$$

If $n \geq N_0$ then $R_n - 1 \geq R - 1 > 0$ and consequently from (3.83) it follows that

$$0 < c_n = \frac{P_n}{R_n - 1} \leq \frac{P_n}{R - 1} \quad \text{where} \quad P_n = (n-1)c_n - nc_{n+1}. \tag{3.84}$$

Thus $P_n > 0$, so by the comparison test in order to show that the series $\sum_{n=1}^{\infty} c_n$ converges it is sufficient to show that the series $\sum_{n=1}^{\infty} P_n$ converges. That $P_n > 0$ means that $(n-1)c_n > nc_{n+1}$, so the sequence nc_{n+1} is a monotonically decreasing sequence of nonnegative real numbers when $n \geq N_0$; hence the limit $A = \lim_{n\to\infty} nc_{n+1}$ exists. For any $N > N_0$

$$\sum_{n=N_0}^{N} P_n = \sum_{n=N_0}^{N} \left((n-1)c_n - nc_{n+1}\right) = (N_0 - 1)c_{N_0} - Nc_{N+1}$$

in view of the obvious cancellations in successive terms in this series; consequently

$$\lim_{N\to\infty} \sum_{n=N_0}^{N} P_n = \lim_{N\to\infty} \left((N_0 - 1)c_{N_0} - Nc_{N+1}\right) = (N_0 - 1)c_{N_0} - A.$$

Therefore the series $\sum_{n=1}^{N} P_n$ converges, and that suffices for the proof.

As for the earlier tests for convergence, Raabe's test can be rephrased and extended when the entries c_n are sufficiently simple.

Corollary 3.51 (Raabe's Test). *If $\sum_{n=0}^{\infty} c_n$ is a series of real or complex numbers where $c_n \neq 0$ for all n, if*

$$R_n = n\left(1 - \frac{|c_{n+1}|}{|c_n|}\right) \tag{3.85}$$

and if the limit $\rho = \lim_{n\to\infty} R_n$ exists then the series is absolutely convergent if $\rho > 1$ and is divergent if $\rho < 1$.

Proof: Again to simplify the notation assume that $c_n > 0$ for all n. If $\rho > 1$ then there is a number N_0 sufficiently large that $R_n \geq R > 1$ for all $n > N_0$, and it follows from the preceding theorem that the series converges. On the other hand if $\rho < 1$ then there is a number N_0 sufficiently large that $R_n - 1 \leq 0$ for all $n \geq N_0$, so from (3.83) in the preceding theorem $(n-1)c_n - nc_{n+1} \leq 0$ for all $n \geq N_0$. Therefore the sequence nc_{n+1} is monotonically increasing for $n \geq N_0$, so in particular $nc_{n+1} \geq N_0 c_{N_0+1}$ for $n \geq N_0$, or equivalently $c_{n+1} \geq \frac{1}{n}N_0 c_{N_0+1}$ for $n \geq N_0$. The harmonic series $\frac{1}{n}$ diverges, as noted in (3.75); thus the partial sums increase without limit hence the series $\sum_{n=1}^{\infty} c_n$ diverges as well, and that suffices for the proof.

Note that Raabe's test, as for the root and ratio tests, is not applicable if the limit $\rho = 1$, so in that case the series may or may not converge. That Raabe's test is stronger than the ratio test is evident from considering the series for which $c_n = \frac{1}{n^2}$. The ratio test conveys no information about the convergence of that series; however for Raabe's test

$$\lim_{n\to\infty} n\left(1 - \frac{c_{n+1}}{c_n}\right) = \lim_{n\to\infty} n\left(1 - \frac{n^2}{(n+1)^2}\right) = \lim_{n\to\infty} \frac{n(2n+1)}{(n+1)^2} = 2$$

so the series converges.

The preceding tests of course can be applied to power series, or more generally to series $\sum_{n=0}^{\infty} f_n(x)$ of functions in a subset U of $\mathbb{R}^n$ or $\mathbb{C}^n$, by considering the corresponding series of constants for each fixed point $x \in U$. However for series of functions another question of interest is whether the series is **uniformly convergnt** in U, meaning that the sequence of partial sums $P_N(x) = \sum_{n=0}^{N} f_n(x)$ converges uniformly in U. That is equivalent to the condition that the sequence of partial sums is uniformly Cauchy, by Theorem 3.19. A standard test for this is the **Weierstrass M-test**, really a special case of the comparison test.

Corollary 3.52. *A sequence $\sum_{n=0}^{\infty} f_n(x)$ of functions $f_n(x)$ in a subset U of $\mathbb{R}^n$ or $\mathbb{C}^n$ is uniformly convergent in U if $|f_n(x)| \leq C_n$ for all $x \in U$, where $\sum_{n=0}^{\infty} C_n$ is a convergent sequence of real numbers $C_n \geq 0$.*

Proof: If $|f_n(x)| \leq C_n$ then the partial sums $P_N(x) = \sum_{n=0}^{N} f_n(x)$ satisfy $|P_N(x) - P_M(x)| \leq \sum_{n=M+1}^{N} |f_n(x)| \leq \sum_{n=M+1}^{N} C_n$; and since the series $\sum_{n=0}^{\infty} C_n$ is convergent then for any $\epsilon > 0$ there is a number N_0 such that $|P_N(x) - P_M(x)| \leq \sum_{n=M+1}^{N} C_n < \epsilon$ for all $x \in U$ whenever $M, N \geq N_0$, showing that the partial sums $P_N(x)$ are a uniformly Cauchy sequence hence that they converge uniformly in U and thereby concluding the proof.

Theorem 3.53. *If a power series $\sum_{n=0}^{\infty} c_n z^n$ converges for a particular value $z_0 \in \mathbb{C}$ then it converges absolutely whenever $|z| < |z_0|$ and it converges uniformly in the set of points $z \in \mathbb{C}$ for which $|z| \leq r$ for any $r < |z_0|$.*

Proof: If the power series converges for $z = z_0$ then $\lim_{n\to\infty} |c_n z_0^n| = 0$ so $|c_n z_0^n| \leq C$ for some $C > 0$ and all $n \geq 0$, hence $|c_n z^n| \leq C(|z|/|z_0|)^n$ for all $n \geq 0$. If $|z| < |z_0|$ then $|z|/|z_0| < 1$ so the series $\sum_{n=0}^{\infty} C(|z|/|z_0|)^n$ converges by the root test hence the series $\sum_{n=0}^{\infty} c_n z^n$ converges absolutely by the comparison test. Actually $|c_n z^n| \leq C(r/|z_0|)^n$ in the entire region $|z| \leq r$ for any $r < |z_0|$; then $(r/|z_0|) < 1$ so the series $\sum_{n=0}^{\infty} C(|z|/|z_0|)^n$ converges by the root test hence the series $\sum_{n=0}^{\infty} c_n z^n$ converges uniformly in that region by the Weierstrass M-test, Corollary 3.52, and that suffices for the proof.

Note particularly that in the preceding theorem the power series is only assumed to converge at z_0; so it may converge conditionally rather than absolutely at that point. For any real or complex power series there is a unique value R, called the **radius of convergence** of the series, such that the series converges whenever $|z| < R$ and diverges whenever $|z| > R$; indeed R is just the supremum of the values $|z|$ for which the series converges at the point z, as an evident consequence of the preceding theorem. Of course it is possible that $R = 0$ so the series converges only when $z = 0$, or that $R = \infty$ so the series converges absolutely for any z. For a finite value of R the series converges absolutely at all points z in the open disc $D_R = \{z \in \mathbb{C}||z| < R\}$ and converges uniformly in any compact subset $\overline{D_r} \subset D_R$ for $r < R$, so the limit function is continuous in the entire open set D_R. The behavior of the function at the boundary $\partial D_R = \{z \in \mathbb{C}||z| = R\}$ of the region of convergence is a rather complicated question in general; but the situation for real power series is much simpler. A real power series with radius of convergence R for instance is continuous in the open interval $(-R, R)$; it may converge at both end points of the interval, just at one end point, or at neither end point. The power series $\sum_{n=1}^{\infty} \frac{1}{n^2} x^n$ will be shown later in this section to have the radius of converge, $R = 1$ and to converge at both end points $x = \pm 1$. The power series $\sum_{n=1}^{\infty} \frac{1}{n} x^n$ has the radius of convergence $R = 1$ and converges (although not absolutely) at the point $x = -1$ but diverges at the point $x = 1$, as already noted. The power series $\sum_{n=1}^{\infty} n x^n$ has the radius of convergence $R = 1$ but obviously diverges at both end points $x = \pm 1$. An interesting and often useful result is that if a real power series with radius of convergence R converges at the end point R then its sum actually is a continous function in the closed interval $[0, R]$, and correspondingly of course at the other end point.

Theorem 3.54 (Abel's Theorem). *If a real power series $f(x) = \sum_{n=0}^{\infty} c_n x^n$ converges at a point $x = R > 0$ then it converges uniformly in the interval $[0, R]$ so its sum is a continuous function on the interval $[0, R]$ and consequently*

$$f(R) = \lim_{x \to R} f(x). \tag{3.86}$$

Proof: To simplify the notation assume $R = 1$. Since the series converges at $x = 1$, if $A_N = c_0 + c_1 + \cdots + c_N$ are the partial sums at $x = 1$ then $f(1) = \lim_{N \to \infty} A_N$; hence for any $\epsilon > 0$ there is a number M such that

$$|f(1) - A_N| < \epsilon \quad \text{whenever} \quad N > M. \tag{3.87}$$

Now following Abel set

$$c_n x^n = (A_n - A_{n-1})x^n = (f(1) - A_{n-1})x^n - (f(1) - A_n)x^n$$

for all n, with the convention that $A_{-1} = 0$; then

$$\sum_{n=\mu}^{\nu} c_n x^n = \sum_{n=\mu}^{\nu}(f(1) - A_{n-1})x^n - \sum_{n=\mu}^{\nu}(f(1) - A_n)x^n$$

$$= \sum_{n=\mu-1}^{\nu-1} (f(1) - A_n)x^{n+1} - \sum_{n=\mu}^{\nu}(f(1) - A_n)x^n$$

$$= \sum_{n=\mu}^{\nu-1} \big(f(1) - A_n\big)(x^{n+1} - x^n)$$

$$+ \big(f(1) - A_{\mu-1}\big)x^{\mu} - \big(f(1) - A_{\nu}\big)x^{\nu}$$

so that

$$\left| \sum_{n=\mu}^{\nu} c_n x^n \right| \leq \sum_{n=\mu}^{\nu-1} \big|f(1) - A_n\big| \big|x^{n+1} - x^n\big| + \big|f(1) - A_{\mu-1}\big| \big|x^{\mu}\big| + \big|f(1) - A_{\nu}\big| \big|x^{\nu}\big|.$$

$$(3.88)$$

If $\mu > M + 1$ then $|f(1) - A_n| < \epsilon$ and $|f(1) - A_{\mu-1}| < \epsilon$ and $|f(1) - A_{\nu}| < \epsilon$ by (3.87); and since $0 \leq x \leq 1$ it follows that $|x| \leq 1$ and $|x^{n+1} - x^n| = x^n - x^{n+1}$ so by the obvious cancellation of terms

$$\sum_{n=\mu}^{\nu-1} \big|x^{n+1} - x^n\big| = \sum_{n=\mu}^{\nu-1} (x^n - x^{n+1}) = x^{\mu} - x^{\nu} \leq 1.$$

Substituting these observations in (3.88) shows that

$$\left| \sum_{n=\mu}^{\nu} c_n x^n \right| \leq \epsilon \sum_{n=\mu}^{\nu-1} \big|x^{n+1} - x^n\big| + \epsilon |x|^{\mu} + \epsilon |x|^{\nu} \leq 3\epsilon,$$

so the partial sums of the series $\sum_{n=0}^{\infty} c_n x^n$ are a uniformly Cauchy sequence in the interval $[0, 1]$ hence the series converges uniformly in that interval and therefore its sum is continuous in that interval. That suffices for the proof.

For a complex power series $\sum_{n=0}^{\infty} c_n z^n$ with the radius of convergence R the preceding theorem can be applied to the restriction of that power series to a straight line segment from the origin in the direction of a complex number ζ with $|\zeta| = 1$, that is, to the series $\sum_{n=0}^{\infty} c_n \zeta^n r^n$ viewed as a power series in the real variable r. If that power series converges at the end point $r = R$ then by the preceding theorem $f(\zeta R) = \lim_{r \to R} f(\zeta r)$; in that sense the sum $f(z)$ converges to its boundary value along each radius of the circle of convergence. That does not imply that the function $f(z)$ actually converges to its boundary value as the point z approaches the boundary point ζ in any other way; the situation is rather more complicated, and is examined rather extensively in the more detailed study of holomorphic functions. On the other hand if a real power series converges in an interval $-R < x < R$ then its sum is a continuously differentiable function in that interval, and its derivative is given by a power series that converges in the same interval.

Theorem 3.55. *If $f(x) = \sum_{n=0}^{\infty} c_n x^n$ in the interval $-R < x < R$ then $f(x)$ is a differentiable function in that interval and its derivative has the power series expansion*

$$f'(x) = \sum_{n=1}^{\infty} n c_n x^{n-1} \quad for \quad -R < x < R \tag{3.89}$$

in which

$$c_n = f^{(n)}(0)/n! \quad for\,all \quad n. \tag{3.90}$$

Proof: The individual terms $c_n x^n$ in the power series can be differentiated, and their derivatives can be taken to define the formal power series

$$\sum_{n=1}^{\infty} n c_n x^{n-1} = \sum_{n=0}^{\infty} (n+1) c_{n+1} x^n. \tag{3.91}$$

Since the original power series converges at the point $x = r$ for any value $r \in (0, R)$ it follows that $\lim_{n\to\infty} |c_n| r^n = 0$ hence that $|c_n| r^n < C$ for some constant $C > 0$ and all n. Then for any value $r_1 \in (0, r)$

$$(n+1)|c_{n+1}| r_1^n = \frac{1}{r}(n+1)|c_{n+1}| r^{n+1} \left(\frac{r_1}{r}\right)^n \leq \frac{1}{r} C(n+1) \epsilon^n \tag{3.92}$$

where $\epsilon = r_1/r < 1$. The power series $\sum_{n=0}^{\infty} \frac{1}{r} C(n+1) \epsilon^n$ converges by the ratio test, since

$$\lim_{n\to\infty} \frac{(n+2)\epsilon^{n+1}}{(n+1)\epsilon^n} = \lim_{n\to\infty} \frac{n+2}{n+1} \epsilon = \epsilon < 1.$$

In view of (3.92) the comparison test shows that the power series (3.91) also converges at $x = r_1$, hence it converges uniformly in the interval $(-r_1, r_1)$. It then follows from Theorem 3.39 that this series converges to the derivative $f'(x)$ of the function $f(x)$. That is the case for any $r_1 < R$, so the function $f(x)$ is differentiable in the full interval $(-R, R)$ and its derivative has the power series expansion (3.89). It is easy to see by induction on k that

$$f^{(k)}(x) = \sum_{n=k}^{\infty} n(n-1)\cdots(n-k+1)\, c_n x^{n-k} \quad for\,any\,\, k \geq 1. \tag{3.93}$$

Indeed (3.89) is the case $k = 1$; and applying the preceding theorem to the series (3.93) shows that

$$f^{(k+1)}(x) = \sum_{n=k+1}^{\infty} n(n-1)\cdots(n-k+1)(n-k)\, c_n x^{n-k-1},$$

which is equation (3.93) for the case $k + 1$. Setting $x = 0$ in (3.93) yields (3.90), and that suffices for the proof.

Corollary 3.56. *If $f(x) = \sum_{n=0}^{\infty} c_n x^n$ in the interval $-R < x < R$ then $f(x)$ is a C^{∞} function in that interval.*

Proof: The preceding theorem shows that the derivative of any function $f(x)$ given by a power series is again a differentiable function given by a power series; the theorem can be applied to the derivative, showing that it too is a differentiable function given by a power series, and by induction it follows that the function $f(x)$ has derivatives of all orders. That suffices for the proof.

Any C^{∞} function $f(x)$ has a unique formal Taylor series (3.67); but that power series is not necessarily convergent,[4] and even if it is convergent its sum is not necessarily equal to the function $f(x)$. One of the problems at the end of this section leads to an example of a nontrivial C^{∞} function $f_0(x)$ for which $f_0^{(n)}(0) = 0$ for all $n \geq 0$; so although the Taylor series converges quite trivially in that case, its sum is not the function $f_0(x)$ but rather the function that is identically zero. If $f(x)$ is any C^{∞} function then the function $f(x) + f_0(x)$ is another C^{∞} function with the same Taylor expansion as $f(x)$. A real analytic function in a region U was defined earlier as a C^{∞} function in U that is equal to its Taylor series in a neighborhood of each point of U. The sum of any convergent power series in a region U is actually a real analytic function in U, as a consequence of the following observation.

Theorem 3.57. *If $f(x) = \sum_{n=0}^{\infty} c_n x^n$ in the interval $-R < x < R$ then for any point $a \in (-R, R)$ the function $f(a + x)$ has a power series expansion*

$$f(a + x) = \sum_{n=0}^{\infty} b_n x^n \quad where \quad b_n = f^{(n)}(a)/n! \; ; \tag{3.94}$$

and this power series converges absolutely in the interval $|x| < R - |a|$.

Proof: The function $f(a + x)$ has the power series expansion

$$f(a + x) = \sum_{n=0}^{\infty} c_n(a + x)^n, \tag{3.95}$$

---

[4]A theorem of Émile Borel, in "Sur quelques points de la théorie des functions," *Ann. Sci. L'École Norm Sup.* 12 (1895): 9–55, asserts that for any real sequence c_ν there is a C^{∞} function $f(x)$ for which $f^{(\nu)}(0) = c_\nu \nu!$; so any power series is the Taylor series of some C^{∞} function at the origin.

and this power series converges absolutely in the region $|a + x| < R$ hence in the interval $|x| < R - |a|$. For values $a > 0, x > 0$ the corresponding power series

$$\widetilde{f}(x) = \sum_{n=0}^{\infty} |c_n|(a + x)^n \tag{3.96}$$

is a convergent series of nonnegative numbers; and when the binomial expansion

$$|c_n|(a + x)^n = \sum_{m=0}^{n} |c_n| \binom{n}{m} a^{n-m} x^m,$$

which is a finite sum of nonnegative numbers, is substituted into the series (3.96) the effect is that of replacing each of the nonnegative terms in the convergent series by a finite sum of nonnegative terms, so the resulting series is a convergent series

$$\widetilde{f}(x) = \sum_{n=0}^{\infty} \sum_{m=0}^{n} |c_n| \binom{n}{m} a^{n-m} x^m$$

$$= |c_0| \binom{0}{0} + |c_1| \binom{1}{0} a + |c_1| \binom{1}{0} x + |c_2| \binom{2}{0} a^2$$

$$+ |c_2| \binom{2}{1} ax + |c_2| \binom{2}{2} x^2 + \cdots \tag{3.97}$$

of nonnegative terms. It follows that the corresponding series

$$f(a + x) = \sum_{n=0}^{\infty} \sum_{m=0}^{n} c_n \binom{n}{m} a^{n-m} x^m$$

$$= c_0 \binom{0}{0} + c_1 \binom{1}{0} a + c_1 \binom{1}{0} x + c_2 \binom{2}{0} a^2$$

$$+ c_2 \binom{2}{1} ax + c_2 \binom{2}{2} x^2 + \cdots \tag{3.98}$$

is an absolutely convergent series. This can be viewed as an absolutely convergent iterated sum of a double series, since the binomial coefficients $\binom{n}{m}$ vanish for $n < m$; and since by Theorem 3.47 (ii) the order of summation of an absolutely convergent double series can be reversed without changing the sum it follows that

$$f(a+x) = \sum_{m=0}^{\infty} \sum_{n=m}^{\infty} c_n \binom{n}{m} a^{n-m} x^m = \sum_{m=0}^{\infty} b_n x^n$$

where

$$b_m = \sum_{n=m}^{\infty} c_n \binom{n}{m} a^{n-m},$$

and that suffices for the proof.

Some of the most common functions in everyday use are defined by power series expansions so are real analytic functions; some examples will indicate the role that the power series expansions have in examining these functions. Consider first the problem of finding differentiable functions $h(x)$ of a real variable x, if there are any, that are solutions of the differential equation $h'(x) = h(x)$, one of the most basic of differential equations. If there is such a function and it is normalized by requiring that $h(0) = 1$ then all its derivatives also satisfy $h^{(n)}(0) = 1$ so it has the Taylor series $h(x) \sim \sum_{n=0}^{\infty} x^n/n!$ at the origin. It follows immediately from the ratio test that this power series has a radius of convergence ∞, hence it converges to a C^∞ function

$$e(x) = \sum_{n=0}^{\infty} \frac{1}{n!} x^n \quad \text{for all} \quad x \in \mathbb{R}. \tag{3.99}$$

By Theorem 3.55 this power series can be differentiated term-by-term; and its derivative is easily seen to be the same power series, so $e'(x) = e(x)$. Thus the function $e(x)$ defined by the power series expansion (3.99) is a C^∞ function on $\mathbb{R}$ that satisfies the differential equation $e'(x) = e(x)$ and the normalizing condition that $e(0) = 1$; it is called the **exponential function**.

It follows from the series expansion (3.99) that $e(x) \geq 1 + x$ for all points $x \geq 0$, so in particular $e(x) > 0$ for all points $x > 0$. The function $f(x) = e(x + a)e(-x)$ is differentiable and clearly $f'(x) = 0$, so actually $f(x)$ is a constant; since $f(-a) = e(a)$ it follows that $e(x + a)e(-x) = f(-a) = e(a)$, and in particular for $a = 0$ since $e(0) = 1$

$$e(x)e(-x) = 1 \quad \text{for all } x \in \mathbb{R} \tag{3.100}$$

so

$$e(x) > 0 \quad \text{and} \quad e(x)^{-1} = e(-x) \quad \text{for all } x \in \mathbb{R}. \tag{3.101}$$

Then from the explicit form of the function $f(x)$ for $a = y$ and the preceding equation it follows that

$$e(x + y) = e(x)e(y) \quad \text{for all } x, y \in \mathbb{R}. \tag{3.102}$$

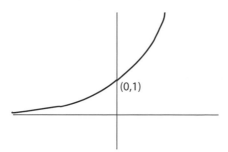

Figure 3.1. The exponential function

A simple consequence of the preceding observations is that $e(x)$ is the only differentiable function on $\mathbb{R}^1$ satisfying $e'(x) = e(x)$ and $e(0) = 1$. Indeed if $f(x)$ is a differentiable function on $\mathbb{R}$ such that $f'(x) = f(x)$ and and $f(0) = 1$ then $g(x) = f(x)e(-x)$ is also a differentiable function on $\mathbb{R}$ and $g'(x) = f'(x)e(-x) - f(x)e'(-x) = 0$ for all $x \in \mathbb{R}$, so $g(x) = c$ is a constant; in particular $c = g(0) = f(0)e(0) = 1$ hence $f(x) = g(x)$. Moreover since $e(x) > 1$ for $x > 0$ it follows from (3.101) that $e(x) < 1$ for $x < 0$; and since $e'(x) = e(x) > 0$ the exponential function is a monotonic increasing function on $\mathbb{R}$, as sketched in the accompanying Figure 3.1. All the terms in the series expansion (3.99) are positive so for $n > 0$ and $x > 0$

$$\frac{e(x)}{x^n} \geq \frac{x^{n+1}}{x^n(n+1)!} = \frac{x}{(n+1)!} \quad \text{hence} \quad \lim_{x \to \infty} e(x)/x^n = \infty \qquad (3.103)$$

showing that the function $e(x)$ increases faster than any power of x as x tends to ∞; equivalently of course $\lim_{x \to \infty} x^n e(x)^{-1} = 0$, showing that the function $e(-x) = e(x)^{-1}$ tends to 0 faster than any power x^{-n} as x tends to ∞.

The value $e(1)$ is an important natural invariant, usually denoted just by e; its explicit value can be calculated approximately, to any desired degree of accuracy, from the power series expansion (3.99), so that for instance

$$e = 2.71828\,18284\,59045\,23536\,02874\ldots \qquad (3.104)$$

with accuracy up to 25 decimal places.[5] Equation (3.101) implies that $e(n) = e(1 + 1 + \cdots + 1) = e(1)^n = e^n$ for any positive integer n; and further it implies that $e(1) = e\left(\frac{1}{n} + \frac{1}{n} + \cdots + \frac{1}{n}\right) = e(\frac{1}{n})^n$ so $e(\frac{1}{n}) = e^{\frac{1}{n}}$, the positive n-th root of the real number e; therefore actually $e(r) = e^r$ for any rational number r. Any real number α can be written as a limit $\alpha = \lim_{n \to \infty} r_n$ of rational numbers r_n, and by the continuity of the exponential function $e(\alpha) = \lim_{n \to \infty} e(r_n) = \lim_{n \to \infty} e^{r_n}$; for this reason the exponential function is customarily denoted by $e^x = e(x)$ for

---

[5]See the *On-Line Encyclopedia of Integer Sequences* (*OEIS*), also cited as *Sloane's Encyclopedia of Integer Sequences*, number A001113, for more extensive approximations.

any real number $x \in \mathbb{R}$. For some purposes, such as considering the composition of functions, it may be clearer to use the notation $e(x)$ though; so both will be used as convenient.

Since the continuous function $e(x)$ is monotonically strictly increasing while $\lim_{x \to -\infty} e(x) = 0$ and $\lim_{x \to \infty} e(x) = \infty$ it follows that the mapping $x \longrightarrow e(x)$ is a homeomorphism $e : \mathbb{R} \longrightarrow \mathbb{R}^+$, where $\mathbb{R}^+ = \{ x \in \mathbb{R} \mid x > 0 \}$; consequently that mapping has a well-defined inverse mapping $\mathbb{R}^+ \longrightarrow \mathbb{R}$, which is called the **logarithm function** and is usually denoted by $\log y$. Thus the continuous real-valued function $\log y$ defined for all $y \in \mathbb{R}^+$ is characterized by the conditions that

$$\log e^x = x \ \text{ for all } x \in \mathbb{R} \ \text{ and } \ e^{\log y} = y \ \text{ for all } y > 0; \tag{3.105}$$

and $\log 1 = 0$ since $e(0) = 1$. If $e(x) = y$ and $e(x + h) = y + k$ then $\log y = x$ and $\log(y + k) = x + h$ hence

$$\frac{\log(y + k) - \log y}{k} = \frac{h}{e(x + h) - e(x)};$$

and since by continuity h tends to 0 as k tends to 0 it follows in the limit that the function $\log y$ is differentiable and

$$\log' y = \frac{1}{e'(x)} = \frac{1}{e(x)} = \frac{1}{y} = y^{-1} \quad \text{for all} \quad y > 0. \tag{3.106}$$

Of course if it is assumed or proved that the function $\log y$ is differentiable the preceding result can be obtained just by differentiating the second identity in (3.105). Equation (3.106) shows that $\log' y$ is a differentiable function, indeed actually is a C^∞ function, hence so is the function $\log y$ itself.

It is usually convenient at least notationally to discuss the local behavior of functions at the origin; so for the local behavior of the logarithm function it is customary to focus on the function $\log(1 + x)$, which is a C^∞ function in a neighborhood of the origin. Applying the chain rule to (3.106) shows that

$$\frac{d}{dy} \log(1 + y) = (1 + y)^{-1}; \tag{3.107}$$

and differentiating this identity further it follows by the obvious induction that

$$\frac{d^n}{dy^n} \log(1 + y) = (-1)^{n-1} \frac{(n - 1)!}{(1 + y)^n}, \tag{3.108}$$

hence there results the Taylor series

$$\log(1 + y) \sim \sum_{n=1}^{\infty} (-1)^{n-1} \frac{1}{n} y^n = y - \frac{1}{2} y^2 + \frac{1}{3} y^3 - \cdots . \qquad (3.109)$$

This power series converges in the interval $-1 < y \le 1$ and diverges for $y = -1$, by comparison with the alternating harmonic series (3.73) and the harmonic series (3.75), so its radius of convergence is $R = 1$; and the derivative of this power series in the interval $-1 < y < 1$ is

$$\frac{d}{dy} \sum_{n=1}^{\infty} (-1)^{n-1} \frac{1}{n} y^n = \sum_{n=1}^{\infty} (-1)^{n-1} y^{n-1} = \frac{1}{1+y},$$

which is summed as a geometric series. Thus the power series appearing in (3.109) and the function $\log(1 + y)$ have the same derivative, and consequently the power series appearing in (3.109) is equal to the function $\log(1 + y)$ near the origin, aside from an additive constant. The power series and the function $\log(1 + y)$ both take the value 0 when $y = 0$ and therefore the power series actually is equal to $\log(1 + y)$ so that

$$\log(1 + y) = \sum_{n=1}^{\infty} (-1)^{n-1} \frac{1}{n} y^n = y - \frac{1}{2} y^2 + \frac{1}{3} y^3 - \cdots \quad \text{for } |y| < 1. \qquad (3.110)$$

Since the power series converges at $y = 1$ it follows from Abel's Theorem, Theorem 3.54, that $\log 2 = \lim_{y \to 1} \log(1 + y)$ is the value of the power series at $y = 1$, yielding one of the classical formulas

$$\log 2 = 1 - \frac{1}{2} + \frac{1}{3} - \frac{1}{4} + \frac{1}{5} - \frac{1}{6} + \cdots . \qquad (3.111)$$

As for the logarithm function itself, if $e(h) = x$ and $e(k) = y$ then $h = \log x$ and $k = \log y$ so $e(h + k) = e(h)e(k) = xy$ hence

$$\log(xy) = \log x + \log y \quad \text{for all} \quad x, y > 0. \qquad (3.112)$$

From this it follows by induction on n that $\log x^n = n \log x$, and upon setting $y = x^n$ it then follows that $\log y^{1/n} = \frac{1}{n} \log y$; altogether then $\log x^\alpha = \alpha \log x$ for any rational $\alpha > 0$ and any $x > 0$, and consequently $x^\alpha = e(\log x^\alpha) = e(\alpha \log x)$ for any $x > 0$ and any rational $\alpha > 0$, which also holds for rational $\alpha < 0$ by noting that $\log(x^{-1}) = -\log x$ for $x > 0$ as a consequence of (3.112). In view of this it is customary to define real powers of positive real numbers by

$$x^y = e(y \log x) = e^{y \log x} \quad \text{for all} \quad x, y \in \mathbb{R}, \ x > 0. \qquad (3.113)$$

Rational powers of positive rationals can be defined by the algebraic operations of powers and roots, so since any real numbers $x, y \in \mathbb{R}$ can be expressed as limits $x = \lim_{m \to \infty} \alpha_m$ and $y = \lim_{n \to \infty} \beta_n$ for rational numbers α_m, β_n then by continuity $x^y = \lim_{m,n \to \infty} (\alpha_m)^{\beta_n}$ so the formal definition (3.113) is compatible with the usual notion for rational numbers. The function of two variables x^y is a differentiable function in both variables, as the composition $x^y = e(y \log x)$ of the differentiable exponential and logarithm functions; and it follows in a straightforward manner from the chain rule for differentiation that

$$\frac{\partial}{\partial x} x^y = y x^{y-1} \quad \text{and} \quad \frac{\partial}{\partial y} x^y = x^y \log x. \tag{3.114}$$

As an occasionally useful application of the relation between the exponential and logarithm functions, the exponential function can be described alternatively by

$$e^x = \lim_{n \to \infty} \left(1 + \frac{x}{n} \right)^n \quad \text{for all } x \in \mathbb{R}. \tag{3.115}$$

Indeed $\lim_{n \to \infty} \left(1 + \frac{x}{n} \right)^n = \lim_{n \to \infty} e \left(n \log(1 + \frac{x}{n}) \right) = e(\lim_{n \to \infty} n \log(1 + \frac{x}{n}))$ since the exponential is a continuous function; and if $t = 1/n$ then by L'Hôpital's rule

$$\lim_{n \to \infty} n \log \left(1 + \frac{x}{n} \right) = \lim_{t \to 0} \frac{\log(1 + tx)}{t} = \lim_{t \to 0} \frac{x}{(1 + tx)} = x,$$

where the limit exists for all $x \in \mathbb{R}$.

The function e^x is defined by a Taylor series that converges absolutely for all real values of the variable x. The series also converges absolutely for all complex values z of the variable, since $|z| \in \mathbb{R}$, so by Theorem 3.43 it converges for all $z \in \mathbb{C}$, thus providing an extension of the exponential function to a complex-valued function e^z of the complex variable $z \in \mathbb{C}$. In the identity $e^{x+y} = e^x \, e^y$ the partial sums of the Taylor series of the two sides are equal polynomials so they have the same value when the variables x, y take complex values; therefore

$$e^{w+z} = e^w \, e^z \quad \text{for all} \quad w, z \in \mathbb{C}. \tag{3.116}$$

When restricted to complex numbers of the form $z = it$ for real values t, the real part of e^{it} is a well-defined function usually denoted by $\cos t$ and called the **cosine** function, while the imaginary part of e^{it} is a well-defined function usually denoted by $\sin t$ and called the **sine** function; thus

$$e^{it} = \cos t + i \sin t. \tag{3.117}$$

When the odd and even terms in the Taylor series (3.99) are separated it follows that

$$e^{it} = \sum_{n=0}^{\infty} \frac{(it)^n}{n!} = \sum_{k=0}^{\infty} \frac{(it)^{2k}}{(2k)!} + \sum_{k=0}^{\infty} \frac{(it)^{2k+1}}{(2k+1)!} \tag{3.118}$$

and consequently that

$$\cos t = \sum_{k=0}^{\infty} \frac{(-1)^k t^{2k}}{(2k)!} = 1 - \frac{t^2}{2!} + \frac{t^4}{4!} - \frac{t^6}{6!} + \cdots \tag{3.119}$$

and

$$\sin t = \sum_{k=0}^{\infty} \frac{(-1)^k t^{2k+1}}{(2k+1)!} = t - \frac{t^3}{3!} + \frac{t^5}{5!} - \frac{t^7}{7!} + \cdots . \tag{3.120}$$

Upon differentiating the preceding two Taylor series or alternatively differentiating (3.117) it follows readily that

$$\cos' t = -\sin t \quad \text{and} \quad \sin' t = \cos t. \tag{3.121}$$

These functions are differentiable in turn, so actually both the sine and cosine are $\mathcal{C}^{\infty}$ functions; and by differentiating again it follows that they satisfy the differential equations

$$\sin'' t = -\sin t \quad \text{and} \quad \cos'' t = -\cos t. \tag{3.122}$$

It is an interesting exercise to show that any $\mathcal{C}^{\infty}$ function $f(t)$ that is a solution of the differential equation $f''(t) = -f(t)$ must be of the form $f(t) = a \sin t + b \cos t$ for some real constants a, b.

Since the coefficients of the Taylor series of the exponential function are real numbers it follows that $\overline{e(it)} = e(-it)$, and consequently $|e(it)|^2 = e(it)\overline{e(it)} = e(it)e(-it) = e(it - it) = 1$ so that

$$|e^{it}|^2 = \sin^2 t + \cos^2 t = 1 \quad \text{for all} \quad t \in \mathbb{R}; \tag{3.123}$$

thus the complex numbers e^{it} for all $t \in \mathbb{R}$ lie on the unit circle $|z| = 1$ in the complex plane, as sketched in Figure 3.2. That identifies the sine and cosine with the familiar trigonometric interpretations as the lengths of the sides of a right triangle with hypotenuse of length 1. From the power series (3.119) it follows that $\cos 0 = 1$ and $\sin 0 = 0$, and from (3.121) it follows that $\sin' 0 = \cos 0 = 1$ so that $\sin t$ is an increasing function of t near the origin; geometrically that means that e^{it} is the point $(1, 0)$ when $t = 0$ and as t increases the imaginary part $\sin t$ increases hence the real part $\cos t$ decreases. That continues

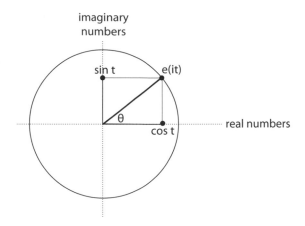

Figure 3.2. $e(iy) = c(y) + i\, s(y)$

until the first value of t at which $\cos t = 0$ so that $e^{it} = (0, 1)$; that particular value of t is customarily denoted by $\pi/2$, so that $e^{\pi i/2} = i$ while $\sin \pi/2 = 1$ and $\cos \pi/2 = 0$. Then since $e^{i(t+\pi/2)} = e^{it}\, e^{\pi i/2} = ie^{it}$ the path traversed by $e(it)$ for $\pi/2 \le t \le \pi$ is just the product of the path traversed by e^{it} for $0 \le t \le \pi/2$ multiplied by i, so is the next quadrant of the circle in Figure 3.2. Continuing that argument shows that the point e^{it} traverses the full circle in Figure 3.2 in the counterclockwise direction as t runs through the interval $0 \le t \le 2\pi$, so that $e^{2\pi i} = 1$ and consequently

$$e^{z+2\pi i} = e^z\, e^{2\pi i} = e^z \quad \text{for all} \quad z \in \mathbb{C}, \tag{3.124}$$

so as t increases further the point e^{it} continues to traverse the unit circle, once in each interval of length 2π in the parameter. Thus the exponential function is a **periodic function** of the complex variable with the period $2\pi i$, so its real and imaginary parts satisfy

$$\sin(z + 2\pi) = \sin z \quad \text{and} \quad \cos(z + 2\pi) = \cos z \quad \text{for all} \quad z \in \mathbb{C}. \tag{3.125}$$

For an arbitrary complex number $z = s + it$ it follows from the functional equation (3.116) that $e^{s+it} = e^s\, e^{it}$ for all $s, t \in \mathbb{R}$; thus for any fixed s the point e^{s+it} traverses a circle of radius $e^s > 0$ as the paramter t varies on the real line, while for any fixed t the point e^{s+it} traverses a straight line segment from (but not including) the origin through the point e^{it} on the unit circle and then ever onwards. The mapping $e : z \longrightarrow e^z$ for $z \in \mathbb{C}$ thus describes a mapping from the complex plane $\mathbb{C}$ onto the complement of the origin in the complex plane $\mathbb{C}$, where the points $z + 2\pi i\, n$ for all $n \in \mathbb{Z}$ have the same image but the otherwise distinct points have distinct images. The inverse mapping $\log : \mathbb{C}^\times \longrightarrow \mathbb{C}$, the natural extension of the logarithm to a complex-valued function of a complex variable, then is a multiple-valued mapping from the complement of the origin in $\mathbb{C}$ onto the entire complex plane $\mathbb{C}$, which is locally single-valued but with the property that when a point $z \in \mathbb{C}^{\times}$ continues along a path that encircles the origin once in a counterclockwise direction and then returns to the same location the image $\log z$ changes to $\log z + 2\pi i$. For many purposes it is convenient to describe points in the complex plane in the form $z = e^{s+it}$ for real numbers s, t, the analogue of polar coordinates in the plane; traditionally $|z| = e^s$ is called the **modulus** of the complex number $z = e^{s+it}$ and t is called the **argument** of z and often is viewed as an angle. With this notation the operation of raising the complex number $z = e^{s+it}$ to a power $z^n = e^{ns+int}$ for some integer $n \in \mathbb{Z}$ is to replace the modulus $|z| = e^s$ by its n-th power $|z|^n = e^{ns}$ and to replace the argument t by its n-th multiple nt. Thus a circle $|z| = r$ centered at the origin is replaced by the circle $|z| = r^n$ and the argument of points in the circle is replaced by n times the argument, so the mapping $z \longrightarrow z^n$ maps each circle centered at the origin n times to its image. Thus any point $z \neq 0$ is the image of n distinct points under the mapping $z \longrightarrow z^n$, while of course the origin is mapped to itself; so every nonzero complex number has precisely n distinct n-th roots.

Another power series expansion that arises quite commonly merits consideration here, as another basic function in analysis. For any real number α the **binomial function**

$$(1 + t)^\alpha = e^{\alpha \log(1+t)} \tag{3.126}$$

is a well-defined function of the variable $t \in \mathbb{R}$ so long as $1 + t > 0$, hence in the interval $-1 < t < \infty$. From the chain rule and the formulas for the derivatives of the exponential and logarithm function it follows readily that $\frac{d}{dt}(1+t)^\alpha = \alpha(1+t)^{\alpha-1}$; iterating this formula shows by induction on n that

$$\frac{d^n}{dt^n}(1+t)^\alpha = \alpha(\alpha-1)\cdots(\alpha-n+1)(1+t)^{\alpha-n} \quad \text{for all} \quad n \geq 1. \tag{3.127}$$

The Taylor series associated to this function hence is given by

$$(1+t)^\alpha \sim \sum_{n=0}^{\infty} \frac{\alpha(\alpha-1)\cdots(\alpha-n+1)}{n!} t^n. \tag{3.128}$$

The coefficients in this Taylor series are called the **binomial coefficients** and usually denoted by

$$\binom{\alpha}{n} = \frac{\alpha(\alpha-1)\cdots(\alpha-n+1)}{n!} \quad \text{for any real } \alpha; \tag{3.129}$$

if α is an integer this reduces to the usual binomial coefficient

$$\binom{\alpha}{n} = \frac{\alpha!}{(\alpha - n)!n!}. \tag{3.130}$$

With this notation the Taylor series (3.128) can be written equivalently as

$$(1+t)^\alpha \sim \sum_{n=0}^{\infty} \binom{\alpha}{n} t^n. \tag{3.131}$$

If α is a nonnegative integer the function (3.126) is a polynomial, as is the Taylor series since the binomial coefficients are zero whenever $n \geq \alpha + 1$; and the Taylor series is a classical polynomial identity. If α is not a positive integer however the binomial series is an infinite series; and that raises the question of its radius of convergence.

Theorem 3.58. *For any real number α the binomial series (3.128) converges absolutely in the open interval $(-1, 1)$; and if $\alpha > 0$ it converges absolutely and uniformly in the closed interval $[-1, 1]$ as well.*

Proof: Setting $c_n = \binom{\alpha}{n}$ for convenience, it follows directly from the definition of the binomial coefficients in (3.129) that

$$\left| \frac{c_{n+1} t^{n+1}}{c_n t^n} \right| = \frac{|\alpha - n|}{n+1} |t|,$$

and since $\lim_{n \to \infty} \frac{|\alpha - n|}{n+1} = 1$ for all real α it follows from the ratio test that the series converges absolutely for $-1 < t < 1$. For the special case that $|t| = 1$ and $\alpha > 0$ it follows from the preceding equation that

$$\left| \frac{c_{n+1}}{c_n} \right| = \frac{n - \alpha}{n+1} \quad \text{if} \quad n > \alpha \tag{3.132}$$

hence

$$\lim_{n \to \infty} n \left(1 - \frac{n - \alpha}{n+1} \right) = \lim_{n \to \infty} \frac{n(1 + \alpha)}{n+1} = 1 + \alpha > 1$$

so by Raabe's test the series converges at $t = \pm 1$; and by Abel's Theorem, Theorem 3.54, it converges uniformly in the closed interval $[-1.1]$, which suffices for the proof.

It follows from the preceding theorem that for all real α the Taylor series (3.131) converges to some function $f(t)$ for $|t| < 1$; but that is not enough yet to show that $f(t) = (1 + t)^\alpha$. However substituting the result that

$$(n + 1)\binom{\alpha}{n + 1} = (n + 1)\frac{\alpha(\alpha - 1) \cdots (\alpha - n)}{1 \cdot 2 \cdots \cdots (n + 1)} = (\alpha - n)\binom{\alpha}{n},$$

into the derivative of the power series expansion of $f(t)$ shows that

$$f'(t) = \sum_{n=1}^{\infty} n\binom{\alpha}{n}t^{n-1} = \sum_{n=0}^{\infty}(n + 1)\binom{\alpha}{n + 1}t^n = \sum_{n=0}^{\infty}(\alpha - n)\binom{\alpha}{n}t^n = \alpha f(t) - tf'(t)$$

so that $(1 + t)f'(t) = \alpha f(t)$ and consequently

$$\frac{d}{dt}\log f(t) = \frac{f'(t)}{f(t)} = \frac{\alpha}{1 + t} = \alpha\frac{d}{dt}\log(1 + t) = \frac{d}{dt}\log(1 + t)^\alpha$$

which implies that $f(t) = c(1 + t)^\alpha$ for some constant c; and setting $t = 0$ shows that $1 = c$, so altogether there is the power series expansion

$$(1 + t)^\alpha = \sum_{n=0}^{\infty}\binom{\alpha}{n}t^n \quad \text{for} \quad -1 < t < 1, \tag{3.133}$$

and if $\alpha > 0$ it converges uniformly in the closed interval $[-1, 1]$.

For an interesting application of the preceding observations, the sine function $\sin x$ is a monotonically strictly increasing function in the interval $(-\frac{\pi}{2}, \frac{\pi}{2})$, varying from $\sin(-\pi/2) = -1$ to $\sin(\pi/2) = +1$. It follows that it has a well defined inverse function $f(x)$ in the interval $(-1, +1)$, for which $f(\sin x) = x$ and $f(0) = 0$. The Inverse Mapping Theorem, to be discussed in Section 5.1, shows that this function f also is differentiable; it then follows from the chain rule that $f'(\sin x) \cos x = 1$, so setting $\sin x = t$ since $\cos^2 x = 1 - \sin^2 x$ it follows that

$$f'(t) = (1 - t^2)^{-1/2} \quad \text{and} \quad f(0) = 0. \tag{3.134}$$

Therefore by the binomial expansion

$$f'(t) = \sum_{n=0}^{\infty}\binom{-\frac{1}{2}}{n}(-t^2)^n$$

and since $f(0) = 0$ the function $f(t)$ itself is given by

$$f(t) = \sum_{n=0}^{\infty}(-1)^n\binom{-\frac{1}{2}}{n}\frac{t^{2n+1}}{2n + 1}$$

$$= t + \frac{1}{2} \cdot \frac{t^3}{3} + \frac{1 \cdot 3}{2 \cdot 4} \cdot \frac{t^5}{5} + \frac{1 \cdot 3 \cdot 5}{2 \cdot 4 \cdot 6} \cdot \frac{t^7}{7} + \cdots. \tag{3.135}$$

This function is called the **arcsine function**, and is denoted either by arcsin t or $\sin^{-1} t$. This series also converges at $t = 1$ by Raabe's test, and since arcsin $1 = \frac{\pi}{2}$ an application of Abel's Theorem, Theorem 3.54, yields the amusing formula

$$\frac{\pi}{2} = 1 + \frac{1}{2} \cdot \frac{1}{3} + \frac{1 \cdot 3}{2 \cdot 4} \cdot \frac{1}{5} + \frac{1 \cdot 3 \cdot 5}{2 \cdot 4 \cdot 6} \cdot \frac{1}{7} + \cdots . \tag{3.136}$$

There are of course many other applications of the binomial series.

There are a number of representations of functions by series other than power series, and these other representations also play important roles in mathematics and its applications. One example is the generalized harmonic series.

Theorem 3.59. *The generalized harmonic series $\sum_{n=1}^{\infty} \frac{1}{n^{\alpha}}$ for a real number α converges if $\alpha > 1$ and diverges if $\alpha \leq 1$. The generalized harmonic series $\sum_{n=1}^{\infty} \frac{1}{n^z}$ for a complex number $z = x + iy \in \mathbb{C}$ is absolutely convergent for $x > 1$ and divergent if $x < 1$.*

Proof: When $\alpha = 1$ the generalized harmonic series reduces to the usual harmonic series, which then diverges; and since $\frac{1}{n^{\alpha}} > \frac{1}{n}$ if $\alpha < 1$ the generalized harmonic series diverges for $\alpha < 1$ by the comparison test. If $c_n = \frac{1}{n^{\alpha}}$ then setting $t = \frac{1}{n}$ and applying L'Hôpital's rule shows that

$$\lim_{n \to \infty} n \left(1 - \frac{c_{n+1}}{c_n} \right) = \lim_{n \to \infty} n \frac{(n+1)^{\alpha} - n^{\alpha}}{(n+1)^{\alpha}} = \lim_{t \to 0} \frac{(1+t)^{\alpha} - 1}{t(1+t)^{\alpha}}$$

$$= \lim_{t \to 0} \frac{\alpha(1+t)^{\alpha-1}}{(1+t)^{\alpha} + t\alpha(1+t)^{\alpha-1}} = \alpha;$$

so if $\alpha > 1$ the series is absolutely convergent by Raabe's test, and that suffices for the proof.

The power series expansion of functions is immensely useful, but holds only for real analytic functions. However any continuous function can be approximated arbitrarily closely by polynomials, although the resulting sequence of approximating polynomials does not have the form of a power series. The basic result for functions of a real variable was established by Weierstrass, and an extensive generalization was proved by Marshall Stone. The following very simple example of the Weierstrass Theorem is basic to the general case and illustrates the sort of result that is possible.

Theorem 3.60. *For any $\epsilon > 0$ there is a polynomial $p(x)$ in the variable $x \in \mathbb{R}$ such that $|p(x) - |x|| < \epsilon$ whenever $|x| \leq M$.*

Proof: By Theorem 3.58 the binomial series $f(x) = (1-x)^{1/2} = \sum_{n=0}^{\infty} c_n x^n$ for the positive square root converges absolutely and uniformly in $[-1, 1]$, so for any $\epsilon > 0$ there is a suitable partial sum that is a polynomial $p(x)$ for which $|p(x) - (1-x)^{1/2}| < \epsilon$ whenever $|x| \leq 1$. Upon setting $x = 1 - t^2$ it follows that

$$\left| p(1-t^2) - (t^2)^{1/2} \right| < \epsilon \quad \text{for} \quad |t| \leq 1.$$

For the positive square root $(t^2)^{1/2} = |t|$ so the preceding equation yields the desired polynomial approximation in the interval $|t| \leq 1$; and setting $t = x/M$ and replacing ϵ by ϵ/M provides the desired polynomial approximation in the interval $|x| \leq M$, which suffices for the proof.

To generalize this, following the ideas of Stone, consider the vector space $\mathcal{C}(S)$ of continuous real-valued functions on a compact topological space S. The functions in $\mathcal{C}(S)$ are all bounded on S, so $\mathcal{C}(S)$ is a complete normed vector space under the ℓ_∞ norm $\|f\|_\infty = \sup_{x \in S} |f(x)|$, since convergence in this norm is just uniform convergence on S. Since this is the only norm that will be considered here, the notation will be simplified by denoting this norm just by $\|f\|$. A vector subspace $\mathcal{A} \subset \mathcal{C}(S)$ is called a **function algebra** if it includes the constant functions and the product fg of any two functions $f, g \in \mathcal{A}$. For example if $S = [0, 1] \subset \mathbb{R}$ the set of all real polynomials is a function algebra. A function algebra is an algebra, as defined in Section 1.3, where the operations of addition and multiplication are the ordinary sum and product of functions; but it is not necessarily a finite dimensional algebra. The closure of the vector subspace $\mathcal{A} \subset \mathcal{C}(S)$ of the normed space is a well-defined closed vector subspace $\overline{\mathcal{A}} \subset \mathcal{C}(S)$; of course it also contains the constant functions, and it is easy to see that it includes the product fg of any two functions $f, g \in \overline{\mathcal{A}}$, so it too is a function algebra on the space S. The quite surprising result is that if it is assumed that the algebra $\mathcal{A}$ **separates points** on S, meaning that for any two distinct points $a, b \in S$ there is a function $f \in \mathcal{A}$ such that $f(a) \neq f(b)$, then actually $\overline{\mathcal{A}} = \mathcal{C}(S)$.

Theorem 3.61 (Stone-Weierstrass Theorem). *If $\mathcal{A} \subset \mathcal{C}(S)$ is a real function algebra on a compact topological space S and $\mathcal{A}$ separates points on S then $\overline{\mathcal{A}} = \mathcal{C}(S)$.*

Proof: A preliminary observation is that if $f \in \overline{\mathcal{A}}$ then $|f| \in \overline{\mathcal{A}}$ as well. To demonstrate this let $M = \|f\|$ and note that by Theorem 3.60 there is a polynomial $p(t)$ such that $|\, p(t) - |t| \,| < \epsilon$ whenever $|t| \leq M$. Since $|f(x)| \leq M$ for any point $x \in S$ it follows that $|\, p(f(x)) - |f(x)| \,| < \epsilon$ for all $x \in S$, or equivalently that $\|p(f) - |f|\,\| < \epsilon$. By the definition of an algebra clearly $p(f) \in \overline{\mathcal{A}}$ whenever $f \in \overline{\mathcal{A}}$; thus $|f| \in \mathcal{C}(S)$ is the limit of points in the closed subspace $\overline{\mathcal{A}} \subset \mathcal{C}(S)$ hence $|f| \in \overline{\mathcal{A}}$.

If $f, g \in \overline{\mathcal{A}}$ then $f \pm g \in \overline{\mathcal{A}}$ and it follows from the preceding observation that $|f \pm g| \in \overline{\mathcal{A}}$ as well; so since the constant function $1/2 \in \overline{\mathcal{A}}$ it further follows that

$$\max(f, g) = \frac{1}{2}(f + g) + \frac{1}{2}|f - g| \in \overline{\mathcal{A}}, \quad \text{and}$$

$$\min(f, g) = \frac{1}{2}(f + g) - \frac{1}{2}|f - g| \in \overline{\mathcal{A}}.$$

If c_1, c_2 are any two distinct points in S by assumption there is a function $g \in \mathcal{A}$ for which $g(c_1) \neq g(c_2)$; and for any specified real numbers a_1, a_2 the function f defined by

$$f(x) = \frac{g(x) - g(c_2)}{g(c_1) - g(c_2)} a_1 + \frac{g(x) - g(c_1)}{g(c_2) - g(c_1)} a_2$$

clearly belongs to the algebra $\mathcal{A}$ and satisfies $f(c_1) = a_1$ and $f(c_2) = a_2$.

With these preliminary observations out of the way, consider a continuous function $h \in \mathcal{C}(S)$ and a real number $\epsilon > 0$. First for any two points $a, b \in S$ there is a function $f_{a,b} \in \overline{\mathcal{A}}$ such that $f_{a,b}(a) = h(a)$ and $f_{a,b}(b) = h(b)$, as in the preceding paragraph if $a \neq b$ and trivially if $a = b$. Since $f_{a,b}(b) = h(b) > h(b) - \epsilon$ then by continuity $f_{a,b}(x) > h(x) - \epsilon$ for all points x in a sufficiently small open neighborhood U_b of the point $b \in S$. The same construction can be carried out for any point $b \in S$, and the open neighborhoods U_b in this construction form an open covering of S. Since S is compact finitely many $U_{b_1}, \ldots, U_{b_n}$ of these neighborhoods will serve to cover S. Then define a function $f_a(x)$ by

$$f_a = \max(f_{a,b_1}, \ldots, f_{a,b_n});$$

this is a function in $\overline{\mathcal{A}}$, also as observed in the preceding paragraph, and it satisfies $f_a(a) = h(a)$ and $f_a(x) > h(x) - \epsilon$ for all points $x \in S$. As in the preceding argument since $f_a(a) = h(a) < h(a) + \epsilon$ it follows by continuity that $f_a(x) < h(x) + \epsilon$ for all points x in a sufficiently small open neighborhood V_a of the point a. The same construction can be carried out for any point $a \in S$, and the open neighborhoods V_a in this construction form an open covering of S. Since S is compact finitely many $V_{c_1}, \ldots, V_{c_m}$ of these neighborhoods will serve to cover S. Then define a function f by

$$f = \min(f_{c_1}, \ldots, f_{c_m});$$

this is a function in $\overline{\mathcal{A}}$, as observed in the preceding paragraph, and by construction $f(x) > h(x) - \epsilon$ and $f(x) < h(x) + \epsilon$ for any points $x \in S$, so that $\|f - h\| < \epsilon$. That is the case for any $\epsilon > 0$, so that h is a limit of functions $f \in \overline{\mathcal{A}}$ and consequently $h \in \overline{\mathcal{A}}$, thereby concluding the proof.

The extension of the preceding theorem to algebras of complex-valued functions requires an additional assumption, that the algebra is **self-adjoint**, meaning that whenever f belongs to the algebra then so does its complex conjugate $\overline{f}$.

Corollary 3.62. *If $A \subset C(S)$ is a self-adjoint complex function algebra on a compact topological space S and A separates points on S then $\overline{A} = C(S)$.*

Proof: Let B be the set of real-valued functions in the algebra A. If $f \in A$ and $f = u + iv$ then $2u = f + \overline{f} \in A$, since A is self-adjoint, hence the real part of f is a function $u \in B$. The real function algebra B also separates points, since for any distinct points $a, b \in S$ there is a function $f \in A$ such that $f(a) = 0$ and $f(b) = 1$, and consequently if $f = u + iv$ then $u \in B$ and $u(a) = 0$ while $u(b) = 1$. Thus the algebra B satisfies the hypothesis of the preceding theorem so its closure is the algebra of real-valued continuous functions on S. In particular if $f = u + iv$ is any continuous complex-valued function on S then $u, v \in \overline{B} \subset \overline{A}$ and consequently $f = u + iv \in \overline{A}$, which suffices for the proof.

Since the algebra of real-valued polynomials in any compact subset $S \subset \mathbb{R}^n$ clearly separates points, it follows from the Stone-Weierstrass Theorem that for any continuous function f on a compact subset $S \subset \mathbb{R}^n$ and any $\epsilon > 0$ there is a polynomial $p(x)$ such that $|f(x) - p(x)| < \epsilon$ for all $x \in S$. The Stone-Weierstrass Theorem however does not imply that these approximating polynomials for successively smaller values of ϵ can be arranged as the partial sums of an infinite sequence of polynomials that sum to the function $f(x)$ on the set S, much as in the case of power series; that caution should be kept in mind.

Another very important application of the Stone-Weierstrass Theorem is to periodic functions. A function f on the real line is **periodic** with period 1 if $f(x + 1) = f(x)$ for all $x \in \mathbb{R}$. The function $e^{2\pi ix}$ is an example of a periodic function with period 1, as are its real and imaginary parts $\cos 2\pi x$ and $\sin 2\pi x$. The function $e^{2\pi ix}$ maps the real line $\mathbb{R}$ to the unit circle S^1 in the complex plane in such a way that the interval $[0, 1]$ is mapped onto the unit circle S^1, with the end points 0 and 1 in $\mathbb{R}$ both mapped to the same point $1 \in S^1$ on the unit circle and the open subset $(0, 1)$ being mapped bijectively to the complement $S^1 \sim \{1\}$ of the point 1 on the unit circle; thus this mapping identifies the quotient $\mathbb{R}/\asymp$ of the real axis by the equivalence relation $\asymp$, where $x \asymp y$ whenever $x - y \in \mathbb{Z}$, with a circle. The unit circle is a compact space, with the induced topology from that of the complex plane, so the Stone-Weierstrass Theorem can be applied to the normed vector space $C(S^1)$ of continuous functions on the unit circle, or equivalently, to the normed vector space of continuous periodic functions on the real line $\mathbb{R}$. The single function $e^{2\pi ix}$ separates points on the circle, hence of course so does the complex algebra generated by this function, the algebra A consisting of all finite linear combinations of the functions $e^{2\pi inx}$ for $n \geq 0$. This algebra is not self-adjoint, since $\overline{e^{2\pi inx}} = e^{2\pi i(-n)x}$; but the algebra $\mathcal{A}^*$ consisting of all finite linear combinations of the functions $e^{2\pi inx}$ for $n \in \mathbb{Z}$ is a self-adjoint algebra, which of course also separates points on S^1. It then follows from the Stone-Weierstrass Theorem that for any continuous periodic function $f(x)$ on $\mathbb{R}$ with the period 1 and for any $\epsilon > 0$ there are complex constants $c_n \in \mathbb{C}$ such that

$$\left| f(x) - \sum_{n=-N}^{N} c_n e^{2\pi inx} \right| < \epsilon. \tag{3.137}$$

These are the finite **Fourier series**. The Stone-Weierstrass Theorem does not ensure though that these finite Fourier series for increasingly small ϵ can be constructed to have the same initial terms, so that the function $f(x)$ is the sum of an infinite series of the form (3.137); that is quite a different matter and will not be pursued further here.

Problems, Group I

1. Either determine the following limits, or show that they do not exist:

   (i) $\displaystyle\lim_{x\to 0} \frac{e^x - \cos x}{\sin x}$, (ii) $\displaystyle\lim_{n\to\infty} \frac{1 - e^{1/n}}{\sin(1/n)}$, (iii) $\displaystyle\lim_{n\to\infty} \frac{1 - \cos(1/n)}{e^{1/n^2} - 1}$.

2. Determine which of the following series converge:

   (i) $\displaystyle\sum_{n=1}^{\infty} \frac{n}{2^n}$, (ii) $\displaystyle\sum_{n=1}^{\infty} \frac{1}{\sqrt{n(n+1)}}$, (iii) $\displaystyle\sum_{n=2}^{\infty} \frac{1}{(\log n)^{(\log n)}}$, (iv) $\displaystyle\sum_{n=1}^{\infty} \frac{n^n}{3^n n!}$.

3. Find the radius of convergence of each of the following complex power series:

   (i) $\displaystyle\sum_{n=2}^{\infty} \frac{1}{\log n} z^n$, (ii) $\displaystyle\sum_{n=1}^{\infty} \frac{2^n}{n!} z^n$, (iii) $\displaystyle\sum_{n=1}^{\infty} \frac{2^n}{n^2} z^n$, (iv) $\displaystyle\sum_{n=1}^{\infty} \frac{n!}{2^{n^2}} z^n$.

4. Find explicit expressions for the sums of the following series in the regions in which they converge:

   (i) $\displaystyle\sum_{n=1}^{\infty} n^2 x^n$, (ii) $\displaystyle\sum_{n=1}^{\infty} \frac{1}{n} x^n$, (iii) $\displaystyle\sum_{n=1}^{\infty} n^2 e^{-nx}$.

5. Show that the series (3.136) converges.

6. Show that if a power series has integer coefficients, infinitely many of which are nonzero, then its radius of convergence is at most 1.

7. Find an example of two convergent real series $\sum x_k$ and $\sum y_k$ such that $\sum x_k y_k$ is divergent. Is there such an example if at least one of the series $\sum x_k$ or $\sum y_k$ is absolutely convergent? Why?

Problems, Group II

8. If $c_n \geq 0$ and the series $\sum_{n=1}^{\infty} c_n$ converges does the series $\sum_{n=1}^{\infty} c_n^2$ necessarily converge? Does the series $\sum_{n=1}^{\infty} \sqrt{c_n}$ necessarily converge?

9. Find the subsets of $\mathbb{R}$ on which sequences (i) and (ii) and series (iii) converge, and the subsets on which the convergence is uniform:

$$\text{(i)} \ \frac{x^{2n}}{1+x^2}, \qquad \text{(ii)} \ \frac{x^2}{1+nx^4}, \qquad \text{(iii)} \ \sum_{n=1}^{\infty} \frac{x^2}{(1+x^2)^n}.$$

10. i) Show that there are no real numbers z for which $e^z = z$, but that there are complex numbers z with that property.
    ii) For what complex numbers $z \in \mathbb{C}$ can the value of $\log z$ be chosen to be a negative real number?

11. If the power series $\sum_{n=1}^{\infty} c_n x^n$ has the radius of convergence R what can you say about the radius of convergence of each of the following power series? (i) $\sum_{n=1}^{\infty} c_n x^{2n}$, (ii) $\sum_{n=1}^{\infty} c_{2n} x^n$, (iii) $\sum_{n=1}^{\infty} c_n x^{n^2}$.

12. By the method used in this book to find the Taylor series of the arcsine function, find the Taylor series at the origin of the arctan function, the inverse of the function $\tan x = \frac{\sin x}{\cos x}$; and by using that expansion show that $\frac{\pi}{4} = 1 - \frac{1}{3} + \frac{1}{5} - \frac{1}{7} + \cdots$.

13. Let $\phi(x) = \begin{cases} 0 & \text{for } x = 0, \\ e^{-1/x^2} & \text{for } x \neq 0, \end{cases}$ and $\psi(x) = \begin{cases} 0 & \text{for } x \leq 0, \\ e^{-1/x^2} & \text{for } x > 0. \end{cases}$

    i) Find $\phi'(0)$.
    ii) Find the Taylor series of $f(x) = \sin x + \psi(x)$ at the origin.
    iii) Does the Taylor series of $f(x)$ converge to $f(x)$?

14. Show that any continuous real-valued function $f(x)$ in the closed interval $[-R, R] \subset \mathbb{R}$ can be approximated within any chosen $\epsilon > 0$ by a function of the form $h(x) = \sum_{n=0}^{N} c_n e^{v_n x}$ where $c_n \in \mathbb{R}$ and $v_n \in \mathbb{Z}$, $v_n \geq 0$.

15. Show that if $f(x) = \sum_{n=0}^{\infty} c_n x^n$ is a real power series that converges in the interval $(-1, 1)$, and if $f(x_n) = 0$ for an infinite sequence of distinct points $x_n > 0$ such that $\lim_{n \to \infty} x_n = 0$, then $f(x) = 0$ in $(-1, 1)$.

16. Consider the Taylor series $\frac{x}{e^x - 1} = \sum_{n=0}^{\infty} \frac{1}{n!} B_n x^n$ where B_n are defined by this expansion. (The numbers B_n, called the Bernoulli numbers, are fascinating numbers with no simple explicit defining formula but a great many applications; for example, $B_{14} = \frac{7}{6}$, $B_{24} = -\frac{236,364,091}{2,730}$, $B_{26} = \frac{8,553,103}{6}$.)

    i) By examining the relation $(e^x - 1)(B_0 + B_1 x + \frac{1}{2!} B_2 x^2 + \cdots) = x$ find explicitly the first four values B_0, B_1, B_2, B_3, and find a recursive formula for the Bernoulli numbers.
    ii) Show that $B_{2n+1} = 0$ for $n > 1$.
    iii) Show that $x \cot x = \sum_{n=0}^{\infty} (-1)^n B_{2n} \frac{2^{2n}}{(2n)!} x^{2n}$ where $\cot(x) = \frac{\cos x}{\sin x}$.

4

Linear Mappings

4.1 Endomorphisms

In Section 1.3 matrices were used to describe linear mappings $A : V \longrightarrow W$ between vector spaces V and W over a field F in terms of bases for the two vector spaces V and W; two matrices describing the same mapping but in terms of different bases for V and for W were called **equivalent matrices**, thereby introducing a useful and natural equivalence relation among matrices over F. The Equivalence Theorem, Corollary 1.17, demonstrated that two $n \times m$ matrices over F are equivalent if and only if they have the same rank, thereby showing that the rank of a matrix is the unique invariant classifying the equivalence of matrices, and obtained normal forms for each equivalence class of matrices.

The special case of linear mappings $A : V \longrightarrow V$ from a vector space V to itself, called **endomorphisms** of the vector space V, leads to different and rather more complicated equivalence relations between matrices. Any endomorphism A of course can be described by a matrix in terms of any basis for the vector space V; but since the mapping is from the vector space V to itself it is only natural to use the same basis for V whether it is viewed as the domain or range of the mapping A. If $\{\mathbf{u}_i\}$ is a basis for the n-dimensional vector space V over a field F then an endomorphism $A : V \longrightarrow V$ takes the basis vector $\mathbf{u}_j$ to the vector $A\mathbf{u}_j = \sum_{i=1}^{n} a_{ij}\mathbf{u}_i$ for some elements $a_{ij} \in F$; so in terms of this basis the endormorphism is described by the $n \times n$ matrix $A = \{a_{ij}\}$, where for convenience the same letter customarily is used both for a matrix and for the linear transformation described by that matrix. If $\mathbf{v}_j$ is any other basis for the vector space V then $\mathbf{u}_j = \sum_{i=1}^{n} s_{ij}\mathbf{v}_i$ for a nonsingular matrix $S = \{s_{ij}\}$, and conversely $\mathbf{v}_j = \sum_{i=1}^{n} \hat{s}_{ij}\mathbf{u}_i$ for the inverse matrix $S^{-1} = \{\hat{s}_{ij}\}$. Therefore

$$A\mathbf{v}_j = A\left(\sum_{i=1}^{n}\hat{s}_{ij}\mathbf{u}_i\right) = \sum_{i=1}^{n}\hat{s}_{ij}A\mathbf{u}_i = \sum_{i,k=1}^{n}\hat{s}_{ij}a_{ki}\mathbf{u}_k = \sum_{i,k,l=1}^{n}\hat{s}_{ij}a_{ki}s_{lk}\mathbf{v}_l \qquad (4.1)$$

$$= \sum_{l=1}^{n}b_{lj}\mathbf{v}_l \quad \text{where} \quad b_{lj} = \sum_{i,k=1}^{n}s_{lk}a_{ki}\hat{s}_{ij} \quad \text{or} \quad B = SAS^{-1}.$$

Matrices A and B related by $B = SAS^{-1}$ for a nonsingular matrix S are said to be **similar** matrices; this is clearly an equivalence relation between $n \times n$ matrices over the field F. In these terms the matrices representing an endomorphism of a vector space V in terms of all bases of that vector space are an equivalence class of similar matrices. It is clear that similar $n \times n$ matrices have the same determinant and rank; but two $n \times n$ matrices of the same determinant and rank are not necessarily similar, as will become apparent in the subsequent discussion.

If $T : V \longrightarrow V$ is an endomorphism of a vector space V over a field F, a nonzero vector $\mathbf{v} \in V$ is said to be an **eigenvector**, or alternatively a **characteristic vector**, of the linear transformation T if $T\mathbf{v} = \lambda \mathbf{v}$ for some $\lambda \in F$; the value $\lambda \in F$ is called the **eigenvalue**, or alternatively the **characteristic value**, of the eigenvector $\mathbf{v}$. The set of all eigenvectors for T with the eigenvalue λ, together with the zero vector, form a linear subspace $V_\lambda \subset V$; for if $\mathbf{v}_1, \mathbf{v}_2 \in V_\lambda$ it follows that $T(\mathbf{v}_1 + \mathbf{v}_2) = \lambda(\mathbf{v}_1 + \mathbf{v}_2)$ and $T(c\mathbf{v}) = c(T\mathbf{v}) = \lambda \cdot c\mathbf{v}$ by the linearity of T, hence $\mathbf{v}_1 + \mathbf{v}_2 \in V_\lambda$ and $c\mathbf{v} \in V_\lambda$. The subspace V_λ is called the **eigenspace** of T for the eigenvalue λ. The eigenspaces for distinct eigenvalues are linearly independent subspaces of V; for if $\lambda_1 \neq \lambda_2$ and $\mathbf{v} \in V_{\lambda_1} \cap V_{\lambda_2}$ then $T\mathbf{v} = \lambda_1 \mathbf{v} = \lambda_2 \mathbf{v}$ hence $(\lambda_1 - \lambda_2)\mathbf{v} = \mathbf{0}$ so $\mathbf{v} = \mathbf{0}$. The dimension of the eigenspace for an eigenvalue λ is called the **multiplicity** of the eigenvalue λ, and the set of all eigenvalues for T is called the **spectrum** of the linear transformation T. Note that the eigenspace for the eigenvalue 0 is just the kernel of the endomorphism T.

An $n \times n$ matrix over a field F describes an endomorphism $A : F^n \longrightarrow F^n$, and the eigenvectors and eigenvalues of this endomorphism are called the eigenvectors and eigenvalues of the matrix A. They can be calculated quite directly in terms of the matrix A by considering the **characteristic polynomial** of that matrix, the polynomial in the variable λ defined by

$$\chi_A(\lambda) = \det(\lambda I - A) \tag{4.2}$$

where I is the $n \times n$ identity matrix. It is clear that similar matrices have the same characteristic polynomial, since if $B = SAS^{-1}$ for a nonsingular matrix S then $\chi_B(\lambda) = \det(\lambda I - B) = \det(\lambda I - SAS^{-1}) = \det(S(\lambda I - A)S^{-1}) = \det(\lambda I - A) = \chi_A(\lambda)$; thus the characteristic polynomial of an endomorphism T can be defined as the characteristic polynomial of any matrix that describes this endomorphism in terms of some basis for the vector space.

Lemma 4.1. *The spectrum of the linear transformation described by a matrix A over a field F is the set of roots, or zeros, of the characteristic polynomial χ_A of that matrix.*

Proof: If the linear transformation described by a matrix A has the nontrivial eigenvector $\mathbf{v}$ for the eigenvalue λ then $A\mathbf{v} = \lambda\mathbf{v}$, or equivalently $(\lambda I - A)\mathbf{v} = \mathbf{0}$; hence the matrix $\lambda I - A$ is necessarily a singular matrix, therefore $\chi_A(\lambda) = \det(\lambda I - A) = 0$ so λ is a root of the polynomial χ_A. Conversely if $\chi_A(\lambda) = \det(\lambda I - A) = 0$ for some $\lambda \in F$ then the matrix $\lambda I - A$ is singular, so there is a nontrivial vector $\mathbf{v}$ such that $(\lambda I - A)\mathbf{v} = \mathbf{0}$, showing that λ is the eigenvalue of the eigenvector $\mathbf{v}$. That suffices for the proof.

For some simple examples, the 5×5 real matrix

$$A = \begin{pmatrix} 0 & 1 & 0 & 0 & 0 \\ -1 & 0 & 0 & 0 & 0 \\ 0 & 0 & 2 & 0 & 0 \\ 0 & 0 & 0 & 3 & 3 \\ 0 & 0 & 0 & 0 & 3 \end{pmatrix}$$

has the characteristic polynomial

$$\chi_A(\lambda) = \det \begin{pmatrix} \lambda & -1 & 0 & 0 & 0 \\ 1 & \lambda & 0 & 0 & 0 \\ 0 & 0 & \lambda - 2 & 0 & 0 \\ 0 & 0 & 0 & \lambda - 3 & -3 \\ 0 & 0 & 0 & 0 & \lambda - 3 \end{pmatrix} = (\lambda^2 + 1)(\lambda - 2)(\lambda - 3)^2.$$

The spectrum of this linear transformation viewed as a real linear transformation consists of the two distinct real roots $\{2, 3\}$ of the characteristic polynomial; however the spectrum of this linear transformation viewed as a complex linear transformation consists of the four distinct complex roots $\{i, -i, 2, 3\}$ of the characteristic polynomial. The eigenspace for the eigenvalue 3 consists of those vectors $\mathbf{x}$ such that

$$\mathbf{0} = \begin{pmatrix} 3 & -1 & 0 & 0 & 0 \\ 1 & 3 & 0 & 0 & 0 \\ 0 & 0 & 1 & 0 & 0 \\ 0 & 0 & 0 & 0 & -3 \\ 0 & 0 & 0 & 0 & 0 \end{pmatrix} \begin{pmatrix} x_1 \\ x_2 \\ x_3 \\ x_4 \\ x_5 \end{pmatrix} = \begin{pmatrix} 3x_1 - x_2 \\ x_1 + 3x_2 \\ x_3 \\ -3x_5 \\ 0 \end{pmatrix}$$

hence such that the coefficients satisfy $3x_1 - x_2 = x_1 + 3x_2 = x_3 = -3x_5 = 0$; this is a set of equations in the variables x_1, x_2, x_3, x_4, x_5 for which the only nontrivial solutions yield the vectors $(0, 0, 0, x_4, 0)$ for any value of x_4. Thus the eigenspace is the one-dimensional linear subspace of $\mathbb{R}^5$ spanned by the basis vector

$$\delta_4 \quad \text{where} \quad \delta_4 = {}^t(0\ 0\ 0\ 1\ 0) \in \mathbb{R}^5.$$

The corresponding calculation shows that the eigenspace for the eigenvalue 2 is the one-dimensional linear subspace of $\mathbb{R}^5$ spanned by the basis vector δ_3 while the eigenspace for the eigenvalue i is the one-dimensional subspace of $\mathbb{C}^5$ spanned by the vector

$$\delta_1 + i\delta_2 = {}^t(1 \ i \ 0 \ 0 \ 0) \in \mathbb{C}^5$$

and the eigenspace for the eigenvalue $-i$ is the one-dimensional subspace of $\mathbb{C}^5$ spanned by the vector

$$i\delta_1 + \delta_2 = {}^t(i \ 1 \ 0 \ 0 \ 0) \in \mathbb{C}^5.$$

For examples in the reverse direction, in a sense, if A is a 2×2 matrix with the characteristic polynomial $\chi_A(\lambda) = (\lambda - a_1)(\lambda - a_2)$ and the eigenvalues a_1, a_2 are distinct then to each of these eigenvalues a_i there corresponds an eigenvector $\mathbf{v}_i$; the two eigenvectors are linearly independent since they lie in distinct eigenspaces of A, so the 2×2 matrix $P = (\mathbf{v}_1 \ \mathbf{v}_2)$ formed with these column vectors is a nonsingular matrix. Then

$$AP = A\,(\mathbf{v}_1 \ \mathbf{v}_2) = (a_1\mathbf{v}_1 \ \ a_2\mathbf{v}_2) = (\mathbf{v}_1 \ \mathbf{v}_2) \begin{pmatrix} a_1 & 0 \\ 0 & a_2 \end{pmatrix} = P \begin{pmatrix} a_1 & 0 \\ 0 & a_2 \end{pmatrix}$$

so $P^{-1}AP = \begin{pmatrix} a_1 & 0 \\ 0 & a_2 \end{pmatrix}$, showing that the matrix A is similar to a diagonal matrix. A matrix that is similar to a diagonal matrix is said to be a **diagonalizable matrix**. The obvious extension of this example shows that if the characteristic polynomial of an $n \times n$ matrix has n distinct roots then that matrix is diagonalizable. On the other hand the matrices $\begin{pmatrix} a_1 & 0 \\ 0 & a_2 \end{pmatrix}$ and $\begin{pmatrix} a & 0 \\ 0 & a \end{pmatrix}$ both have the characteristic polynomial $(\lambda - a)^2$; the first is not diagonalizable, since it is easily seen to have but a single eigenvector, while the second one is already diagonal. Thus the condition that the characteristic polynomial of an $n \times n$ matrix A has n distinct roots is a sufficient but not necessary condition that the matrix is diagonalizable; a condition that is both necessary and sufficient will be given later in the Diagonalizability Theorem, Corollary 4.8.

If $p(\lambda) \in F[\lambda]$, so $p(\lambda) = \sum_{j=0}^{N} c_j \lambda^j$ for some $c_j \in F$, then $p(A)$ for any $n \times n$ matrix A with coefficients in the field F is defined by $p(A) = \sum_{j=0}^{N} c_j A^j$; thus $p(A)$ is another $n \times n$ matrix with coefficients in the field F. Any relation involving sums and products of polynomials $p(\lambda)$ yields in this way the corresponding relation among the matrices $p(A)$. Some care must be taken when considering polynomials in several different variables though; for instance $xyx = x^2y$ for any elements x, y in a field, but $XYX \neq X^2Y$ for some matrices, for instance for $X = \begin{pmatrix} 0 & 1 \\ 0 & 0 \end{pmatrix}$ and $Y = \begin{pmatrix} 0 & 1 \\ 1 & 0 \end{pmatrix}$. However so long as the matrices X and Y **commute**, meaning that $XY = YX$, polynomial identities also hold for polynomials in these matrices.

Lemma 4.2. *If A is an $n \times n$ matrix with coefficients in a field F and $A\mathbf{v} = \lambda\mathbf{v}$ then $p(A)\mathbf{v} = p(\lambda)\mathbf{v}$ for any polynomial p with coefficients in F.*

Proof: If $p(\lambda) = \sum_{j=0}^{N} c_j \lambda^j$ and $A\mathbf{v} = \lambda\mathbf{v}$ then since $A^n\mathbf{v} = \lambda^n\mathbf{v}$ it follows that $p(A)\mathbf{v} = \sum_{j=0}^{N} c_j A^j \mathbf{v} = \sum_{j=0}^{N} c_j \lambda^j \mathbf{v} = p(\lambda)\mathbf{v}$, which suffices for the proof.

The preceding result is really rather simple, but worth listing as a separate lemma for emphasis since it is quite commonly used. The value of a polynomial in a scalar $\lambda \in F$ when λ is replaced by a matrix is rather clear; but there also arise situations in which the coefficients of polynomials are matrices. If $P(\lambda) = \sum_i P_i \lambda^i$ and $Q(\lambda) = \sum_i Q_j \lambda^j$ are two matrices in a scalar $\lambda \in F$, where the coefficients P_i, Q_j are $n \times n$ matrices, then the product of these two polynomials is defined to be the polynomial $R(\lambda) = P(\lambda)Q(\lambda) = \sum_{i,j} P_i Q_j \lambda^{i+j}$; note that it is not necessarily the case that $P(\lambda)Q(\lambda) = Q(\lambda)P(\lambda)$, since it is quite possible that $P_i Q_j \neq Q_j P_i$ for some indices i, j. When the variable λ is replaced by a matrix A some care must be taken when considering the product $P(A)Q(A)$, since it is not necessarily the case that $P(A)Q(A) = R(A)$. However if the matrix A commutes with all of the matrices Q_j then it is clear that $P(A)Q(A) = R(A)$. With this in mind it is possible to turn to the following result, which will play a significant role in the subsequent discussion.

Theorem 4.3 (Cayley-Hamilton Theorem). *If A is an $n \times n$ matrix over a field F then $\chi_A(A) = 0$.*

Proof: As in the discussion of Cramer's rule, the adjugate $X^\dagger$ of an $n \times n$ matrix X is an $n \times n$ matrix, the entries of which are polynomials in the entries of the matrix X; and it has the property that $(\det X)\, I = X^\dagger \cdot X$. In particular for the matrix $X = \lambda I - A$ for which $\det X = \chi_A(\lambda)$ this takes the form

$$\chi_A(\lambda)\, I = (\lambda I - A)^\dagger \cdot (\lambda I - A). \tag{4.3}$$

The entries in the matrix $B(\lambda) = (\lambda I - A)^\dagger$ are polynomials in λ, so that matrix can be written out explicitly as a sum of terms of the form $B_j \lambda^j$ for some matrices B_j; then (4.3) takes the form of the polynomial identity

$$\chi_A(\lambda)\, I = \left(\sum_j B_j \lambda^j \right) (\lambda I - A).$$

When λ is replaced by the matrix A this reduces to

$$\chi_A(A)\, I = \left(\sum_j B_j A^j \right) (A - A) = 0,$$

by the discussion in the preceding paragraph, and that suffices for the proof.

The preceding result actually is more complicated, and somewhat more surprising, than might be expected from the simplicity of the proof; an example might illustrate that. The characteristic polynomial for the matrix $A = \begin{pmatrix} a & b \\ c & d \end{pmatrix}$ over a field F is $\chi_A(\lambda) = \det \begin{pmatrix} \lambda-a & -b \\ -c & \lambda-d \end{pmatrix} = \lambda^2 - (a+d)\lambda + (ad - bc)$ and consequently $\chi_A(A) = A^2 - (a+d)A + (ad - bc)I$, which when written out explicitly has the form

$$\chi_A(A) = \begin{pmatrix} a^2 + bc & (a+d)b \\ (a+d)c & bc + d^2 \end{pmatrix} - (a+d)\begin{pmatrix} a & b \\ c & d \end{pmatrix} + (ad - bc)\begin{pmatrix} 1 & 0 \\ 0 & 1 \end{pmatrix}.$$

A straightforward calculation shows that actually $\chi_A(A) = 0$. It is clear that a corresponding explicit calculation in the general case would be rather a challenge; and it is evident that it is really not so obvious that $\chi_A(A) = 0$.

In the further discussion of endomorphisms, polynomials other than the characteristic polynomial play significant roles; and it is necessary to discuss some further general properties of polynomials that are used in that discussion. The set of polynomials in a variable x with coefficients in a field F traditionally is denoted by $F[x]$; clearly $F[x]$ is a ring. Particularly convenient for many purposes are **monic polynomials**, polynomials of the form $x^n + a_{n-1}x^{n-1} + \cdots + a_1 x + a_0$, that is, polynomials with leading coefficient 1. If $f(x) = a_m x^m + \cdots$ and $p(x) = b_n x^n + \cdots$ are polynomials of degrees $m \geq n$ then $f(x) - \frac{a_m}{b_n}x^{m-n}p(x) = r_1(x)$ is a polynomial of degree at most $m - 1$; if $\deg r_1(x) \geq n$ the same process can be applied to the polynomial $r_1(x)$, and so on, so that eventually $f(x) = p(x)q(x) + r(x)$ for uniquely determined polynomials $q(x)$ and $r(x)$ where the degree of $r(x)$ is strictly less than the degree of $p(x)$ or $r(x) = 0$. This is the classical **division algorithm** for polynomials over a field.[1] A polynomial $p(x)$ is said to **divide** a polynomial $f(x)$, or equivalently $f(x)$ is said to be **divisible** by $p(x)$, if $f(x) = p(x)q(x)$ for some polynomial $q(x)$; that $p(x)$ divides $f(x)$ is denoted by $p(x)|f(x)$. Note that $p(x)|f(x)$ if and only if the remainder $r(x) = 0$ in the division algorithm $f(x) = p(x)q(x) + r(x)$. A nonconstant polynomial $p(x)$ is said to be **reducible** if it can be written as the product $p(x) = f(x)g(x)$ of two nonconstant polynomials; and a nonconstant polynomial that is not reducible is said to be **irreducible**, so a nonconstant polynomial $p(x)$ is irreducible if and only if it cannot be written as a product $p(x) = f(x)g(x)$ of two nonconstant polynomials. For example the polynomial $x + 1$ is irreducible over any field; the polynomial $x^2 + 1$ is irreducible as a real polynomial, but not irreducible as a complex polynomial since it can be written as the product $x^2 + 1 = (x + i)(x - i)$. The **greatest common divisor** of two polynomials $f(x), g(x) \in F[x]$ is the monic

---

[1] The division algorithm is often called the **Euclidean algorithm**, since it appears in Euclid's *Elements* as a tool for calculating the greatest common divisor of two integers. Rings in which there is an analogue of the Euclidean algorithm are called **Euclidean rings** or **Euclidean domains** and occur frequently in algebra.

polynomial $d(x) \in F[x]$ of the greatest degree such that $d(x)|f(x)$ and $d(x)|g(x)$. The **least common multiple** of two polynomials $f(x), g(x) \in F[x]$ is the monic polynomial $m(x) \in F[x]$ of least degree such that $f(x)|m(x)$ and $g(x)|m(x)$. A **polynomial ideal** $\mathcal{I}$, usually just called an **ideal** when it is understood that what is involved is a collection of polynomials, is a set of polynomials such that whenever $p(x), q(x) \in \mathcal{I}$ then $p(x) + q(x) \in \mathcal{I}$ and $f(x)p(x) \in \mathcal{I}$ for any arbitrary polynomial $f(x) \in F[x]$. Note carefully that this is not the condition that whenever $p(x), q(x) \in \mathcal{I}$ then $p(x) + q(x) \in \mathcal{I}$ and $p(x)q(x) \in \mathcal{I}$; that is the condition that the set $\mathcal{I}$ is a **subrng**[2] of the ring of polynomials. An ideal in $F[x]$ is a subrng of the ring $F[x]$; but not every subrng of $F[x]$ is an ideal in $F[x]$. For example, the polynomial x^2 generates the subrng of $F[x]$ consisting of all polynomials in x^2, hence the subrng consisting of all polynomials $\sum_{n=0}^{N} a_n x^{2n}$; on the other hand x^2 generates the ideal consisting of all polynomials $F[x]x^2$, hence the ideal of all polynomials $\sum_{n=2}^{N} a_n x^n$, all polynomials having a double zero at the origin. The set of polynomials that vanish at a particular element $a \in F$ is another standard example of an ideal in $F[x]$. For any finite set of polynomials $p_1(x), \ldots, p_n(x)$ with coefficients in a field F the set of all polynomials that can be written as sums $\sum_{i=1}^{m} f_i(x)p_i(x)$ for any polynomials $f_i(x) \in F[x]$ is another polynomial ideal, called the polynomial ideal **generated** by the polynomials $p_i(x)$.

Theorem 4.4 (Basic Theorem). (i) *If a polynomial ideal $\mathcal{I} \subset F[x]$ contains a nonzero constant then $\mathcal{I} = F[x]$.*

(ii) *A nonzero polynomial ideal $\mathcal{I}$ contains a unique monic polynomial $d(x)$ of least possible degree; and $d(x)|p(x)$ for any other polynomial $p(x) \in \mathcal{I}$, so $\mathcal{I}$ is the polynomial ideal generated by $d(x)$.*

(iii) *If $p(x) \in F[x]$ is irreducible and if $p(x)|f(x)g(x)$ for some polynomials $f(x), g(x) \in F[x]$ then either $p(x)|f(x)$ or $p(x)|g(x)$.*

(iv) *Any nonconstant polynomial $f(x) \in F[x]$ can be written as a product of irreducible polynomials. This factorization is unique up to multiplication of each factor by a nonzero scalar and up to order.*

Proof: (i) If $c \in \mathcal{I}$ for some constant $c \neq 0$ then for any polynomial $f(x)$ the product $cf(x) \in \mathcal{I}$ hence $f(x) = \frac{1}{c}(cf(x)) \in \mathcal{I}$ also.

(ii) A nonzero polynomial ideal $\mathcal{I}$ necessarily contains a monic polynomial $d(x)$ of least possible degree. If $\deg d(x) = 0$ so that $d(x) = 1$ then it follows from (i) that $\mathcal{I}$ consists of all polynomials, and any polynomial $p(x)$ is the product $p(x) = p(x) \cdot 1$. If $\deg d(x) > 0$ and $p(x) \in \mathcal{I}$ then $p(x) = q(x)d(x) + r(x)$ for a unique polynomial $r(x)$ for which $\deg r(x) < \deg d(x)$ by the division algorithm for polynomials. Since $p(x) \in \mathcal{I}$ and $d(x) \in \mathcal{I}$ it follows that $r(x) = p(x) - q(x)d(x) \in \mathcal{I}$; but since $\deg r(x) < \deg d(x)$ and $\deg d(x)$ is the smallest

---

[2]In the earlier discussion a subring was required to contain the identity element 1, so here a subrng means a subring but one that does not necessarily contain the identity element 1.

degree of any nonzero polynomial in $\mathcal{I}$ necessarily $r(x) = 0$, hence $d(x)|p(x)$. If $d_0(X)$ is another monic polynomial in $\mathcal{I}$ of the least possible degree then by what has just been proved $d(x)|d_0(x)$ and $d_0(x)|d(x)$ so that $d_0(x) = cd(x)$ for some constant $c \in F$; and since $d(x)$ and $d_0(x)$ are both monic polynomials necessarily $c = 1$ so that $d_0(x) = d(x)$.

(iii) If $p(x)$ is irreducible and $p(x)|f(x)g(x)$ for some polynomials $f(x)$ and $g(x)$ then $f(x)g(x) = p(x)q(x)$ for some polynomial $q(x)$. The polynomials $p(x)$ and $f(x)$ generate a polynomial ideal $\mathcal{I}$. If $\mathcal{I}$ does not consist of all polynomials then by (ii) there is a nonconstant polynomial $d(x)$ such that $d(x)|p(x)$ and $d(x)|f(x)$; since $p(x)$ is irreducible necessarily $p(x) = c\,d(x)$ for some $c \in F$ so that $p(x)|f(x)$. On the other hand if $\mathcal{I}$ consists of all polynomials then in particular $1 \in \mathcal{I}$ hence $1 = a(x)p(x) + b(x)f(x)$ for some polynomials $a(x)$ and $b(x)$; and then $g(x) = g(x) \cdot 1 = g(x)a(x)p(x) + g(x)b(x)f(x) = g(x)a(x)p(x) + b(x)q(x)p(x)$ showing that $p(x)|g(x)$.

(iv) If $f(x)$ is not irreducible it can be written as a product $f(x) = f_1(x)f_2(x)$ of two nonconstant polynomials, each of which will have degree strictly less than the degree of $f(x)$. The same argument can be applied to each of the factors $f_1(x)$ and $f_2(x)$, so each is either irreducible or is a product of two nonconstant polynomials of strictly lower degrees. In this way the initial polynomial eventually can be written as a product $f(x) = c\,p_1(x) \cdots p_r(x)$ of irreducible monic polynomials. If there is another such decomposition then there will be a monic irreducible polynomial $p(x)$ other than $p_1(x), \ldots, p_r(x)$ that divides $f(x)$. By (iii) either $p(x)|p_1(x)$, which is impossible since $p(x) \neq p_1(x)$ and both are irreducible, or $p(x)|p_2(x) \cdots p_r(x)$. Repeating this argument clearly leads eventually to the conclusion that $p(x)$ must be one of the polynomials $p_i(x)$, a contradiction which suffices to conclude the proof.

The characteristic polynomial $\chi_A(\lambda)$ of a matrix A is a monic polynomial such that $\chi_A(A) = 0$, by the Cayley-Hamilton Theorem; and of course any polynomial in the ideal generated by $\chi_A(\lambda)$ also vanishes at the matrix A. However there may be still other polynomials that vanish at the matrix A.

Theorem 4.5 (Minimal Polynomial Theorem). *Let A be an $n \times n$ matrix over a field F.*

(i) *There is a unique monic polynomial $\mu_A(\lambda) \in F[\lambda]$ of minimal degree for which $\mu_A(A) = 0$.*

(ii) *A polynomial $p(\lambda)$ satisfies $p(A) = 0$ if and only if $\mu_A(\lambda)|p(\lambda)$.*

(iii) *The roots of the monic polynomial $\mu_A(\lambda)$ are precisely the eigenvalues of the matrix A.*

Proof: (i) and (ii) It is clear that the set of all polynomials $p(\lambda)$ such that $p(A) = 0$ is an ideal in $F[\lambda]$, which is a nonzero ideal since it contains $\chi_A(\lambda)$. The conclusions of (i) and (ii) follow immediately from Theorem 4.4 (ii).

(iii) Finally since $\chi_A(A) = 0$ by the Cayley-Hamilton Theorem it follows from part (ii) that $\chi_A(\lambda) = \mu_A(\lambda)q(\lambda)$ for some polynomial $q(\lambda)$ and consequently that the roots of the polynomial $\mu_A(\lambda)$ must be among the roots of the polynomial $\chi_A(\lambda)$, which are the eigenvalues of the matrix A. Conversely if λ is an eigenvalue of A then $A\mathbf{v} = \lambda\mathbf{v}$ for an eigenvector $\mathbf{v} \neq \mathbf{0}$ and it follows from Lemma 4.2 that $\mu_A(A)\mathbf{v} = \mu_A(\lambda)\mathbf{v}$; but $\mu_A(A) = 0$ so $\mu_A(\lambda)\mathbf{v} = \mathbf{0}$, and since $\mathbf{v} \neq \mathbf{0}$ necessarily $\mu_A(\lambda) = 0$, which demonstrates (iii) and thereby concludes the proof.

The polynomial $\mu_A(\lambda)$ is called the **minimal polynomial** of the matrix A. If $p(A) = 0$ for a polynomial $p(\lambda)$ then $p(C^{-1}AC) = C^{-1}p(A)C = 0$ for any nonsingular matrix C since for instance $C^{-1}A^2C = C^{-1}AC \cdot C^{-1}AC = (C^{-1}AC)^2$; hence similar matrices have the same minimal polynomial, so the minimal polynomial of a linear transformation T on a vector space V can be defined to be the minimal polynomial of any matrix representing the linear transformation T in terms of a basis for the vector space V. As a consequence of the preceding theorem the characteristic and minimal polynomials for any matrix A are related by $\mu_A(\lambda)|\chi_A(\lambda)$. To find the minimal polynomial for a matrix A it thus suffices just to look at those polynomials $\mu(\lambda)$ such that $\mu(\lambda)|\chi_A(\lambda)$, a finite set of polynomials, and to see for which of these polynomials $\mu(A) = 0$. For example, the matrix $A = \begin{pmatrix} 2 & 0 \\ 0 & 2 \end{pmatrix}$ has the characteristic polynomial $\chi_A(x) = (x-2)^2$, so to find the minimal polynomial it suffices just to look at the polynomials $(x-2)$ and $(x-2)^2$; and since $(A - 2I) = 0$ it follows that $\mu_A(x) = (x-2)$. On the other hand the matrix $B = \begin{pmatrix} 2 & 1 \\ 0 & 2 \end{pmatrix}$ has the same characteristic polynomial $\chi_B(x) = (x-2)^2$; but $(B - 2I) = \begin{pmatrix} 0 & 1 \\ 0 & 0 \end{pmatrix} \neq 0$ while $(B - 2I)^2 = \begin{pmatrix} 0 & 1 \\ 0 & 0 \end{pmatrix}\begin{pmatrix} 0 & 1 \\ 0 & 0 \end{pmatrix} = 0$ so $\mu_B(x) = (x-2)^2$. The matrices A and B thus have the same characteristic polynomials but different minimal polynomials; consequently they are not similar matrices.

Theorem 4.6 (Direct Sum Theorem). *If $C = A \oplus B$ where A, B are square matrices with coefficients in a field F then*
(i) the characteristic polynomial $\chi_C(\lambda)$ of the matrix C is the product $\chi_C(\lambda) = \chi_A(\lambda)\chi_B(\lambda)$, and
(ii) the minimal polynomial $\mu_C(\lambda)$ of the matrix C is the least common multiple of the minimal polynomials $\mu_A(\lambda)$ and $\mu_B(\lambda)$.

Proof: (i) This follows immediately from the observation that

$$\det\big((\lambda I - A) \oplus (\lambda I - B)\big) = \det(\lambda I - A) \cdot \det(\lambda I - B).$$

(ii) If $C = A \oplus B$ and $f(\lambda)$ is any polynomial it is clear that $f(C) = f(A) \oplus f(B)$ hence that $f(C) = 0$ if and only if $f(A) = f(B) = 0$, or equivalently if and only if $\mu_A(\lambda)|f(\lambda)$ and $\mu_B(\lambda)|f(\lambda)$. Therefore the minimal polynomial $\mu_C(x)$ is the monic polynomial of least degree that is divisible by both $\mu_A(\lambda)$ and $\mu_B(\lambda)$ so $\mu_C(\lambda)$ is the least common multiple of $\mu_A(\lambda)$ and $\mu_B(\lambda)$. That suffices for the proof.

Theorem 4.7 (Primary Decomposition Theorem). *If the minimal polynomial $\mu_A(\lambda)$ of a matrix A with coefficients in a field F is the product $\mu_A(\lambda) = \prod_{i=1}^{m} p_i(\lambda)^{r_i}$ for distinct irreducible polynomials $p_i(\lambda)$ then the matrix A is similar to the direct sum $A_1 \oplus A_2 \oplus \cdots \oplus A_m$ of matrices A_i where $p_i(\lambda)^{r_i} = \mu_{A_i}(\lambda)$ is the minimal polynomial of the matrix A_i, and the matrices A_i are unique up to similarity.*

Proof: Let $f_i(\lambda) = \mu_A(\lambda)/p_i(\lambda)^{r_i}$ for $1 \le i \le m$, so that $f_i(\lambda)$ also are polynomials with coefficients in F; and let $\mathcal{I}$ be the polynomial ideal generated by the polynomials $f_i(\lambda)$. Theorem 4.4 shows that $\mathcal{I}$ contains a polynomial $d(\lambda)$ of least degree and that $d(\lambda)|f_i(\lambda)$ for $1 \le i \le m$; but since $p_i(\lambda)$ are $m > 1$ distinct irreducible polynomials there are no polynomials of positive degree that divide all the polynomials $f_i(\lambda)$, hence $d(\lambda) = 1$. Thus $1 \in \mathcal{I}$ so $1 = \sum_{i=1}^{m} f_i(\lambda)g_i(\lambda)$ for some polynomials $g_i(\lambda)$; consequently the matrices $E_i = f_i(A)g_i(A)$ satisfy

$$I = \sum_{i=1}^{m} E_i \qquad \text{where } I \text{ is the identity matrix.} \qquad (4.4)$$

The matrices E_i commute with one another and with the matrix A since all of them are polynomials in the matrix A; and if $i \ne j$ then since

$$f_i(\lambda)f_j(\lambda) = \frac{\mu_A(\lambda)\mu_A(\lambda)}{p_i(\lambda)^{r_i}p_j(\lambda)^{r_j}} = \mu_A(\lambda)q(\lambda)$$

for some polynomial $q(\lambda)$ it follows that

$$E_iE_j = f_i(A)g_i(A)f_j(A)g_j(A) = \mu_A(A)q(A)g_i(A)g_j(A) = 0 \qquad \text{if} \qquad i \ne j. \qquad (4.5)$$

Furthermore it follows from (4.4) and (4.5) that

$$E_j = E_jI = \sum_{i=1}^{m} E_jE_i = E_j^2. \qquad (4.6)$$

In terms of these $n \times n$ matrices E_i with coefficients in the field F introduce the subspaces

$$V_i = E_iF^n = \left\{ E_i\mathbf{v} \,\middle|\, \mathbf{v} \in F^n \right\} \subset F^n, \qquad (4.7)$$

noting that $AV_i = AE_iF^n = E_iAF^n \subset E_iF^n = V_i$ so that

$$AV_i \subset V_i \quad \text{for} \quad 1 \le i \le m. \qquad (4.8)$$

It follows from (4.4) that any vector $\mathbf{v} \in F^n$ can be written as a sum $\mathbf{v} = \sum_{i=1}^{m} E_i \mathbf{v}$ of vectors $E_i \mathbf{v} \in V_i$; and by (4.6) and (4.5) this is a unique decomposition, since if $\mathbf{v} = \sum_{i=1}^{m} E_i \mathbf{v}_i = \sum_{i=1}^{m} E_i \mathbf{w}_i$ for some vectors $\mathbf{v}_i, \mathbf{w}_i$ then $\mathbf{0} = \sum_{i=1}^{m} E_i(\mathbf{v}_i - \mathbf{w}_i)$ hence $\mathbf{0} = E_j\left(\sum_{i=1}^{m} E_i(\mathbf{v}_i - \mathbf{w}_i)\right) = E_j E_j(\mathbf{v}_j - \mathbf{w}_j) = E_j(\mathbf{v}_j - \mathbf{w}_j)$ so $E_j \mathbf{v}_j = E_j \mathbf{w}_j$ for all j. Thus the vector space F^n is the direct sum

$$F^n = \bigoplus_{i=1}^{m} V_i \tag{4.9}$$

where the subspaces V_i are preserved under the action of A as in (4.8). Finally from the preceding definitions it further follows that $p_i^{r_i}(A)V_i = p_i^{r_i}(A)E_i F^n = p_i^{r_i}(A)f_i(A)g_i(A)F^n = \mu_A(A)g_i(A)F^n$, so since $\mu_A(A) = 0$

$$p_i^{r_i}(A)V_i = \mathbf{0} \quad \text{for} \quad 1 \le i \le m. \tag{4.10}$$

These observations essentially yield the proof of the theorem. Indeed after a suitable nonsingular linear transformation of the vector space F^n, which amounts to replacing the matrix A by a similar matrix which also will be labeled A to simplify the notation, it can be assumed that the direct sum decomposition (4.9) has the form

$$F^n = \begin{pmatrix} V_1 \\ V_2 \\ \cdots \\ V_m \end{pmatrix}. \tag{4.11}$$

After this change of variables (4.8) amounts to the condition that the matrix A has the corresponding direct sum decomposition

$$A = \begin{pmatrix} A_1 & 0 & \cdots & 0 \\ 0 & A_2 & \cdots & 0 \\ & & \cdots & \\ 0 & 0 & \cdots & A_m \end{pmatrix} \tag{4.12}$$

and (4.10) amounts to the condition that $p_i^{r_i}(A_i) = 0$, so that $p_i(\lambda)^{r_i}$ must be divisible by the minimal polynomial of the matrix A_i. If $p_i^{r_i}(\lambda)$ is not itself the minimal polynomial for the matrix A_i then the minimal polynomial must be $p_i^{s_i}(\lambda)$ where $s_i < r_i$; but then by Theorem 4.6 the minimal polynomial of the matrix A would be $\prod_{i=1}^{m} p_i(\lambda)^{s_i}$, which is not the case.

Suppose there is another direct sum decomposition

$$A = \begin{pmatrix} A'_1 & 0 & \cdots & 0 \\ 0 & A'_2 & \cdots & 0 \\ & & \cdots & \\ 0 & 0 & \cdots & A'_m \end{pmatrix} \tag{4.13}$$

satisfying the conclusions of the theorem, so that $\mu_{A'_i}(\lambda) = p_i(\lambda)^{r_i}$; and consider the corresponding decomposition

$$F^n = \begin{pmatrix} V'_1 \\ V'_2 \\ \cdots \\ V'_m \end{pmatrix} \tag{4.14}$$

of the vector space F^n. Since $\mu_{A'_i}(\lambda) = p_i(\lambda)^{r_i}$ then as in the preceding argument $E_i V'_i = \mathbf{0}$; therefore for any vector $\mathbf{v}'_i \in V'_i$ it follows that $\mathbf{v}'_i = I\mathbf{v}'_i = \sum_{j=1}^m E_j \mathbf{v}'_i = E_i \mathbf{v}'_i \in V_i$ so that $V'_i \subset V_i$, and since $\dim V'_i = \dim V_i$ necessarily $V'_i = V_i$, which suffices for the proof.

The preceding theorem provides a simple characterization of diagonalizable matrices.

Corollary 4.8 (Diagonalizability Theorem). *A matrix A with coefficients in a field F is diagonalizable if and only if its minimal polynomial has the form $\mu_A(\lambda) = \prod_{j=1}^m (\lambda - b_j)$ where b_j are distinct elements of the field F.*

Proof: A diagonal matrix with coefficients in the field F is just the direct sum $A = \oplus_i a_i$ of one-dimensional matrices $a_i \in F$. It is evident from the definitions that the characteristic and minimal polynomials of a 1×1 matrix a_i are equal and are explicitly $\chi_{a_i}(\lambda) = \mu_{a_i}(\lambda) = (\lambda - a_i)$. It then follows immediately from Theorem 4.6 that the characteristic polynomial of the matrix $A = \bigoplus_{i=1}^n a_i$ is $\chi_A(\lambda) = \prod_{i=1}^n (\lambda - a_i)$ and that the minimal polynomial of A is $\mu_A(\lambda) = \prod_{j=1}^m (\lambda - b_j)$ where b_j are the distinct terms among the diagonal entries a_i $\mu_A(\lambda) = \prod_{j=1}^m (\lambda - b_j)$. Conversely if the minimal polynomial of a matrix A has the form $\mu_A(\lambda) = \prod_{j=1}^m (\lambda - b_j)$ where b_j are distinct elements of the field F the factors $(\lambda - b_j)$ are distinct irreducible polynomials, so by the preceding theorem the matrix A is similar to the direct sum $A_1 \oplus A_2 \oplus \cdots \oplus A_m$ where the minimal polynomial of the matrix A_i is $\mu_{a_i}(\lambda) = (\lambda - b_i)$; then $0 = \mu_{a_i}(A_i) = A_i - b_i I$ so $A_i = b_i I$ is a diagonal matrix, and that concludes the proof.

Not all matrices are diagonalizable; so it is necessary to find other normal forms to which non-diagonalizable matrices can be reduced. For the purposes of the discussion here it is enough to consider just one special class of non-diagonalizable matrices. A matrix $N \in F^{n \times n}$ is said to be **nilpotent** if $N^r = 0$ for some integer $r > 0$; the least integer r for which $N^r = 0$ is called the **degree** of the nilpotent matrix N. It is clear that any matrix similar to a nilpotent matrix of degree r is also a nilpotent matrix of degree r, since if $N^r = 0$ and P is a nonsingular matrix then $(PNP^{-1})^r = PN^rP^{-1} = P0P^{-1} = 0$.

Theorem 4.9 (Nilpotent Matrix Theorem). *A nilpotent $n \times n$ matrix N with coefficients in a field F is similar to a direct sum*

$$N = N_{\rho_1} \oplus N_{\rho_2} \oplus \cdots \oplus N_{\rho_k} \tag{4.15}$$

where N_ρ is a $\rho \times \rho$ matrix of the form

$$N_\rho = \begin{pmatrix} 0 & 1 & 0 & 0 & \cdots & 0 & 0 \\ 0 & 0 & 1 & 0 & \cdots & 0 & 0 \\ 0 & 0 & 0 & 1 & \cdots & 0 & 0 \\ & & & \cdots & & & \\ 0 & 0 & 0 & 0 & \cdots & 0 & 1 \\ 0 & 0 & 0 & 0 & \cdots & 0 & 0 \end{pmatrix}; \tag{4.16}$$

the indices ρ_i are uniquely determined up to order, and two nilpotent $n \times n$ matrices are similar if and only if they have the same set of indices ρ_i.

Proof: Note that if δ_i is the Kronecker basis for the vector space F^ρ the condition that a matrix N has the form (4.16) can be restated as the condition that

$$N\delta_1 = \mathbf{0} \quad \text{while} \quad N\delta_i = \delta_{i-1} \quad \text{for} \quad 2 \leq i \leq \rho; \tag{4.17}$$

it is evident from this that the matrix N is a nilpotent matrix of degree ρ, and since a direct sum of nilpotent matrices is again nilpotent it follows that any direct sum of matrices of the form (4.16) is nilpotent.

To show conversely that any nilpotent matrix is similar to a direct sum of matrices satisfying (4.17), let N be a nilpotent $n \times n$ matrix of degree r over the field F and introduce the subspaces

$$V_i = \ker N^i = \left\{ \mathbf{v} \in F^n \,\middle|\, N^i\mathbf{v} = \mathbf{0} \right\} \subset F^n \quad \text{for} \quad 0 \leq i \leq r. \tag{4.18}$$

Clearly

$$\mathbf{0} = V_0 \subset V_1 \subset V_2 \subset \cdots \subset V_{r-1} \subset V_r = F^n \tag{4.19}$$

since $N^0 = I$ and $N^r = 0$; and if $d_i = \dim V_i$ it follows that $0 = d_0 \le d_1 \le d_2 \le \cdots \le d_{r-1} \le d_r = n$. It is also clear that $NV_i \subset V_{i-1}$, since if $\mathbf{v} \in V_i$ then $N^i\mathbf{v} = \mathbf{0}$ hence $N^{i-1}N\mathbf{v} = N^i\mathbf{v} = \mathbf{0}$ so $N\mathbf{v} \in V_{i-1}$. The principal step in the proof is to show that if vectors $\mathbf{v}_1, \ldots, \mathbf{v}_k \in V_i \subset F^n$ represent linearly independent vectors in the quotient space V_i/V_{i-1} then the vectors $N\mathbf{v}_1, \ldots, N\mathbf{v}_k \in V_{i-1} \subset F^n$ represent linearly independent vectors in the quotient space V_{i-1}/V_{i-2}. To demonstrate that, suppose that vectors $\mathbf{v}_1, \ldots, \mathbf{v}_k \in V_i$ represent linearly independent vectors in the quotient space V_i/V_{i-1} and that $\sum_{j=1}^k c_j N\mathbf{v}_j \in V_{i-2}$ for some $c_j \in F$; then $\mathbf{0} = N^{i-2}\left(\sum_{j=1}^k c_j N\mathbf{v}_j\right) = N^{i-1}\left(\sum_{j=1}^k c_j \mathbf{v}_j\right)$ hence $\sum_{j=1}^k c_j \mathbf{v}_j \in V_{i-1}$, so since the vectors $\mathbf{v}_j$ represent linearly independent vectors in V_i/V_{i-1} necessarily $c_j = 0$ for all indices j. Consequently the vectors $N\mathbf{v}_j \in V_{i-1}$ represent linearly independent vectors in V_{i-1}/V_{i-2}.

With these auxiliary results established, choose linearly independent vectors $\mathbf{v}_{r,j_r} \in V_r = F^n$ for $1 \le j_r \le d_r - d_{r-1}$ so that

$$\{\mathbf{v}_{r,j_r}\} \quad \text{represent a basis for } V_r/V_{r-1}. \tag{4.20}$$

The vectors $N\mathbf{v}_{r,j_r} \in V_{r-1} \subset F^n$ then represent linearly independent vectors in the quotient space V_{r-1}/V_{r-2}; if they do not span that vector space choose additional linearly independent vectors $\mathbf{v}_{r-1,j_{r-1}} \in V_{r-1}$ for $1 \le j_{r-1} \le (d_{r-1} - d_{r-2}) - (d_r - d_{r-1})$ so that

$$\{N\mathbf{v}_{r,j_r}\}, \quad \{\mathbf{v}_{r-1,j_{r-1}}\} \quad \text{represent a basis for } V_{r-1}/V_{r-2}. \tag{4.21}$$

Continue this process until finally there are vectors $\mathbf{v}_{1,j_1} \in V_1$ so that the vectors

$$\{N^{r-1}\mathbf{v}_{r,j_r}\}, \quad \{N^{r-2}\mathbf{v}_{r-1,j_{r-1}}\}, \ldots, \{N\mathbf{v}_{2,j_2}\}, \quad \{\mathbf{v}_{1,j_1}\} \tag{4.22}$$

$$\text{represent a basis for } V_1/V_0 = V_1.$$

For each index j_r the vectors

$$\mathbf{v}_{r,j_r}, N\mathbf{v}_{r,j_r}, N^2\mathbf{v}_{r,j_r}, \ldots, N^{r-1}\mathbf{v}_{r,j_r} \tag{4.23}$$

are linearly independent. Indeed if $\sum_{i=0}^{r-1} c_i N^i \mathbf{v}_{r,j_r} = \mathbf{0}$ for some scalars $c_i \in F$, not all of which are zero, and if s is the least integer for which $c_s \ne 0$ then

$$\mathbf{0} = N^{r-1-s} \sum_{i=s}^{r-1} c_i N^i \mathbf{v}_{r,j_r} = c_s N^{r-1} \mathbf{v}_{r,j_r}$$

which is a contradiction since $N^{r-1}\mathbf{v}_{r,j_r} \ne \mathbf{0}$ and $c_s \ne 0$. The vectors (4.23) thus are a basis for a linear subspace $W_{r,j_r} \subset F^n$ of dimension r. Relabel these vectors by setting $\delta_i = N^{r-i}\mathbf{v}_{r,j_r}$; then $N\delta_1 = NN^{r-1}\mathbf{v}_{r,j_r} = \mathbf{0}$ while $N\delta_i = NN^{r-i}\mathbf{v}_{r,j_r} = N^{r-(i-1)}\mathbf{v}_{r,j_r} = \delta_{i-1}$ for $i > 1$, which is condition (4.17) for the restriction $N|W_{r,j_r}$

where $\rho = r$. This construction can be repeated for each of the vectors $\mathbf{v}_{r,j_r}$; the number of such vectors is the dimension $\dim V_r/V_{r-1} = d_r - d_{r-1}$, so in terms of an appropriate basis there result $d_r - d_{r-1}$ separate $r \times r$ matrix blocks of the form (4.16) for $\rho = r$. The process can be continued beginning with the vectors $\mathbf{v}_{r-1,j_{r-1}}$ to obtain a further collection of $(d_{r-1} - d_{r-2}) - (d_r - d_{r-1})$ separate $(r - 1) \times (r - 1)$ matrix blocks of the form (4.16), and so on for the rest of the vectors $\mathbf{v}_{i,j_i}$. This shows altogether that the matrix N is similar to a direct sum of matrix blocks of the form (4.16); and since the number of blocks of different sizes is determined by the dimensions d_i this direct sum is uniquely determined up to the order of the separate blocks, which suffices to conclude the proof.

For matrices over a field F with minimal polynomials which factor fully into linear factors over that field there is a convenient normal form, the **Jordan Normal Form**, that serves to provide a complete classification up to similarity. The Jordan normal form is a representation of a matrix as a direct sum of **Jordan blocks**, which are $\rho \times \rho$ matrices of the form

$$B(\rho; a) = \begin{pmatrix} a & 1 & 0 & 0 & \cdots & 0 & 0 \\ 0 & a & 1 & 0 & \cdots & 0 & 0 \\ 0 & 0 & a & 1 & \cdots & 0 & 0 \\ & & & \cdots & & & \\ & & & \cdots & & & \\ 0 & 0 & 0 & 0 & \cdots & a & 1 \\ 0 & 0 & 0 & 0 & \cdots & 0 & a \end{pmatrix} \tag{4.24}$$

for a scalar $a \in F$ that clearly is the eigenvalue of the matrix $B(\rho; a)$.

Theorem 4.10. *The characteristic and minimal polynomials of the Jordan block $B(\rho; a)$ are*

$$\chi_{B(\rho;a)}(\lambda) = \mu_{B(\rho;a)}(\lambda) = (\lambda - a)^\rho;$$

and the eigenspace of the Jordan block $B(\rho; a)$ is one dimensional.

Proof: It is a straightforward calculation that a matrix $B(\rho; a)$ of the form (4.24) has the determinant $\chi_{B(\rho;a)}(\lambda) = \det B(\rho; a) = (\lambda - a)^\rho$. Then since $B(\rho; a) - aI = N$ is a $\rho \times \rho$ matrix of the form (4.16) it is a nilpotent matrix of degree ρ, so the minimal polynomial for the matrix $B(\rho; a)$ is the polynomial $\mu_{B(\rho;a)}(\lambda) = (\lambda - a)^\rho$. The only eigenvalue of the matrix $B(\rho; a)$ is a, the unique root of the polynomial $\chi_{B(\rho;a)}(\lambda)$; and it is another straightforward calculation to verify that the eigenvectors of the matrix $B(\rho; a)$ for this eigenvalue are scalar multiples of the vector

$$\mathbf{v} = \begin{pmatrix} 1 \\ 0 \\ \cdots \\ 0 \end{pmatrix},$$

so the eigenspace for the eigenvector a is one dimensional. That suffices for the proof.

Theorem 4.11 (Jordan Normal Form Theorem). *A matrix A over a field F with minimal polynomial* $\mu_A(\lambda) = \prod_{i=1}^{m}(\lambda - a_i)^{\mu_i}$ *for some scalars* $a_i \in F$ *is similar to a direct sum*

$$A_1 \oplus A_2 \oplus \cdots \oplus A_m \tag{4.25}$$

of matrices A_i *with minimal polynomials* $\mu_{A_i}(\lambda) = (\lambda - a_i)^{\mu_i}$*; and each matrix* A_i *in turn is similar to a direct sum*

$$B(\rho_{i_1}; a_i) \oplus B(\rho_{i_2}; a_i) \oplus \cdots \oplus B(\rho_{i_{k(i)}}; a_i) \tag{4.26}$$

of Jordan blocks $B(\rho_j; a_i)$*. This decomposition is unique up to the order of the Jordan blocks.*

Proof: If the matrix A has the minimal polynomial $\mu_A(\lambda) = \prod_{i=1}^{m}(\lambda - a_i)^{\mu_i}$ then by the Primary Decomposition Theorem, Theorem 4.7, the matrix A is similar to a direct sum $A_1 \oplus A_2 \oplus \cdots \oplus A_m$ where $\mu_{A_i}(\lambda) = (\lambda - a_i)^{\mu_i}$, and this decomposition is unique up to similarity and to the order of the matrices A_i. Each matrix A_i is a root of its minimal polynomial so $(A_i - a_iI)^{\mu_i} = 0$; hence the matrix $N_i = A_i - a_iI$ is a $\nu_i \times \nu_i$ nilpotent matrix of degree μ_i and $A_i = a_iI + N_i$. By the Nilpotent Matrix Theorem, Theorem 4.9, the matrix N_i is similar to a direct sum $N_{i,\rho_1} \oplus N_{i,\rho_2} \oplus \cdots \oplus N_{i,\rho_k}$ of matrices N_{i,ρ_j} of the form (4.16). Since a scalar matrix remains a scalar matrix under similarity of matrices it follows that the matrix A_i is similar to a direct sum of matrices of the form $B(\rho, a) = aI + N_\rho$, and that suffices for the proof.

The preceding theorem provides the complete classification up to similarity of matrices for which the minimal polynomial factors completely. In particular it provides the complete classification up to similarity of complex matrices as a consequence of the Fundamental Theorem of Algebra, that any complex polynomial has roots. The invariants characterizing a similarity class of matrices are the unordered sets of pairs (ρ_{i_j}, a_i), where each pair consists of an integer $\rho_{i_j} \geq 1$ and a complex number a_i; and two complex matrices are similar if and only if they have the same set of these pairs, in any order.

If A is an $n \times n$ matrix in the Jordan normal form (4.25) then

$$n = \sum_{i=1}^{k} \rho_i \qquad (4.27)$$

and by the Direct Sum Theorem, Theorem 4.6 (i), the characteristic polynomial for the matrix A is

$$\chi_A(\lambda) = \prod_{i=1}^{k} (\lambda - a_i)^{\rho_i}. \qquad (4.28)$$

The roots a_i are not necessarily distinct, for several different Jordan blocks may have the same eigenvalue. If b_j for $1 \leq j \leq l$ are the distinct roots a_i and if σ_j is the maximum of the sizes ρ_i of the Jordan blocks for which $a_i = b_j$ it follows from the Direct Sum Theorem, Theorem 4.6 (ii), that the minimal polynomial for the matrix A is

$$\mu_A(\lambda) = \prod_{j=1}^{l} (\lambda - b_j)^{\sigma_j}. \qquad (4.29)$$

The eigenvalue of a Jordan block $B(\rho_i, a_i)$ is a_i and the eigenspace is one-dimensional so it is determined up to a scalar multiple by the choice of a single eigenvector; and that vector extends to an eigenvector of the matrix A with the eigenvalue a_i. For example if

$$A = B(2,3) \oplus B(1,3) \oplus B(1,4),$$

or more explicitly

$$A = \begin{pmatrix} 3 & 1 & 0 & 0 \\ 0 & 3 & 0 & 0 \\ 0 & 0 & 3 & 0 \\ 0 & 0 & 0 & 4 \end{pmatrix},$$

then the matrix A has the characteristic and minimal polynomials

$$\chi_A(\lambda) = (\lambda - 3)^3(\lambda - 4), \quad \text{and} \quad \mu_A(\lambda) = (\lambda - 3)^2(\lambda - 4)$$

so the eigenvalues are 3 and 4. The eigenvector $^t(1,0)$ of the Jordan block $B(2,3)$ extends to an eigenvector $^t(1,0,0,0)$ of the matrix A with eigenvalue 3, the eigenvector 1 of the Jordan block $B(1,3)$ extends to an eigenvector $^t(0,0,1,0)$ of the matrix A with the eigenvalue 3, and the eigenvector 1 of the Jordan block $B(1,4)$ extends to an eigenvector $^t(0,0,0,1)$ of the matrix A with the eigenvalue 4; altogether the eigenspace V_3 of the matrix A with the eigenvalue 3 is the two-dimensional linear subspace of $\mathbb{C}^4$ spanned by the two vectors $^t(1,0,0,0)$ and $^t(0,0,1,0)$ while the eigenspace V_4 of the matrix A with the eigenvalue 4 is the one-dimensional linear subspace of $\mathbb{C}^4$ spanned by the vector $^t(0,0,0,1)$.

To determine the normal form under similarity of a complex matrix A the first step is to calculate the characteristic polynomial $\chi_A(\lambda) = \det(\lambda I - A)$ and to find its roots. Calculating the roots actually is the hardest part of the determination of the normal form, and the only part that is not fairly straightforward and mechanical. Indeed there is no simple procedure or explicit algebraic formula for finding the roots of a general polynomial, and the roots usually can be determined only approximately. For that reason the whole procedure of finding the Jordan normal form is more a theoretical than a practical calculation. However if the roots of the characteristic polynomial can be found, that polynomial can be written in terms of the distinct roots as the product (4.28); and the minimal polynomial then is the product (4.29) for the minimal values σ_i for which $\mu_A(A) = 0$, a straightforward calculation. The next step is to determine the eigenvectors for A, another straightforward algebraic calculation. The set of eigenvectors for any particular eigenvalue a form the eigenspace $V_a \subset \mathbb{C}^n$; and the dimension $\dim V_a$ is the number of Jordan blocks with the eigenvalue a appearing in the Jordan normal form of the matrix A. The sizes of the various Jordan blocks are not always determined uniquely just by the minimal and characteristic polynomials and the number of eigenvectors though, so a further calculation is required. There are two alternative approaches to this part of the calculation. On the one hand it is possible to proceed as in the proof of the Jordan Normal Form Theorem, Theorem 4.11. The Primary Decomposition Theorem, Theorem 4.7, shows that if b_i are the roots of the minimal polynomial $\mu_A(\lambda)$ then the matrix A can be decomposed into a direct sum of matrices A_i for which the minimal polynomial is $\mu_i(\lambda) = (\lambda - b_i)^{\sigma_i}$. The matrices $(A_i - b_i I)$ then are nilpotent matrices, and their normal form can be calculated as in the proof of the Nilpotent Matrix Theorem, Theorem 4.9; and that yields the Jordan normal form for the matrix A_i. On the other hand an alternative, and in many ways a simpler and more direct procedure, is just to work with the eigenvectors. For a Jordan block such as

$$B(3, a) = \begin{pmatrix} a & 1 & 0 \\ 0 & a & 1 \\ 0 & 0 & a \end{pmatrix}$$

and for a vector $\mathbf{x} = {}^t(x_1, x_2, x_3) \in \mathbb{C}^3$ since

$$B(3, a)\mathbf{x} = \begin{pmatrix} a & 1 & 0 \\ 0 & a & 1 \\ 0 & 0 & a \end{pmatrix} \begin{pmatrix} x_1 \\ x_2 \\ x_3 \end{pmatrix} = \begin{pmatrix} ax_1 + x_2 \\ ax_2 + x_3 \\ ax_3 \end{pmatrix}$$

the single eigenvector for this matrix block is $\mathbf{x}_1 = {}^t(1, 0, 0)$, so

$$A\mathbf{x}_1 = a\mathbf{x}_1.$$

The vector $\mathbf{x}_2 = {}^t(0, 1, 0)$ satisfies $B(3, a)\mathbf{x}_2 = a\mathbf{x}_2 + \mathbf{x}_1$ so $\mathbf{x}_2$ is a solution of the equation

$$\left(B(3, a) - aI\right)\mathbf{x}_2 = \mathbf{x}_1;$$

and the vector $\mathbf{x}_3 = {}^t(0, 0, 1)$ satisfies $B(3, a)\mathbf{x}_3 = a\mathbf{x}_3 + \mathbf{x}_2$ so $\mathbf{x}_3$ is a solution of the equation

$$\left(B(3, a) - aI\right)\mathbf{x}_3 = \mathbf{x}_2.$$

If $B(3, a)$ is part of the Jordan normal form for a matrix A then the vectors $\mathbf{x}_i$ associated to the Jordan block $B(3, a)$ naturally extend to vectors $\mathbf{v}_i$ in the vector space on which the matrix A acts; and these extensions satisfy the system of linear equations

$$\begin{aligned}
(A - aI)\mathbf{v}_1 &= \mathbf{0} \\
(A - aI)\mathbf{v}_2 &= \mathbf{v}_1 \qquad\qquad\qquad (4.30) \\
(A - aI)\mathbf{v}_3 &= \mathbf{v}_2.
\end{aligned}$$

The vectors $\mathbf{v}_i$ then span the vector space on which the matrix A acts as the Jordan block $B(3, a)$; so a linear change of variables transforming the vectors $\mathbf{v}_i$ to the Kronecker basis vectors reduces the matrix A to the Jordan normal form. For larger or smaller Jordan blocks the corresponding results hold.

For a more explicit example of the determination of the Jordan normal form, if

$$A = \begin{pmatrix} -5 & 9 & -4 \\ -4 & 8 & -4 \\ 1 & 0 & 0 \end{pmatrix}$$

it is a simple calculation to verify that $\det(\lambda I - A) = \lambda^3 - 3\lambda^2 + 4 = (\lambda + 1)(\lambda - 2)^2$; hence the eigenvalues are $-1, 2$ and the eigenvectors are solutions of the system of linear equations $(A - \lambda I)\mathbf{v} = \mathbf{0}$. A straightforward calculation shows that the eigenvectors of this matrix are

$$\mathbf{v}_1' = {}^t(\ -1 \quad 0 \quad 1\ ) \quad \text{for the eigenvalue} \quad \lambda = -1,$$

$$\mathbf{v}_1'' = {}^t(\ 2 \quad 2 \quad 1\ ) \quad \text{for the eigenvalue} \quad \lambda = 2.$$

The factor $(\lambda + 1)$ corresponds to a 1×1 submatrix of the normal form of the matrix A, with entry the eigenvalue -1. The factor $(\lambda - 2)^2$ corresponds to a 2×2 submatrix of the normal form of the matrix A; and since there is only one eigenvector $\mathbf{v}_1''$ for the eigenvalue $\lambda = 2$, the associated Jordan block must be of the form $\left(\begin{smallmatrix} 2 & 1 \\ 0 & 2 \end{smallmatrix}\right)$. Therefore the Jordan normal form for the matrix A thus is

$$B = \begin{pmatrix} -1 & 0 & 0 \\ 0 & 2 & 1 \\ 0 & 0 & 2 \end{pmatrix}.$$

To reduce the matrix A to this normal form, associated to the 2×2 block there must be a vector $\mathbf{v}_2''$ for which $(A - 2I)\mathbf{v}_2'' = \mathbf{v}_1''$; a straightforward calculation shows that it is the vector

$$\mathbf{v}_2'' = {}^t(\ 1 \quad 1 \quad 0\ ).$$

The vectors $\mathbf{v}_1'$, $\mathbf{v}_1''$, $\mathbf{v}_2''$ form a basis for the vector space, so the matrix

$$C = (\mathbf{v}_1' \ \mathbf{v}_1'' \ \mathbf{v}_2'') = \begin{pmatrix} -1 & 2 & 1 \\ 0 & 2 & 1 \\ 1 & 1 & 0 \end{pmatrix}$$

is nonsingular; and since $C\delta_j = \mathbf{v}_j$ it follows that $C^{-1}AC = B$. In this simple case there is a single eigenvector for each eigenvalue so a single candidate for the vector $\mathbf{v}_1$ in the set of formulas (4.30) for each eigenvalue. If the eigenspace V_a for a particular eigenvalue a has dimension $\dim V_a > 1$ then there are at least 2 different Jordan blocks with that eigenvalue and correspondingly two different choices for the eigenvectors $\mathbf{v}_1$ in formulas (4.30). These choices cannot be made randomly; it is necessary to try to solve at least the first two equations in (4.30) simultaneously to determine the vectors to be used.

As already noted, the Jordan normal form of a matrix is not always determined uniquely just by its characteristic and minimal polynomials, although there are only finitely many different possible Jordan normal forms with given characteristic and minimal polynomials. For instance the two matrices

$$A_1 = \begin{pmatrix} a & 1 \\ 0 & a \end{pmatrix} \oplus \begin{pmatrix} a & 1 \\ 0 & a \end{pmatrix} \oplus \begin{pmatrix} a & 0 \\ 0 & a \end{pmatrix}$$

and

$$A_2 = \begin{pmatrix} a & 1 \\ 0 & a \end{pmatrix} \oplus \begin{pmatrix} a & 0 \\ 0 & a \end{pmatrix} \oplus \begin{pmatrix} a & 0 \\ 0 & a \end{pmatrix}$$

both have the characteristic polynomial $\chi_{a_i}(\lambda) = (\lambda - a)^6$ and the minimal polynomial $\mu_{a_i}(\lambda) = (\lambda - a)^2$, in view of the preceding discussion and Theorem 4.6; but they are not similar matrices since they do not have the same Jordan normal forms, or even more simply, because they have different numbers of eigenvectors. Actually in this example the eigenspace of the matrix A_1 has dimension 4 while the eigenspace of the matrix A_2 has dimension 5, so that serves to distinguish them.

Any real matrix M can be viewed as a complex matrix; so M is similar to a complex matrix SMS^{-1} in Jordan normal form, although the matrix S exhibiting this similarity can be a complex matrix. The eigenvalues of a real matrix are either real, in which case the arguments can proceed as with complex matrices, or are complex conjugate pairs, in which case it is possible to choose normal forms that are not diagonal but are formed from 2×2 matrix blocks. For instance, for any real values a, b the following two matrices are similar:

$$\begin{pmatrix} a + ib & 0 \\ 0 & a - ib \end{pmatrix} \qquad \begin{pmatrix} a & b \\ -b & a \end{pmatrix};$$

and generalized Jordan normal forms can be taken with diagonal terms of either real numbers or 2×2 real matrices of the above forms. This topic will not be discussed any further here, though.

The discussion of Jordan normal forms of matrices is really of more theoretical than practical use; it can be used in a theoretical examination of matrices or in proofs of general properties of matrices, but the explicit calculation of the roots of the characteristic polynomial is impracticable in nontrival cases. The topic of direct calculations with large matrices is really an entirely different, difficult, and important branch of mathematics.

Problems, Group I

1. Which of the following sets are ideals in the polynomial ring $F[x]$ for a field F? For each set that is an ideal find its monic generator.

    (i) The set of $f \in F[x]$ such that $\deg f \geq 6$.
    (ii) The set of $f \in F[x]$ such that $f(2) = f(4) = 0$.
    (iii) The set of $f \in F[x]$ such that $f(x) = x^2 p(x) + x^5 q(x)$ for some $p, q \in F[x]$.
    (iv) The set of $f \in F[x]$ that involve only even powers of the variable x.

2. Show that two polynomials in $\mathbb{C}[x]$ have the greatest common divisor equal to 1 if and only if the two polynomials have no common roots. Show that this is not the case for two polynomials in $\mathbb{R}[x]$.

3. Let A and B be $n \times n$ matrices over a field F.

    (i) Show that AB and BA always have the same eigenvalues.
    (ii) Do AB and BA always have the same characteristic polynomial?
    (iii) Do AB and BA always have the same minimal polynomial?

4.   Find a 3×3 real matrix A with the minimal polynomial $\mu_A(\lambda) = \lambda^2$.

5.   Show that the following matrix over a field F has the same minimal and

characteristic polynomials: $\begin{pmatrix} 0 & 0 & a \\ 1 & 0 & b \\ 0 & 1 & c \end{pmatrix}$.

Problems, Group II

6.   If A is a matrix with the characteristic polynomial

$$\chi_A(\lambda) = (\lambda - 2)^3(\lambda - 3)^4$$

and the minimal polynomial $\mu_A(\lambda)$, how many eigenvectors can A have in the
following cases? (More precisely, describe the possible eigenspaces.) Justify
your answer.
   (i) $\deg \mu_A(\lambda) = 7$
   (ii) $\deg \mu_A(\lambda) = 6$
   (iii) $\deg \mu_A(\lambda) = 5$
   (iv) $\deg \mu_A(\lambda) = 2$

7.   Find the Jordan normal form for the complex matrix

$$A = \begin{pmatrix} 2 & 0 & 1 & -3 \\ 0 & 2 & 10 & 4 \\ 0 & 0 & 2 & 0 \\ 0 & 0 & 0 & 3 \end{pmatrix}$$

and find a matrix P such that PAP^{-1} is the Jordan normal form.

8.   Show that 3×3 complex matrices are similar if and only if they have the same
characteristic and minimal polynomials. Show that is false for 4×4 matrices.

9.   If $A : V \longrightarrow V$ is a linear mapping between two vector spaces over a field show
that there is a direct sum decomposition $V = V_1 \oplus V_2$ which is preserved by A,
in the sense that $AV_1 \subset V_1$ and $AV_2 \subset V_2$, and is such that A is nilpotent on V_1
and A is nonsingular on V_2.

10.   (i) If $\mathcal{I} \subset F[x]$ is the ideal in the polynomial ring over a field F generated by
       two polynomials $f(x), g(x) \in F[x]$ and if $\mathcal{I} = F[x] d(x)$ for a monic
       polynomial $d(x) \in F[x]$ show that $d(x)$ is the greatest common divisor of
       the polynomials $f(x)$ and $g(x)$.

(ii) Show that the greatest common divisor $d(x)$ of two polynomials $f(x), g(x) \in F[x]$ can be written as the sum $d(x) = p(x)f(x) + q(x)g(x)$ for some polynomials $p(x), q(x) \in F[x]$.

(iii) Show that if $p(x)$ is the greatest common divisor and $q(x)$ is the least common multiple of two polynomials $f(x), g(x) \in F[x]$ then $f(x)g(x) = cp(x)q(x)$ for some $c \in F$.

11. Let V be the vector space of $n \times n$ matrices over a field F, let $T : V \longrightarrow V$ be the linear mapping defined by $T(X) = AX$ for any $X \in V$ and for a fixed matrix $A \in V$, and let A also denote the linear transformation from F^n to F^n defined by the matrix A.

(i) Do the linear transformations A and T have the same eigenvalues?

(ii) Do the linear transformations A and T have the same characteristic polynomial?

(iii) Do the linear transformations A and T have the same minimal polynomial?

12. Show that if an $n \times n$ complex matrix A satisfies $A^r = I$ and if A has just one eigenvalue λ then $A = \lambda I$. What are the possible values for λ?

13. Show that $\det(I + N) = 1$ for any nilpotent matrix N.

14. Classify up to similarity all 3×3 complex matrices A such that $A^2 = I$.

15. Show that if A is an invertible $n \times n$ matrix over a field F there is a polynomial $p(x) \in F[x]$ such that $A^{-1} = p(A)$.

16. Show that any complex matrix is similar to its transpose.

4.2 Inner Product Spaces

There are other equivalence relations between matrices, associated to other uses of matrices in linear algebra. A **bilinear function** on a finite dimensional vector space V over a field F is defined to be a mapping $f : V \times V \longrightarrow F$ such that $f(\mathbf{u}, \mathbf{v})$ is a linear function of the variable $\mathbf{u} \in V$ for each fixed $\mathbf{v} \in V$ and also is a linear function of the variable $\mathbf{v} \in V$ for each fixed $\mathbf{u} \in V$; a bilinear function is a special case of multilinear functions. In terms of a basis $\{\mathbf{v}_i\}$ for the n-dimensional vector space V a bilinear function f is described by the matrix $A \in F^{n \times n}$ with the entries $a_{ij} = f(\mathbf{v}_i, \mathbf{v}_j)$; for if $\mathbf{u} = \sum_{i=1}^n x_i \mathbf{v}_i$ and $\mathbf{v} = \sum_{j=1}^n y_j \mathbf{v}_j$

then

$$f(\mathbf{u}, \mathbf{v}) = f\left(\sum_{i=1}^{n} x_i \mathbf{v}_i, \sum_{i=1}^{n} y_j \mathbf{v}_j\right) = \sum_{i,j=1}^{n} x_i y_j f(\mathbf{v}_i, \mathbf{v}_j) = \sum_{i,j=1}^{n} a_{ij} x_i y_j. \qquad (4.31)$$

When the coefficients $\{x_i\}$ and $\{y_j\}$ are viewed as describing column vectors $\mathbf{x}$ and $\mathbf{y}$ in F^n the preceding equation can be written in matrix form as

$$f(\mathbf{x}, \mathbf{y}) = {}^t\mathbf{x} A \mathbf{y}, \qquad (4.32)$$

where ${}^t\mathbf{x}$ is the transpose of the column vector $\mathbf{x}$ so is a row vector. Any other basis $\mathbf{v}'_i$ for the vector space V has the form $\mathbf{v}'_i = \sum_{j=1}^{n} p_{ij} \mathbf{v}_j$ for some nonsingular matrix $P = \{p_{ij}\} \in \mathrm{Gl}(n, F)$; so a vector $\mathbf{v} \in V$ has the descriptions $\mathbf{v} = \sum_{i=1}^{n} x'_i \mathbf{v}'_i = \sum_{i,j=1}^{n} x'_i p_{ij} \mathbf{v}_j = \sum_{j=1}^{n} x_j \mathbf{v}_j$ where the coordinates are related by $x_j = \sum_{i=1}^{n} x'_i p_{ij}$, or alternatively when the set of coordinates is viewed as a column vector, by $\mathbf{x} = {}^t P \mathbf{x}'$. Consequently the explicit descriptions (4.31) in terms of the two bases are related by

$$f(\mathbf{u}, \mathbf{v}) = \sum_{i,j=1}^{n} a_{ij} x_i y_j = \sum_{i,j,k,l=1}^{n} a_{ij} x'_k p_{ki} \, y'_l p_{lj} = \sum_{k,l=1}^{n} a'_{kl} x'_k y'_l \qquad (4.33)$$

where

$$a'_{kl} = \sum_{i,j=1}^{n} a_{ij} p_{ki} p_{lj} \quad \text{or in matrix terms} \quad A' = PA\,{}^t P. \qquad (4.34)$$

Two $n \times n$ matrices A, A' over a field F that are related as in (4.34) for a nonsingular matrix $P \in \mathrm{Gl}(n, F)$ are said to be **congruent** matrices; note particularly that this relation involves only square matrices. It is easy to see that this is an equivalence relation between $n \times n$ matrices over the field F. The set of matrices describing a bilinear function $f : V \times V \longrightarrow F$ as in (4.31) for all choices of bases for V thus is an equivalence class of congruent matrices. If P is a nonsingular $n \times n$ matrix then rank $(PX) = $ rank $(XP) = $ rank X for any other $n \times n$ matrix X, by Theorem 1.16; it follows that all congruent matrices have the same rank, which is defined to be the **rank** of the bilinear function described by any of these congruent matrices.

A bilinear function $f(\mathbf{x}, \mathbf{y})$ on a vector space V over a field F is **trivial** if $f(\mathbf{x}, \mathbf{y}) = 0$ for all $\mathbf{x}, \mathbf{y} \in V$, and otherwise is **nontrivial**. It is said to be **symmetric** if $f(\mathbf{x}, \mathbf{y}) = f(\mathbf{y}, \mathbf{x})$ for all vectors $\mathbf{x}, \mathbf{y} \in V$, and is said to be **skew symmetric** if $f(\mathbf{x}, \mathbf{y}) = -f(\mathbf{y}, \mathbf{x})$ for all vectors $\mathbf{x}, \mathbf{y} \in V$. If $f(\mathbf{x}, \mathbf{y})$ is a symmetric bilinear function described by a matrix A then since $a_{ij} = f(\mathbf{v}_i, \mathbf{v}_j)$ for basis vectors $\mathbf{v}_i, \mathbf{v}_j$ it follows that $a_{ij} = f(\mathbf{v}_i, \mathbf{v}_j) = f(\mathbf{v}_j, \mathbf{v}_i) = a_{ji}$ so A is a symmetric matrix, in the sense that $A = {}^tA$; and conversely if A is a symmetric matrix then $f(\mathbf{x}, \mathbf{y}) = {}^t\mathbf{x}A\mathbf{y} = {}^t\mathbf{y}\,{}^tA\mathbf{x} = {}^t\mathbf{y}A\mathbf{x} = f(\mathbf{y}, \mathbf{x})$. Correspondingly a bilinear function $f(\mathbf{x}, \mathbf{y})$ is skew-symmetric if and only if the matrix representing it is skew-symmetric, in the sense that $A = -{}^tA$. Any matrix congruent to a symmetric matrix is easily seen to be symmetric, while any matrix that is congruent to a skew-symmetric matrix is easily seen to be skew-symmetric. It is clear that any $n \times n$ matrix A over a field for which $2 \neq 0$ can be written uniquely as the sum $A = A' + A''$ of the symmetric matrix $A' = \frac{1}{2}(A + {}^tA)$ and the skew-symmetric matrix $A'' = \frac{1}{2}(A - {}^tA)$. On the other hand if $2 = 0$ in the field then $-a_{ij} = a_{ij}$ so a matrix is symmetric if and only if it is skew-symmetric. The discussion in this section will be limited to the consideration of symmetric bilinear functions over fields in which $2 \neq 0$. The inner or dot product is an example of a symmetric bilinear function, and generalizations of this occur frequently in analysis. It is quite common to simplify the notation by setting $f(\mathbf{x}, \mathbf{y}) = (\mathbf{x}, \mathbf{y})$ for any symmetric bilinear function $f(\mathbf{x}, \mathbf{y})$; but when doing so it is important to keep in mind that a general symmetric bilinear function does not necessarily satisfy the positivity condition (2.5), so it is perhaps simpler just to continue with the notation $f(\mathbf{x}, \mathbf{y})$.

A basis $\{\mathbf{v}_i\}$ for a vector space V with a symmetric bilinear function $f(\mathbf{v}, \mathbf{w})$ is said to be an **orthogonal basis** if $f(\mathbf{v}_i, \mathbf{v}_j) = 0$ whenever $i \neq j$; this notion thus depends on the choice of a symmetric bilinear function on V. It is perhaps worth pointing out that there are no such bases for a nontrivial skew-symmetric bilinear function, to emphasize that only symmetric bilinear functions will be considered further here. Note that orthogonality does not specify the actual values of $f(\mathbf{v}_i, \mathbf{v}_i)$; some may be zero, although not all can be zero if the bilinear function is nontrivial. The special case of an orthogonal basis for which $f(\mathbf{v}_i, \mathbf{v}_i) = 1$ for all the basis vectors $\mathbf{v}_i$ is called an **orthonormal** basis; alternatively an orthonormal basis is characterized by the condition that $f(\mathbf{v}_i, \mathbf{v}_j) = \delta_j^i$, in terms of the Kronecker symbol.

Lemma 4.12. *If $f(\mathbf{x}, \mathbf{y})$ is a nontrivial symmetric bilinear function in a vector space V over a field F for which $2 \neq 0$ then there is at least one vector $\mathbf{v} \in V$ such that $f(\mathbf{v}, \mathbf{v}) \neq 0$.*

Proof: For a bilinear function on a vector space V to be nontrivial there must be vectors $\mathbf{v}_1, \mathbf{v}_2 \in V$ for which $f(\mathbf{v}_1, \mathbf{v}_2) \neq 0$. It may be the case that $f(\mathbf{v}_1, \mathbf{v}_1) \neq 0$ or that $f(\mathbf{v}_2, \mathbf{v}_2) \neq 0$, in which case there is nothing to prove. However if $f(\mathbf{v}_1, \mathbf{v}_1) = f(\mathbf{v}_2, \mathbf{v}_2) = 0$ then by bilinearity and symmetry $f(\mathbf{v}_1 + \mathbf{v}_2, \mathbf{v}_1 + \mathbf{v}_2) = f(\mathbf{v}_1, \mathbf{v}_1) + f(\mathbf{v}_1, \mathbf{v}_2) + f(\mathbf{v}_2, \mathbf{v}_1) + f(\mathbf{v}_2, \mathbf{v}_2) = 0 + 2f(\mathbf{v}_1, \mathbf{v}_2) + 0 \neq 0$, and that suffices for the proof.

It is worth noting that the assumption in the preceding lemma that $2 \neq 0$ is critical; indeed the bilinear function on the vector space $(\mathbb{Z}/2\mathbb{Z})^2$ over the field $\mathbb{Z}/2\mathbb{Z}$ described by the matrix $\begin{pmatrix} 0 & 1 \\ 1 & 0 \end{pmatrix}$ is nontrivial but $f(\mathbf{v}, \mathbf{v}) = 0$ for every vector $\mathbf{v}$. The usefulness of such vectors is clarified by the following observation.

Theorem 4.13 (Decomposition Theorem). *If $f(\mathbf{x}, \mathbf{y})$ is a symmetric bilinear function on a vector space V of dimension n over a field F for which $2 \neq 0$ and if $\mathbf{v}_1 \in V$ is a vector for which $f(\mathbf{v}_1, \mathbf{v}_1) \neq 0$ then the vector space V has a direct sum decomposition $V = V' \oplus V''$ where $V' \subset V$ is the one-dimensional linear subspace spanned by the vector $\mathbf{v}_1$ and $V'' \subset V$ is the $(n-1)$- dimensional linear subspace consisting of those vectors $\mathbf{v}'' \in V$ for which $f(\mathbf{v}'', \mathbf{v}_1) = 0$.*

Proof: If a vector $\mathbf{v} \in V$ can be written as the sum $\mathbf{v} = c\mathbf{v}_1 + \mathbf{v}''$ for some scalar $c \in F$ where $\mathbf{v}'' \in V''$ then $0 = f(\mathbf{v}'', \mathbf{v}_1) = f(\mathbf{v} - c\mathbf{v}_1, \mathbf{v}_1) = f(\mathbf{v}, \mathbf{v}_1) - cf(\mathbf{v}_1, \mathbf{v}_1)$ so it must be the case that $c = f(\mathbf{v}, \mathbf{v}_1)/f(\mathbf{v}_1, \mathbf{v}_1)$; and conversely if $c = f(\mathbf{v}, \mathbf{v}_1)/f(\mathbf{v}_1, \mathbf{v}_1)$ then $f(\mathbf{v} - c\mathbf{v}_1, \mathbf{v}_1) = 0$ so $\mathbf{v} - c\mathbf{v}_1 = \mathbf{v}'' \in V''$. Thus any vector $\mathbf{v} \in V$ can be written uniquely as the sum $\mathbf{v} = c\mathbf{v}_1 + \mathbf{v}''$ for some scalar $c \in F$ where $\mathbf{v}'' \in V''$; and that is just the condition that $V = V' \oplus V''$, from which it follows further that dim $V'' = n - 1$. That suffices for the proof.

Corollary 4.14 (Gram-Schmidt Theorem). *For any nontrivial symmetric bilinear function on a finite dimensional vector space V over a field F in which $2 \neq 0$ there is an orthogonal basis for V.*

Proof: The theorem will be demonstrated by induction on the dimension n of the vector space V. The theorem is quite obviously true if $n = 1$; so let V be a vector space of dimension $n > 1$ and assume that the result has been demonstrated for all vector spaces of dimension strictly less than n. Lemma 4.12 shows that there is a vector $\mathbf{v}_1 \in V$ for which $f(\mathbf{v}_1, \mathbf{v}_1) \neq 0$; and the preceding theorem then shows that there is a direct sum decomposition $V = V' \oplus V''$ where $V' = F\mathbf{v}_1$ is the one-dimensional subspace spanned by the vector $\mathbf{v}_1$ and V'' is the $(n-1)$-dimensional subspace consisting of vectors $\mathbf{v}''$ such that $f(\mathbf{v}'', \mathbf{v}_1) = 0$. By the induction hypothesis there is an orthogonal basis $\mathbf{v}''_1, \ldots, \mathbf{v}''_{n-1}$ for the subspace $V'' \subset V$; then $\mathbf{v}_1, \mathbf{v}''_1, \ldots, \mathbf{v}''_{n-1}$ is a basis for the n-dimensional vector space V, and since $f(\mathbf{v}_1, \mathbf{v}''_j) = 0$ for $1 \leq j \leq n - 1$ while $f(\mathbf{v}''_i, \mathbf{v}''_j) = 0$ whenever $i \neq j$ this is an orthogonal basis for the vector space V. That suffices for the proof.

It is perhaps worth noting that the preceding theorem does not assert that there is an orthonormal basis; for instance it may be the case that $f(\mathbf{v}_i, \mathbf{v}_i) = 0$ for all but one of the basis vectors. The decomposition of vectors in Theorem 4.13 can be used to construct an explicit orthogonal basis for a vector space beginning with any given basis $\mathbf{u}_1, \ldots, \mathbf{u}_n$. Indeed after modifying the first basis vector $\mathbf{u}_1$ as necessary it can be assumed that $f(\mathbf{u}_1, \mathbf{u}_1) \neq 0$, so there is the direct sum decomposition $V = F\mathbf{u}_1 + V''$ as in Theorem 4.13.

Set $\mathbf{v}_1 = \mathbf{u}_1$ and decompose the remaining vectors as $\mathbf{u}_j = c_j\mathbf{v}_1 + \mathbf{u}_j''$ where $\mathbf{u}_j'' \in V''$ for $2 \le j \le n$. Repeat the preceding process for the basis $\mathbf{u}_2'', \ldots, \mathbf{u}_n''$ of the vector space V'', and continue in the same manner; that leads to an orthogonal basis for the vector space V. This procedure is known as the **Gram-Schmidt Orthogonalization Process** and is quite commonly used either theoretically or actually explicitly.

Corollary 4.15. *Any symmetric matrix over a field F in which $2 \ne 0$ is congruent to a diagonal matrix.*

Proof: To a symmetric matrix A there is associated the symmetric bilinear function $f(\mathbf{u}, \mathbf{v})$ as in (4.31). By the preceding theorem there is an orthogonal basis $\{\mathbf{v}_i\}$ for this bilinear function, and the matrix representing the bilinear function in terms of this basis is of the form $A' = PA\,{}^tP$ for some nonsingular matrix P, so is congruent to the matrix A. The entries in the matrix A' are $a_{ij}' = f(\mathbf{v}_i, \mathbf{v}_j) = 0$ unless $i = j$, showing that A' is a diagonal matrix.

The preceding corollary can be interpreted as providing a normal form for symmetric matrices under the equivalence relation of congruence. However this normal form is not uniquely determined, since for instance if D is a diagonal matrix with diagonal entries d_i and if P is a nonsingular diagonal matrix with entries p_i then $PD\,{}^tP = D'$ is a diagonal matrix congruent to D but with the entries $d_i' = p_i^2 d_i$. The further examination of congruence for general fields leads to rather difficult algebraic questions; but only the fields $\mathbb{R}$ and $\mathbb{C}$ will be considered in the remainder of this section. For a real diagonal matrix $D = \{d_i\}$ there is as just observed a congruent diagonal matrix D' with the entries $d_i' = p_i^2 d_i$; and if $p_i = 1/\sqrt{|d_i|}$ then $d_i' = \pm 1$ or 0, so any symmetric real matrix is congruent to a diagonal matrix with entries ± 1 or 0 along the diagonal. If P is a permutation matrix then $PD'\,{}^tP = D''$ is a diagonal matrix obtained from D' by rearranging the rows and columns, that is, by rearranging the diagonal elements. Altogether then any real symmetric matrix is congruent to one of the form

$$D = \begin{pmatrix} I_{n_+} & 0 & 0 \\ 0 & -I_{n_-} & 0 \\ 0 & 0 & 0 \end{pmatrix}, \tag{4.35}$$

where I_n denotes the identity $n \times n$ matrix and 0 denotes the zero matrix. The diagonal matrix D has n_+ positive diagonal elements, each of which is $+1$, and n_- negative diagonal elements, each of which is -1, while the remaining diagonal elements are 0. The rank of the matrix D is $r = n_+ + n_-$, so the matrix D is determined fully by its rank r and the difference $s = n_+ - n_-$ between the number of real and negative diagonal elements, called the **signature** of the matrix D. A basic result[3] of matrix theory is that two congruent symmetric real matrices also have the same signature.

Theorem 4.16 (Sylvester's Law of Inertia). *Congruent symmetric real matrices have the same rank and signature.*

Proof: The theorem is equivalent to the assertion that if $f(\mathbf{u}, \mathbf{v})$ is a symmetric bilinear function on a finite dimensional real vector space V, and if $\{\mathbf{u}_i\}$ and $\{\mathbf{v}_i\}$ are two orthogonal bases for V for which the inner products $f(\mathbf{u}_i, \mathbf{u}_i)$ and $f(\mathbf{v}_i, \mathbf{v}_i)$ are either $+1$ or -1 or 0, then the number of basis vectors for which the inner product is $+1$ is the same for the two bases; for since congruent matrices have the same rank the number of basis vectors for which the inner product is 0 will be the same for the two bases, hence the number of basis vectors for which the inner product is -1 also will be the same for the two bases. Relabel the basis vectors $\mathbf{u}_i$ so that $U_+ \subset V$ is the subspace of V spanned by the basis vectors $\mathbf{u}_i^+$ for which the inner product is $+1$, for $1 \le i \le n^+$; that $U_- \subset V$ is the subspace spanned by the basis vectors $\mathbf{u}_i^-$ for which the inner product is -1, for $1 \le i \le n^-$; and that U_0 is the subspace spanned by the basis vectors $\mathbf{u}_i^0$ for which the inner product is 0, for $1 \le i \le n^0$. Let V_+, V_-, V_0 be the corresponding subspaces in terms of the basis vectors $\mathbf{v}_i$. The vector space V thus can be written as a direct sum of subspaces in two ways, namely $V = U_+ \oplus U_- \oplus U_0 = V_+ \oplus V_- \oplus V_0$. If $\mathbf{u} \in U_+$ and $\mathbf{u} \ne \mathbf{0}$ then $\mathbf{u} = \sum_{i=1}^{n^+} a_i^+ \mathbf{u}_i^+$ where not all the coefficients a_i^+ are zero; and since the basis is orthogonal $f(\mathbf{u}, \mathbf{u}) = \sum_{i,j=1}^{n^+} f(a_i^+ \mathbf{u}_i^+, a_j^+ \mathbf{u}_j^+) = \sum_{i=1}^{n^+} (a_i^+)^2 > 0$. The corresponding argument shows that $f(\mathbf{u}, \mathbf{u}) < 0$ whenever $\mathbf{u} \in U_-$ and $f(\mathbf{u}, \mathbf{u}) = 0$ whenever $\mathbf{u} \in U_0$, and similarly for the basis $\{\mathbf{v}_i\}$. Now suppose that contrary to what is to be proved $\dim U_+ > \dim V_+$; then $U_+ \cap (V_- \oplus V_0) \ne \mathbf{0}$, so there will be a nontrivial vector $\mathbf{u}$ in that intersection. Since $\mathbf{u} \in U_+$ it follows that $f(\mathbf{u}, \mathbf{u}) > 0$; but since $\mathbf{u} \in V_- \oplus V_0$ it also follows that $f(\mathbf{u}, \mathbf{u}) \le 0$, a contradiction which concludes the proof.

Corollary 4.17. *Two real symmetric matrices are congruent if and only if they have the same rank and signature; and any real symmetric matrix is congruent to a unique matrix of the form* (4.35).

Proof: This is really a restatement of the results of Corollary 4.15 and Theorem 4.16, as in the preceding discussion, so no further proof is required.

This gives the complete classification of symmetric real matrices under congruence. The corresponding result for complex symmetric matrices is somewhat simpler.

---

[3]This result was first proved by J. J. Sylvester in 1852, and customarily has been named after him in this rather dramatic way.

Corollary 4.18. *Two complex symmetric matrices are congruent if and only if they have the same rank; and any complex symmetric matrix is congruent to a unique matrix of the form*

$$D = \begin{pmatrix} I_r & 0 \\ 0 & 0 \end{pmatrix}. \tag{4.36}$$

Proof: As in the preceding discussion, any diagonal matrix $D = \{d_i\}$ over a field F is congruent to the diagonal matrix $D' = \{d_i'\}$ where $d_i' = p_i^2 d_i$ for any $p_i \neq 0$ in that field. Since any complex number has a square root, it follows that when F is the field of complex numbers the diagonal entries of the matrix D' can be taken to be either 1 or 0; and a further congruence $D'' = PD'\,{}^tP$ for a permutation matrix P can rearrange the diagonal elements, so altogether any complex symmetric matrix is congruent to one of the form (4.36), where r is the rank of that matrix. Since two congruent matrices have the same rank, it follows that the rank and hence the normal form (4.36) are uniquely determined, and that suffices for the proof.

There can be associated to any symmetric bilinear function $f(\mathbf{u}, \mathbf{v})$ on a finite dimensional vector space over a field F the **quadratic function** $q(\mathbf{v}) = f(\mathbf{v}, \mathbf{v})$. When the bilinear function is described by a matrix A over the field F in terms of a choice of coordinates x_i the quadratic function can be written

$$q(\mathbf{x}) = \sum_{i,j=1}^{n} a_{ij} x_i x_j, \tag{4.37}$$

so it is actually a homogeneous polynomial of degree 2 in the coordinates x_i. Under changes of coordinates in F^n the matrix A is replaced by congruent matrices, the matrices $PA\,{}^tP$ for nonsingular matrices P over the field F. Of course for a skew-symmetric bilinear function the associated quadratic function vanishes identically; so this is of interest only for symmetric bilinear functions. The discussion of these quadratic functions here will be limited to real quadratic functions, those over the field $\mathbb{R}$ of real numbers. In that case the quadratic function $q(\mathbf{x})$ is said to be **positive definite** if $q(\mathbf{x}) > 0$ for any nonzero vector $\mathbf{x} \in \mathbb{R}^n$ and is said to be **negative definite** if $q(\mathbf{x}) < 0$ for any nonzero vector $\mathbf{x} \in \mathbb{R}^n$. If the quadratic function described by a symmetric real matrix A is positive definite the matrix A is said to be a **positive definite matrix**, and correspondingly if the quadratic function described by a symmetric real matrix A is negative definite the matrix A is said to be a **negative definite matrix**. It is evident from the definition that a matrix A is negative definite if and only if the matrix $-A$ is positive definite; and that a diagonal real matrix is positive definite if and only if all its diagonal terms are strictly positive and is negative definite if and only if all its diagonal terms are strictly negative. Since the condition that a quadratic function be positive definite clearly is independent of the choice of basis, a symmetric real matrix A is positive definite if and only if it is congruent to the identity matrix, and correspondingly it is negative definite if and only if it is congruent to the negative of the identity matrix. To determine whether a given symmetric real matrix A is positive definite it is not necessary to go through the whole Gram-Schmidt orthogonalization process though, for there is a simpler test that is expressed in terms of the **principal minors** of the matrix A, the determinants of the square matrices formed from the first v rows and columns of the matrix A for the successive values $v = 1, 2, \ldots, n$. Thus the principal minors of a matrix $A = \{a_{ij}\}$ are the values

$$a_{11}, \quad \det \begin{pmatrix} a_{11} & a_{12} \\ a_{21} & a_{22} \end{pmatrix}, \quad \det \begin{pmatrix} a_{11} & a_{12} & a_{13} \\ a_{21} & a_{22} & a_{23} \\ a_{31} & a_{32} & a_{33} \end{pmatrix}, \cdots . \tag{4.38}$$

Theorem 4.19 (Sylvester's Criterion). *A symmetric real matrix A is positive definite if and only if all principal minors are strictly positive; and A is negative definite if and only if all the odd principal minors are strictly negative and all the even principal minors are strictly positive.*

Proof: First, to show that the principal minors of a positive definite real matrix are positive, consider a symmetric positive definite $n \times n$ matrix A. By Corollary 4.15 the matrix A is congruent to a diagonal matrix D, which as noted can be assumed to be of the form (4.35); and the matrix D also is positive definite, so it can be assumed that D is the identity matrix hence that $A = {}^t\!PP$ for a nonsingular matrix P. Then $\det A = \det {}^t\!P \cdot \det P = (\det P)^2 > 0$, so at least that subdeterminant is positive. If A_r is the leading $r \times r$ minor of the matrix A and $\mathbf{x} = \begin{pmatrix} \mathbf{x}_r \\ 0 \end{pmatrix} \in \mathbb{R}^n$ where $\mathbf{x}_r \in \mathbb{R}^r$ then since the matrix A is positive definite it follows that $0 < {}^t\!\mathbf{x}A\mathbf{x} = {}^t\!\mathbf{x}_r A_r \mathbf{x}_r$; that means that the matrix A_r is positive definite, hence in view of the special case already demonstrated it follows that $\det A_r > 0$.

Next, it will be demonstrated by induction on n that if the principal minors of an $n \times n$ real symmetric matrix are positive then that matrix is positive definite; that is trivially true for $n = 1$, so assume it is true for some particular $n \geq 1$ and consider an $(n + 1) \times (n + 1)$ matrix A for which all the principal minors are strictly positive. Decompose the matrix A into blocks of the form

$$A = \begin{pmatrix} A_0 & \mathbf{a} \\ {}^t\!\mathbf{a} & a \end{pmatrix}$$

where A_0 is an $n \times n$ matrix, $\mathbf{a} \in \mathbb{R}^n$ is a column vector and $a \in \mathbb{R}$ is a real number. By assumption $\det A_0 > 0$ so A_0 is a nonsingular matrix. Consider then the auxiliary $(n+1) \times (n+1)$ matrices

$$A' = \begin{pmatrix} A_0 & 0 \\ 0 & a' \end{pmatrix} \quad \text{and} \quad B = \begin{pmatrix} I_n & 0 \\ {}^t\mathbf{a}A_0^{-1} & 1 \end{pmatrix},$$

where $a' = a - {}^t\mathbf{a}A_0^{-1}\mathbf{a} \in \mathbb{R}$ and I_n is the $n \times n$ identity matrix. A straightforward calculation shows that

$$A = B \, A' \, {}^tB. \tag{4.39}$$

Since the principal minors of A_0 are also principal minors of A, and all are strictly positive by hypothesis, it follows from the inductive assumption that A_0 is positive definite. Clearly $\det B = 1$ so $\det A = \det A' = a' \det A_0$, and since $\det A > 0$ and $\det A_0 > 0$ it follows that $a' > 0$. The matrix A' therefore is positive definite, hence the congruent matrix A is positive definite, which establishes the inductive step for this part of the argument, so the matrix A is positive definite.

Finally a symmetric real matrix B is negative definite if and only if $-B$ is positive definite; hence it follows from the result just established for positive definite matrices that B is negative definite if and only if the principal minors of the matrix $-B$ are all positive. If B_r is the leading $r \times r$ submatrix of B then the associated principal minor is $\det(-B_r) = (-1)^r \det B_r$, which is positive if and only if the sign of $\det B_r$ is $(-1)^r$, and that suffices to conclude the proof.

As an application of Sylvester's Criterion, if $f(\mathbf{x})$ is a $\mathcal{C}^2$ function in an open subset $U \subset \mathbb{R}^n$ and if $\mathbf{a} \in U$ is a critical point of that function then its Taylor expansion in local coordinates centered at the point $\mathbf{a}$ has the form

$$f(\mathbf{a} + \mathbf{x}) = f(\mathbf{a}) + \frac{1}{2} \sum_{i,j=1}^{n} \partial_{ij}f(\xi_\mathbf{x})x_ix_j \tag{4.40}$$

where $\xi_\mathbf{x}$ is a point on the line segment connecting the point $\mathbf{a}$ and the point $\mathbf{x}$. The $n \times n$ matrix $\{\partial_{ij}f(\mathbf{x})\}$ is known as the **hessian** or the **hessian matrix** of the function $f(\mathbf{x})$. If the hessian of $f(\mathbf{x})$ is positive definite at the point $\mathbf{a}$ then all the principal minors are positive at the point $\mathbf{a}$; and since the second partial derivatives are continuous the principal minors will remain positive in an open neighborhood V of the point $\mathbf{a}$ hence the hessian itself will be positive definite in the open neighborhood V of the point $\mathbf{a}$. Consequently $f(\mathbf{x}) \geq f(\mathbf{a})$ for all points $\mathbf{x} \in V$, so the point $\mathbf{a}$ is a local minimum of the function $f(\mathbf{x})$. If the hessian matrix is negative definite the corresponding argument shows that the point $\mathbf{a}$ is a local maximum of the function $f(\mathbf{x})$. It should be noted though that this is a sufficient but not necessary condition for a local maximum or minimum, since a local extremum may occur at a point at which all the second derivatives vanish. For functions of a single variable the only principal minor is just $f''(a)$, so this is a simple test to see whether a critical point is a local maximum or minimum. For functions of two variables there are just two principal minors, the values $\partial_{11}f(\mathbf{a})$ and $\partial_{11}f(\mathbf{a})\partial_{22}f(\mathbf{a}) - \partial_{12}f(\mathbf{a})\partial_{21}f(\mathbf{a})$; for functions of more variables the calculation becomes more complicated.

The special case of positive definite bilinear functions leads back to the real inner product spaces considered in Section 2.1, a topic of particular interest in analysis; but it is important in analysis also to consider the complex analogue of real inner product spaces. A real inner product space was defined in (2.5) to be a real vector space U together with a mapping which assigns to any vectors $\mathbf{x}, \mathbf{y} \in U$ a real number $(\mathbf{x}, \mathbf{y}) \in \mathbb{R}$ that satisfies the conditions of linearity, symmetry, and positivity. Analogously a **complex inner product space** is defined to be a complex vector space V together with a mapping which assigns to any vectors $\mathbf{v}, \mathbf{w} \in V$ a complex number $(\mathbf{v}, \mathbf{w}) \in \mathbb{C}$ that satisfies the following conditions:

(i) *sesquilinearity*: $(c_1\mathbf{v}_1 + c_2\mathbf{v}_2, \ \mathbf{w}) = c_1(\mathbf{v}_1, \mathbf{w}) + c_2(\mathbf{v}_2, \mathbf{w})$
$(\mathbf{v}, \ c_1\mathbf{w}_1 + c_2\mathbf{w}_2) = \bar{c}_1(\mathbf{v}, \mathbf{w}_1) + \bar{c}_2(\mathbf{v}, \mathbf{w}_2)$
for any $c_1, c_2 \in \mathbb{C}$,

(ii) *hermitian symmetry*: $(\mathbf{v}, \mathbf{w}) = \overline{(\mathbf{w}, \mathbf{v})}$,

(iii) *positivity*: $(\mathbf{v}, \mathbf{v}) \geq 0$ and $(\mathbf{v}, \mathbf{v}) = 0$ if and only if $\mathbf{v} = \mathbf{0}$,

where $\bar{c}$ denotes the complex conjugate of $c \in \mathbb{C}$. Condition (i) means that the function $(\mathbf{v}, \mathbf{w})$ is linear in the first variable but **conjugate linear** in the second variable; so it is called **sesquilinearity** rather than bilinearity. Condition (ii) is a modification of symmetry, under which interchanging the entries changes the value of the function $(\mathbf{v}, \mathbf{w})$ to its complex conjugate; so it is called **hermitian symmetry** rather than symmetry. An immediate consequence of hermitian symmetry is that $(\mathbf{v}, \mathbf{v})$ is a real number, so that condition (iii) does make sense. A complex inner product space has an associated norm $\|\mathbf{v}\|_2 = \sqrt{(\mathbf{v}, \mathbf{v})}$ defined by the nonnegative square root of the real number $(\mathbf{v}, \mathbf{v}) \geq 0$, in parallel to the case of a real inner product space. The Cauchy-Schwarz inequality holds in a complex inner product space with essentially the same proof as given in Theorem 2.9 for a real inner product space; the details of verifying that inequality will be left to the reader. An **isometry** or an **isomorphism** between two complex inner product spaces V, W, with the inner products $(\mathbf{u}, \mathbf{v})_V$ and $(\mathbf{u}, \mathbf{v})_W$, respectively, is defined to be a bijective complex linear mapping $T : V \longrightarrow W$ between these two vector spaces that preserves the complex inner product, in the sense that

$$(T\mathbf{u}, T\mathbf{v})_W = (\mathbf{u}, \mathbf{v})_V \quad \text{for all } \mathbf{u}, \mathbf{v} \in V; \tag{4.41}$$

the spaces V and W are then said to be **isometric** or **isomorphic** complex inner product spaces. The two vectors spaces are also isometric as normed vector spaces, with the norm associated to the inner product. This is easily seen to be an equivalence relation between complex inner product spaces that can be viewed as identifying them for many purposes.

A real inner product space can be viewed as a special case of a complex inner product space, in which vectors and the field are real so that the complex conjugate of any vector or scalar is just that vector or scalar. The results already established for real inner product spaces normally extend to complex inner product spaces, with suitable modifications as necessary; and the results to be considered subsequently for complex inner product spaces normally specialize to corresponding results for real inner product spaces, also with suitable modifications as necessary. In both cases the modifications will be noted in the course of the discussion.

Any choice of a basis $\{\mathbf{u}_i\}$ for an n-dimensional complex vector space V establishes an isomorphism $\phi : \mathbb{C}^n \longrightarrow V$ between the complex vector spaces $\mathbb{C}^n$ and V, so it can be viewed as establishing a complex **coordinate system** on V; explicitly the image of the vector $\mathbf{x} = (x_1, \dots, x_n) \in \mathbb{C}^n$ is the vector $\mathbf{v} = \phi(\mathbf{x}) = \sum_{i=1}^n x_i \mathbf{u}_i \in V$ described by the coordinates x_i. If V is a complex inner product space this linear isomorphism then can be used to define an inner product structure on the complex vector space $\mathbb{C}^n$ for which ϕ is an isometry. Indeed if $\mathbf{x} = (x_1, \dots, x_n) \in \mathbb{C}^n$ and $\mathbf{y} = (y_1, \dots, y_n) \in \mathbb{C}^n$ while $\phi(\mathbf{x}) = \sum_{i=1}^n x_i \mathbf{u}_i = \mathbf{v} \in V$ and $\phi(\mathbf{y}) = \sum_{j=1}^n y_j \mathbf{u}_j = \mathbf{w}$ then the complex inner product on $\mathbb{C}^n$ must be given by

$$(\mathbf{x}, \mathbf{y}) = \left(\phi(\mathbf{x}), \phi(\mathbf{y})\right) = \left(\sum_{i=1}^n x_i \mathbf{u}_i, \sum_{j=1}^n y_j \mathbf{u}_j\right) = \sum_{i,j=1}^n x_i \bar{y}_j (\mathbf{u}_i, \mathbf{u}_j) \tag{4.42}$$

$$= \sum_{i,j=1}^n a_{ij} x_i \bar{y}_j \quad \text{where } a_{ij} = (\mathbf{u}_i, \mathbf{u}_j) \in \mathbb{C}$$

or in matrix terms

$$(\mathbf{x}, \mathbf{y}) = {}^t\mathbf{x} A \bar{\mathbf{y}} \quad \text{for the matrix } A = \{a_{ij}\}. \tag{4.43}$$

Since V is a complex inner product space $a_{ij} = (\mathbf{u}_i, \mathbf{u}_j) = \overline{(\mathbf{u}_j, \mathbf{u}_i)} = \bar{a}_{ji}$ by hermitian symmetry. Equivalently in matrix terms $A = {}^t\bar{A}$; a matrix A satisfying this condition is said to be a **hermitian matrix**. Conversely if A is a hermitian matrix then $(\mathbf{x}, \mathbf{y}) = {}^t\mathbf{x}A\overline{\mathbf{y}} = {}^t\overline{\mathbf{y}}\,{}^tA\mathbf{x} = \overline{{}^t\mathbf{y}\,{}^t\overline{A}\overline{\mathbf{x}}} = \overline{{}^t\mathbf{y}A\overline{\mathbf{x}}} = \overline{(\mathbf{y}, \mathbf{x})}$ so the expression (4.43) is hermitian symmetric. Also since V is a complex inner product space $\sum_{i,j} a_{ij} x_i \overline{x}_j = (\mathbf{u}, \mathbf{u}) > 0$ for any nonzero complex vector $\mathbf{x} = \{x_i\}$ by positivity; a hermitian matrix A such that ${}^t\mathbf{x}A\overline{\mathbf{x}} = \sum_{i,j} a_{ij} x_i \overline{x}_j > 0$ for any nonzero complex vector $\mathbf{x} = \{x_i\}$ is called a **positive definite hermitian matrix**. Thus any positive definite hermitian matrix A defines the structure of a complex inner product space on $\mathbb{C}^n$ with the inner product (4.43).

Any finite dimensional complex inner product space V has an orthogonal basis, a basis $\{\mathbf{u}_i\}$ for which $(\mathbf{u}_i, \mathbf{u}_j) = 0$ if $i \neq j$, by an argument similar to that in the proof of the Gram-Schmidt Theorem; again the details of verifying that the proof can be carried through in the complex case will be left to the reader. By positivity $\|\mathbf{u}_i\| > 0$ for any nonzero vector $\mathbf{u}_i$; therefore the vectors $\mathbf{u}_i^* = \mathbf{u}_i/\|\mathbf{u}_i\|$ actually are an orthonormal basis for V, a basis satisfying the condition that $(\mathbf{u}_i, \mathbf{u}_j) = \delta_j^i$ for all indices i, j. The coordinate system on V defined by the basis $\{\mathbf{u}_i^*\}$ then exhibits an isometry of the complex inner product space V with the complex inner product space $\mathbb{C}^n$ for which the complex inner product is defined by the positive definite hermitian matrix I, the identity matrix. This inner product space is called the **standard complex inner product space**, and the vector space $\mathbb{C}^n$ with this inner product will be denoted by $\mathbb{C}_I^n$; thus $\mathbb{C}_I^n$ is the vector space $\mathbb{C}^n$ with the complex inner product

$$(\mathbf{x}, \mathbf{y}) = {}^t\mathbf{x} \cdot \overline{\mathbf{y}} = \sum_{i=1}^n x_i \overline{y}_i \quad \text{for any vectors } \mathbf{x} = \{x_i\}, \quad \mathbf{y} = \{y_i\} \in \mathbb{C}^n, \qquad (4.44)$$

and it is the analogue of the vector space $\mathbb{R}^n$ with the standard real inner product. For many purposes it is sufficient just to consider the standard complex inner product space $\mathbb{C}^n$ rather than a general complex inner product space, just as in the earlier case of the real inner product space.

An isometry between the complex inner product space $\mathbb{C}_I^n$ and itself is a linear mapping $P : \mathbb{C}^n \longrightarrow \mathbb{C}^n$ between these two vector spaces that preserves the complex inner product structure, in the sense that $(P\mathbf{x}, P\mathbf{y}) = (\mathbf{x}, \mathbf{y})$ for any vectors $\mathbf{x}, \mathbf{y} \in \mathbb{C}^n$. More explicitly the inner product is given by (4.44) so the isometry is characterized by ${}^t\mathbf{x} \cdot \overline{\mathbf{y}} = (\mathbf{x}, \mathbf{y}) = (P\mathbf{x}, P\mathbf{y}) = {}^t(P\mathbf{x}) \cdot \overline{P\mathbf{y}} = {}^t\mathbf{x}\,{}^tP \cdot \overline{P}\overline{\mathbf{y}}$ for all $\mathbf{x}, \mathbf{y} \in \mathbb{C}^n$; and that is an identity in the variables $\mathbf{x}, \mathbf{y} \in \mathbb{C}^n$ if and only if ${}^tP\overline{P} = I$, the identity matrix. A matrix P satisfying this condition is called a **unitary matrix**; so a unitary matrix is characterized by the condition that

$$ {}^tP\overline{P} = I, \text{ the identity matrix, or equivalently } P^{-1} = \overline{P}. \qquad (4.45) $$

The set of all $n \times n$ unitary matrices clearly is a group under multiplication; this group is called the **unitary group** and is often denoted by $U(n)$. A real matrix A is hermitian precisely when $A = {}^tA$, that is, when the matrix A is symmetric. A real matrix A is unitary precisely when $A\,{}^{t}A = I$, or equivalently when $A^{-1} = {}^{t}A$. A real unitary matrix is called an **orthogonal** matrix; and the set of all $n \times n$ orthogonal matrices is a group under multiplication, called the **orthogonal group** and often denoted by $O(n)$.

If $A : \mathbb{C}_I^n \longrightarrow \mathbb{C}_I^n$ is an endomorphism of the complex vector space $\mathbb{C}_I^n$ and if $P : \mathbb{C}_I^n \longrightarrow \mathbb{C}_I^n$ is the isometry defined by a unitary matrix P then when P is viewed as a change of coordinates on the complex vector space $\mathbb{C}_I^n$ the linear transformation A in the new coordinates introduced by P is the linear transformation PAP^{-1}. This is a matrix that is similar to the matrix A, under the equivalence relation of similarity of matrices as in (4.1). However if P is a unitary matrix then $P^{-1} = \overline{{}^{t}P}$ so $PAP^{-1} = PA\,\overline{{}^{t}P}$ and the linear transformation A in terms of the new coordinates introduced by P also can be considered as the matrix $PA\,\overline{{}^{t}P}$. This is a matrix that is a complex analogue of a matrix congruent to the matrix A. Thus this is really a new equivalence relation, a combination of similarity and congruence with a complex twist, called **hermitian congruence**. In the analogous situation for real matrices, where P is an orthogonal matrix, this is the relation called **orthogonal congruence**. The examination of this equivalence relation can be carried out either in terms of orthonormal bases for inner product spaces or in terms of properties of unitary or orthogonal matrices; the subsequent discussion will focus on the first of these approaches.

Theorem 4.20. *For any endomorphism $T : V \longrightarrow V$ of a finite dimensional complex inner product space V there is a unique endomorphism $T^* : V \longrightarrow V$ such that*

$$(T\mathbf{u}, \mathbf{v}) = (\mathbf{u}, T^*\mathbf{v}) \quad \text{for all } \mathbf{u}, \mathbf{v} \in V. \tag{4.46}$$

If the endomorphism T is described by a matrix T in terms of an orthonormal basis for V then the endomorphism T^* is described by the matrix $T^* = \overline{{}^{t}T}$.

Proof: The choice of an orthonormal basis establishes an isometry between the inner product space V and the standard complex inner product space $\mathbb{C}_I^n$, so it suffices to prove the theorem just for the case of an endomorphism T of the inner product space $\mathbb{C}_I^n$, where the inner product has the form (4.44). For any vectors $\mathbf{x}, \mathbf{y} \in \mathbb{C}^n$

$$(T\mathbf{x}, \mathbf{y}) = {}^{t}(T\mathbf{x}) \cdot \overline{\mathbf{y}} = {}^{t}\mathbf{x}\,{}^{t}T\overline{\mathbf{y}} = {}^{t}\mathbf{x} \cdot \overline{(\overline{{}^{t}T}\mathbf{y})} = (\mathbf{x}, \overline{{}^{t}T}\mathbf{y}),$$

and that suffices for the proof.

The endomorphism T^* for any complex inner product space is called the **adjoint** of the endomorphism T, and the matrix $T^* = \overline{{}^{t}T}$ is called the **adjoint** of the matrix T. For a real vector space the adjoint of an endomorphism T also is defined by the equation (4.46); and when the adjoint of a real transformation is described by a matrix T in terms of an orthonormal basis the adjoint is described by the transpose $T^* = {}^tT$. Adjoint endomorphisms are quite basic tools in the study of inner product spaces; some of their standard properties are summarized in the following theorem. There are actually two different approaches to the proof of assertions about adjoint matrices for finite dimensional inner product spaces: proofs can be expressed either in terms of the basic defining equation (4.46) for the adjoint transformation or in terms of the matrix description $T^* = {}^t\overline{T}$ of endomorphisms in terms of an orthonormal basis. Generally the former approach will be used here, unless the alternative explicit calculation actually is more convenient.

Theorem 4.21. *For a finite dimensional complex inner product space V the adjoint operation has the following properties:*
(i) $(S + T)^* = S^* + T^*$ *for any endomorphisms S, T.*
(ii) $I^* = I$ *for the identity mapping I.*
(iii) $\lambda^* = \overline{\lambda}$ *for the mapping defined as multiplication by $\lambda \in \mathbb{C}$.*
(iv) $(ST)^* = T^*S^*$ *for any endomorphisms S, T.*
(v) $(T^*)^* = T$ *for any endomorphism T.*

Proof: It is clear from the defining equation (4.46) that (i) and (ii) hold. Since $(\lambda\mathbf{u}, \mathbf{v}) = \lambda(\mathbf{u}, \mathbf{v}) = (\mathbf{u}, \overline{\lambda}\mathbf{v})$ for any $\lambda \in \mathbb{C}$ and any vectors $\mathbf{u}, \mathbf{v} \in V$ that demonstrates (iii). From (4.46) it follows that $(\mathbf{u}, (ST)^*\mathbf{v}) = (ST\mathbf{u}, \mathbf{v}) = (T\mathbf{u}, S^*\mathbf{v}) = (\mathbf{u}, T^*S^*\mathbf{v})$ for any vectors $\mathbf{u}, \mathbf{v} \in V$, which shows that (iv) holds. Finally $((T^*)^*\mathbf{u}, \mathbf{v}) = \overline{(\mathbf{v}, (T^*)^*\mathbf{u})} = \overline{(T^*\mathbf{v}, \mathbf{u})} = (\mathbf{u}, T^*\mathbf{v}) = (T\mathbf{u}, \mathbf{v})$ for all vectors $\mathbf{u}, \mathbf{v} \in V$, which demonstrates (v) and thereby concludes the proof.

Actually the explicit proof of the preceding theorem for matrices, where $T^* = {}^t\overline{T}$, is possibly even simpler than the proof given here. For real inner product spaces the adjoint operation has the same properties, except that $\lambda^* = \lambda$ for the operation of multiplying by a real number $\lambda \in \mathbb{R}$. It is clear from the preceding theorem that the defining equation (4.46) for the adjoint operator can be rephrased equivalently in the reverse order, so as the condition that

$$(\mathbf{u}, T\mathbf{v}) = (T^*\mathbf{u}, \mathbf{v}) \quad \text{for all } \mathbf{u}, \mathbf{v} \in V. \tag{4.47}$$

Two subspaces W_1, W_2 of an inner product space V are said to be **orthogonal** if $(\mathbf{u}_1, \mathbf{u}_2) = 0$ for any vectors $\mathbf{u}_1 \in W_1$ and $\mathbf{u}_2 \in W_2$; and the **orthogonal complement** $W^\perp$ of a linear subspace $W \subset V$ is defined to be the subset

$$W^\perp = \left\{ \mathbf{u} \in V \,\middle|\, (\mathbf{u}, \mathbf{w}) = 0 \text{ for all } \mathbf{w} \in W \right\}, \tag{4.48}$$

which is easily seen also to be a linear subspace of V.

Theorem 4.22. *If V is a finite dimensional complex inner product space then*
(i) $\ker T = (T^*V)^\perp$ *and* $\ker T^* = (TV)^\perp$ *for any endomorphism T of V;*
(ii) $(W^\perp)^\perp = W$ *for any subspace* $W \subset V;$
(iii) $V = W \oplus W^\perp$ *for any subspace* $W \subset V;$ *and*
(iv) $T^*V = (\ker T)^\perp$ *and* $TV = (\ker T^*)^\perp$ *for any endomorphism T of V.*

Proof: Clearly $T\mathbf{u} = \mathbf{0}$ for a vector $\mathbf{u} \in V$ if and only if $0 = (T\mathbf{u}, \mathbf{v}) = (\mathbf{u}, T^*\mathbf{v})$ for all $\mathbf{v} \in V$, which implies the first assertion in (i), and the second assertion follows from the first assertion as an immediate consequence of Theorem 4.21 (v). If $\mathbf{v}_1, \ldots, \mathbf{v}_r$ is an orthonormal basis for the subspace $W \subset V$ where $r = \dim W$ this basis can be extended to an orthonormal basis $\mathbf{v}_1, \ldots, \mathbf{v}_n$ for the full vector space V where $n = \dim V$. It is evident that the vectors $\mathbf{v}_{r+1}, \ldots, \mathbf{v}_n$ form a basis for the subspace $W^\perp \subset V$; correspondingly the vectors $\mathbf{v}_1, \ldots, \mathbf{v}_r$ form a basis for $(W^\perp)^\perp$ as well as for W, from which (ii) and (iii) clearly follow. Finally (iv) follows directly from (i) and (ii), and that suffices for the proof.

The same theorem holds for the special case of real inner product spaces. A very important class of endomorphisms of either a real or a complex inner product space V are the **normal** endomorphisms, defined as those endomorphisms T such that

$$TT^* = T^*T; \tag{4.49}$$

although this may seem a somewhat peculiar condition, it does turn out to be both natural and important. Several more naturallydefined families of endomorphisms are special cases of normal endomorphisms. For example, an endomorphism T is said to be **self-adjoint** if $T^* = T$; obviously a self-adjoint endomorphism is a normal endomorphism. A complex self-adjoint matrix T is one for which $\,^t\overline{T} = T$ while a real self-adjoint matrix T is one for which $\,^tT = T$ so is just a symmetric matrix; thus these natural families of matrices are normal matrices. For other examples, a unitary matrix was defined to be a matrix T for which $\,^t\overline{T} = T^{-1}$, and an orthogonal matrix was defined as a real unitary matrix, so as a matrix T for which $\,^tT = T^{-1}$; and both unitary and orthogonal matrices are normal matrices, so have the following special properties.

Theorem 4.23. *A normal endomorphism T of a finite dimensional complex inner product space V satisfies the following conditions:*
(i) $\|T\mathbf{u}\| = \|T^*\mathbf{u}\|$ *for any vector* $\mathbf{u} \in V$.
(ii) $\ker T^* = \ker T$.
(iii) $TV = T^*V$.
(iv) $(T - \lambda I)$ *is also normal for any* $\lambda \in \mathbb{C}$.
(v) *If* $T\mathbf{v} = \lambda\mathbf{v}$ *for a vector* $\mathbf{v} \in V$ *and a complex constant* λ *then* $T^*\mathbf{v} = \overline{\lambda}\mathbf{v}$.

(vi) *If* $T\mathbf{u}_1 = \lambda_1\mathbf{u}_1$ *and* $T\mathbf{u}_2 = \lambda_2\mathbf{u}_2$ *for vectors* $\mathbf{u}_1, \mathbf{u}_2$ *and complex constants* λ_1, λ_2 *where* $\lambda_1 \neq \lambda_2$ *then* $(\mathbf{u}_1, \mathbf{u}_2) = 0$.

Proof: (i) For any vector $\mathbf{u} \in V$ it follows from the properties of adjoint endomorphisms and the definition of a normal endomorphism that $(T\mathbf{u}, T\mathbf{u}) = (\mathbf{u}, T^*T\mathbf{u}) = (\mathbf{u}, TT^*\mathbf{u}) = (T^*\mathbf{u}, T^*\mathbf{u})$, which is equivalent to (i).

(ii) This is an immediate consequence of (i).

(iii) This is an immediate consequence of (ii) and Theorem 4.22 (iv).

(iv) Since $(\lambda I)^* = \bar{\lambda}I$ for any complex constant λ by Theorem 4.21 (ii) and Theorem 4.21 (iii) it follows from Theorem 4.21 (i) that $(T - \lambda I)^* = T^* - \bar{\lambda}I$; therefore $(T - \lambda I)(T - \lambda I)^* = TT^* - \lambda T^* - \bar{\lambda}T + |\lambda|^2 I$ and similarly $(T - \lambda I)^*(T - \lambda I) = T^*T - \lambda T^* - \bar{\lambda}T + |\lambda|^2 I$, so since $TT^* = T^*T$ it follows that $(T - \lambda I)(T - \lambda I)^* = (T - \lambda I)^*(T - \lambda I)$ hence $(T - \lambda I)$ is normal.

(v) If $T\mathbf{v} = \lambda\mathbf{v}$ then $\mathbf{v} \in \ker(T - \lambda I)$, and since $T - \lambda I$ is normal by (iv) it follows from (ii) that $\ker(T - \lambda I)^* = \ker(T - \lambda I)$; and as already noted $(T - \lambda I)^* = T^* - \bar{\lambda}I$ hence $T^*\mathbf{v} = \bar{\lambda}\mathbf{v}$.

(vi) If $T\mathbf{u}_1 = \lambda_1\mathbf{u}_1$ and $T\mathbf{u}_2 = \lambda_2\mathbf{u}_2$ then $T^*\mathbf{u}_i = \bar{\lambda}_i\mathbf{u}_i$ by (v). Therefore $\lambda_1(\mathbf{u}_1, \mathbf{u}_2) = (T\mathbf{u}_1, \mathbf{u}_2) = (\mathbf{u}_1, T^*\mathbf{u}_2) = (\mathbf{u}_1, \bar{\lambda}_2\mathbf{u}_2) = \lambda_2(\mathbf{u}_1, \mathbf{u}_2)$; so if $\lambda_1 \neq \lambda_2$ then $(\mathbf{u}_1, \mathbf{u}_2) = 0$, and that suffices for the proof.

Again the same results hold for real inner product spaces. It is an interesting exercise to demonstrate the converse of (i) of the preceding theorem, that is, to show that if $\|T\mathbf{u}\| = \|T^*\mathbf{u}\|$ for any vector $\mathbf{u} \in V$ then the endomorphism T is normal. However the true significance of normal endomorphisms really lies in the following theorem.

Theorem 4.24 (Spectral Theorem). *An endomorphism T of a finite dimensional complex inner product space V is normal if and only if V has an orthonormal basis consisting of eigenvectors of T.*

Proof: If $\mathbf{v}_1, \ldots, \mathbf{v}_n$ is an orthonormal basis for the vector space V consisting of eigenvectors for the endomorphism T, so that $T\mathbf{v}_i = \lambda_i\mathbf{v}_i$, then the matrix representation of T in terms of this basis is the diagonal matrix with the entries λ_i along the diagonal; the adjoint matrix is just the complex matrix with the diagonal entries $\bar{\lambda}_i$, and these two matrices of course commute so T is a normal linear transformation. Conversely if T is a normal linear transformation of the complex vector space V it has nontrivial eigenvectors and the eigenvectors for any particular eigenvalue form a finite dimensional vector subspace; any orthonormal basis of that subspace of course consists of eigenvectors of T. The eigenvectors for distinct eigenvalues are orthogonal vectors by Theorem 4.23 (vi); so the set of all eigenvectors of T form a vector subspace $W \subset V$ spanned by a collection of orthonormal eigenvectors $\mathbf{v}_i$ with eigenvalues λ_i. If W is not the entire vector space V its orthogonal complement $W^\perp$ will be a nontrivial subspace of V, consisting of vectors $\mathbf{u} \in V$ such that $(\mathbf{u}, \mathbf{v}_i) = 0$ for $1 \leq i \leq r$. If

$\mathbf{u} \in W^{\perp}$ then since $T^*\mathbf{v}_i = \overline{\lambda_i}\mathbf{v}_i$ by Theorem 4.23 (v) it follows that $(T\mathbf{u}, \mathbf{v}_i) = (\mathbf{u}, T^*\mathbf{v}_i) = (\mathbf{u}, \overline{\lambda_i}\mathbf{v}_i) = \lambda_i(\mathbf{u}, \mathbf{v}_i) = 0$, hence $T\mathbf{u} \in W^{\perp}$ as well; therefore $TW^{\perp} \subset W^{\perp}$. But the endomorphism T of this vector space $W^{\perp}$ must also have an eigenvector, which is not in the space W of all eigenvectors. That contradiction shows that $W = V$, hence that V is spanned by the eigenvectors of T, thereby concluding the proof.

Corollary 4.25 (Spectral Theorem for Complex Matrices). *A complex matrix T is hermitian congruent to a diagonal matrix if and only if the matrix T is normal.*

Proof: The preceding theorem shows that the endomorphism described by a matrix T is normal if and only if the inner product vector space has a basis consisting of eigenvectors of the endomorphism T. A change of basis in the inner product space $\mathbb{C}^n$ has the effect of replacing the matrix T by a hermitian congruent matrix; and a matrix is diagonal if and only if the basis vectors are eigenvectors of the linear transformation described by the matrix. That suffices for the proof.

The remarkable part of the preceding corollary is that there is such a simple and readily calculable test to determine whether a matrix is hermitian congruent to a diagonal matrix. Unitary and self-adjoint matrices are particular examples of normal matrices, so any such matrices are hermitian congruent to diagonal matrices. On the other hand there are normal matrices that are neither unitary nor self-adjoint, so the theorem does cover still other cases. For example, it is a straightforward calculation to verify that the matrix

$$\begin{pmatrix} 1 & 1 & 0 \\ 0 & 1 & 1 \\ 1 & 0 & 1 \end{pmatrix} \tag{4.50}$$

is a normal matrix but is neither unitary nor self-adjoint.

The situation for real matrices is rather different, since real normal matrices do not necessarily have any eigenvectors at all. For example the real matrix $\begin{pmatrix} 0 & -1 \\ 1 & 0 \end{pmatrix}$ is easily seen to be normal, but it has no real eigenvalues hence no real eigenvectors; actually that is the usual situation for real 2×2 normal matrices. The proof of the spectral theorem for real inner product spaces thus fails just at the point of using the existence of eigenvectors. However what is in some senses a rather simpler corresponding result still holds for real inner product spaces, based on the following observation.

Lemma 4.26. *The eigenvalues of any self-adjoint endomorphism of a real or complex inner product space are real numbers.*

Proof: If $T : V \longrightarrow V$ is a self-adjoint endomorphism of an inner product space V and $T\mathbf{v} = \lambda\mathbf{v}$ for some nonzero vector $\mathbf{v} \in V$ and some complex number λ then $\lambda(\mathbf{v}, \mathbf{v}) = (T\mathbf{v}, \mathbf{v}) = (\mathbf{v}, T^*\mathbf{v}) = (\mathbf{v}, T\mathbf{v}) = (\mathbf{v}, \lambda\mathbf{v}) = \bar{\lambda}(\mathbf{v}, \mathbf{v})$ and since $(\mathbf{v}, \mathbf{v}) \neq 0$ it follows that $\lambda \in \mathbb{R}$, which suffices for the proof.

Theorem 4.27. *An endomorphism T of a finite dimensional real inner product space is self-adjoint if and only if V has an orthonormal basis consisting of eigenvectors of T.*

Proof: If V has an orthonormal basis consisting of real eigenvectors of T then V is isometric to the standard real inner product space for which the endomorphism T is represented by a real diagonal matrix, hence T is self-adjoint. Conversely suppose that T is a self-adjoint endomorphism of a finite dimensional real vector space V. The vector space V is isometric to the standard real inner product space $\mathbb{R}_I^n$, and the endomorphism T is then described by a self-adjoint matrix T. This matrix always has a complex eigenvalue; but the preceding lemma shows that its eigenvalues are all real hence have real eigenvectors. Therefore the endomorphism T actually has real eigenvectors, so the proof of the spectral theorem for complex matrices carries over to a corresponding result for the vector space V, to conclude the proof.

Corollary 4.28. *A real matrix T is orthogonally congruent to a diagonal matrix if and only if the matrix T is self-adjoint.*

Proof: The preceding theorem shows that the endomorphism described by a matrix T is self-adjoint if and only if the inner product space has an orthonormal basis consisting of eigenvectors of the endomorphism T. A change of basis in the inner product space $\mathbb{R}^n$ has the effect of replacing the matrix T by an orthogonally congruent matrix; and a matrix is diagonal if and only if the basis vectors are eigenvectors of the linear transformation described by the matrix. That suffices for the proof.

It is worth mentioning here that the spectral theorem and the corresponding theorem for real inner product spaces generally require the hypothesis that the vector spaces are finite dimensional. Infinite dimensional real and complex inner product spaces play important roles in analysis, and a major topic is the investigation of conditions under which the spectral theorem holds in some form or other for these vector spaces.

Problems, Group I

1. Verify that congruence is an equivalence relation among matrices.

2. Find an orthonormal basis for the vector space $\mathbb{R}^3$ in terms of the usual symmetric bilinear function $f(\mathbf{x}, \mathbf{y}) = \sum_{i=1}^{3} x_i y_i$ by applying the Gram-Schmidt process beginning with the basis $\mathbf{v}_1 = \{1, 0, 1\}$, $\mathbf{v}_2 = \{1, 0, -1\}$, $\mathbf{v}_3 = \{0, 3, 4\}$.

3. Find an orthonormal basis for the vector space $\mathbb{R}^3$ in terms of the symmetric bilinear function described by the matrix $\begin{pmatrix} 1 & 0 & 1 \\ 0 & 2 & 1 \\ 1 & 1 & 2 \end{pmatrix}$.

4. In $\mathbb{R}^4$ in terms of the usual symmetric bilinear function $f(\mathbf{x}, \mathbf{y})$ let $V \subset \mathbb{R}^4$ be the subspace consisting of all vectors $\mathbf{x}$ such that $f(\mathbf{x}, \mathbf{u}) = f(\mathbf{x}, \mathbf{v}) = 0$ for the vectors ${}^t\mathbf{u} = \{1, 0, -1, 1\}$ and ${}^t\mathbf{v} = \{2, 3, -1, 2\}$. Find an orthonormal basis for V.

5. For what real values x is the matrix

$$\begin{pmatrix} x & 1 & 2 \\ 1 & 4 & 5 \\ 2 & 5 & 8 \end{pmatrix}$$

positive definite? Negative definite? What are the rank and signature of this matrix for the special case in which $x = 0$?

6. Show that the Cauchy-Schwarz inequality holds in any finite dimensional complex inner product space.

7. Show that if an endomorphism T of an inner product space satisfies $\|T\mathbf{u}\| = \|T^*\mathbf{u}\|$ for all vectors $\mathbf{u}$ then T is normal. (This is the converse of (i) in Theorem 4.23.)

8. On the inner product space $\mathbb{C}^2$ with the usual inner product show that the matrix

$$A = \begin{pmatrix} 1 & i \\ i & 1 \end{pmatrix}$$

is normal and find an orthonormal basis of $\mathbb{C}^2$ consisting of eigenvectors of A.

9. Show that for any vector space the only endomorphism that is both normal and nilpotent is the zero mapping.

Problems, Group II

10. (i) Show that the mapping $f : \mathbb{R}^{2\times2} \times \mathbb{R}^{2\times2} \longrightarrow \mathbb{R}$ from pairs of 2×2 real matrices to $\mathbb{R}$, defined by $f(A, B) = \text{tr}(AB)$, is a symmetric bilinear function, where $\text{tr}(X)$ is the trace of the matrix X.

(ii) Find a matrix representation M of the mapping f for the basis consisting of the matrices e_{ij} having entry 1 in row i column j and entries 0 otherwise. Determine the rank and signature of M.

(iii) Find a matrix representation M_1 of the restriction of f to the subspace consisting of pairs of skew-symmetric matrices, for some choice of a basis for this subspace. Determine the rank and signature of the matrix M_1.

11. Show that if V is a finite dimensional vector space over a field F in which $2 \neq 0$ and if $f(\mathbf{x}, \mathbf{y})$ is a nontrivial skew-symmetric bilinear function on V then there is no basis for V in terms of which the matrix describing the mapping f is upper triangular.

12. Show that if the columns of an $n \times n$ real matrix form an orthonormal basis for $\mathbb{R}^n$ in terms of the usual symmetric bilinear function then so do the rows (where both rows and columns are viewed as vectors in $\mathbb{R}^n$).

13. Show that if A is a symmetric $n \times n$ real matrix and if $\mathbf{a} \in \mathbb{R}^n$ is a critical point for the function $g(\mathbf{x})$ on $\mathbb{R}^n$ defined by $g(\mathbf{x}) = f(A\mathbf{x}, \mathbf{x})/f(\mathbf{x}, \mathbf{x})$ for $\mathbf{x} \neq \mathbf{0}$ and $g(\mathbf{0}) = 0$, where $f(\mathbf{x}, \mathbf{y})$ is the usual symmetric bilinear function on $\mathbb{R}^n$, then $A\mathbf{a} = \alpha\, \mathbf{a}$ for some real number α. What can you say about the value of α?

14. Show that a complex matrix A is normal if and only if $A = A_1 + i\, A_2$ where A_1, A_2 are commuting self-adjoint matrices.

15. Find an example of a complex 2×2 matrix A such that A^2 is normal but A is not normal.

16. Show that a complex matrix A is normal if and only if A^* can be written as a polynomial in A.

17. Show that the matrix $A = \begin{pmatrix} 1 & 2 & 3 \\ 2 & 3 & 4 \\ 3 & 4 & 5 \end{pmatrix}$ is normal and find a real orthogonal matrix P so that $PA\,{}^t P$ is a diagonal matrix.

5

Geometry of Mappings

A continuous one-to-one mapping $\phi : U \longrightarrow V$ between two open subsets $U, V \subset \mathbb{R}^n$ with a continuous inverse $\phi^{-1} : V \longrightarrow U$ is called a **homeomorphism** between these two sets. A homeomorphism really identifies the two sets as far as topological properties are concerned. Indeed the mapping ϕ takes open subsets of U to open subsets of V, since the image $\phi(E) \subset V$ of any open subset $E \subset U$ is the inverse image $(\phi^{-1})^{-1}(E)$ of the set E under the continuous mapping $\phi^{-1} : V \longrightarrow U$, and the inverse mapping ϕ^{-1} takes open subsets of V to open subsets of U; consequently the mapping ϕ identifies continuous functions $f : U \longrightarrow \mathbb{R}$ on U with continuous functions $f \circ \phi^{-1} : V \longrightarrow \mathbb{R}$ on V, and correspondingly for the inverse mapping ϕ^{-1}. If the mapping and its inverse are both continuously differentiable, in which case the mapping is called a $\mathcal{C}^1$ **homeomorphism** or sometimes a $\mathcal{C}^1$ **diffeomorphism**, this mapping also identifies $\mathcal{C}^1$ functions on the two sets; and if the mapping ϕ and its inverse are $\mathcal{C}^r$ mappings they identify $\mathcal{C}^r$ functions on U with those on V. It is a familiar result from calculus in one variable that if $f : I \longrightarrow \mathbb{R}^1$ is a continuously differentiable function in an open interval $I \subset \mathbb{R}^1$ such that $f'(x) \neq 0$ for all $x \in I$ then the mapping f is a $\mathcal{C}^1$ homeomorphism from I to its image $f(I)$. The proof of this result is often skipped or treated casually; but it is a model for one proof of the corresponding result in higher dimensions, so it may be helpful to review that special case first. Suppose that $f : [a, b] \longrightarrow \mathbb{R}^1$ is a continuous mapping from a closed interval in $\mathbb{R}^1$ into $\mathbb{R}$ and that f is $\mathcal{C}^1$ in the open interval (a, b) and $f'(x) \neq 0$ at each point $x \in (a, b)$; since the image $f([a, b])$ of the compact connected set $[a, b]$ is necessarily also compact and connected it must be a closed interval, so it can be assumed that $f([a, b]) = [c, d]$. The set of points where $f'(x) > 0$ and the set of points at which $f'(x) < 0$ are open subsets of (a, b), the union of which is the entire interval (a, b); and since an interval is connected it must be the case that one of these two sets is empty, so either $f'(x) > 0$ at all points $x \in (a, b)$ or $f'(x) < 0$ at all points $x \in (a, b)$. Suppose first that $f'(x) > 0$ at all points $x \in (a, b)$. It follows from the Mean Value Theorem that whenever $a \leq x_1 < x_2 \leq b$ there is a point ξ such that $x_1 < \xi < x_2$ and $f(x_2) - f(x_1) = f'(\xi)(x_2 - x_1) > 0$; thus f is a strictly increasing function in $[a, b]$, so it is a one-to-one mapping from the closed interval $[a, b]$ onto the closed interval $[c, d]$ and has a well-defined inverse mapping $g : [c, d] \longrightarrow [a, b]$. For any $\delta > 0$ the continuous positive function

$f'(x)$ attains its minimum value at a point in the compact subset $[a + \delta, b - \delta]$, and that minimum must be a positive number $m > 0$ since $f'(x) > 0$ everywhere; so $f'(x) \geq m > 0$ for all $x \in [a + \delta, b - \delta]$. Therefore whenever $a + \delta \leq x_1 < x_2 \leq b - \delta$ it follows from the Mean Value Theorem that there is a point ξ such that $x_1 < \xi < x_2$ and $f(x_2) - f(x_1) = f'(\xi)(x_2 - x_1) \geq m(x_2 - x_1)$. If $y_i = f(x_i)$ then $x_i = g(y_i)$ and the preceding inequality can be written $y_2 - y_1 \geq m(g(y_2) - g(y_1))$. That is the case for any points y_i for which $f(a + \delta) \leq y_1 < y_2 \leq f(b - \delta)$, which shows that the inverse function g is continuous in the interval $[f(a + \delta), f(b - \delta)]$; and since that is the case for any $\delta > 0$ it follows that the function g is continuous in (c, d). Furthermore in the interval $c \leq y_1 < y_2 \leq d$ if $x_i = g(y_i)$ then

$$\frac{g(y_2) - g(y_1)}{y_2 - y_1} = \frac{x_2 - x_1}{f(x_2) - f(x_1)};$$

since g is continuous, x_2 tends to x_1 as y_2 tends to y_1, so in the limit the preceding equation reduces to $g'(y_1) = 1/f'(x_1)$, showing that the function g is differentiable and that its derivative is also continuous. The corresponding result of course holds if $f'(x) < 0$ in (a, b). This argument critically used the hypothesis that the derivative $f'(x)$ is a continuous function, and the result does not hold without that hypothesis. Indeed the function

$$h(x) = \frac{1}{2}x + x^2 \sin \frac{1}{x} \tag{5.1}$$

is differentiable at all points $x \in \mathbb{R}^1$ and it is easy to see that $h'(0) = \frac{1}{2}$; but the derivative of $h(x)$ alternates in sign in any open neighborhood of 0 so it cannot be a one-to-one mapping in any open neighborhood of 0.

If $\mathbf{f} : U \longrightarrow V$ is a $\mathcal{C}^1$ homeomorphism between two open subsets $U, V \subset \mathbb{R}^n$ and if $\mathbf{g} : V \longrightarrow U$ is its inverse then $\mathbf{g}(\mathbf{f}(\mathbf{x})) = \mathbf{x}$, and an application of the chain rule shows that $\mathbf{g}'(\mathbf{f}(\mathbf{x}))\mathbf{f}'(\mathbf{x}) = I$, the identity matrix, and consequently that $\det \mathbf{f}'(\mathbf{x}) \neq 0$ at each point $\mathbf{x} \in U$; thus a necessary condition for the mapping $\mathbf{f}$ to be a $\mathcal{C}^1$ homeomorphism is that $\det \mathbf{f}'(\mathbf{x}) \neq 0$ for all $\mathbf{x} \in U$. The matrix $\mathbf{f}'$ is called the **Jacobian matrix** of the mapping $\mathbf{f}$, and its determinant is called the **Jacobian** of the mapping $\mathbf{f}$. This necessary condition is sufficient locally. It is convenient first to establish a special case of that result, from which the general case follows easily.

Theorem 5.1. *If $\mathbf{f} : W \longrightarrow \mathbb{R}^n$ is a continuously differentiable mapping defined in an open neighborhood $W \subset \mathbb{R}^n$ of the origin such that $\mathbf{f}(\mathbf{0}) = \mathbf{0}$ and $\mathbf{f}'(\mathbf{0}) = I$, the identity matrix, then for any sufficiently small open subneighborhood $U \subset W$ of the origin the restriction of the mapping $\mathbf{f}$ to U is a $\mathcal{C}^1$ homeomorphism $\mathbf{f} : U \longrightarrow V$ between U and an open neighborhood $V \subset \mathbb{R}^n$ of the origin.*

Proof: The mapping $\mathbf{r}(\mathbf{x}) = \mathbf{f}(\mathbf{x}) - \mathbf{x}$ is a continuously differentiable mapping such that $\mathbf{r}(\mathbf{0}) = \mathbf{0}$ and $\mathbf{r}'(\mathbf{0}) = \mathbf{0}$; therefore there is a closed cell $\overline{\Delta} \subset W \subset \mathbb{R}^n$ centered at the origin $\mathbf{0}$ sufficiently small that the following conditions are satisfied:

(i) $\|\mathbf{r}'(\mathbf{x})\|_o \leq \frac{1}{2}$ for all $\mathbf{x} \in \overline{\Delta}$, for the operator norm for ℓ_2 norms; and

(ii) $\det \mathbf{f}'(\mathbf{x}) \neq 0$ for all $\mathbf{x} \in \overline{\Delta}$.

The first step in the proof is to establish a basic inequality on which the remainder of the proof rests. From the Mean Value Inequality (3.56) it follows that for any points $\mathbf{a}, \mathbf{b} \in \overline{\Delta}$ there is a point $\mathbf{c}$ on the line connecting those two points for which $\|\mathbf{r}(\mathbf{b}) - \mathbf{r}(\mathbf{a})\|_2 \leq \|\mathbf{r}'(\mathbf{c})\|_o \|\mathbf{b} - \mathbf{a}\|_2$ in terms of the operator norm $\|\mathbf{r}'(\mathbf{c})\|_o$; it follows from the assumption (i) that $\|\mathbf{r}(\mathbf{b}) - \mathbf{r}(\mathbf{a})\|_2 \leq \frac{1}{2} \|\mathbf{b} - \mathbf{a}\|_2$. Then from this inequality and the triangle inequality it follows that

$$\|\mathbf{b} - \mathbf{a}\|_2 \leq \underbrace{\|(\mathbf{f}(\mathbf{b}) - \mathbf{f}(\mathbf{a})) - (\mathbf{b} - \mathbf{a})\|_2}_{=\mathbf{r}(\mathbf{b}) - \mathbf{r}(\mathbf{a})} + \|(\mathbf{f}(\mathbf{b}) - \mathbf{f}(\mathbf{a}))\|_2$$

$$\leq \frac{1}{2} \|\mathbf{b} - \mathbf{a}\|_2 + \|(\mathbf{f}(\mathbf{b}) - \mathbf{f}(\mathbf{a}))\|_2$$

hence that

$$\|\mathbf{f}(\mathbf{b}) - \mathbf{f}(\mathbf{a})\|_2 \geq \frac{1}{2} \|\mathbf{b} - \mathbf{a}\|_2, \tag{5.2}$$

which is the basic inequality.

It follows from (5.2) that the mapping $\mathbf{f}: \overline{\Delta} \longrightarrow \mathbb{R}^n$ is an injective mapping; for if $\mathbf{f}(\mathbf{b}) = \mathbf{f}(\mathbf{a})$ that inequality shows that $\|\mathbf{b} - \mathbf{a}\|_2 = 0$. In particular $\mathbf{f}(\mathbf{x}) \neq \mathbf{f}(\mathbf{0}) = \mathbf{0}$ for any point $\mathbf{x} \in \partial \Delta$, the boundary of the cell $\overline{\Delta}$; so since $\partial \Delta$ is compact there is a positive number $d > 0$ such that

$$\|\mathbf{f}(\mathbf{x})\|_2 \geq d > 0 \quad \text{for all} \quad \mathbf{x} \in \partial \Delta. \tag{5.3}$$

The next step is to show that the open neighborhood V of the origin defined by

$$V = \left\{ \mathbf{y} \in \mathbb{R}^n \,\middle|\, \|\mathbf{y}\|_2 < \frac{1}{2} d \right\} \tag{5.4}$$

is contained in the image $\mathbf{f}(\Delta)$. To demonstrate that, for any point $\mathbf{b} \in V$ the function $\psi(\mathbf{x}) = \|\mathbf{f}(\mathbf{x}) - \mathbf{b}\|_2$ is a continuous function on $\overline{\Delta}$, and it is just necessary to show that $\psi(\mathbf{a}) = 0$ for some point $\mathbf{a} \in \Delta$. Since $\mathbf{b} \in V$ then $\psi(\mathbf{0}) = \|\mathbf{b}\|_2 < \frac{1}{2} d$ by (5.4), while if $\mathbf{x} \in \partial \Delta$ then in view of (5.3) and the triangle inequality

$$d \leq \|\mathbf{f}(\mathbf{x})\|_2 \leq \|\mathbf{f}(\mathbf{x}) - \mathbf{b}\|_2 + \|\mathbf{b}\|_2 = \psi(\mathbf{x}) + \|\mathbf{b}\|_2 < \psi(\mathbf{x}) + \frac{d}{2}$$

so $\psi(\mathbf{x}) > \frac{d}{2}$. Thus $\psi(\mathbf{0}) < \psi(\mathbf{x})$ for any point $\mathbf{x} \in \partial \Delta$, so the function $\psi(\mathbf{x})$ must take its minimum value at some interior point $\mathbf{a} \in \Delta$; and that point is the obvious candidate to be a zero of the function $\psi(\mathbf{x})$. The square of this function, the function $\psi(\mathbf{x})^2 = \sum_{i=1}^{n} \left(f_i(\mathbf{x}) - b_i\right)^2$, then also takes its minimum value at the point $\mathbf{a} \in \Delta$, hence that point is a critical point of $\psi(\mathbf{x})^2$, a zero of all of its partial derivatives $\partial_j \psi(\mathbf{x})^2 = \sum_{i=1}^{n} 2\left(f_i(\mathbf{x}) - b_i\right) \partial_j f_i(\mathbf{x})$; thus $0 = \sum_{i=1}^{n} 2\left(f_i(\mathbf{a}) - b_i\right) \partial_j f_i(\mathbf{a})$ for $1 \leq j \leq n$, which can be written equivalently in terms of the matrix $\mathbf{f}'(\mathbf{a})$ as

$$2\,\mathbf{f}'(\mathbf{a})\big(\mathbf{f}(\mathbf{a}) - \mathbf{b}\big) = 0.$$

However the matrix $\mathbf{f}'(\mathbf{a})$ is nonsingular by assumption (ii), hence $\mathbf{f}(\mathbf{a}) - \mathbf{b} = \mathbf{0}$ as desired.

What has been shown so far is that the mapping $\mathbf{f} : \overline{\Delta} \longrightarrow \mathbb{R}^n$ is an injective mapping, and that $V \subset \mathbf{f}(\Delta)$. Consequently the subset $U = \mathbf{f}^{-1}(V) \subset \Delta$ is an open neighborhood of the origin for which $\mathbf{f}$ is a one-to-one mapping $\mathbf{f} : U \longrightarrow V$ from U onto V; so there is a well-defined inverse mapping $\mathbf{g} = \mathbf{f}^{-1} : V \longrightarrow U$ from V onto U. To show that this mapping $\mathbf{g}$ is continuously differentiable, for any points $\mathbf{y}, \mathbf{y} + \mathbf{k} \in V$ let $\mathbf{x}, \mathbf{x} + \mathbf{h}$ be their images under the mapping $\mathbf{g} : V \longrightarrow U$, so that

$$
\begin{aligned}
\mathbf{x} &= \mathbf{g}(\mathbf{y}) \quad \text{and} \quad \mathbf{x} + \mathbf{h} = \mathbf{g}(\mathbf{y} + \mathbf{k}) \\
\mathbf{y} &= \mathbf{f}(\mathbf{x}) \quad \text{and} \quad \mathbf{y} + \mathbf{k} = \mathbf{f}(\mathbf{x} + \mathbf{h}).
\end{aligned}
\tag{5.5}
$$

By the basic inequality (5.2)

$$
\begin{aligned}
\|\mathbf{g}(\mathbf{y} + \mathbf{k}) - \mathbf{g}(\mathbf{y})\|_2 &= \|(\mathbf{x} + \mathbf{h}) - \mathbf{x}\|_2 \leq 2\|\mathbf{f}(\mathbf{x} + \mathbf{h}) - \mathbf{f}(\mathbf{x})\|_2 \\
&= 2\|(\mathbf{y} + \mathbf{k}) - \mathbf{y}\|_2 = 2\|\mathbf{k}\|_2,
\end{aligned}
$$

hence $\mathbf{g}$ is continuous. Since $\mathbf{f}$ is differentiable

$$\mathbf{k} = \mathbf{f}(\mathbf{x} + \mathbf{h}) - \mathbf{f}(\mathbf{x}) = \mathbf{f}'(\mathbf{x})\mathbf{h} + \epsilon_{\mathbf{f}}(\mathbf{h}) \quad \text{where} \lim_{\mathbf{h} \to 0} \frac{\|\epsilon_{\mathbf{f}}(\mathbf{h})\|_2}{\|\mathbf{h}\|_2} = 0,$$

from which it follows that

$$\mathbf{g}(\mathbf{y} + \mathbf{k}) - \mathbf{g}(\mathbf{y}) = \mathbf{h} = \mathbf{f}'(\mathbf{x})^{-1}\mathbf{k} + \epsilon_{\mathbf{g}}(\mathbf{k}) \tag{5.6}$$

where $\epsilon_{\mathbf{g}}(\mathbf{k}) = -\mathbf{f}'(\mathbf{x})^{-1}\epsilon_{\mathbf{f}}(\mathbf{h})$. Since $\det \mathbf{f}'(\mathbf{x}) \neq 0$ for all $\mathbf{x} \in \overline{\Delta}$ and $\overline{\Delta}$ is compact it follows $\|\mathbf{f}'(\mathbf{x})^{-1}\|_o \leq M$ for some constant M and all points $\mathbf{x} \in \overline{\Delta}$; hence $\|\mathbf{f}'(\mathbf{x})^{-1}\mathbf{k}\|_2 \leq M\|\mathbf{k}\|_2$ and similarly $\|\epsilon_{\mathbf{g}}(\mathbf{k})\|_2 \leq M\|\epsilon_{\mathbf{f}}(\mathbf{h})\|_2$. Altogether then

$$\frac{\|\epsilon_{\mathbf{g}}(\mathbf{k})\|_2}{\|\mathbf{k}\|_2} \leq M \frac{\|\epsilon_{\mathbf{f}}(\mathbf{h})\|_2}{\|\mathbf{h}\|_2} \cdot \frac{\|\mathbf{h}\|_2}{\|\mathbf{k}\|_2} \leq 2M \frac{\|\epsilon_{\mathbf{f}}(\mathbf{h})\|_2}{\|\mathbf{h}\|_2} \tag{5.7}$$

since by the fundamental inequality (5.2) yet again

$$\frac{\|\mathbf{h}\|_2}{\|\mathbf{k}\|_2} = \frac{\|(\mathbf{x} + \mathbf{h}) - \mathbf{x}\|_2}{\|\mathbf{f}(\mathbf{x} + \mathbf{h}) - \mathbf{f}(\mathbf{x})\|_2} \leq 2.$$

Since $\lim_{\mathbf{h} \to \mathbf{0}} \frac{\|\epsilon_{\mathbf{f}}(\mathbf{h})\|_2}{\|\mathbf{h}\|_2} = 0$ while $\lim_{\mathbf{k} \to \mathbf{0}} \mathbf{h} = \mathbf{0}$ as a consequence of the continuity of the mapping $\mathbf{g}$, it follows from (5.7) that $\lim_{\mathbf{k} \to \mathbf{0}} \frac{\|\epsilon_{\mathbf{g}}(\mathbf{k})\|_2}{\|\mathbf{k}\|_2} = 0$ and consequently from (5.6) that the mapping $\mathbf{g}$ is differentiable, and moreover that $\mathbf{g}'(\mathbf{y}) = \mathbf{f}'(\mathbf{x})^{-1} = \mathbf{f}'(\mathbf{g}(\mathbf{y}))^{-1}$ so the derivative $\mathbf{g}'(\mathbf{y})$ is also continuous; that suffices for the proof.

Theorem 5.2 (Inverse Mapping Theorem). *If* $\mathbf{f} : W \longrightarrow \mathbb{R}^n$ *is a continuously differentiable mapping in an open neighborhood* $W \subset \mathbb{R}^n$ *of a point* $\mathbf{a} \in \mathbb{R}^n$ *such that* $\det \mathbf{f}'(\mathbf{a}) \neq 0$ *then for any sufficiently small open subneighborhood* $U \subset W$ *of the point* $\mathbf{a}$ *the restriction of the mapping* $\mathbf{f}$ *to* U *is a* $\mathcal{C}^1$ *homeomorphism* $\mathbf{f} : U \longrightarrow V$ *between* U *and an open neighborhood* V *of the image point* $\mathbf{b} = \mathbf{f}(\mathbf{a})$. *If* $\mathbf{g} : V \longrightarrow U$ *is the inverse mapping then* $\mathbf{g}'(\mathbf{y}) = \mathbf{f}'(\mathbf{g}(\mathbf{y}))^{-1}$ *for each point* $\mathbf{y} \in V$.

Proof: Introduce the invertible affine mappings $\phi, \psi : \mathbb{R}^n \longrightarrow \mathbb{R}^n$ defined by

$$\phi(\mathbf{x}) = \mathbf{f}'(\mathbf{a})^{-1}\mathbf{x} + \mathbf{a}, \quad \psi(\mathbf{x}) = \mathbf{x} - \mathbf{b}.$$

The composition $\mathbf{F} = \psi \circ \mathbf{f} \circ \phi$ is then a continuously differentiable mapping from an open neighborhood of the origin $\mathbf{0} \in \mathbb{R}^n$ into $\mathbb{R}^n$ such that $\mathbf{F}(\mathbf{0}) = \mathbf{0}$, and by the chain rule $\mathbf{F}'(\mathbf{0}) = \psi'(\mathbf{b})\mathbf{f}'(\mathbf{a})\phi'(\mathbf{0}) = I \cdot \mathbf{f}'(\mathbf{a})\mathbf{f}'(\mathbf{a})^{-1} = I$. The preceding theorem shows that the mapping $\mathbf{F}$ is invertible, and that its inverse $\mathbf{G}$ is a continuously differentiable mapping from an open neighborhood of the origin into $\mathbb{R}^n$. The composition $\mathbf{g} = \phi \circ \mathbf{G} \circ \psi$ is a continuously differentiable mapping from an open neighborhood of $\mathbf{b}$ into $\mathbb{R}^n$, and since $\mathbf{f} = \psi^{-1} \circ \mathbf{F} \circ \phi^{-1}$ it follows that $\mathbf{g} \circ \mathbf{f} = \phi \circ \mathbf{G} \circ \psi \circ \psi^{-1} \circ \mathbf{F} \circ \phi^{-1}$ is the identity mapping so that $\mathbf{g}$ is the inverse mapping to $\mathbf{f}$. Finally since $\mathbf{y} = \mathbf{f}(\mathbf{g}(\mathbf{y}))$ for any point $\mathbf{y} \in V$ it follows from the chain rule that $I = \mathbf{f}'(\mathbf{g}(\mathbf{y}))\mathbf{g}'(\mathbf{y})$ where I is the identity matrix, so that $\mathbf{g}'(\mathbf{y}) = \mathbf{f}'(\mathbf{g}(\mathbf{y}))^{-1}$, and that concludes the proof.

One consequence of the Inverse Mapping Theorem is that if $\mathbf{f} : U \longrightarrow \mathbb{R}^n$ is a continuously differentiable mapping defined in an open subset $U \subset \mathbb{R}^n$ such that $\det \mathbf{f}'(\mathbf{x}) \neq 0$ for all points $\mathbf{x} \in U$ then $\mathbf{f}$ is an **open mapping**, in the sense that the image of any open subset $U_0 \subset U$ is an open subset $\mathbf{f}(U_0) \subset \mathbb{R}^n$. The inverse image of any open set under a continuous mapping is always open; but the image of an open subset under a general continuous mapping is not necessarily open, so open continuous mappings are a special subclass of continuous mappings. If $\mathbf{f} : U \longrightarrow \mathbb{R}^n$ is a $\mathcal{C}^1$ homeomorphism from an open subset $U \subset \mathbb{R}^n$ with coordinates $(t_1, \ldots, t_n)$ onto an open subset $V \subset \mathbb{R}^n$ with coordinates $(x_1, \ldots, x_n)$, points in V can be described either in terms of the coordinates $(x_1, \ldots, x_n)$ in $\mathbb{R}^n$ or alternatively in terms of the coordinates $(t_1, \ldots, t_n) \in U \subset \mathbb{R}^n$; for this reason the parameters $(t_1, \ldots, t_n)$ are often described as being another **local coordinate system** in V, and the mapping $\mathbf{f}$ is called a $\mathcal{C}^1$ **local**

change of coordinates in V. Of course it is also possible just to consider continuous changes of coordinates through a homeomorphism that is not necessarily given by a C^1 mapping. The usefulness of a change of coordinates is that it may be possible to describe the local geometry much more simply in terms of the coordinates $(t_1, \ldots, t_n)$ than in terms of the initial coordinates $(x_1, \ldots, x_n)$. For example, an arc of a circle of radius 1 centered at the origin in the plane with a coordinate system (x_1, x_2) can be described locally as a straight line segment $r = 1$ in terms of polar coordinates (r, θ).

There is a more abstract and general form for changes of coordinates, which opens an important area of mathematics. A Hausdorff topological space M is called a **manifold**, or more precisely a **topological manifold of dimension** n, if it has an open covering $\{V_\alpha\}$, each set of which is homeomorphic to an open subset $U_\alpha \subset \mathbb{R}^n$. Thus if M is manifold then $M \subset \bigcup_\alpha V_\alpha$ and there are homeomorphisms $\phi_\alpha : U_\alpha \longrightarrow V_\alpha$ for some subsets $U_\alpha \subset \mathbb{R}^n$. Customarily the subsets U_α are assumed to be pairwise disjoint. The collection $\{V_\alpha, \phi_\alpha\}$ of open subsets of M and homeomorphisms from open subsets of $\mathbb{R}^n$ is called a **coordinate covering** of the topological space M; the subsets V_α are called **coordinate neighborhoods** of the manifold M, and the coordinates in U_α are considered **local coordinates** on the manifold M through the homeomorphism ϕ_α. If $\{V_\alpha, \phi_\alpha\}$ and $\{W_\beta, \psi_\beta\}$ are two coordinate coverings of M it is clear that their union is also a coordinate covering of M. For example a circle can be viewed as a one-dimensional manifold, for if it is described in terms of polar coordinates (r, θ) in a plane $\mathbb{R}^2$ then the angle θ can be taken as a local coordinate in any segment of the circle other than the full circle; similarly a sphere in $\mathbb{R}^3$, the surface of a solid ball $B_r \in \mathbb{R}^3$, can be viewed as a two-dimensional manifold by using spherical coordinates (r, θ, ϕ) in $\mathbb{R}^3$. What is interesting is that there are a great many examples of manifolds, some of which can be described naturally as subsets of Euclidean spaces as in these examples and others that are naturally described more abstractly.

Functions defined in a neighborhood $V_\alpha \subset M$ of a manifold M can be viewed as functions of the local coordinates in the open subset $U_\alpha \subset \mathbb{R}^n$, so it is possible to consider those that are continuously differentiable or C^1 functions of these local coordinates; but of course it is not necessarily the case that C^1 functions in V_α are also C^1 in the intersection $V_\alpha \cap V_\beta$ when viewed as functions of the local coordinates in V_β, unless the homeomorphisms ϕ_α and ϕ_β are differentiably compatible. A **differentiable coordinate covering** $\{V_\alpha, \phi_\alpha\}$ of a Hausdorff space M, or more precisely a C^1- **coordinate covering** of M, is a coordinate covering such that for any nonempty intersection $V_\alpha \cap V_\beta \subset M$ the mapping

$$\phi_{\beta\alpha} = \phi_\beta^{-1} \circ \phi_\alpha : \quad \phi_\alpha^{-1}(V_\alpha \cap V_\beta) \longrightarrow \phi_\beta^{-1}(V_\alpha \cap V_\beta)$$

is a C^1 mapping. The situation may be clarified by considering Figure 5.1. It is clear that if a function $f : M \longrightarrow \mathbb{R}$ is a C^1 mapping in a coordinate neighborhood $V_\alpha \subset M$ of a differentiable coordinate covering $\{V_\alpha, \phi_\alpha\}$ of M, meaning that

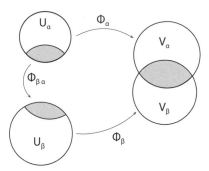

Figure 5.1. Local coordinate coverings of a manifold

the composition $f \circ \phi_\alpha : U_\alpha \longrightarrow \mathbb{R}$ is a C^1 mapping in $U_\alpha \subset \mathbb{R}^n$, then it is also a C^1 mapping in the intersection $V_\alpha \cap V_\beta$ in terms of the coordinates in V_β; consequently a function $f : M \longrightarrow \mathbb{R}$ is defined to be a C^1 **function** on M in terms of a differentiable coordinate covering $\{V_\alpha, \phi_\alpha\}$ if $f \circ \phi_\alpha$ is a C^1 mapping for each coordinate neighborhood V_α. That does not necessarily imply that the function is a C^1 function in terms of a different differentiable coordinate covering of the topological space M. Two C^1 coordinate coverings $\{V_\alpha, \phi_\alpha\}$ and $\{W_\beta, \psi_\beta\}$ are called **equivalent C^1 coordinate coverings** if their union is a C^1 coordinate covering; that means that the local coordinates in intersections $V_\alpha \cap W_\beta$ also must be compatible. The collection of all equivalent C^1 coordinate coverings of M is called a C^1 **structure** on the topological space M. Of course there are corresponding definitions for C^r-structures and C^∞ structures. If M and M' are two C^1 manifolds of dimensions n and n', with C^1 coordinate coverings $\{V_\alpha, \phi_\alpha\}$ and $\{V'_\beta, \phi'_\beta\}$, then just as for real-valued functions on manifolds a continuous mapping $f : M \longrightarrow M'$ can be expressed in terms of the local coordinates on the two manifolds; explicitly if $f(V_\alpha) \subset V'_\beta$ then the mapping f induces the mapping $(\phi'_\beta)^{-1} \circ f \circ \phi_\alpha$ from an open subset of $\mathbb{R}^n$ to an open subset of $\mathbb{R}^{n'}$; and the mapping f is said to be a C^1 mapping if these representations of f in terms of the local coordinates on M and M' are C^1 homeomorphisms. A differentiable mapping is a C^1 homeomorphism, or a C^1 diffeomorphism, if these local representations are C^1 mappings. The corresponding definitions again hold for C^r-structures and C^∞ structures. The investigation of these structures and mappings is the topic of differential topology. The question whether a topological manifold admits a differentiable structure or not, and if it admits a differentiable structure whether or not there are additional differentiable structures, has been investigated extensively; but that leads too far afield to be discussed further here.[1]

---

[1]The first example of a topological manifold admitting distinct differentiable structures was found by John Milnor, who showed that there are precisely 28 distinct differentiable structures on a seven-dimensional sphere. For further information see for example the *Princeton Companion to Mathematics*.

Figure 5.2. The mapping $\mathbf{f} : (r, \theta) \longrightarrow (r\cos\theta, r\sin\theta)$ in an open neighborhood of the origin

It should be emphasized that the Inverse Mapping Theorem is only a local result; a mapping $\mathbf{f} : U \longrightarrow \mathbb{R}^m$ for which $\det \mathbf{f}'(\mathbf{x}) \neq 0$ at all points $\mathbf{x} \in U \subset \mathbb{R}^n$ need not be invertible if $n > 1$. For example the derivative of the mapping $\mathbf{f} : \mathbb{R}^2 \longrightarrow \mathbb{R}^2$ defined by $\mathbf{f}(r, \theta) = (r\cos\theta, r\sin\theta)$ is the matrix

$$\mathbf{f}'(r, \theta) = \begin{pmatrix} \dfrac{\partial x}{\partial r} & \dfrac{\partial x}{\partial \theta} \\ \dfrac{\partial y}{\partial r} & \dfrac{\partial y}{\partial \theta} \end{pmatrix} = \begin{pmatrix} \cos\theta & -r\sin\theta \\ \sin\theta & r\cos\theta \end{pmatrix} \tag{5.8}$$

for which $\det \mathbf{f}'(r, \theta) = r$. The Inverse Mapping Theorem asserts that the mapping $\mathbf{f}$ is locally invertible near any point $(r, \theta) \in \mathbb{R}^2$ for which $r \neq 0$; but $\mathbf{f}$ is clearly not a one-to-one mapping in the open subset $U = \{ (r, \theta) \mid r > 0 \}$ since $\mathbf{f}(r, \theta) = \mathbf{f}(r, \theta + 2n\pi)$ for any integer n. It is very difficult to find general conditions ensuring that a mapping is globally invertible.[2] Points at which the Jacobian of a differentiable mapping $\mathbf{f} : \mathbb{R}^n \longrightarrow \mathbb{R}^n$ vanishes are known as **singular points** or **singularities** of the mapping. Mappings can be very complicated indeed near their singular points. For a simple example, consider again the mapping $\mathbf{f} : \mathbb{R}^2 \longrightarrow \mathbb{R}^2$ given by $\mathbf{f}(r, \theta) = (r\cos\theta, r\sin\theta)$; the origin $r = \theta = 0$ is a singular point since $\mathbf{f}'(0, 0) = \left(\begin{smallmatrix} 1 & 0 \\ 0 & 0 \end{smallmatrix}\right)$. The mapping is sketched in the accompanying Figure 5.2. A cell centered at the origin is mapped to the

---

[2]The "Jacobian Problem" is whether a mapping $\mathbf{f} : \mathbb{R}^n \longrightarrow \mathbb{R}^n$ from the n-dimensional vector space to itself for $n > 1$ defined by polynomial equations for which $\det \mathbf{f}'(\mathbf{x}) = 1$ at all points $\mathbf{x} \in \mathbb{R}^n$ is necessarily a homeomorphism from $\mathbb{R}^n$ onto $\mathbb{R}^n$. This is a long-standing open problem in mathematics, having been raised by O. H. Keller in 1939 for polynomial mappings $\mathbf{f} : \mathbb{C}^n \longrightarrow \mathbb{C}^n$ over the complex numbers. It is known to be true for polynomial mappings $\mathbf{f} : \mathbb{R}^2 \longrightarrow \mathbb{R}^2$ of degree at most 100, but not in general, at least at the time of this writing. It is known to be false for polynomial mappings over finite fields. It is also false for mappings involving functions more general than polynomials; for instance the mapping $\mathbf{f} : \mathbb{C}^2 \longrightarrow \mathbb{C}^2$ defined by $\mathbf{f}(x, y) = (e^x, e^{-x}y)$ has the Jacobian $\det \mathbf{f}'(x, y) = 1$ at all points (x, y) but its image excludes the axis $(0, y)$ and the mapping is not one-to-one. A famous example due to Fatou and Bieberbach, and discussed for example in the book *Several Complex Variables* by S. Bochner and W. T. Martin (Princeton University Press, 1948), is an injective mapping $\mathbf{f} : \mathbb{C}^2 \longrightarrow \mathbb{C}^2$ given by functions that are everywhere analytic and for which the Jacobian $\det \mathbf{f}'(\mathbf{x}) = 1$ at all points $\mathbf{x} \in \mathbb{C}^2$, but the image omits an open subset of $\mathbb{C}^2$. A survey of the problem and some approaches, and examples of some false proofs, can be found in the paper by H. Bass, E. H. Connell, and D. Wright in the *Bulletin of the American Mathematical Society* 7 (1982). If it is just assumed that $\det \mathbf{f}'(\mathbf{x}) \neq 0$ for all points $\mathbf{x} \in \mathbb{C}^2$, the "strong Jacobian problem," there is a counterexample for mappings $\mathbf{f} : \mathbb{R}^2 \longrightarrow \mathbb{R}^2$ by S. Pinchuk in *Mathematische Zeitschrift* 217 (1994).

bow-tie region, and the mapping is locally one-to-one except along the axis $r = 0$; that entire axis is mapped to a single point, the origin. The general study of the geometry of singularities of mappings is quite extensive, including what is known as the theory of catastrophes.[3]

An important and sometimes somewhat confusing concept is that of the **orientation** of a real vector space V. For any two bases $\mathbf{u}_i$ and $\mathbf{v}_i$ of an n-dimensional real vector space V there is a unique nonsingular real $n \times n$ matrix $A = \{a_{ij}\}$ such that $\mathbf{u}_i = \sum_{j=1}^{n} a_{ij}\mathbf{v}_j$ for $1 \le i \le n$. The two bases $\mathbf{u}_i$ and $\mathbf{v}_i$ are said to have the **same orientation** if $\det A > 0$. Clearly this determines an equivalence relation among bases, considering two bases equivalent if they have the same orientation; an equivalence class is called an **orientation** of the vector space V. It is customary to say that two bases that are not equivalent under this relation have the **opposite orientation**; so the bases $\mathbf{u}_i$ and $\mathbf{v}_i$ have the opposite orientation if $\det A < 0$. There are thus two possible orientations of any real vector space; but there is not a unique natural orientation of a vector space, so a choice must be made by selecting a basis that defines the orientation if the vector space is to be viewed as an oriented vector space. A proper subspace W of an oriented vector space V does not inherit a natural orientation from that of V; the orientation of W must be specified separately by the choice of a basis for W. For instance a general line in an oriented plane has two orientations, the two directions in which a vector along the line can point; but neither is determined by the orientation of the plane.

An orientation of the vector space $\mathbb{R}^n$ also can be described by an ordering of the coordinates in $\mathbb{R}^n$, specifying the coordinates as $x_1, x_2, \ldots, x_n$ for example and taking as the associated basis for $\mathbb{R}^n$ the derivatives $dx_1, dx_2, \ldots, dx_n$ of the coordinate functions, the unit vectors $dx_j = \delta_j$ in terms of the Kronecker vectors. A **permutation** π of the variables in $\mathbb{R}^n$, a reordering of the variables, can be described by a **permutation matrix** M, a matrix that has a single entry of 1 in each row and column and all other entries 0. For example

$$\begin{pmatrix} 0 & 1 & 0 & 0 \\ 0 & 0 & 0 & 1 \\ 1 & 0 & 0 & 0 \\ 0 & 0 & 1 & 0 \end{pmatrix} \begin{pmatrix} x_1 \\ x_2 \\ x_3 \\ x_4 \end{pmatrix} = \begin{pmatrix} x_2 \\ x_4 \\ x_1 \\ x_3 \end{pmatrix}$$

---

[3]See for example the books *Catastrophe Theory and Its Applications* by Tim Poston and Ian Stewart (New York: Dover, 1998) or *Catastrophe Theory* by Vladimir Arnold (Berlin: Springer-Verlag, 1992).

so this matrix describes the permutation $\pi(x_1, x_2, x_3, x_4) = (x_2, x_4, x_1, x_3)$, and more generally

$$
\begin{pmatrix} {}^t\delta_{i_1} \\ {}^t\delta_{i_2} \\ \cdots \\ {}^t\delta_{i_n} \end{pmatrix} \begin{pmatrix} x_1 \\ x_2 \\ \cdots \\ x_n \end{pmatrix} = \begin{pmatrix} x_{i_1} \\ x_{i_2} \\ \cdots \\ x_{i_n} \end{pmatrix}
$$

for any permutation $(i_1, i_2, \ldots, i_n)$ of the integers $(1, 2, \ldots, n)$, where ${}^t\delta_i$ is the transpose of the Kronecker vector δ_i. The **sign** of the permutation π described by a permutation matrix M, denoted by $\mathrm{sgn}(\pi)$, is defined by $\mathrm{sgn}(\pi) = \det M$, which is ± 1. If $x_{i_j} = \sum_{k=1}^n m_{jk} x_k$ is a permutation of the variables in $\mathbb{R}^n$ described by a permutation matrix $M = \{m_{ij}\}$ then $dx_{i_j} = \sum_{k=1}^n m_{jk} dx_k$ describes the relation between the bases in terms of the two orders of the variables; the bases dx_i and dx_{i_j} have the same orientation for any permutation π for which $\mathrm{sgn}(\pi) = +1$, hence two orders of the coordinates in $\mathbb{R}^n$ describe the same orientation of $\mathbb{R}$ precisely when they differ by a permutation π of sign $\mathrm{sgn}(\pi) = +1$. A permutation that interchanges any two consecutive variables, called a **transposition**, is orientation reversing, since $\det(\begin{smallmatrix} 0 & 1 \\ 1 & 0 \end{smallmatrix}) = -1$. Any permutation clearly can be written as the result of successively applying transpositions; and the sign of a succession of n transpositions is $(-1)^n$, so a permutation is orientation preserving precisely when it can be written as the successive application of an even number of transpositions. This provides an alternative to calculating a determinant to determine the sign of a permutation. There is often no natural orientation associated to the choice of coordinates in $\mathbb{R}^n$ though. For instance the orientation expressed by polar coordinates r, θ in $\mathbb{R}^2$ depends upon the order in which these coordinates are listed, as is the case for a set of coordinates $a, \alpha, \aleph$ in $\mathbb{R}^3$. It is customary to consider an orientation in $\mathbb{R}^2$ as either clockwise or counterclockwise, depending on the order of the variables; and to consider an orientation in $\mathbb{R}^3$ as that specified by either the order of fingers on the right hand or on the left hand, again a choice of the order of the variables.

If $\mathbf{f} : U \longrightarrow \mathbb{R}^n$ is a continuously differentiable mapping from an open subset $U \subset \mathbb{R}^n$ into $\mathbb{R}^n$ and if $\det \mathbf{f}'(\mathbf{a}) \neq 0$ at a point $\mathbf{a} \in U$ then either $\det \mathbf{f}'(\mathbf{a}) > 0$, in which case the mapping $\mathbf{f}$ is said to be **orientation preserving** at the point $\mathbf{a}$, or $\det \mathbf{f}'(\mathbf{a}) < 0$, in which case the mapping $\mathbf{f}$ is said to be **orientation reversing** at the point $\mathbf{a}$. A $\mathcal{C}^1$ mapping thus is orientation preserving at a point $\mathbf{a}$ precisely when its determinant matrix $\mathbf{f}'(\mathbf{a})$ is an orientation preserving linear transformation. If U is a connected open set and the Jacobian of the mapping $\mathbf{f}$ is nowhere vanishing then the mapping $\mathbf{f}$ is either orientation preserving at all points of U or orientation reversing at all points of U. For example the

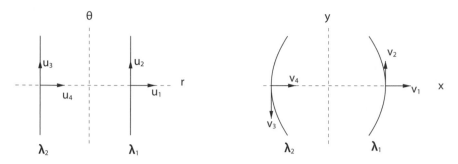

Figure 5.3. The mapping $\mathbf{f}: (r, \theta) \longrightarrow (r\cos\theta, r\sin\theta)$ preserves the orientation on the set where $r > 0$ but reverses the orientation on the set where $r < 0$

mapping $\mathbf{f}: \mathbb{R}^2(r, \theta) \longrightarrow \mathbb{R}^2(x, y)$ that takes a point (r, θ) to the point $\mathbf{f}(r, \theta) = (r\cos\theta, r\sin\theta)$ has the Jacobian $\det \mathbf{f}'(r, \theta) = r$, and consequently

if $r > 0$ then $\det \mathbf{f}'(r, \theta) > 0$ and $\mathbf{f}$ is orientation preserving,
if $r < 0$ then $\det \mathbf{f}'(r, \theta) < 0$ and $\mathbf{f}$ is orientation reversing,
if $r = 0$ then $\det \mathbf{f}'(r, \theta) = 0$ and $\mathbf{f}$ is singular,

as indicated in the accompanying Figure 5.3.

For a $\mathcal{C}^1$ manifold M with the differentiable structure defined by a coordinate covering $\{V_\alpha, \phi_\alpha\}$ the coordinates $z_\alpha \in U_\alpha$ and $z_\beta \in U_\beta$ are related by a $\mathcal{C}^1$ mapping $z_\beta = \phi_{\beta\alpha}(z_\alpha)$ for any nonempty intersection $V_\alpha \cap V_\beta$ as in Figure 5.3. The orientations of U_α and U_β defined by the orders of the coordinates in these neighborhoods are the same precisely when $\det \phi'_{\beta\alpha}(z_\alpha) > 0$. The coordinate covering $\{V_\alpha, \phi_\alpha\}$ is said to define an **orientation** of the manifold M if the orientations in intersecting coordinate neighborhoods $V_\alpha \cap V_\beta$ are the same, that is, if $\det \phi'_{\beta\alpha}(z_\alpha) > 0$ for every nonempty intersection $V_\alpha \cap V_\beta$; and in that case the manifold M is said to be an **oriented manifold**. If the orders of the coordinates in each coordinate neighborhood of an oriented manifold are changed by an orientation reversing permutation the result is still an oriented manifold, but with its orientation reversed. Manifolds such that it is possible to choose a coordinate covering defining an orientation are said to be **orientable manifolds**; and each such has two choices of an orientation. However there are manifolds, such as the Möbius strip, for which it is not possible to find a coordinate covering for which the orientations agree in all intersections of coordinate neighborhoods; such manifolds are said to be **non-orientable manifolds**.

Locally any $\mathcal{C}^1$ homeomorphism can be written as a composition of simpler homeomorphisms, which permits the local study of $\mathcal{C}^1$ homeomorphisms to be reduced to the examination of simpler mappings when that is helpful. A particularly simple homeomorphism is one that changes only a single variable, a mapping $\mathbf{f} : \mathbb{R}^n \longrightarrow \mathbb{R}^n$ for which the coordinate functions $f_i(\mathbf{x})$ have the form $f_i(\mathbf{x}) = x_i$ for $i \neq j$ while $f_j(\mathbf{x})$ is a function of all the variables.

Theorem 5.3. *A $\mathcal{C}^1$ mapping $\mathbf{f} : U \longrightarrow \mathbb{R}^n$ defined in an open neighborhood U of the origin in $\mathbb{R}^n$, that takes the origin to the origin and has a nonzero Jacobian at the origin, can be written in any sufficiently small open subneighborhood $V \subset U$ of the origin as the composition of $\mathcal{C}^1$ homeomorphisms that change only a single variable.*

Proof: For convenience a $\mathcal{C}^1$ homeomorphism $\mathbf{f} : U \longrightarrow \mathbb{R}^n$ defined in an open neighborhood $U \subset \mathbb{R}^n$ of the origin will be said to be of type r if $\mathbf{f}(\mathbf{0}) = \mathbf{0}$ and its coordinate functions f_i have the form $f_i(\mathbf{x}) = x_i$ for at least $r - 1$ of the indices i; but this convention will be used just for the course of the present proof. Any $\mathcal{C}^1$ homeomorphism taking the origin to the origin is of type 1 of course, since that imposes no conditions on its coordinate functions; but it is of type $n + 1$ if and only if it is the identity mapping. The theorem will be demonstrated by showing that if $\mathbf{f}$ is any $\mathcal{C}^1$ homeomorphism of type r in an open neighborhood U of the origin then after restricting that neighborhood sufficiently there is a homeomorphism $\boldsymbol{\phi} : U \longrightarrow U$ such that $\boldsymbol{\phi}$ changes just one variable and the composition $\mathbf{f} \circ \boldsymbol{\phi}$ is of type $r + 1$. Indeed if that result is demonstrated then for any homeomorphism $\mathbf{f} : U \longrightarrow \mathbb{R}^n$ for which $\mathbf{f}(\mathbf{0}) = \mathbf{0}$ after restricting the neighborhood U sufficiently there will be a sequence of $\mathcal{C}^1$ homeomorphisms $\boldsymbol{\phi}_i : U \longrightarrow \mathbb{R}^n$ each of which changes just a single variable such that $\mathbf{f} \circ \boldsymbol{\phi}_1 \circ \boldsymbol{\phi}_2 \circ \cdots \circ \boldsymbol{\phi}_k = I$, the identity mapping, and then $\mathbf{f} = \boldsymbol{\phi}_k^{-1} \circ \cdots \circ \boldsymbol{\phi}_2^{-1} \circ \boldsymbol{\phi}_1^{-1}$, thus exhibiting $\mathbf{f}$ as the composition of $\mathcal{C}^1$ homeomorphisms each of which changes just a single variable.

Suppose therefore that $\mathbf{f} : U \longrightarrow \mathbb{R}^n$ is a $\mathcal{C}^1$ homeomorphism of type r. For convenience of notation the variables can be relabeled so that $f_i(\mathbf{x}) = x_i$ for $1 \leq i < r$. In that case points $\mathbf{x} \in \mathbb{R}^n$ can be written in the form

$$\mathbf{x} = \begin{pmatrix} \mathbf{x}_I \\ x_r \\ \mathbf{x}_{II} \end{pmatrix} \quad \text{where} \quad \begin{cases} \mathbf{x}_I \in \mathbb{R}^{r-1} & I = (1, 2, \ldots, r-1) \\ x_r \in \mathbb{R}^1 \\ \mathbf{x}_{II} \in \mathbb{R}^{n-r} & II = (r+1, r+2, \ldots, n) \end{cases}$$

and the mapping $\mathbf{f}$ can be written correspondingly in the form

$$\mathbf{f}(\mathbf{x}) = \begin{pmatrix} \mathbf{x}_I \\ f_r(\mathbf{x}) \\ \mathbf{f}_{II}(\mathbf{x}) \end{pmatrix} \quad \text{where} \quad \begin{cases} \mathbf{x}_I \in \mathbb{R}^{r-1} & I = (1, 2, \ldots, r-1) \\ f_r(\mathbf{x}) \in \mathbb{R}^1 \\ \mathbf{f}_{II}(\mathbf{x}) \in \mathbb{R}^{n-r} II = (r+1, r+2, \ldots, n). \end{cases}$$

In these terms the derivative of the mapping $\mathbf{f}$ is the $n \times n$ matrix

$$\mathbf{f}'(\mathbf{x}) = \begin{pmatrix} I_{r-1} & 0 & 0 \\ \partial_I f_r(\mathbf{x}) & \partial_r f_r(\mathbf{x}) & \partial_{II} f_r(\mathbf{x}) \\ \partial_I \mathbf{f}_{II}(\mathbf{x}) & \partial_r \mathbf{f}_{II}(\mathbf{x}) & \partial_{II} \mathbf{f}_{II}(\mathbf{x}) \end{pmatrix} \qquad (5.9)$$

in which I_{r-1} is the $(r-1) \times (r-1)$ identity matrix while

$$\partial_I f_r(\mathbf{x}) = \big(\partial_1 f_r(\mathbf{x}) \cdots \partial_{r-1} f_r(\mathbf{x})\big), \quad a \ 1 \times (r-1) \text{ matrix,}$$
$$\partial_I \mathbf{f}_{II}(\mathbf{x}) = \big\{\, \partial_j f_i(\mathbf{x}) \,\big|\, r+1 \leq i \leq n,\ 1 \leq j \leq r-1 \,\big\}, \text{ an } (n-r) \times (r-1) \text{ matrix,}$$

and similarly for the other components of the matrix (5.9). Since $\det \mathbf{f}'(\mathbf{0}) \neq 0$ not all the entries in column r of this matrix are zero; so after a further relabeling of the coordinate functions it can be assumed that $\partial_r f_r(\mathbf{0}) \neq 0$. Let $\boldsymbol{\psi}_r : U \longrightarrow \mathbb{R}^n$ be the mapping defined by

$$\boldsymbol{\psi}_r \begin{pmatrix} \mathbf{x}_I \\ x_r \\ \mathbf{x}_{II} \end{pmatrix} = \begin{pmatrix} \mathbf{x}_I \\ f_r(\mathbf{x}) \\ \mathbf{x}_{II} \end{pmatrix},$$

for which

$$\boldsymbol{\psi}'_r(\mathbf{x}) = \begin{pmatrix} I_{r-1} & 0 & 0 \\ \partial_I f_r(\mathbf{x}) & \partial_r f_r(\mathbf{x}) & \partial_{II} f_r(\mathbf{x}) \\ 0 & 0 & I_{n-r}. \end{pmatrix} \qquad (5.10)$$

Since $\det \boldsymbol{\psi}'_r(\mathbf{0}) = \partial_r f_r(\mathbf{0}) \neq 0$ it follows from the Inverse Mapping Theorem that the mapping $\boldsymbol{\psi}_r$ is invertible in a sufficiently small open neighborhood $U_r \subset U$ of the origin. The inverse mapping $\boldsymbol{\phi}_r = \boldsymbol{\psi}_r^{-1}$ of course has the corresponding form

$$\boldsymbol{\phi}_r \begin{pmatrix} \mathbf{x}_I \\ x_r \\ \mathbf{x}_{II} \end{pmatrix} = \begin{pmatrix} \mathbf{x}_I \\ g_r(\mathbf{x}) \\ \mathbf{x}_{II} \end{pmatrix}$$

for some function $g_r(\mathbf{x})$; and since

$$\mathbf{x} = \boldsymbol{\psi}_r(\boldsymbol{\phi}_r(\mathbf{x})) = \boldsymbol{\psi}_r \begin{pmatrix} \mathbf{x}_I \\ g_r(\mathbf{x}) \\ \mathbf{x}_{II} \end{pmatrix} = \begin{pmatrix} \mathbf{x}_I \\ f_r(\boldsymbol{\phi}_r(\mathbf{x})) \\ \mathbf{x}_{II} \end{pmatrix}$$

it follows that $f_r(\phi_r(\mathbf{x})) = x_r$ and therefore

$$\mathbf{f}(\phi_r(\mathbf{x})) = \mathbf{f}\begin{pmatrix} \mathbf{x}_I \\ g_r(\mathbf{x}) \\ \mathbf{x}_{II} \end{pmatrix} = \begin{pmatrix} \mathbf{x}_I \\ f_r(\phi_r(\mathbf{x})) \\ \mathbf{f}_{II}(\phi_r(\mathbf{x})) \end{pmatrix} = \begin{pmatrix} \mathbf{x}_I \\ x_r \\ \mathbf{f}_{II}(\phi_r(\mathbf{x})) \end{pmatrix},$$

so the composition $\mathbf{f} \circ \phi_r$ is of type $r + 1$. As already observed, that suffices to conclude the proof.

Corollary 5.4 (Decomposition Theorem). *A C^1 mapping $\mathbf{f} : U \longrightarrow \mathbb{R}^n$ defined in an open neighborhood U of a point $\mathbf{a} \in \mathbb{R}^n$ with a nonzero Jacobian at the point $\mathbf{a}$ can be written in any sufficiently small open subneighborhood $V \subset U$ of the point $\mathbf{a}$ as the composition of C^1 homeomorphisms that change only a single variable.*

Proof: If $\mathbf{f}(\mathbf{a}) = \mathbf{b}$ then $\mathbf{g}(\mathbf{x}) = \mathbf{f}(\mathbf{a} + \mathbf{x}) - \mathbf{b}$ is a C^1 mapping defined in the open neighborhood U of the origin that takes the origin to the origin and has a nonzero Jacobian at the origin; by the preceding theorem the homeomorphism $\mathbf{g}$ can be written in some open subneighborhood $V \subset U$ as a composition of C^1 homeomorphisms that change only a single variable, and $\mathbf{f}(\mathbf{x}) = \mathbf{g}(\mathbf{x} - \mathbf{a}) + \mathbf{b} = T_{\mathbf{b}} \circ \mathbf{g} \circ T_{-\mathbf{a}}$ in terms of the translation mappings $T_{\mathbf{c}}(\mathbf{x}) = \mathbf{x} + \mathbf{c}$. Any translation can be written as a composition of translations in a single variable, so altogether then $\mathbf{f}$ is written as a composition of C^1 homeomorphisms that change only a single variable, which suffices for the proof.

Problems, Group I

1. Show that the function (5.1) is differentiable at all points $x \in \mathbb{R}^1$ but that it does not describe a one-to-one mapping in any open neighborhood of the origin.

2. Consider the mapping $\mathbf{f} : \mathbb{R}^2 \longrightarrow \mathbb{R}^2$ defined by

$$\mathbf{f}(x_1, x_2) = \begin{pmatrix} e^{x_1} \cos x_2 \\ e^{x_1} \sin x_2 \end{pmatrix}.$$

  i) Find the Jacobian matrix of the mapping $\mathbf{f}$.

  ii) Find formulas for the local inverse of the mapping $\mathbf{f}$ at those points at which the local inverse exists.

  iii) Is the mapping $\mathbf{f}$ injective?

  iv) Is the mapping $\mathbf{f}$ surjective?

  v) At which points is the mapping $\mathbf{f}$ nonsingular and orientation preserving?

3.   i) Which permutations of the coordinate axes in $\mathbb{R}^4$ are orientation preserving mappings?
     ii) What are the permutation matrices representing these mappings?

4.   i) Find the Jacobian matrix and the Jacobian of the mapping $\mathbf{f} : \mathbb{R}^2 \longrightarrow \mathbb{R}^2$ defined by $\mathbf{f}(\mathbf{x}) = \begin{pmatrix} \sin x_1 \\ \cos x_2 \end{pmatrix}$.
     ii) Find the singularities of the mapping and describe the mapping at those singular points.

Problems, Group II

5.   Introduce an equivalence relation in $\mathbb{R}^2$ by setting $(x_1, x_2) \asymp (y_1, y_2)$ if and only if $(x_1 - y_1) \in \mathbb{Z}$ and $(x_2 - y_2) \in \mathbb{Z}$; let $T = \mathbb{R}^2 / \asymp$ be the space of equivalence classes and let $\phi : \mathbb{R}^2 \longrightarrow T$ be the natural mapping that associates to any point in $\mathbb{R}^2$ its equivalence class in T. Introduce the topology on T in which a set $U \subset T$ is open if and only if $\phi^{-1}(U)$ is an open subset of $\mathbb{R}^2$. Show that the set T can be given the structure of a compact two-dimensional $\mathcal{C}^\infty$ manifold.

6.   Consider the mapping $f : \mathbb{R}^2 \longrightarrow \mathbb{R}^2$ defined by

$$f(x_1, x_2) = \left(-x_1 + \sqrt{x_1^2 + x_2^2}, -x_1 - \sqrt{x_1^2 + x_2^2} \right).$$

   i) Find the Jacobian of this mapping.
   ii) Find the local inverse of this mapping at those points at which it is locally bijective. At which of these points is the mapping orientation preserving?
   iii) Show that lines parallel to the coordinate axes in the image space of this mapping correspond to parabolas $x_2^2 = 2p \left(x_1 + \frac{p}{2} \right)$. (These coordinates in the image space are called parabolic coordinates.)

7.   Consider the mapping $\mathbf{f} : \mathbb{R}^{2 \times 2} \longrightarrow \mathbb{R}^{2 \times 2}$ defined by $\mathbf{f}(X) = X^2$.

   i) Find the Jacobian matrix of the mapping $\mathbf{f}$ and evaluate it at the identity matrix I.
   ii) Show that the mapping $\mathbf{f}$ is a bijective mapping from an open neighborhood of the identity matrix I to an open neighborhood of the identity matrix I.
   iii) The local inverse $\mathbf{f}^{-1}$ of the mapping $\mathbf{f}$ in an open neighborhood of the identity matrix can be viewed as the mapping $\mathbf{f}^{-1}(X) = \sqrt{X}$. What is the derivative of this function $\sqrt{X}$?

    iv) Can you find an explicit formula for the function $\sqrt{X}$ by using the binomial expansion?

8.   Consider the mapping $\phi : \mathbb{R}^3 \longrightarrow \mathbb{R}^3$ defined by

$$\mathbf{f}(r, \phi, \theta) = (r \sin \phi \cos \theta, r \sin \phi \sin \theta, r \cos \phi).$$

(The coordinates (r, ϕ, θ) are called spherical coordinates in $\mathbb{R}^3$.)

    i) Find the Jacobian of the mapping $\mathbf{f}$.
    ii) Determine those points at which the mapping $\mathbf{f}$ is locally bijective, and find a local inverse mapping.
    iii) Is the mapping $\mathbf{f}$ orientation preserving at those points at which it is locally bijective?
    iv) What is the image of the mapping $\mathbf{f}$?

9.   A cylinder is constructed from a strip of paper by gluing the two ends together in the same order; the Möbius strip is constructed from a strip of paper by gluing the two ends of the strip together in the reverse order. Show that a cylinder is an orientable manifold but that the Möbius strip is a non-orientable manifold. It is amusing to note as well that while a cylinder has two edges and two sides a Möbius strip has only one side and one edge.

5.2   Implicit Function Theorem

The Inverse Mapping Theorem can be applied to derive a simple local normal form at a point $\mathbf{a} \in \mathbb{R}^m$ for mappings $\mathbf{f} : \mathbb{R}^m \longrightarrow \mathbb{R}^n$ when $m > n$ and rank $\mathbf{f}'(\mathbf{a}) = n$, the maximal possible rank, showing that in suitable local coordinates near the point $\mathbf{a}$ the mapping is the simplest linear mapping between spaces of different dimensions; the general result will be discussed in Section 5.3, but a special case of particular importance and usefulness will be discussed in this section.

Theorem 5.5. *If $\mathbf{f} : U \longrightarrow \mathbb{R}^n$ is a C^1 mapping defined in an open neighborhood U of the origin $\mathbf{0} \in \mathbb{R}^m$ where $m > n$, and if $\mathbf{f}(\mathbf{0}) = \mathbf{0}$ and the $n \times n$ matrix consisting of the first n columns of the $n \times m$ matrix $\mathbf{f}'(\mathbf{0})$ is of rank n, then after shrinking the neighborhood U if necessary there is a C^1 homeomorphism $\phi : V \longrightarrow U$ between an open neighborhood V of the origin $\mathbf{0} \in \mathbb{R}^m$ and the open neighborhood U, where*

$$\phi(t_1, \ldots, t_m) = \{ \underbrace{g_1(\mathbf{t}), \ldots, g_n(\mathbf{t})}_{\mathbf{g(t)}}, t_{n+1}, \ldots, t_m \} \tag{5.11}$$

for some C^1 functions $g_i(\mathbf{t})$ of the variables $\mathbf{t} \in V$, such that $g_i(\mathbf{0}) = 0$ and

$$(\mathbf{f} \circ \boldsymbol{\phi})(t_1 \ldots, t_m) = \{t_1, \ldots, t_n\}. \tag{5.12}$$

Proof: The assertion of the theorem can be summarized as the existence of the mapping $\boldsymbol{\phi}$ in the diagram

$$
\begin{array}{ccccc}
\mathbb{R}^m & & \mathbb{R}^m & & \mathbb{R}^n \\
\cup & & \cup & & \cup \\
V & \xrightarrow{\;\boldsymbol{\phi}\;} & U & \xrightarrow{\;\mathbf{f}\;} & \mathbb{R}^n
\end{array}
$$

with the properties (5.11) and (5.12). To simplify the notation write points $\mathbf{x} \in \mathbb{R}^m$ in the form

$$\mathbf{x} = \begin{pmatrix} \mathbf{x}_I \\ \mathbf{x}_{II} \end{pmatrix} \quad \text{where} \quad \begin{cases} \mathbf{x}_I \in \mathbb{R}^n & I = (1, 2, \ldots, n) \\ \mathbf{x}_{II} \in \mathbb{R}^{m-n} & II = (n+1, n+2, \ldots, m). \end{cases}$$

The $n \times m$ matrix $\mathbf{f}'(\mathbf{x})$ can be written

$$\mathbf{f}'(\mathbf{x}) = \big(\partial_I \mathbf{f}(\mathbf{x}) \partial_{II} \mathbf{f}(\mathbf{x})\big)$$

in terms of the $n \times n$ matrix $\partial_I \mathbf{f}(\mathbf{x})$ having the entries $(\partial_I(\mathbf{f}(\mathbf{x})))_{ij} = \partial_j f_i(\mathbf{x})$ for $1 \le i, j \le n$ and the $n \times (m-n)$ matrix $\partial_{II} \mathbf{f}(\mathbf{x})$ with entries $(\partial_{II}(\mathbf{f}(\mathbf{x})))_{ij} = \partial_j f_i(\mathbf{x})$ for $1 \le i \le n$ and $n+1 \le j \le m$. Introduce the C^1 mapping $\boldsymbol{\psi} : U \longrightarrow \mathbb{R}^m$ defined by

$$\boldsymbol{\psi}(\mathbf{x}) = \boldsymbol{\psi}\begin{pmatrix} \mathbf{x}_I \\ \mathbf{x}_{II} \end{pmatrix} = \begin{pmatrix} \mathbf{f}(\mathbf{x}) \\ \mathbf{x}_{II} \end{pmatrix}$$

and note that

$$\boldsymbol{\psi}'(\mathbf{x}) = \begin{pmatrix} \partial_I \mathbf{f}(\mathbf{x}) & \partial_{II} \mathbf{f}(\mathbf{x}) \\ \partial_I \mathbf{x}_{II} & \partial_{II} \mathbf{x}_{II} \end{pmatrix} = \begin{pmatrix} \partial_I \mathbf{f}(\mathbf{x}) & \partial_{II} \mathbf{f}(\mathbf{x}) \\ 0 & I_{m-n} \end{pmatrix}$$

where I_{m-n} is the identity $(m-n) \times (m-n)$ matrix; consequently $\det \boldsymbol{\psi}'(\mathbf{0}) = \det \partial_I \mathbf{f}(\mathbf{0}) \ne 0$, so it follows from the Inverse Mapping Theorem that after shrinking the neighborhood U if necessary the mapping $\boldsymbol{\psi}$ is a one-to-one mapping from U onto an open neighborhood V of the image $\boldsymbol{\psi}(\mathbf{0}) = \mathbf{0} \in \mathbb{R}^m$ with a C^1 inverse $\boldsymbol{\phi} = \boldsymbol{\psi}^{-1}$, which of course has the form

$$\boldsymbol{\phi}(\mathbf{t}) = \begin{pmatrix} \mathbf{g}(\mathbf{t}) \\ \mathbf{t}_{II} \end{pmatrix}$$

for a mapping $\mathbf{g} : V \longrightarrow \mathbb{R}^n$ for which $\mathbf{g}(\mathbf{0}) = \mathbf{0}$ since $\boldsymbol{\phi}(\mathbf{0}) = \mathbf{0}$; so $\boldsymbol{\phi}$ is a mapping of the form (5.11). Then

$$\begin{pmatrix} \mathbf{t}_I \\ \mathbf{t}_{II} \end{pmatrix} = (\boldsymbol{\psi} \circ \boldsymbol{\phi}) \begin{pmatrix} \mathbf{t}_I \\ \mathbf{t}_{II} \end{pmatrix} = \boldsymbol{\psi} \begin{pmatrix} \mathbf{g}(\mathbf{t}) \\ \mathbf{t}_{II} \end{pmatrix} = \begin{pmatrix} \mathbf{f}(\boldsymbol{\phi}(\mathbf{t})) \\ \mathbf{t}_{II} \end{pmatrix}$$

so $\mathbf{f}(\boldsymbol{\phi}(\mathbf{t})) = \mathbf{t}_I$, which demonstrates (5.12) and thereby concludes the proof.

Corollary 5.6. *If $\mathbf{f} : U \longrightarrow \mathbb{R}^n$ is a $\mathcal{C}^1$ mapping defined in an open subset $U \subset \mathbb{R}^m$ where $m > n$, and if $\operatorname{rank} \mathbf{f}'(\mathbf{a}) = n$ at a point $\mathbf{a} \in U$, then there is a $\mathcal{C}^1$ change of coordinates in an open neighborhood of the point $\mathbf{a}$ such that the mapping $\mathbf{f} - \mathbf{f}(\mathbf{a})$ is a linear mapping described by a matrix of rank n in terms of these coordinates.*

Proof: The mapping $\tilde{\mathbf{f}}(\mathbf{x}) = \mathbf{f}(\mathbf{x} + \mathbf{a}) - \mathbf{f}(\mathbf{a})$ is $\mathcal{C}^1$ in an open neighborhood of the origin, and rank $\tilde{\mathbf{f}}'(\mathbf{0}) = \operatorname{rank} \mathbf{f}'(\mathbf{a}) = n$ while $\mathbf{f}(\mathbf{0}) = \mathbf{0}$. By a permutation of the coordinates it can be supposed that the first n columns of the $n \times m$ matrix $\tilde{\mathbf{f}}'(\mathbf{0})$ are linearly independent. The mapping $\tilde{\mathbf{f}}$ thus satisfies the hypotheses of the preceding theorem, so after a further $\mathcal{C}^1$ change of coordinates $\boldsymbol{\phi}$ near the origin the composition $L\mathbf{t} = \tilde{\mathbf{f}}(\boldsymbol{\phi}(\mathbf{t})) = \mathbf{f}(\boldsymbol{\phi}(\mathbf{t}) + a) - \mathbf{f}(\mathbf{a})$ is a linear mapping of the form (5.12) of rank n; so if $\boldsymbol{\psi}(\mathbf{t}) = \boldsymbol{\phi}(\mathbf{t}) + \mathbf{a}$ then $\mathbf{f}(\boldsymbol{\psi}(\mathbf{t})) - \mathbf{f}(\mathbf{a}) = L\mathbf{t}$, and that suffices for the proof.

Corollary 5.7. *If $\mathbf{f} : U \longrightarrow \mathbb{R}^n$ is a continuously differentiable mapping defined in an open subset $U \subset \mathbb{R}^m$ where $m > n$, and if $\operatorname{rank} \mathbf{f}'(\mathbf{a}) = n$ at a point $\mathbf{a} \in U$, then there is a $\mathcal{C}^1$ change of coordinates in an open neighborhood $U_{\mathbf{a}}$ of the point $\mathbf{a}$ in terms of which the set $\{ \mathbf{x} \in U_{\mathbf{a}} \mid \mathbf{f}(\mathbf{x}) = \mathbf{f}(\mathbf{a}) \}$ is a piece of a linear subspace of dimension $m - n$.*

Proof: It follows from the preceding Corollary 5.6 that there is a local coordinate system in an open neighborhood $U_{\mathbf{a}}$ of the point $\mathbf{a}$ such that the mapping $\mathbf{f} - \mathbf{f}(\mathbf{a})$ is a linear mapping of rank n in terms of these coordinates; consequently the set of points in $U_{\mathbf{a}}$ at which the mapping $\mathbf{f} - \mathbf{f}(\mathbf{a})$ vanishes is a linear subspace of dimension $m - n$, which suffices for the proof.

One of the most important consequences of Theorem 5.5 involves a more precise assumption about the form of the derivative $\mathbf{f}'(\mathbf{0})$.

Theorem 5.8 (Implicit Function Theorem). *If $\mathbf{f} : U \longrightarrow \mathbb{R}^n$ is a $\mathcal{C}^1$ mapping defined in an open neighborhood U of the origin $\mathbf{0} \in \mathbb{R}^m$ where $m > n$, and if $\mathbf{f}(\mathbf{0}) = \mathbf{0}$ and $\operatorname{rank} \begin{pmatrix} \partial_1 \mathbf{f}(\mathbf{0}) & \cdots & \partial_n \mathbf{f}(\mathbf{0}) \end{pmatrix} = n$, then in a sufficiently small open cell $\Delta = \Delta_I \times \Delta_{II} \subset U$ containing the origin, written as the product of a cell $\Delta_I \subset \mathbb{R}^n$ in the space of the first n variables $\mathbf{x}_I = \{x_1, \ldots, x_n\}$ and a cell $\Delta_{II} \subset \mathbb{R}^{m-n}$ in the space of the last $m - n$ variables $\mathbf{x}_{II} = \{x_{n+1}, \ldots, x_m\}$, there is a $\mathcal{C}^1$ mapping $\mathbf{h} : \Delta_{II} \longrightarrow \Delta_I$ such that $\mathbf{f}(\mathbf{x}) = \mathbf{0}$ for a point $\mathbf{x} = \{\mathbf{x}_I, \mathbf{x}_{II}\}$ in $\Delta_I \times \Delta_{II}$ if and only if $\mathbf{x}_I = \mathbf{h}(\mathbf{x}_{II})$.*

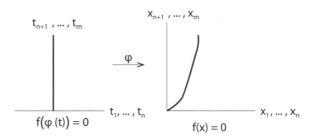

Figure 5.4. Illustration of the proof of the Implicit Function Theorem

Proof: It follows from Theorem 5.5 that if the neighborhood U is sufficiently small there is a C^1 homeomorphism $\phi : V \longrightarrow U$ from an open neighborhood V of the origin in the space $\mathbb{R}^m$ of the variables $t_1, \ldots, t_m$ onto U such that (5.11) and (5.12) are satisfied. Any point $\mathbf{x} \in U$ is the image $\mathbf{x} = \phi(\mathbf{t})$ of a unique point $\mathbf{t} \in V$, and it follows from (5.12) that $\mathbf{f}(\mathbf{x}) = \mathbf{0}$ if and only if $(\mathbf{f} \circ \phi)(\mathbf{t}) = \{t_1, \ldots, t_n\} = \mathbf{0}$, or equivalently if and only if $\mathbf{t}_I = \mathbf{0}$ when the point $\mathbf{t} \in \mathbb{R}^m$ is written $\mathbf{t} = \{\mathbf{t}_I, \mathbf{t}_{II}\}$ where $\mathbf{t}_I \in \mathbb{R}^n$ and $\mathbf{t}_{II} \in \mathbb{R}^{m-n}$; and in that case it follows from (5.11) that $\{\mathbf{x}_I, \mathbf{x}_{II}\} = \mathbf{x} = \phi(\mathbf{t}) = \{\mathbf{g}(\mathbf{0}, \mathbf{t}_{II}), \mathbf{t}_{II}\}$, so that actually $\mathbf{t}_{II} = \mathbf{x}_{II}$ and consequently $\mathbf{x} = \{\mathbf{g}(\mathbf{0}, \mathbf{x}_{II}), \mathbf{x}_{II}\}$ or equivalently $\mathbf{x}_I = \mathbf{g}(\mathbf{0}, \mathbf{x}_{II})$, which suffices for the proof.

The argument in the proof of the preceding theorem is illustrated in the accompanying Figure 5.4. The zeros of the function $\mathbf{f}$ form the image of the linear subspace defined by the equations $t_1 = \cdots = t_n = 0$ under the C^1 mapping $\phi :$ $\mathbb{R}^m \longrightarrow \mathbb{R}^m$ that describes the change of coordinates in an open neighborhood of the origin. The mapping ϕ does not change the last $m - n$ coordinates but only the first n coordinates of the points; so the mapping preserves the "height" of points above the coordinate axis of the first n coordinates, but moves the points by changing their first n coordinates. Points on the linear subspace are described by the parameters $t_i = x_i$ for $n + 1 \leq i \leq m$ while the coordinates t_i are held constant for $1 \leq i \leq n$; the zeros of the function $\mathbf{f}$ are also described by the parameters $t_i = x_i$ for $n + 1 \leq i \leq m$.

In the Implicit Function Theorem, the equation $\mathbf{f}(\mathbf{x}) = \mathbf{0}$ can be viewed as a description of relations among the coordinates $x_1, \ldots, x_m$ in $\mathbb{R}^m$, which implicitly describe some of the coordinates as functions of the remaining coordinates; the theorem provides conditions determining which coordinates really can be viewed as explicit functions of the remaining coordinates, at least locally. For example, if $f : U \longrightarrow \mathbb{R}^1$ is a continuously differentiable function in an open subset $U \subset \mathbb{R}^m$ and if $\partial_1 f(\mathbf{a}) \neq 0$ at some point $\mathbf{a} \in U$ then by the Implicit Function Theorem in an open cell $\Delta \subset U$ containing the point $\mathbf{a}$ there is a function $g(x_2, \ldots, x_n)$ such that $f(x_1, \ldots, x_m) = 0$ for a point $\{x_1, \ldots, x_m\} \in \Delta$ if and only if $x_1 = g(x_2, \ldots, x_m)$. The equation $f(x_1, x_2, \ldots, x_m) = 0$ defines x_1

implicitly as a function of the remaining variables, and in an open neighborhood of any point $\mathbf{a}$ at which $\partial_1 f(\mathbf{a}) \neq 0$ this equation also defines x_1 explicitly as a function $x_1 = g(x_2, \ldots, x_m)$ of the remaining variables. It is obvious that some condition on the function f is necessary for this to be true; for instance without any condition the function $f(x_1, \ldots, x_m)$ might be independent of the variable x_1, in which case it can say nothing at all about that variable. The example in which $f(x_1, x_2, x_3) = x_1^2 + x_2^2 + x_3^2 - 1$ illustrates another way in which some condition on the function f is necessary. The equation $f(\mathbf{x}) = 0$ can be viewed as defining x_1 as the function $x_1 = \pm\sqrt{1 - x_2^2 - x_3^2}$, where the square root has two well-defined locally continuous values so long as $x_2^2 + x_3^2 < 1$; but x_1 is not a well-defined function in an open neighborhood of any point at which $x_2^2 + x_3^2 = 1$, or equivalently at which $\partial_1 f(x_1, x_2, x_3) = 0$. If the variable x_1 is defined explicitly by $x_1 = g(x_2, x_3)$ then $f\big(g(x_2, x_3), x_2, x_3\big) = 0$ identically in the variables x_2, x_3; and the chain rule can be applied to calculate the derivatives of the function $g(x_2, x_3)$, since

$$
0 = \frac{\partial}{\partial x_2} f\big(g(x_2, x_3), x_2, x_3\big)
$$

$$
= \partial_1 f\big(g(x_2, x_3), x_2, x_3\big) \partial_1 g(x_2, x_3) + \partial_2 f\big(g(x_2, x_3), x_2, x_3\big).
$$

Thus the partial derivatives of g can be expressed in terms of the partial derivatives of f, and similarly for higher derivatives. Of course the same argument can be applied to another variable x_i, expressing it as a function of the remaining variables provided that $\partial_i f(\mathbf{x}) \neq 0$. If f_1, f_2 are continuously differentiable functions in an open subset $U \subset \mathbb{R}^4$ for which $f_1(\mathbf{a}) = f_2(\mathbf{a}) = 0$ and

$$
\det \begin{pmatrix} \partial_1 f_1(\mathbf{a}) & \partial_2 f_1(\mathbf{a}) \\ \partial_1 f_2(\mathbf{a}) & \partial_2 f_2(\mathbf{a}) \end{pmatrix} \neq 0 \tag{5.13}
$$

then the set of points at which $f_1(\mathbf{x}) = f_2(\mathbf{x}) = 0$ near $\mathbf{a}$ can be described explicitly as the set of points at which x_1 and x_2 are explicitly given functions of the remaining variables, say $x_1 = g_1(x_3, x_4)$ and $x_2 = g_2(x_3, x_4)$. Thus there are the identities $f_1\big(g_1(x_3, x_4), g_2(x_3, x_4), x_3, x_4\big) = f_2\big(g_1(x_3, x_4), g_2(x_3, x_4), x_3, x_4\big) = 0$; and the partial derivatives of these identities yield two linear equations relating the partial derivatives of the functions $g_1(x_3, x_4)$ and $g_2(x_3, x_4)$, which can be solved explicitly in view of (5.13).

The set $V = \{\, \mathbf{x} \in U \mid \mathbf{f}(\mathbf{x}) = \mathbf{0} \,\}$ of zeros of a mapping $\mathbf{f} : U \longrightarrow \mathbb{R}^n$ defined in an open subset $U \subset \mathbb{R}^m$ is called the **zero locus** of the mapping $\mathbf{f}$. If the mapping is continuous its zero locus is a relatively closed subset of U, since it is the inverse image of the closed set $\mathbf{0} \in \mathbb{R}^n$ under a continuous mapping. A relatively closed subset $V \subset U$ of an open set $U \subset \mathbb{R}^n$ is called a k-**dimensional submanifold** of U if for any point $\mathbf{a} \in V$ there are an open neighborhood

$U_{\mathbf{a}} \subset U$ and a continuous local change of coordinates in $U_{\mathbf{a}}$ such that $V \cap U_{\mathbf{a}}$ is a linear subspace of dimension k in terms of these coordinates. If the change of coordinates is described by a continuously differentiable mapping the subset V is said somewhat more precisely to be a C^1 **submanifold**; similarly it is said to be a C^2 submanifold if the change of coordinates is described by a C^2 mapping, and so on. Corollary 5.7 can be interpreted as the assertion that if $\mathbf{f} : U \longrightarrow \mathbb{R}^n$ is a C^1 mapping from an open subset $U \subset \mathbb{R}^m$ into $\mathbb{R}^n$ for $m > n$ with the zero locus $V \subset U$, and if rank $\mathbf{f}'(\mathbf{x}) = n$ at each point $\mathbf{x} \in V$, then V is a C^1 $(m-n)$-dimensional submanifold of U. It is quite common to rephrase this result just in terms of the coordinate functions $f_1, \ldots, f_n$ of the mapping $\mathbf{f}$ and to speak of the set of common zeros of these functions. In these terms, Corollary 5.7 asserts that if the gradient vectors $\nabla f_1, \ldots, \nabla f_n$ are linearly independent vectors at each point of the set V of the common zeros of these functions then V is a C^1 submanifold of dimension $m-n$ in U. For example the circle

$$V = \left\{ (x_1, x_2) \in \mathbb{R}^2 \ \middle| \ x_1^2 + x_2^2 = 1 \right\} \subset \mathbb{R}^2 \tag{5.14}$$

can be defined as the zero locus of the function $f(\mathbf{x}) = x_1^2 + x_2^2 - 1$ in the plane, and since $f'(\mathbf{x}) = (2x_1 \, 2x_2)$ is a matrix of rank 1 at each point $\mathbf{x} \in V$, at which not both $x_1 = 0$ and $x_2 = 0$, the zero locus V is a one-dimensional submanifold of the plane. Indeed it can be described locally as the linear space $r = 1$ in terms of polar coordinates (r, θ) in the plane in a neighborhood of each point of V; but of course there is no single coordinate neighborhood in which the entire circle can be described by the equation $r = 1$, since the other coordinate θ is not a single-valued function on the full circle. For some purposes an alternative characterization of submanifolds is quite convenient.

Theorem 5.9. *A relatively closed subset $V \subset U$ of an open set $U \subset \mathbb{R}^m$ is an $(m-n)$-dimensional C^1 submanifold of U if and only if for any point $\mathbf{a} \in V$, after a permutation of the coordinates in $\mathbb{R}^m$ if necessary, there is an open cell $\Delta_{\mathbf{a}} = \Delta_{I,\mathbf{a}} \times \Delta_{II,\mathbf{a}}$ centered at the point $\mathbf{a} \in V$, a product of cells $\Delta_{I,\mathbf{a}} \subset \mathbb{R}^n$ and $\Delta_{II,\mathbf{a}} \subset \mathbb{R}^{m-n}$, such that the intersection $V \cap \Delta_{\mathbf{a}}$ is described by the equations*

$$V \cap \Delta_{\mathbf{a}} = \left\{ \mathbf{x} \in \Delta_{\mathbf{a}} \ \middle| \ x_i = g_i(x_{n+1}, \ldots, x_m), \quad 1 \le i \le n \right\} \tag{5.15}$$

for some C^1 functions g_i in the cell $\Delta_{II,\mathbf{a}} \subset \mathbb{R}^{m-n}$.

Proof: If V is an $(m-n)$-dimensional C^1 submanifold of U it follows from the Implicit Function Theorem that V can be described in an open neighborhood of any point $\mathbf{a} \in V$ by the equations (5.15). Conversely if V is defined by the equations (5.15) these equations can be rewritten $f_i(\mathbf{x}) = x_i - g_i(x_{n+1}, \ldots, x_m) = 0$ for $1 \le i \le n$, and since $\nabla f_i(\mathbf{x})$ are clearly linearly independent vectors it follows that V is an $(m-n)$-dimensional C^1 submanifold. That suffices for the proof.

The condition that a subset $V \subset U$ of an open set $U \subset \mathbb{R}^n$ be a submanifold is a local condition; so if a relatively closed subset V is a submanifold in a neighborhood of each of its points it is a submanifold in U. Two $\mathcal{C}^1$ submanifolds $V_1 \subset U_1, V_2 \subset U_2$ of open sets $U_1, U_2 \subset \mathbb{R}^m$ are called **homeomorphic** $\mathcal{C}^1$ **submanifolds** if there is a $\mathcal{C}^1$ homeomorphism $\phi : U_1 \longrightarrow U_2$ such that $\phi(V_1) = V_2$; this is clearly an equivalence relation that can be viewed as identifying two submanifolds that are essentially the same. The corresponding notion holds for topological submanifolds, for which the equivalence mapping is merely a homeomorphism, or for $\mathcal{C}^r$ submanifolds for $r > 1$. This equivalence relation really involves not just the point set V but the particular way in which that set is imbedded in the open set $U \subset \mathbb{R}^m$. For example a circle in $\mathbb{R}^3$ and a knotted curve in $\mathbb{R}^3$ are both submanifolds of $\mathbb{R}^3$; but they are not homeomorphic topological submanifolds, since there is no homeomorphism of $\mathbb{R}^3$ to itself that transforms a circle into a knotted curve.[4]

The set of common zeros of a collection of functions $f_1, \ldots, f_n$ for which the gradients $\nabla f_1(\mathbf{x}), \ldots, \nabla f_n(\mathbf{x})$ are not linearly independent at some points is not necessarily a submanifold at these exceptional points but may have **singularities** of one sort or another. Of course if the gradients of these functions are linearly independent the squares $f_i(\mathbf{x})^2$ of these functions have the same set of common zeros as do the original functions, forming a submanifold; but the gradients of the squares of these functions vanish at all their common zeros. The study of singularities of differentiable mappings has been and still is a very active area of mathematical research.[5] A polynomial function $p(z_1, z_2)$ of two complex variables amounts to a pair of real functions of four real variables, the real and imaginary parts of the variables and of the function; so the set V of points at which $p(z_1, z_2) = 0$ is a candidate to be a submanifold of dimension 2 in $\mathbb{R}^4$. For a linear polynomial $p(z_1, z_2) = a_1 z_1 + a_2 z_2$ the set V is a linear submanifold, and the intersection $V \cap S_\epsilon^3$ of the set V and the sphere $S_\epsilon^3 = \{ (z_1, z_2) \mid |z_1|^2 + |z_2|^2 = \epsilon \}$ is a one-dimensional submanifold of the sphere, a linear slice of the sphere so just an ordinary circle. If the polynomial $p(z_1, z_2)$ has no constant term but a nontrivial linear term then again the set V is a two-dimensional submanifold near the origin, and the intersection $V \cap S_\epsilon^3$ is a submanifold of the sphere that is $\mathcal{C}^1$ homeomorphic to a circle in the sphere. The set V of zeros of the polynomial $p(z_1, z_2) = z_1^2 + z_2^2$

---

[4]The study of knots is an old subject in mathematics, going back to the work of P. G. Tait in the 1870s, but it is still a very active subject, with recent ties to mathematical physics and representation theory among other topics. Classical knot theory focuses on the classification of knots and the algebraic invariants that distinguish different knots. Further information can be found in the book by R. H. Crowell and R. H. Fox, *Introduction to Knot Theory* (Springer, 1963).

[5]Some classical examples of singularities are isolated singularities of algebraic curves, the topological properties of which were investigated by K. Brauner (*Abh. Math. Sem. Hamburg* 6 (1928): 8–54).

is a two-dimensional submanifold near the origin except at the origin itself, which is a singularity; actually this polynomial splits as the product $p(z_1, z_2) = (z_1 + iz_2)(z_1 - iz_2)$ of two linear functions both of which vanish at the origin, so near the origin the set V is the union of two linear spaces of dimension 2, the linear spaces $z_1 + iz_2 = 0$ and $z_1 - iz_2 = 0$, which meet only at a single point, the origin. In this case the intersection $V \cap S_\epsilon^3$ is a submanifold consisting of two disjoint circles which are not linked in the three-dimensional sphere. For the polynomial $p(z_1, z_2) = z_1^2 + z_2^3$ the origin is again a singularity, and in this case the intersection $V \cap S_\epsilon^3$ as a submanifold of the sphere is a nontrivial knot in the sphere, which is fairly easy to describe. In general the intersection $V \cap S_\epsilon^3$ is a submanifold of the sphere that is a knot of a special type, and its topological structure determines a good deal of the structure of the set V at its singular point. There are analogous topological sets that arise in the examination of isolated singularities of the sets of zeros of polynomials in several variables, a fascinating subject in its own right; these examples can be used to introduce the exotic differentiable structures on higher dimensional spheres, as discovered by J. W. Milnor.[6]

Problems, Group I

1. Consider the following functions in $\mathbb{R}^3$:

$$f_1(\mathbf{x}) = e^{x_1} + e^{x_2} + e^{x_3} + x_1 x_2 x_3 - 3,$$

$$f_2(\mathbf{x}) = x_1 + x_2^2 + x_3^3 + \sin\left(e^{x_1 x_2 x_3} - 1\right).$$

   i) Show that the equations $f_1(\mathbf{x}) = f_2(\mathbf{x}) = 0$ define a one-dimensional submanifold V in an open neighborhood of the origin.
   ii) Show that the submanifold V can be described explicitly in an open neighborhood of the origin by equations of the form

$$x_1 = g_1(x_3) \quad \text{and} \quad x_2 = g_2(x_3),$$

   for some C^1 functions $g_1(x_3), g_2(x_3)$ in a neighborhood of the origin in $\mathbb{R}^1$ such that $g_1(0) = 0$ and $g_2(0) = 0$.
   iii) Find the derivatives $g_1'(0)$ and $g_2'(0)$.
   iv) Show that the equation $f_2(\mathbf{x}) = 0$ defines a two-dimensional submanifold W in an open neighborhood of the origin.

---

[6] A more detailed discussion can be found in Milnor's book *Singular Points of Complex Hyper surfaces*, Annals of Mathematics Studies 61 (Princeton University Press, 1968).

2.   i) Show that the two equations

$$\sin(x_1 + 2x_2) + \cos(3x_3 + 4x_4) = 1, \quad e^{3x_1+5x_3} + e^{2x_2+4x_4} = 2$$

define x_1 and x_3 as functions of x_2 and x_4 near the origin $\mathbf{0} = (0,0,0,0)$, and determine the value of $\dfrac{\partial x_1}{\partial x_2}$ at that point.

ii) Show that these two equations also define x_2 and x_4 as functions of x_1 and x_3 near the origin, and determine the value of $\dfrac{\partial x_2}{\partial x_1}$.

3.   Show that 2×2 orthogonal matrices form a $\mathcal{C}^1$ submanifold in $\mathbb{R}^{2 \times 2}$. What is its dimension?

Problems, Group II

4.   i) Does the equation $x_1^2 + 2x_2^2 + 3x_3^2 = 3\cos^2 x_3$ determine x_2 explicitly as a function of x_1 and x_3 in an open neighborhood of the point $(1, 1, 0)$? Why?

ii) Does that equation determine x_3 explicitly as a function of x_1 and x_2 in an open neighborhood of the point $(1,1,0)$? Why?

5.   Let $f_1(\mathbf{x})$ and $f_2(\mathbf{x})$ be $\mathcal{C}^1$ functions in an open subset $U \subset \mathbb{R}^2$, set $f_3(\mathbf{x}) = f_1(\mathbf{x})f_2(\mathbf{x})$, and let V_j be the zero locus of the function $f_j(\mathbf{x})$ in U.

i) Show that $V_3 = V_1 \cup V_2$.

ii) Show that if $\mathbf{a} \in V_1 \cap V_2$ then $(\nabla f_3)(\mathbf{a}) = 0$.

6.   i) If $f(x_1, x_2, x_3) = x_1^2 x_2 + e^{x_1} + x_3$ show that there is a differentiable function g of two variables in an open neighborhood of the point $(1, -1)$ such that $g(1, -1) = 0$ and $f(g(x_2, x_3), x_2, x_3) = 0$ in an open neighborhood of the point $(1, -1)$.

ii) Find $\partial_1 g(1, -1)$ and $\partial_2 g(1, -1)$.

7.   Find sufficient conditions on a $\mathcal{C}^1$ function $f(x_1, x_2)$ defined in an open neighborhood of the origin in $\mathbb{R}^2$ and vanishing at the origin so that there exists a $\mathcal{C}^1$ function $g(x)$ in an open neighborhood of the origin $0 \in \mathbb{R}^1$ such that $g(0) = 0$, $g'(0) \neq 0$, and $f(f(x, g(x)), g(x)) = 0$ for all values x near 0.

8.   Show that the union of the two positive coordinate axes in $\mathbb{R}^2$, the set $\{(x_1, x_2) \in \mathbb{R}^2 \mid x_1 x_2 = 0, \ x_1 \geq 0, \ x_2 \geq 0\}$, is a continuous submanifold of $\mathbb{R}^2$ but is not a $\mathcal{C}^1$ submanifold.

9.   i) Show that there is a C^1 function $f(x_1, x_2)$ in an open neighborhood of the origin in $\mathbb{R}^2$ such that $f(0, 0) = 1$ and

$$f(x_1, x_2)^2 = \sin\left(x_1 x_2 f(x_1, x_2)\right) + 1.$$

ii) Show that the derivative $\partial_1^2 f(0, 0)$ exists.

5.3 Rank Theorem

The Inverse Mapping Theorem can be applied to describe mappings between spaces in situations more general than those to which the Implicit Function Theorem applies, in particular to describe mappings $\mathbf{f} : \mathbb{R}^m \longrightarrow \mathbb{R}^n$ where $m < n$.

Theorem 5.10 (Rank Theorem). *If $\mathbf{f} : U \longrightarrow \mathbb{R}^n$ is a continuously differentiable mapping defined in an open subset $U \subset \mathbb{R}^m$ and if rank $\mathbf{f}'(\mathbf{x}) = r$ for all points in an open neighborhood of a point $\mathbf{a} \in U$ then after a C^1 change of coordinates in open neighborhoods of the points $\mathbf{a} \in \mathbb{R}^m$ and $\mathbf{f}(\mathbf{a}) \in \mathbb{R}^n$ the mapping $\mathbf{f}$ is a linear mapping of rank r.*

Proof: This theorem involves a change of local coordinates in both the domain and range of the mapping $\mathbf{f}$, so requires the existence at least locally of C^1 homeomorphisms ψ and θ in the diagram

$$\mathbb{R}^m \xrightarrow{\;\psi\;} \mathbb{R}^m \xrightarrow{\;\mathbf{f}\;} \mathbb{R}^n \xrightarrow{\;\theta\;} \mathbb{R}^n$$

so that the composition $\theta \circ \mathbf{f} \circ \psi : \mathbb{R}^m \longrightarrow \mathbb{R}^n$ is a linear mapping of rank r. If rank $\mathbf{f}'(\mathbf{a}) = r$ then by suitably relabeling the variables in the domain $\mathbb{R}^m$ and range $\mathbb{R}^n$ of the mapping $\mathbf{f}$ it can be assumed that the leading $r \times r$ submatrix of $\mathbf{f}'(\mathbf{a})$ is a nonsingular $r \times r$ matrix; and by hypothesis that is also the case for the matrix $\mathbf{f}'(\mathbf{x})$ for all points $\mathbf{x} \in U_{\mathbf{a}} \subset \mathbb{R}^m$ for a sufficiently small open neighborhood $U_{\mathbf{a}}$ of the point $\mathbf{a}$. As before it is convenient to write points $\mathbf{x} \in \mathbb{R}^m$ in the form $\mathbf{x} = \left(\begin{smallmatrix} \mathbf{x}_I \\ \mathbf{x}_{II} \end{smallmatrix}\right)$ where $\mathbf{x}_I \in \mathbb{R}^r$ and $\mathbf{x}_{II} \in \mathbb{R}^{m-r}$, and correspondingly to write points $\mathbf{y} \in \mathbb{R}^n$ in the form $\mathbf{y} = \left(\begin{smallmatrix} \mathbf{y}_I \\ \mathbf{y}_{II} \end{smallmatrix}\right)$ where $\mathbf{y}_I \in \mathbb{R}^r$ and $\mathbf{y}_{II} \in \mathbb{R}^{n-r}$. The mapping $\mathbf{f}$ can then be written

$$\mathbf{f}\begin{pmatrix} \mathbf{x}_I \\ \mathbf{x}_{II} \end{pmatrix} = \begin{pmatrix} \mathbf{f}_I(\mathbf{x}) \\ \mathbf{f}_{II}(\mathbf{x}) \end{pmatrix}$$

where $\mathbf{x}_I, \mathbf{f}_I(\mathbf{x}) \in \mathbb{R}^r$ while $\mathbf{x}_{II} \in \mathbb{R}^{m-r}$ and $\mathbf{f}_{II}(\mathbf{x}) \in \mathbb{R}^{n-r}$; and the matrix $\mathbf{f}'(\mathbf{x})$ then can be decomposed into the submatrices

$$\mathbf{f}'(\mathbf{x}) = \begin{pmatrix} \partial_I \mathbf{f}_I(\mathbf{x}) & \partial_{II} \mathbf{f}_I(\mathbf{x}) \\ \partial_I \mathbf{f}_{II}(\mathbf{x}) & \partial_{II} \mathbf{f}_{II}(\mathbf{x}) \end{pmatrix}$$

where $\partial_I \mathbf{f}_I(\mathbf{x})$ is an $r \times r$ matrix which is nonsingular for all points $\mathbf{x} \in U_\mathbf{a}$. In these terms, introduce the mapping $\phi : U_\mathbf{a} \longrightarrow \mathbb{R}^m$ defined by

$$\phi \begin{pmatrix} \mathbf{x}_I \\ \mathbf{x}_{II} \end{pmatrix} = \begin{pmatrix} \mathbf{f}_I(\mathbf{x}) \\ \mathbf{x}_{II} \end{pmatrix},$$

so that

$$\phi'(\mathbf{x}) = \begin{pmatrix} \partial_I \mathbf{f}_I(\mathbf{x}) & \partial_{II} \mathbf{f}_I(\mathbf{x}) \\ 0 & I_{m-r} \end{pmatrix}$$

where I_{m-r} is the $(m-r) \times (m-r)$ identity matrix; since the matrix $\partial_I \mathbf{f}_I(\mathbf{x})$ is nonsingular for all points $\mathbf{x} \in U_\mathbf{a}$ it follows that the matrix $\phi'(\mathbf{x})$ also is nonsingular for all points $\mathbf{x} \in U_\mathbf{a}$. If the neighborhood $U_\mathbf{a}$ is sufficiently small it follows from the Inverse Mapping Theorem that the restriction of the mapping ϕ to $U_\mathbf{a}$ is a $\mathcal{C}^1$ homeomorphism $\phi : U_\mathbf{a} \longrightarrow V$ between the open sets $U_\mathbf{a}$ and $V \subset \mathbb{R}^m$. Clearly the inverse mapping $\psi : V \longrightarrow U_\mathbf{a}$ has the form

$$\psi \begin{pmatrix} \mathbf{t}_I \\ \mathbf{t}_{II} \end{pmatrix} = \begin{pmatrix} \mathbf{g}(\mathbf{t}) \\ \mathbf{t}_{II} \end{pmatrix} \quad \text{for a } \mathcal{C}^1 \text{ mapping } \mathbf{g} : V \longrightarrow \mathbb{R}^r,$$

and then

$$\begin{pmatrix} \mathbf{t}_I \\ \mathbf{t}_{II} \end{pmatrix} = \mathbf{t} = \phi(\psi(\mathbf{t})) = \begin{pmatrix} \mathbf{f}_I(\psi(\mathbf{t})) \\ \mathbf{t}_{II} \end{pmatrix} \quad \text{hence } \mathbf{f}_I(\psi(\mathbf{t})) = \mathbf{t}_I;$$

consequently the composite mapping $\mathbf{h}(\mathbf{t}) = (\mathbf{f} \circ \psi)(\mathbf{t})$ has the form

$$\mathbf{h}(\mathbf{t}) = \mathbf{f}(\psi(\mathbf{t})) = \begin{pmatrix} \mathbf{f}_I(\psi(\mathbf{t})) \\ \mathbf{f}_{II}(\psi(\mathbf{t})) \end{pmatrix} = \begin{pmatrix} \mathbf{t}_I \\ \mathbf{h}_{II}(\mathbf{t}) \end{pmatrix} \quad \text{where } \mathbf{h}_{II}(\mathbf{t}) = \mathbf{f}_{II}(\psi(\mathbf{t})) \in \mathbb{R}^{n-r},$$

$$(5.16)$$

hence

$$\mathbf{h}'(\mathbf{t}) = \begin{pmatrix} I_r & 0 \\ \partial_I \mathbf{h}_{II}(\mathbf{t}) & \partial_{II} \mathbf{h}_{II}(\mathbf{t}) \end{pmatrix} \qquad (5.17)$$

where I_r is the $r \times r$ identity matrix. On the other hand since $\mathbf{h} = \mathbf{f} \circ \psi$ then by the chain rule $\mathbf{h}'(\mathbf{t}) = \mathbf{f}'(\psi(\mathbf{t})) \cdot \psi'(\mathbf{t})$; the $m \times m$ matrix $\psi'(\mathbf{t})$ is nonsingular for all $\mathbf{t} \in V$, while by hypothesis rank $\mathbf{f}'(\psi(\mathbf{t})) = r$ for all $\mathbf{t} \in V$, and consequently rank $\mathbf{h}'(\mathbf{t}) = r$ for all points $\mathbf{t} \in V$ as well. However it is evident from (5.17) that rank $\mathbf{h}'(\mathbf{t}) = r$ if and only if $\partial_{II}\mathbf{h}_{II}(\mathbf{t}) = 0$ for all points $\mathbf{t} \in V$; so the mapping $\mathbf{h}_{II}$ is a function just of the variables $\mathbf{t}_I \in \mathbb{R}^r$. The mapping $\theta : \mathbf{f}_I\big(\psi(V)\big) \times \mathbb{R}^{n-r} \longrightarrow \mathbb{R}^n$ defined by

$$\theta \begin{pmatrix} \mathbf{y}_I \\ \mathbf{y}_{II} \end{pmatrix} = \begin{pmatrix} \mathbf{y}_I \\ \mathbf{y}_{II} - \mathbf{h}_{II}(\mathbf{y}_I) \end{pmatrix}$$

has the derivative

$$\theta'(\mathbf{y}) = \begin{pmatrix} I_r & 0 \\ -\partial_I\mathbf{h}_{II}(\mathbf{y}_I) & I_{n-r} \end{pmatrix},$$

so after restricting the domain of θ if necessary this mapping can be viewed as a local change of coordinates in some neighborhood W of $\mathbf{f}(\mathbf{a})$ by the Inverse Mapping Theorem, and after restricting the neighborhood $U_{\mathbf{a}}$ further if necessary it can be assumed that $\mathbf{f}(U_{\mathbf{a}}) \subset W$. It is evident from the definition of this mapping and the explicit form (5.16) of the mapping $\mathbf{h}$, in which $\mathbf{h}_{II}(\mathbf{t})$ is a function of the variables $\mathbf{t}_I$ alone, that

$$(\theta \circ \mathbf{f} \circ \psi)(\mathbf{t}) = \theta\big(\mathbf{h}(\mathbf{t})\big) = \theta \begin{pmatrix} \mathbf{t}_I \\ \mathbf{h}_{II}(\mathbf{t}) \end{pmatrix} = \begin{pmatrix} \mathbf{t}_I \\ \mathbf{h}_{II}(\mathbf{t}) - \mathbf{h}_{II}(\mathbf{t}) \end{pmatrix} = \begin{pmatrix} \mathbf{t}_I \\ 0 \end{pmatrix},$$

which is a linear mapping of rank r. That suffices for the proof.

Theorem 5.11. *If* $\mathbf{f} : U \longrightarrow \mathbb{R}^n$ *is a continuously differentiable mapping defined in an open subset* $U \subset \mathbb{R}^m$ *and if* rank $\mathbf{f}'(\mathbf{x}) = r$ *for all points* $\mathbf{x} \in U$ *then*
(i) *the image* $\mathbf{f}(U_{\mathbf{a}})$ *of any sufficiently small open neighborhood* $U_{\mathbf{a}} \subset U$ *of a point* $\mathbf{a} \in U$ *is a* C^1 *submanifold of dimension* r *of an open neighborhood of the point* $\mathbf{f}(\mathbf{a}) \in \mathbb{R}^n$;
(ii) *for any point* $\mathbf{b} \in \mathbf{f}(U) \subset \mathbb{R}^n$ *the inverse image* $\mathbf{f}^{-1}(\mathbf{b})$ *is a* C^1 *submanifold of dimension* $m - r$ *in* U.

Proof: It follows from the preceding theorem that after suitable changes of coordinates in open neighborhoods $U_{\mathbf{a}}$ of the point $\mathbf{a} \in U \subset \mathbb{R}^m$ and $V_{\mathbf{b}}$ of the point $\mathbf{b} = \mathbf{f}(\mathbf{a}) \in \mathbb{R}^n$ the mapping $\mathbf{f}$ is a linear mapping of rank r; consequently the image $\mathbf{f}(U_{\mathbf{a}})$ is a linear subspace of dimension r in terms of the new coordinates in $\mathbb{R}^n$ and the inverse image $\mathbf{f}^{-1}(\mathbf{y})$ of any point $\mathbf{y} \in \mathbf{f}(U_{\mathbf{a}}) \subset \mathbb{R}^n$ is a linear subspace of dimension $m - r$ in terms of the new coordinates in $\mathbb{R}^m$. Consequently the image $\mathbf{f}(U_{\mathbf{a}})$ is an r-dimensional submanifold of $V_{\mathbf{b}}$ in terms of the original coordinates in $\mathbb{R}^n$, and the inverse image $\mathbf{f}^{-1}(\mathbf{y}) \cap U_{\mathbf{a}}$ of any point $\mathbf{y} \in \mathbf{f}(U_{\mathbf{a}})$ is an $(m - r)$-dimensional C^1 submanifold of $U_{\mathbf{a}}$ in terms of the original coordinates in $\mathbb{R}^m$. The latter statement holds for any point in $\mathbf{f}^{-1}(\mathbf{y})$; so since the inverse image $\mathbf{f}^{-1}(\mathbf{y}) \subset U$ is a relatively closed subset of U which is locally a submanifold of U it must be a submanifold of U, and that suffices for the proof.

Part (i) of the preceding theorem can be expected to hold only locally. For instance it is quite possible that the image of an open interval $I \subset \mathbb{R}^1$ under a continuously differentiable mapping $\phi : I \longrightarrow \mathbb{R}^n$ is a curve that intersects itself at some points, so that although the image is locally a one-dimensional submanifold the full image has singularities; and the image of the open interval need not even be a closed subset of $\mathbb{R}^n$. The situation in higher dimensions of course is even more complicated; but the local result is of considerable use. For the Inverse Mapping Theorem and the Implicit Function Theorem the hypothesis only required that the mapping be of maximal rank at a point $\mathbf{a}$, or equivalently that there is a square submatrix of maximal size in the Jacobian matrix of the mapping so that the determinant of that matrix at the point $\mathbf{a}$ is nonzero; and since the Jacobian of a $\mathcal{C}^1$ mapping is a continuous function it is necessarily nonzero at all points near enough to $\mathbf{a}$. However if the rank of the Jacobian matrix is not maximal at a point $\mathbf{a}$ its rank at nearby points may well be greater at points arbitrarily close to $\mathbf{a}$. That is the reason that the hypothesis in the Rank Theorem requires that the rank be locally constant. In some applications of the Rank Theorem it is still the case that the rank is maximal; so if the rank is maximal at a point it is maximal at all nearby points.

Theorem 5.12. *If $\phi : U \longrightarrow \mathbb{R}^n$ is a continuously differentiable mapping from an open set $U \subset \mathbb{R}^r$ into $\mathbb{R}^n$ where $r < n$ and if $\operatorname{rank} \phi'(\mathbf{a}) = r$ at some point $\mathbf{a} \in U$ then the image under ϕ of any sufficiently small open neighborhood of the point $\mathbf{a} \in U$ is a $\mathcal{C}^1$ submanifold of dimension r in an open neighborhood of the point $\phi(\mathbf{a}) \in \mathbb{R}^n$.*

Proof: If $\operatorname{rank} \phi'(\mathbf{a}) = r$ at a point $\mathbf{a} \in U$ for a mapping $\phi : \mathbb{R}^r \longrightarrow \mathbb{R}^n$ where $r < n$ then $\operatorname{rank} \phi'(\mathbf{x}) = r$ for all points $\mathbf{x}$ near $\mathbf{a}$, so the desired result follows immediately from the preceding theorem, and that suffices for the proof.

For example a continuous mapping $\mathbf{f} : [0, 1] \longrightarrow \mathbb{R}^m$ can be viewed as a **parametrized curve** in the vector space $\mathbb{R}^m$, beginning at the point $\mathbf{f}(0)$ and ending at the point $\mathbf{f}(1)$; and if $\mathbf{f}'(t) \neq 0$ at each point $t \in (0, 1)$ then by Theorem 5.12 the image of an open neighborhood of any point $t_0 \in (0, 1)$ is a one-dimensional submanifold of an open neighborhood of the point $\mathbf{f}(t_0) \in \mathbb{R}^m$. That the mapping $\mathbf{f}$ is differentiable at the point t_0 means that

$$\mathbf{f}(t) = \mathbf{f}(t_0) + \mathbf{f}'(t_0)(t - t_0) + \boldsymbol{\epsilon}(t) \quad \text{where} \quad \lim_{t \to t_0} \frac{\|\boldsymbol{\epsilon}(t)\|}{\|t - t_0\|} = 0$$

for all points t sufficiently near t_0. The parametrized curve

$$\mathbf{f}_0(t) = \mathbf{f}(t_0) + \mathbf{f}'(t_0)(t - t_0)$$

is a line in $\mathbb{R}^m$ passing through the point $\mathbf{f}(t_0)$, called the **tangent line** to the parametrized curve $\mathbf{f}$ at the point $\mathbf{f}(t_0) = \mathbf{a}_0$, and its image is of course another submanifold of $\mathbb{R}^m$; the vector $\mathbf{f}'(t_0)$ is called the **tangent vector** to the parametrized curve $\mathbf{f}$ at the point $\mathbf{f}(t_0) = \mathbf{a}_0$. It follows from the definition and the uniqueness of the derivative that the tangent line to the parametrized curve $\mathbf{f}$ at the point $\mathbf{a}_0$ is the parametrized line through that point that is the best linear approximation to the $\mathbf{f}$ near that point. It is not uncommon to change the parametrization of the tangent line so that the point $\mathbf{a}_0$ corresponds to the parameter value $t = 0$, so to describe the tangent line as the parametrized curve $\mathbf{f}_0(t) = \mathbf{f}(t_0) + \mathbf{f}'(t_0)t$. A continuously differentiable mapping $\phi : U \longrightarrow \mathbb{R}^n$ from an open subset $U \subset \mathbb{R}^m$ into $\mathbb{R}^n$ takes a parametrized curve $\mathbf{f}$ to the parametrized curve $\phi \circ \mathbf{f}$, and by the chain rule the tangent vector to the image $\phi \circ \mathbf{f}$ at the point $\phi(\mathbf{a}_0)$ is the vector $(\phi \circ \mathbf{f})'(t_0) = \phi'(\mathbf{a}_0)\mathbf{f}'(t_0)$. Thus the derivative $\phi'(\mathbf{a}_0)$ of the mapping ϕ at the point $\mathbf{a}_0$ can be interpreted as the linear mapping that takes tangent vectors to parametrized curves through the point $\mathbf{a}_0$ to tangent vectors to the image parametrized curves through the point $\phi(\mathbf{a}_0)$.

If V is a $\mathcal{C}^1$ submanifold of an open subset $U \subset \mathbb{R}^m$ then through each point $\mathbf{a} \in V$ it is possible to pass a continuously differentiable parametrized curve contained entirely in the set V near that point; for in some local coordinates near the point $\mathbf{a}$ the subset V is a linear subspace, which contains straight lines through that point. Moreover if $\dim V = r$ there are parametrized curves through that point, the tangent vectors to which span a linear subspace of dimension r; for in some local coordinates near the point $\mathbf{a}$ the subset V is a linear subspace of dimension r, and there are r linearly independent straight lines through any point of a linear space of dimension r. For a submanifold $V \subset U$ of an open set $U \subset \mathbb{R}^m$ defined by equations

$$V = \left\{ \mathbf{x} \in U \,\middle|\, f_1(\mathbf{x}) = \cdots = f_{m-r}(\mathbf{x}) = 0 \right\} \tag{5.18}$$

where the vectors $\nabla f_i(\mathbf{x})$ are linearly independent at each point $\mathbf{x} \in V$, the linear subspace of dimension r consisting of the tangent vectors at the point $\mathbf{a}$ of all continuously differentiable parametrized curves contained in V and passing through that point is called the **tangent space** to the submanifold V at the point $\mathbf{a}$ and is denoted by $\mathcal{T}_{\mathbf{a}}(V)$. The terminology here is a bit casual; what is called the tangent space is actually a translate of a linear subspace from the origin to a point on the submanifold. If $\phi : I \longrightarrow V$ is a $\mathcal{C}^1$ parametrized curve for which $\phi(0) = \mathbf{a}$, then since $f_i(\phi(t)) = 0$ identically in t it follows from the chain rule that $f_i'(\mathbf{a})\phi'(0) = 0$, or equivalently that $\nabla f_i(\mathbf{a}) \cdot \phi'(0) = 0$; thus the tangent vector to the parametrized curve ϕ is perpendicular to all of the gradient vectors

$\nabla f_i(\mathbf{a})$, hence the tangent space to V at the point $\mathbf{a} \in V$ is the linear subspace consisting of all points $\mathbf{x} \in \mathbb{R}^m$ such that the vector $\mathbf{x} - \mathbf{a}$ is perpendicular to the vectors $\nabla f_i(\mathbf{a})$. The equation of the tangent space therefore is

$$\mathcal{T}_{\mathbf{a}}(V) = \left\{ \mathbf{x} \in \mathbb{R}^m \,\middle|\, (\mathbf{x} - \mathbf{a}) \cdot \nabla f_i(\mathbf{a}) = 0 \ \text{ for } \ 1 \le i \le m - r \right\}. \tag{5.19}$$

The tangent space is the linear subspace of $\mathbb{R}^n$ that is the best local approximation to the submanifold V near that point, just as is the tangent line to a parametrized curve. For example, if $V \subset \mathbb{R}^3$ is the surface defined by

$$V = \left\{ \mathbf{x} \in \mathbb{R}^3 \,\middle|\, f(\mathbf{x}) = x_1^2 + x_2^3 + x_3^4 - 3 = 0 \right\}$$

then $f'(1,1,1) = (2\ 3\ 4)$ so the equation of the tangent plane to V at the point $\{1, 1, 1\}$ is $(2, 3, 4) \cdot (x_1 - 1, x_2 - 1, x_3 - 1) = 0$ or equivalently $2x_1 + 3x_2 + 4x_3 = 9$.

To determine the extrema of a $\mathcal{C}^1$ function g defined on an r-dimensional submanifold $V \subset \mathbb{R}^m$ it is always possible theoretically to introduce local coordinates in an open neighborhood U of any point $\mathbf{a} \in V$ so that $V \cap U$ is an r-dimensional linear subspace in terms of those coordinates; the restriction of the function g to V then is locally just a function on an r-dimensional linear subspace of $\mathbb{R}^m$, so can be viewed as a function on an open subset of $\mathbb{R}^r$ and its critical points are candidates for points at which the function has an extremum. However that can be a rather arduous calculation in all but the simplest situations, since determining the local coordinates in terms of which the submanifold is linear can be quite difficult to do explicitly. However the Implicit Function Theorem provides a much simpler and widely used method for finding local extrema on a submanifold. For the submanifold V defined by (5.18) where the vectors $\nabla f_i(\mathbf{x})$ are linearly independent at all points $\mathbf{x} \in V$, a point $\mathbf{a} \in V$ will be a local extremum of the restriction of the continuously differentiable function $g(\mathbf{x})$ to the submanifold V if and only if it is a local extremum of the restriction of the function $g(\mathbf{x})$ to any continuously differentiable parametrized curve lying in V and passing through the point $\mathbf{a}$. Then for any $\mathcal{C}^1$ parametrized curve $\boldsymbol{\phi} : U \longrightarrow \mathbb{R}^m$ in an open neighborhood $U \subset \mathbb{R}^1$ of the origin $t = 0$ passing through the point $\mathbf{a}$, so that $\boldsymbol{\phi}(0) = \mathbf{a}$, and contained in V, so that $f_i(\boldsymbol{\phi}(t)) = 0$ for all $t \in U$, the origin $t = 0$ is a local extremum for the composite function $g(\boldsymbol{\phi}(t))$ and consequently

$$0 = \frac{d}{dt} g(\boldsymbol{\phi}(t))\Big|_{t=0} = g'(\mathbf{a})\boldsymbol{\phi}'(0) = \nabla g(\mathbf{a}) \cdot \boldsymbol{\phi}'(0);$$

thus $\nabla g(\mathbf{a})$ must be perpendicular to the tangent line to the parametrized curve, and since that is the case for all parametrized curves in V the vector $\nabla g(\mathbf{a})$ must be perpendicular to the tangent space $\mathcal{T}_a(V)$ to the submanifold V at the point $\mathbf{a}$.

However the tangent space is characterized in (5.19) as the plane perpendicular to the gradients $\nabla f_i(\mathbf{a})$, so the vector $\nabla g(\mathbf{a})$ must be contained in the span of the vectors $\nabla f_i(\mathbf{a})$, consequently must be expressible as a linear combination

$$\nabla g(\mathbf{a}) = \sum_{i=1}^{m-r} \lambda_i \nabla f_i(\mathbf{a}) \tag{5.20}$$

for some parameters $\lambda_i \in \mathbb{R}$. This set of equations, combined with the equations $f_i(\mathbf{a}) = 0$ for the submanifold V, are an explicit set of equations that can determine the local extrema of the function $g(\mathbf{x})$, a technique called the method of **Lagrange multipliers**.

To apply this method to examine the extrema of a function g on a subset $V \subset \mathbb{R}^m$ defined by the vanishing of finitely many $\mathcal{C}^1$ functions f_i, the first step is to see whether the set V is a $\mathcal{C}^1$ submanifold; the method only provides candidates for extrema at points at which V is a submanifold, so if the set V has any singularities those points must be examined separately. The next step is to examine the equations (5.20) together with the equations $f_i(\mathbf{x}) = 0$, to find candidates for points at which the function g has local maxima or minima. It is often clear from the problem or can be determined fairly readily whether any of these points actually is a local maxima or minima. For instance, if M is a compact submanifold and there are only two candidates for local maxima and minima, one of these must be a global maximum and the other a global minimum, and which is determined simply by the values of the function g at these points. If it is not clear whether a point $\mathbf{a}$ is a local extrema or not, the manifold V can be parametrized in an open neighborhood of that point as in (5.15) in Theorem 5.9, and the restriction of the function g to that neighborhood is then a function of the variables $x_{m-r+1}, \ldots, x_m$ to which the usual test in terms of the hessian of the function g in these coordinates can be tried; it is not necessary to determine the local parametrization (5.15) explicitly, of course, since the chain rule can be used to determine the second derivatives of the function g in terms of these variables.

As an example of the use of this method, find the minimum of $x + y + z$ subject to the restrictions that $\frac{a}{x} + \frac{b}{y} + \frac{c}{z} = 1$ where $x, y, z > 0$. The submanifold V is defined by the equation $f(x, y, z) = \frac{a}{x} + \frac{b}{y} + \frac{c}{z} - 1 = 0$, and the function of interest is $g(x, y, z) = x + y + z$, where

$$\nabla g = (1, 1, 1) \qquad \nabla f = \left(\frac{-a}{x^2}, \frac{-b}{y^2}, \frac{-c}{z^2} \right).$$

The Lagrange formula is the equation $\nabla g = \lambda \cdot \nabla f$; the first component of this vector equation is $\lambda = -x^2/a$, and upon substituting this into the other two components there result the equations

$$y = x\sqrt{\frac{b}{a}} \quad \text{and} \quad z = x\sqrt{\frac{c}{a}}.$$

Since $a, b,$ and c are fixed there results a one-parameter set of solutions, parametrized by x. When restricted to the set $S = \{(x, y, z) : f(x, y, z) = 0, x > 0, y > 0, z > 0\}$ this is

$$\frac{a}{x} + \frac{b}{x\sqrt{b/a}} + \frac{c}{x\sqrt{c/a}} = \frac{\sqrt{a}}{x}(\sqrt{a} + \sqrt{b} + \sqrt{c}) = 1$$

$$\Rightarrow (x, y, z) = (\sqrt{a} + \sqrt{b} + \sqrt{c})\left(\sqrt{a}, \sqrt{b}, \sqrt{c}\right).$$

Thus the extremum associated to this point is $g(x, y, z) = (\sqrt{a} + \sqrt{b} + \sqrt{c})^2$. To see that this is the minimum, the set S is noncompact but the function g must have a minimum on any compact subset of S. The function g increases whenever one of the coordinates $x, y,$ or z becomes very large. Since S is contained entirely in the first octant, it is only necessary to minimize g on a reasonable compact subset near the origin, which is the solution found.

When finding the extrema of a continuously differentiable function $g(\mathbf{x})$ on a set such as the closed semidisc

$$D = \left\{ \mathbf{x} \in \mathbb{R}^2 \,\middle|\, x_1^2 + x_2^2 \le 1, x_2 \ge 0 \right\}$$

in the plane, the procedure is first to examine critical points in the interior of the semidisc, then to examine the two pieces of the boundary of the semidisc that are local submanifolds, the semicircle $x_1^2 + x_2^2 = 1$, $x_2 > 0$ and the axis $-1 < x_1 < 1$, $x_2 = 0$, and then the two remaining points $(-1, 0)$ and $(1, 0)$ that are points where the boundary fails to be a submanifold. In general, the procedure is to examine critical points in the interior of the region, and then to examine the pieces of the boundary that are submanifolds and apply the method of Lagrange multipliers in all these cases.

If $\mathbf{f} : U \longrightarrow \mathbb{R}^n$ is a continuously differentiable mapping defined in an open subset $U \subset \mathbb{R}^m$ where $m > n$ then rank $\mathbf{f}'(\mathbf{a}) \le n$ at any point $\mathbf{a} \in U$; and if rank $\mathbf{f}'(\mathbf{x}) = n$ at a point $\mathbf{x} = \mathbf{a} \in U$ then it is also the case that rank $\mathbf{f}'(\mathbf{x}) = n$ for all points $\mathbf{x} \in V$ for an open neighborhood V of the point $\mathbf{a}$, since if some $n \times n$ submatrix of $\mathbf{f}'(\mathbf{a})$ has a nonzero determinant then by continuity that submatrix of $\mathbf{f}'(\mathbf{x})$ will also have a nonzero determinant for all points $\mathbf{x}$ near enough $\mathbf{a}$. The Rank Theorem asserts that the mapping $\mathbf{f}$ is equivalent to a linear mapping of rank n in an open neighborhood of each point $\mathbf{x} \in V$, and consequently the image of the mapping $\mathbf{f}$ is a full open neighborhood of the point $\mathbf{f}(\mathbf{a})$. In particular the image of the mapping $\mathbf{f}$ is not contained within any proper linear subspace of the image space $\mathbb{R}^n$, so the coordinate functions f_i of the mapping $\mathbf{f}$ are linearly independent functions at each point $\mathbf{a} \in U$ for which rank $\mathbf{f}'(\mathbf{a}) = n$. On the other hand if rank $\mathbf{f}'(\mathbf{x}) = r$ at all points $\mathbf{x} \in U$ where $r < n$, then by the Rank Theorem the mapping $\mathbf{f}$ will be a linear mapping of rank r in suitable local coordinates in open neighborhoods of each point $\mathbf{x} \in U \subset \mathbb{R}^m$ and its image $\mathbf{f}(\mathbf{a}) \in \mathbb{R}^n$; that means that the image of the restriction of the mapping $\mathbf{f}$ to an open neighborhood $V_{\mathbf{a}}$ of any point $\mathbf{a} \in U$ is contained in a submanifold of dimension r in an open neighborhood $V_{\mathbf{f}(\mathbf{a})} \subset \mathbb{R}^n$ of the image point $\mathbf{f}(\mathbf{a})$. This submanifold is described as the set of common zeros of $n - r$ linear functions in some local coordinates near the point $\mathbf{f}(\mathbf{a})$, and these functions will at least be some continuously differentiable functions h_i near the point $\mathbf{f}(\mathbf{a})$ in the initial coordinates in $\mathbb{R}^n$. Therefore the component functions f_i of the mapping $\mathbf{f}$ satisfy a collection of $n - r$ nontrivial equations $h_v(f_1(\mathbf{x}), f_2(\mathbf{x}), \ldots, f_n(\mathbf{x})) = 0$ for $1 \leq v \leq n - r$; that is usually expressed by saying that the functions f_i are **functionally dependent**.

Problems, Group I

1. Find the tangent vector to the curve $C \subset \mathbb{R}^3$ defined parametrically by the mapping $\mathbf{f} : \mathbb{R} \longrightarrow \mathbb{R}^3$ where

$$\mathbf{f}(t) = \left\{ e^{(t-1)(t-2)}, t^2 - 4, t^2 - 1 \right\}$$

at the point on the curve corresponding to the parameter value $t = 1$; and find the equation of the tangent line to the curve C at that point.

2. Consider the plane curve C defined by the equation

$$f(x_1, x_2) = x_1^3 + x_2^3 - 3x_1 x_2 = 0.$$

i) Show that the portion of the curve C in the half-plane $x_2 > 0$ is a submanifold.
ii) Find the tangent vector to the curve C at the point $(\frac{3}{2}, \frac{3}{2})$ and the equation of the tangent line to C at that point.

3. Find a normal vector and the equation of the tangent plane to the surface in $\mathbb{R}^4$ defined by the equation $x_1^2 x_2 + x_2 x_3 + x_4 = 3$ at the point $(1, 1, 1, 1) \in \mathbb{R}^4$.

4. Consider the function $f(x_1, x_2, x_3) = x_1^2 + x_1 x_2 + x_2^2 + x_2 x_3 + x_3^2$ in $\mathbb{R}^3$.

i) By using the method of Lagrange multipliers find the maximum and minimum values of this function on the spherical surface S defined by $x_1^2 + x_2^2 + x_3^2 = 1$, and the points at which these extreme values are attained.

    ii) By using the method of Lagrange multipliers find the maximum and minimum values of the function f on the intersection $S \cap T$ of the spherical surface S and the plane T defined by $c_1 x_1 + c_2 x_2 + c_3 x_3 = 0$, where $(c_1, c_2, c_3) \in S$ is a point at which the function f takes its maximum value on S.

5.   What are the dimensions of the cylinder in $\mathbb{R}^3$ with the least total area (including both top and bottom and the circular sides) and with the volume 1.

6.   What is the largest possible volume of a box in $\mathbb{R}^3$ for which the sum of the areas of the base and of the four sides is a specified value A? What are the lengths of the sides of this maximal box?

Problems, Group II

7.   Consider the hyperboloid surface $S \subset \mathbb{R}^3$ defined by the equation

$$x_1^2 + x_2^2 - x_3^2 = 25.$$

    i) Find a unit normal vector to S at a point $(a_1, a_2, 5)$.
    ii) Find the tangent plane to S at a point $(a_1, a_2, 5)$.
    iii) Show that the line segment ℓ joining the points $(a_1, a_2, 5)$ and $(a_2, -a_1, -5)$ is contained in the intersection of the surface S and its tangent plane at the point $(a_1, a_2, 5)$.

8.   Find the minimuum value of the function $f(\mathbf{x}) = x_1 + x_2 + x_3$ of variables $x_j > 0$ in $\mathbb{R}^3$ on the submanifold defined by the equation $\frac{a_1}{x_1} + \frac{a_2}{x_2} + \frac{a_3}{x_3} = 1$ for some constants $a_j > 0$.

9.   By considering the maximal value of the product $x_1 \cdot x_2 \cdots x_n$ for real numbers $x_j > 0$ satisfying the condition that $x_1 + x_2 + \cdots + x_n = c$, derive another proof of the inequality of the arithmetic and geometric means.

10.   If A is a symmetric $n \times n$ matrix show that the maximum and minimum values of the inner product $f(\mathbf{x}) = A\mathbf{x} \cdot \mathbf{x}$ on the unit sphere in $\mathbb{R}^n$ are the largest and smallest eigenvalues of the matrix A.

11.   Suppose that S_1 and S_2 are two-dimensional $\mathcal{C}^1$ submanifolds in $\mathbb{R}^3$ and that ℓ is a line segment from a point $\mathbf{a}_1 \in S_1$ to $\mathbf{a}_2 \in S_2$ that has the minimal length among all line segments joining points of these two surfaces. Show that the

line ℓ is perpendicular to the tangent planes to the surfaces S_1 and S_2 at the points $\mathbf{a}_1$ and $\mathbf{a}_2$.

12.   Suppose that M is a compact two-dimensional $\mathcal{C}^1$ submanifold in $\mathbb{R}^3$, such as the surface of a sphere or doughnut for example. Show that for every two-dimensional linear subspace $L \subset \mathbb{R}^3$ there is a point $a \in M$ such that the tangent space to M at the point a is the plane $a + L$, the translate of the linear subspace L to the point $a \in M$.

6

Integration

6.1 Riemann Integral

The basic operation of the Riemann integral is the integration of functions over closed cells $\overline{\Delta} \subset \mathbb{R}^n$. The **content** of a closed cell

$$\overline{\Delta} = \left\{ \mathbf{x} \in \mathbb{R}^n \,\middle|\, a_j \leq x_j \leq b_j \text{ for } 1 \leq j \leq n \right\}$$

is defined by

$$|\overline{\Delta}| = \prod_{j=1}^{n} (b_j - a_j), \tag{6.1}$$

and is also considered as the content $|\Delta|$ of the corresponding open cell Δ. The content of a cell or interval in $\mathbb{R}^1$ is its length, the content of a cell in $\mathbb{R}^2$ is its area, and the content of a cell in $\mathbb{R}^3$ is its volume; so the content of a cell is just a generalization of the length or area or volume to arbitrary dimensions. The cell can be decomposed as a union of smaller closed cells by adjoining finitely many points of subdivision

$$a_j = c_{j,0} < c_{j,1} < \cdots < c_{j,n_j} = b_j$$

of each of its sides and writing $\overline{\Delta}$ as the union of the closed subcells

$$\left\{ \mathbf{x} \in \mathbb{R}^n \,\middle|\, c_{j,k-1} \leq x_j \leq c_{j,k} \text{ for } 1 \leq j \leq n, 1 \leq k \leq n_j \right\};$$

such a decomposition $\overline{\Delta} = \bigcup_i \overline{\Delta}_i$ of the cell is called a **partition** $\mathcal{P}$ of the cell. The subcells of the partition are not disjoint, but only have portions of their boundaries in common. For the case $n = 1$ this is the decomposition of an interval, a cell in $\mathbb{R}^1$, as a union of subintervals; and for $n = 2$ it is illustrated in the accompanying Figure 6.1. It is clear that $|\overline{\Delta}| = \sum_i |\overline{\Delta}_i|$ for any partition of a cell $\overline{\Delta}$. If f is a bounded function on a cell $\overline{\Delta} \subset \mathbb{R}^n$, its least upper bound and greatest lower bound in the set $\overline{\Delta}$ are well-defined finite real numbers; the **upper sum** of the function f for the partition $\mathcal{P}$ is defined by

$$S^*(f, \mathcal{P}) = \sum_i \sup_{\mathbf{x} \in \overline{\Delta}_i} f(\mathbf{x}) |\overline{\Delta}_i|, \tag{6.2}$$

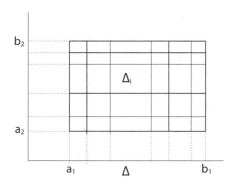

Figure 6.1. Partition of cell $\Delta \subset \mathbb{R}^2$

and similarly the **lower sum** is defined by

$$S_*(f, \mathcal{P}) = \sum_i \inf_{\mathbf{x} \in \overline{\Delta}_i} f(\mathbf{x}) |\overline{\Delta}_i|. \tag{6.3}$$

It is clear that

$$S_*(f, \mathcal{P}) \leq S^*(f, \mathcal{P}) \tag{6.4}$$

for any partition $\mathcal{P}$ of the cell $\overline{\Delta}$. Adding further points of subdivision to the sides of the cell leads to a finer decomposition of the cell as a union of subcells, called a **refinement** of the partition. If $\mathcal{P}_2$ is a refinement of a partition $\mathcal{P}_1$ of a cell Δ then

$$S^*(f, \mathcal{P}_2) \leq S^*(f, \mathcal{P}_1) \quad \text{and} \quad S_*(f, \mathcal{P}_2) \geq S_*(f, \mathcal{P}_1); \tag{6.5}$$

for if a cell $\overline{\Delta}_i$ in the partition $\mathcal{P}_1$ is decomposed into a union $\overline{\Delta}_i = \bigcup_j \overline{\Delta}_{ij}$ of cells $\overline{\Delta}_{ij}$ in the partition $\mathcal{P}_2$ then

$$\sup_{\mathbf{x} \in \overline{\Delta}_{ij}} f(\mathbf{x}) \leq \sup_{\mathbf{x} \in \overline{\Delta}_i} f(\mathbf{x}) \quad \text{and} \quad \sum_j |\overline{\Delta}_{ij}| = |\overline{\Delta}_i|$$

and consequently

$$S^*(f, \mathcal{P}_2) = \sum_{i,j} \sup_{\mathbf{x} \in \overline{\Delta}_{ij}} f(\mathbf{x}) |\overline{\Delta}_{ij}|$$

$$\leq \sum_{i,j} \sup_{\mathbf{x} \in \overline{\Delta}_i} f(\mathbf{x}) |\overline{\Delta}_{ij}| = \sum_i \sup_{\mathbf{x} \in \overline{\Delta}_i} f(\mathbf{x}) |\Delta_i| = S^*(f, \mathcal{P}_1),$$

and correspondingly for the lower sums with the reversed inequalities. It follows that if $\mathcal{P}_1$ and $\mathcal{P}_2$ are any two partitions of a cell $\overline{\Delta}$ then

$$S_*(f, \mathcal{P}_1) \leq S^*(f, \mathcal{P}_2); \tag{6.6}$$

for by considering all the points of subdivision of the sides of the cell $\overline{\Delta}$ from both partitions as defining a further partition $\mathcal{P}'$, a common refinement of both of these partitions, then by (6.4) and (6.5)

$$S_*(f,\mathcal{P}_1) \le S_*(f,\mathcal{P}') \le S^*(f,\mathcal{P}') \le S^*(f,\mathcal{P}_2). \tag{6.7}$$

In view of (6.7) the upper sums $S^*(f,\mathcal{P})$ for all of the partitions $\mathcal{P}$ of a cell $\overline{\Delta}$ are bounded below by any lower sum $S_*(f,\mathcal{P}_1)$, and the lower sums $S_*(f,\mathcal{P})$ for all of the partitions $\mathcal{P}$ of a cell $\overline{\Delta}$ are bounded above by any upper sum $S^*(f,\mathcal{P}_2)$; so it is possible to define the **upper integral** and **lower integral** of the function f over the cell $\overline{\Delta}$ by

$$\int_{\overline{\Delta}}^* f = \inf_{\mathcal{P}} S^*(f,\mathcal{P}) \quad \text{and} \quad \int_{*\overline{\Delta}} f = \sup_{\mathcal{P}} S_*(f,\mathcal{P}), \tag{6.8}$$

where the infimum and supremum are extended over all partitions $\mathcal{P}$ of the cell $\overline{\Delta}$, and clearly

$$\int_{*\overline{\Delta}} f \le \int_{\overline{\Delta}}^* f. \tag{6.9}$$

It may be the case though that the upper integral is strictly greater than the lower integral. For example, if f is the function on the real line $\mathbb{R}$ defined by

$$f(\mathbf{x}) = \begin{cases} 0 & \text{if } \mathbf{x} \text{ is rational,} \\ 1 & \text{if } \mathbf{x} \text{ is irrational,} \end{cases} \tag{6.10}$$

then $\sup_{x\in\overline{\Delta}_i} f(\mathbf{x}) = 1$ and $\inf_{x\in\overline{\Delta}_i} f(\mathbf{x}) = 0$ for any partition $\mathcal{P}$ of a cell $\overline{\Delta}$ into a union $\overline{\Delta} = \bigcup_i \overline{\Delta}_i$, and consequently $S^*(f,\mathcal{P}) = |\overline{\Delta}|$ and $S_*(f,\mathcal{P}) = 0$; and since that is the case for any partition $\mathcal{P}$ it follows that $\int_{\overline{\Delta}}^* f = |\overline{\Delta}|$ and $\int_{*\overline{\Delta}} f = 0$, so $\int_{\overline{\Delta}}^* f > \int_{*\overline{\Delta}} f$ in this case. On the other hand, it is clear that if $f(\mathbf{x}) = c$ at all points $\mathbf{x} \in \overline{\Delta}$ then $\int_{\overline{\Delta}}^* f = \int_{*\overline{\Delta}} f = c|\overline{\Delta}|$. More generally if f is a continuous function in a cell $\overline{\Delta}$ then $\int_{\overline{\Delta}}^* f = \int_{*\overline{\Delta}} f$. Indeed if f is continuous in $\overline{\Delta}$ then it is uniformly continuous since $\overline{\Delta}$ is compact, so for any $\epsilon > 0$ there is a partition $\mathcal{P}$ of the cell as a union $\overline{\Delta} = \bigcup_i \overline{\Delta}_i$ where the cells $\overline{\Delta}_i$ are sufficiently small that $\sup_{\mathbf{x}\in\overline{\Delta}_i} f(\mathbf{x}) - \inf_{\mathbf{x}\in\overline{\Delta}_i} f(\mathbf{x}) \le \epsilon$ for each cell $\overline{\Delta}_i$. Then

$$S^*(f,\mathcal{P}) - S_*(f,\mathcal{P}) = \sum_i \left(\sup_{\mathbf{x}\in\overline{\Delta}_i} f(\mathbf{x})|\overline{\Delta}_i| - \inf_{\mathbf{x}\in\overline{\Delta}_i} f(\mathbf{x})|\overline{\Delta}_i| \right)$$

$$\le \sum_i \epsilon|\overline{\Delta}_i| = \epsilon|\overline{\Delta}|; \tag{6.11}$$

since $S_*(f,\mathcal{P}) \le \int_{*\overline{\Delta}} f \le \int_{\overline{\Delta}}^* f \le S^*(f,\mathcal{P})$ it follows that $\int_{\overline{\Delta}}^* f - \int_{*\overline{\Delta}} f \le \epsilon|\overline{\Delta}|$ for any $\epsilon > 0$ and consequently that $\int_{\overline{\Delta}}^* f = \int_{*\overline{\Delta}} f$. This proof can be tweaked to yield a more general result that is a complete characterization of those functions for which the upper and lower integrals are actually equal; but that requires a digression to discuss some further properties of functions and point sets.

The **oscillation** of a bounded function f in a subset $S \subset \mathbb{R}^n$ was defined in (3.15) by

$$o_f(S) = \sup_{\mathbf{x} \in S} f(\mathbf{x}) - \inf_{\mathbf{x} \in S} f(\mathbf{x}), \qquad (6.12)$$

and the oscillation at a point $\mathbf{a} \in S$ was defined in (3.16) by

$$o_f(\mathbf{a}) = \lim_{r \to 0} o_f(\mathcal{N}_r(\mathbf{a})). \qquad (6.13)$$

In view of (6.11) it is clear that

$$S^*(f, \mathcal{P}) - S_*(f, \mathcal{P}) = \sum_i o_f(\overline{\Delta}_i)|\overline{\Delta}_i| \qquad (6.14)$$

for any partition $\mathcal{P}$ of the cell Δ, so the notion of oscillation is closely related to the question whether the upper and lower integrals of a function are equal. The relevant properties of the notion of oscillation for this purpose are that a bounded function f in an open neighborhood of a point $\mathbf{a} \in \mathbb{R}^n$ is continuous at $\mathbf{a}$ if and only if $o_f(\mathbf{a}) = 0$, as demonstrated in Theorem 3.14, and that if f is a bounded function in a subset $S \subset \mathbb{R}^n$ then the set

$$E = \left\{ \mathbf{x} \in S \,\middle|\, o_f(\mathbf{x}) < \epsilon \right\} \qquad (6.15)$$

is a relatively open subset of S for any $\epsilon > 0$, as demonstrated in Theorem 3.15.

A subset $S \subset \mathbb{R}^n$ has **content zero** if for any $\epsilon > 0$ there are *finitely many open cells* Δ_i such that $S \subset \bigcup_i \Delta_i$ and $\sum_i |\Delta_i| < \epsilon$; the set S has **measure zero** if for any $\epsilon > 0$ there are *countably many open cells* Δ_i such that $S \subset \bigcup_i \Delta_i$ and $\sum_i |\Delta_i| < \epsilon$. Clearly any set of content zero is also a set of measure zero; and a compact set of measure zero is also of content zero, since if S is compact and $S \subset \bigcup_i \Delta_i$ where $\sum_i |\Delta_i| < \epsilon$ then S is already contained in the union of finitely many of these cells Δ_i. It is worth pointing out here that both definitions can just as well be phrased in terms of closed cells $\overline{\Delta}$; for given any open cell Δ and any $\epsilon > 0$ there are closed cells $\overline{\Delta}_-$ and $\overline{\Delta}_+$ such that $\overline{\Delta}_- \subset \Delta \subset \overline{\Delta}_+$ while $|\Delta| - |\overline{\Delta}_-| < \epsilon$ and $|\overline{\Delta}_+| - |\Delta| < \epsilon$.

Lemma 6.1. *If countably many subsets $E_i \subset \mathbb{R}^n$ each have measure zero then their union $E = \bigcup_{i=1}^{\infty} E_i$ also has measure zero.*

Proof: Since E_i has measure zero then for any $\epsilon > 0$ there are countably many open cells Δ_{ij} for $j = 1, 2, \ldots$ such that $E_i \subset \bigcup_{j=1}^{\infty} \Delta_{ij}$ and $\sum_{j=1}^{\infty} \Delta_{ij} < \epsilon/2^i$; then

$E = \bigcup_{i=1}^{\infty} E_i \subset \bigcup_{i,j=1}^{\infty} \Delta_{ij}$, which is also a countable union of open cells, and $\sum_{i,j=1}^{\infty} |\Delta_{ij}| = \sum_{i=1}^{\infty} \sum_{j=1}^{\infty} |\Delta_{ij}| < \sum_{i=1}^{\infty} \frac{1}{2^i} \epsilon = \epsilon$, which suffices for the proof.

For example, the subset $\mathbb{Q} \subset \mathbb{R}^1$ of rational numbers does not have content zero, since no finite number of finite open intervals can cover all the rational numbers; but it does have measure zero as a consequence of the preceding lemma, since each individual point of course has measure zero. There are uncountable sets of measure zero as well; for instance $\mathbb{R}^1 \subset \mathbb{R}^2$ is a subset of measure zero in $\mathbb{R}^2$. To see that, the set $\mathbb{R}^1$ can be decomposed into the union $\mathbb{R}^1 = \bigcup_{n=-\infty}^{\infty} I_n$ where $I_n = [n, n+1]$, and for any $\epsilon > 0$ the segment I_n is a subset of the open cell

$$\Delta_n = \left\{ (x_1, x_2) \in \mathbb{R}^2 \mid n - \epsilon/2 < x_1 < n+1+\epsilon/2, \ -\epsilon/2 < x_2 < \epsilon/2 \right\}$$

for which $|\Delta_n| = (1 + \epsilon)\epsilon$, showing that I_n is a set of measure zero; and $\mathbb{R}^1 = \bigcup_n I_n$ then is a set of measure zero by Lemma 6.1. Moreover if $V \subset U$ is a C^1 submanifold of dimension k in an open subset $U \subset \mathbb{R}^n$ where $1 \leq k < n$ then by Theorem 5.9 for any point $\mathbf{a} \in V$, after relabeling the coordinates if necessary, the intersection $V \cap \Delta_{\mathbf{a}}$ of the submanifold V with a sufficiently small cell $\Delta_{\mathbf{a}} = \Delta'_{\mathbf{a}} \times \Delta''_{\mathbf{a}}$ centered at the point $\mathbf{a}$, where $\Delta'_{\mathbf{a}}$ is in the subspace $\mathbb{R}^{n-k}$ consisting of the first $n - k$ coordinates $x_1, \ldots, x_{n-k}$ and $\Delta''_{\mathbf{a}}$ is in the subspace $\mathbb{R}^k$ consisting of the last k coordinates $x_{n-k+1}, \ldots, x_n$, is described by the equations

$$V \cap \Delta_{\mathbf{a}} = \left\{ \mathbf{x} \in \Delta_{\mathbf{a}} \mid x_i = g_i(x_{n-k+1}, \ldots, x_n) \text{ for } 1 \leq i \leq n-k \right\}$$

for some continuously differentiable functions $g_i(\mathbf{x}'')$ of the variables $\mathbf{x}'' \in \Delta''_{\mathbf{a}}$. Each point $\mathbf{x} \in V \cap \overline{\Delta_{\mathbf{a}}}$ can be enclosed in a small cell $\Delta_i = \Delta'_i \times \Delta''_i$ where $|\Delta'_i| < \epsilon$, and since $V \cap \overline{\Delta_{\mathbf{a}}}$ is compact finitely many of these small cells cover $V \cap \overline{\Delta_{\mathbf{a}}}$; and $\sum_i |\Delta_i| < \epsilon |\Delta''_{\mathbf{a}}|$, from which it is evident that the set $V \cap \Delta_{\mathbf{a}}$ is of content 0. The entire submanifold V can be covered by countably many such cells $\Delta_{\mathbf{a}}$, so by Lemma 6.1 the set V has measure 0.

Lemma 6.2. *If $E \subset \mathbb{R}^n$ has content zero its closure $\overline{E}$ also has content zero; but if $E \subset \mathbb{R}^n$ has measure zero its closure $\overline{E}$ does not necessarily have measure zero.*

Proof: If $E \subset \mathbb{R}^n$ has content zero then for any $\epsilon > 0$ there are finitely many open cells Δ_i such that $E \subset \cup_i \Delta_i$ and $\sum_i |\Delta_i| < \epsilon$. Since $\overline{E} \subset \cup_i \overline{\Delta}_i$ and $\sum_i |\overline{\Delta}_i| < \epsilon$ it follows that $\overline{E}$ also is a set of content zero, for as noted earlier the definition of a set of content zero can be phrased equivalently in terms of open or closed cells. The set of rational numbers $\mathbb{Q}$ is a set of measure zero in $\mathbb{R}$ but its closure is the full set $\overline{\mathbb{Q}} = \mathbb{R}$, which is not a set of measure zero. That suffices for the proof.

A function f is **continuous almost everywhere** in a set $S \subset \mathbb{R}^n$ if the subset $E \subset S$ of points where f fails to be continuous is a set of measure zero.

For example, the function f on $\mathbb{R}$ in example (3.6), defined by

$$f(x) = \begin{cases} \frac{1}{q} & \text{if } x = \frac{p}{q} \text{ where } p, q \text{ are coprime integers,} \\ 0 & \text{if } x \text{ is irrational,} \end{cases} \tag{6.16}$$

is continuous at the irrational numbers but is discontinuous only at the rational numbers, a set of measure zero, so it is continuous almost everywhere.

Theorem 6.3 (Riemann-Lebesgue Theorem). *If f is a bounded function in a cell $\overline{\Delta} \subset \mathbb{R}^n$ then $\int_{\overline{\Delta}}^* f = \int_{*\overline{\Delta}} f$ if and only if f is continuous almost everywhere in $\overline{\Delta}$.*

Proof: For any bounded function f in a cell $\overline{\Delta}$ and for any $\epsilon > 0$ let

$$E_\epsilon = \left\{ \mathbf{x} \in \overline{\Delta} \,\Big|\, o_f(\mathbf{x}) < \epsilon \right\} \quad \text{and} \quad F_\epsilon = \left\{ \mathbf{x} \in \overline{\Delta} \,\Big|\, o_f(\mathbf{x}) \geq \epsilon \right\},$$

so $F_\epsilon = \overline{\Delta} \sim E_\epsilon$. By Lemma 3.15 the set E_ϵ is relatively open in $\overline{\Delta}$, so its complement F_ϵ is a relatively closed subset of $\overline{\Delta}$; and since $\overline{\Delta}$ is compact it follows that the closed subset $F_\epsilon \subset \overline{\Delta}$ is compact. By Lemma 3.14 the function f is continuous at a point $\mathbf{a} \in \overline{\Delta}$ if and only if $o_f(\mathbf{a}) = 0$; so the set of points in $\overline{\Delta}$ at which f is not continuous is the set $F = \{ \mathbf{x} \in \overline{\Delta} \mid o_f(\mathbf{x}) > 0 \}$, and $F_\epsilon \subset F$ for any $\epsilon > 0$ while $F = \bigcup_{n=1}^\infty F_{\frac{1}{n}}$.

To turn to the proof itself, first suppose that the function f is bounded and continuous almost everywhere in $\overline{\Delta}$. The set F then is of measure zero, as are the subsets F_ϵ; and since the sets F_ϵ are compact, they even are of content zero. Therefore for any $\epsilon > 0$ there are finitely many open cells $\Delta'_j \subset \mathbb{R}^n$ such that $F_\epsilon \subset \bigcup_j \Delta'_j$ and $\sum_j |\Delta'_j| < \epsilon$. The complement $K = \overline{\Delta} \sim \bigcup_j \Delta'_j$ is a closed and consequently compact subset of $\overline{\Delta}$, and if $\mathbf{a} \in K$ then $\mathbf{a} \notin F_\epsilon$ so $o_f(\mathbf{a}) < \epsilon$; hence for any point $\mathbf{a} \in K$ there is an open cell $\Delta''_{\mathbf{a}}$ containing that point such that $o_f(\overline{\Delta}''_{\mathbf{a}}) < \epsilon$. Since K is compact finitely many of the open cells $\Delta''_{\mathbf{a}_k}$ cover the compact set K. Let $\mathcal{P}$ be the partition $\overline{\Delta} = \bigcup_i \overline{\Delta}_i$ of the cell $\overline{\Delta}$ determined by the points of subdivision of the finitely many cells Δ'_j and $\Delta''_{\mathbf{a}_k}$. Let I' be the set of those indices i such that the cell $\overline{\Delta}_i$ is contained within some of the cells $\overline{\Delta}'_j$, and let I'' be the remaining set of indices. Thus $F_\epsilon \subset \bigcup_{i \in I'} \overline{\Delta}_i$ and $\sum_{i \in I'} |\overline{\Delta}_i| < \epsilon$, while if $i \in I''$ then $\overline{\Delta}_i \subset \overline{\Delta}''_{\mathbf{a}_k}$ for some of the cells $\overline{\Delta}''_{\mathbf{a}_k}$ so $o_f(\overline{\Delta}_i) < \epsilon$. If $|f(\mathbf{x})| \leq M$ for all points $\mathbf{x} \in \overline{\Delta}$ then $o_f(D) \leq 2M$ for any subset $D \subset \overline{\Delta}$ and consequently

$$S^*(f, \mathcal{P}) - S_*(f, \mathcal{P}) = \sum_i o_f(\overline{\Delta}_i)|\overline{\Delta}_i| = \sum_{i \in I'} o_f(\overline{\Delta}_i)|\overline{\Delta}_i| + \sum_{i \in I''} o_f(\overline{\Delta}_i)|\overline{\Delta}_i|$$

$$\leq 2M \sum_{i \in I'} |\overline{\Delta}_i| + \epsilon \sum_{i \in I''} |\overline{\Delta}_i| \leq 2M\epsilon + |\overline{\Delta}|\epsilon.$$

Since that is the case for any ϵ it follows that

$$\int_{\overline{\Delta}}^* f = \int_{*\overline{\Delta}} f.$$

For the converse direction, suppose that f is a bounded function in $\overline{\Delta}$ such that $\int_{\overline{\Delta}}^* f = \int_{*\overline{\Delta}}$. For any integer $n > 0$ and any $\epsilon > 0$ there is a partition $\mathcal{P}$ of the cell $\overline{\Delta}$ as a union $\overline{\Delta} = \bigcup_i \overline{\Delta}_i$ such that

$$\sum_i o_f(\overline{\Delta}_i)|\overline{\Delta}_i| = S^*(f, \mathcal{P}) - S_*(f, \mathcal{P}) < \epsilon/n.$$

Let I' be the set of those indices i such that $\Delta_i \cap F_{1/n} \neq \emptyset$ and let I'' be the remaining indices; thus $F_{1/n} \subset \bigcup_{i \in I'} \overline{\Delta}_i \cup \bigcup_{i \in I} \partial \overline{\Delta}_i$, and if $i \in I'$ there is a point $\mathbf{a}_i \in \Delta_i$ such that $o_f(\mathbf{a}_i) \geq 1/n$ and consequently $o_f(\overline{\Delta}_i) \geq 1/n$. Then

$$\epsilon/n > \sum_{i \in I'} o_f(\overline{\Delta}_i)|\overline{\Delta}_i| + \sum_{i \in I''} o_f(\overline{\Delta}_i)|\overline{\Delta}_i| \geq \sum_{i \in I'} o_f(\overline{\Delta}_i)|\overline{\Delta}_i| \geq \sum_{i \in I'} (1/n)|\overline{\Delta}_i|;$$

thus $F_{1/n} \subset \bigcup_{i \in I'} \overline{\Delta}_i \cup \bigcup_{i \in I} \partial \overline{\Delta}_i$ where $\sum_{i \in I'} |\overline{\Delta}_i| < \epsilon$ and $\bigcup_{i \in I} \partial \overline{\Delta}_i$ has measure zero. Since that is the case for any $\epsilon > 0$ it follows that $F_{1/n}$ is a set of measure zero; and therefore the countable union $F = \bigcup_{n=1}^{\infty} F_{1/n}$ also is a set of measure zero, which concludes the proof.

It is convenient to say that a function f is **Riemann integrable** over a subset $S \subset \mathbb{R}^n$ if it is bounded and continuous almost everywhere in S. If two functions f_1 and f_2 are Riemann integrable in a subset $S \subset \mathbb{R}^n$ their sum $f_1 + f_2$ and product $f_1 f_2$ are also Riemann integrable in S; for if $|f_j(\mathbf{x})| \leq M_j$ for all $\mathbf{x} \in S$ then $|f_1(\mathbf{x}) + f_2(\mathbf{x})| \leq (M_1 + M_2)$ and $|f_1(\mathbf{x})f_2(\mathbf{x})| \leq M_1 M_2$ for all $\mathbf{x} \in S$ while if f_j is continuous in the complement of a subset $E_j \subset S$ of measure zero then $f_1 + f_2$ and $f_1 f_2$ are continuous in the the complement of the set $E_1 \cup E_2$ which also is of measure zero. The preceding theorem shows that the Riemann integrable functions in a cell $\overline{\Delta}$ are precisely those functions for which the upper and lower integrals over $\overline{\Delta}$ exist and are equal.[1] The common value of the upper and lower integrals of a Riemann integrable function f is called the **Riemann integral** of the function f over the cell $\overline{\Delta}$ and is denoted by $\int_{\overline{\Delta}} f$, so that

$$\int_{\overline{\Delta}} f = \int_{\overline{\Delta}}^* f = \int_{*\overline{\Delta}} f; \tag{6.17}$$

---

[1] It is perhaps more common to say that a function is Riemann integrable over a cell if and only if the upper and lower integrals exist and are equal; but that property rests on the results of rather nontrivial constructions involving the function rather than on somewhat more intrinsic properties of the function itself. Riemann integrability as defined here makes equal sense in any subset of $\mathbb{R}^n$ without the need of defining the integral over more general sets.

for convenience this is often called just the integral of the function f, although some caution is necessary since there are other common notions of integrals, such as the Lebesgue or Denjoy integrals. It is perhaps more common and certainly more traditional to denote the Riemann integral by

$$\int_{\overline{\Delta}} f(\mathbf{x}) \, d\mathbf{x} = \int_{\overline{\Delta}} f, \tag{6.18}$$

a somewhat redundant notation but one that is convenient in specifying the variables involved when considering the effect of changes of variables on the integral, as for example in Theorem 6.19. Note that if f, g are Riemann integrable functions in a cell $\overline{\Delta} \subset \mathbb{R}^n$ and if $f(\mathbf{x}) \geq g(\mathbf{x})$ for all $\mathbf{x} \in \overline{\Delta}$ then $S^*(f, \mathcal{P}) \geq S^*(g, \mathcal{P})$ for all partitions $\mathcal{P}$ of $\overline{\Delta}$ hence $\int_{\overline{\Delta}} f \geq \int_{\overline{\Delta}} g$; this simple observation, the integration of inequalities, is quite useful and will be extended further in the subsequent discussion.

The explicit value of a given Riemann integral can be calculated directly from the definition of the integral; but in practice there is another standard technique for calculating integrals in one dimension that is almost universally used for fairly simple functions.

Theorem 6.4 (Second Fundamental Theorem of Calculus). *If $f(x)$ is a C^1 function in an open subset $U \subset \mathbb{R}^1$ then*

$$\int_{[a,b]} f' = f(b) - f(a) \tag{6.19}$$

for any cell $[a, b] \subset U$.

Proof: For any partition

$$\mathcal{P} = a = x_0 < x_1 < \cdots < x_{n-1} < x_n = b$$

of the cell $[a, b] \subset U$ and for any points ξ_i such that $x_i \leq \xi_i \leq x_{i+1}$ clearly

$$\inf_{x_i \leq x \leq x_{i+1}} f'(x) \leq f'(\xi_i) \leq \sup_{x_i \leq x \leq x_{i+1}} f'(x)$$

so it follows that

$$S_*(f', \mathcal{P}) \leq \sum_{i=0}^{n-1} f'(\xi_i)(x_{i+1} - x_i) \leq S^*(f', \mathcal{P}). \tag{6.20}$$

Since f is C^1 in U then by the Mean Value Theorem for functions of a single variable, Theorem 3.29, there are points ξ_i in the intervals $x_i < \xi_i < x_{i+1}$ for which $f(x_{i+1}) - f(x_i) = f'(\xi_i)(x_{i+1} - x_i)$; for these points

$$\sum_{i=0}^{n-1} f'(\xi_i)(x_{i+1} - x_i) = \sum_{i=0}^{n-1} \left(f(x_{i+1}) - f(x_i) \right) = f(x_n) - f(x_0) = f(b) - f(a).$$

Substituting this observation in (6.20) shows that

$$S_*(f', \mathcal{P}) \leq f(b) - f(a) \leq S^*(\mathcal{P}, f'),$$

and since that is the case for any partition $\mathcal{P}$ it follows that

$$\int_{* \ [a,b]} f' \leq f(b) - f(a) \leq \int_{[a,b]}^* f';$$

however the function f' is Riemann integrable, as a continuous function on $[a, b]$, so $\int_{* \ [a,b]} f' = \int_{[a,b]}^* f' = \int_{[a,b]} f'$ in the preceding equation, which leads immediately to (6.19) and thereby concludes the proof.

It is perhaps worth noting that the hypotheses in the preceding theorem can be weakened; to apply the Mean Value Theorem it is only required that the function f be continuous in the closed interval $[a, b]$ and differentiable in the interior (a, b), and to show the equality of the upper and lower integrals it is only required that the function f' be Riemann integrable over $[a, b]$, so be bounded and continuous almost everywhere in that interval. However in practice this theorem is usually applied to quite regular functions, those that can be expressed as derivatives of continuously differentiable functions, such as polynomials and some expressions involving exponentials, logarithms, and the trigonometric functions; for instance $\int_{[0, \pi/2]} \sin x = -\cos(\pi/2) + \cos(0) = 1$ since $\cos'(x) = -\sin(x)$. The following special case of the Second Fundamental Theorem is also quite commonly used and is well worth keeping in mind.

Corollary 6.5 (Integration by Parts). *If f, g are C^1 functions in an interval $[a, b]$ then*

$$\int_{[a,b]} fg' = f(b)g(b) - f(a)g(a) - \int_{[a,b]} f'g. \tag{6.21}$$

Proof: This follows immediately from an application of the preceding theorem to the product fg, using the linearity of integrals which will be established in Theorem 6.8.

Of course the Second Fundamental Theorem can be used to evaluate the integrals of a continuous function f only if it is possible to find an **antiderivative** of f, a C^1 function F such that $F' = f$; and there are quite simple functions,[2] such as $f(x) = \sqrt{P(x)}$ for some polynomials $P(x)$ of degree at least 3, for which there are no elementary antiderivatives $F(x)$. That any continuous function actually has some antiderivative, however complicated it may be, is the content of the **First Fundamental Theorem of Calculus**; and the First and Second Theorems taken together are usually referred to as the **Fundamental Theorem of Calculus**. The antiderivatives are most easily expressed by using the **oriented integral** $\int_a^b f$ of a Riemann integrable function f, defined by

$$\int_a^b f = \begin{cases} \int_{[a,b]} f & \text{if} \quad a \le b, \\ -\int_{[b,a]} f & \text{if} \quad a \ge b; \end{cases} \tag{6.22}$$

thus oriented integrals have the basic property that

$$\int_a^b f = -\int_b^a f \quad \text{for any} \quad a, b \in \mathbb{R}. \tag{6.23}$$

It is easy to see, as a direct consequence of this definition, that

$$\int_a^b f + \int_b^c f = \int_a^c f \quad \text{for any} \quad a, b, c \in \mathbb{R}; \tag{6.24}$$

this is of course trivial if $a < b < c$ but interesting in the other possible cases.

Theorem 6.6 (First Fundamental Theorem of Calculus). *If f is a continuous function in an interval $[a, b]$ the oriented integral $F(x) = \int_a^x f$ is a C^1 function in (a, b) such that $F'(x) = f(x)$ for all $x \in (a, b)$.*

Proof: If $F(x) = \int_a^x f$ it follows from (6.24) that $F(x + h) - F(x) = \int_x^{x+h} f$ whenever $x, x + h \in [a, b]$, where h can be either positive or negative. Since f is continuous then for any fixed point $x \in (a, b)$ and any $\epsilon > 0$ if h is sufficiently small

$$f(x) - \epsilon \le f(t) \le f(x) + \epsilon \quad \text{for all } t \text{ between } x \text{ and } x + h.$$

The oriented integral of this inequality over the interval between x and $x + h$ yields the inequalities

$$\big(f(x) - \epsilon\big)h \le F(x + h) - F(x) \le \big(f(x) + \epsilon\big)h \quad \text{if } h > 0$$

or

$$\big(f(x) - \epsilon\big)h \ge F(x + h) - F(x) \ge \big(f(x) + \epsilon\big)h \quad \text{if } h < 0;$$

---

[2]For examples see the discussion of elliptic functions in *A Course of Modern Analysis* by Whittaker and Watson.

and dividing by h and taking the limit as h tends to 0 shows that

$$f(x) - \epsilon \leq F'(x) \leq f(x) + \epsilon.$$

Since that is the case for all $\epsilon > 0$ it follows that $F'(x) = f(x)$ in both cases, and that suffices for the proof.

Although only the integral of a function over a cell has been defined so far, it is possible to extend the integral to more general sets remarkably easily. A bounded subset $D \subset \mathbb{R}^n$ is called a **Jordan measurable set** if its boundary ∂D is a set of measure zero; since ∂D is a closed bounded set, and hence a compact set, then actually ∂D is a set of content zero for any Jordan measurable set D. It is easier to show that a set is a Jordan measurable set by using the characterization that the boundary is a set of measure zero; but for many results about Jordan measurable sets it is more convenient to use the equivalent characterization that the boundary is a set of content zero. The equivalence of these two characterizations should be kept in mind in the subsequent discussion. The definition of a Jordan measurable set does not impose any further restrictions on the set D; so a Jordan measurable set may be open or closed or even a more general point set. An open or closed cell is a Jordan measurable set, since as noted earlier $\partial \Delta$ is a set of measure zero for any cell Δ.

Lemma 6.7. *If $D_1, D_2 \subset \mathbb{R}^n$ are Jordan measurable sets then so are both their union $D_1 \cup D_2$ and their intersection $D_1 \cap D_2$.*

Proof: Obviously if $D_1, D_2 \subset \mathbb{R}^n$ are bounded sets then so are their union and intersection. Suppose then that their boundaries $\partial D_1, \partial D_2$ are sets of measure zero. For any subsets A, B of a topological space since $A \cup B \subset \overline{A} \cup \overline{B}$ and $\overline{A} \cup \overline{B}$ is a closed set it follows that $\overline{A \cup B} \subset (\overline{A} \cup \overline{B})$; and similarly $\overline{A \cap B} \subset (\overline{A} \cap \overline{B})$. Therefore

$$\partial(D_1 \cup D_2) = \overline{(D_1 \cup D_2)} \bigcap \overline{\sim (D_1 \cup D_2)} \subset \overline{(D_1 \cup D_2)} \bigcap \overline{(\sim D_1 \cap \sim D_2)}$$
$$\subset \left(\overline{D_1} \cap \overline{\sim D_1} \cap \overline{\sim D_2} \right) \bigcup \left(\overline{D_2} \cap \overline{\sim D_1} \cap \overline{\sim D_2} \right)$$
$$\subset \left(\overline{D_1} \cap \overline{\sim D_1} \right) \bigcup \left(\overline{D_2} \cap \overline{\sim D_2} \right) = \partial D_1 \cup \partial D_2,$$

which is a union of two sets of measure zero so is a set of measure zero. Similarly

$$\partial(D_1 \cap D_2) = \overline{(D_1 \cap D_2)} \bigcap \overline{\sim (D_1 \cap D_2)} \subset \overline{(D_1 \cap D_2)} \bigcap \overline{(\sim D_1 \cup \sim D_2)}$$
$$\subset \left(\overline{D_1} \cap \overline{D_2} \cap \overline{\sim D_1} \right) \bigcup \left(D_1 \cap \overline{D_2} \cap \overline{\sim D_2} \right)$$
$$\subset \left(\overline{D_1} \cap \overline{\sim D_1} \right) \bigcup \left(\overline{D_2} \cap \overline{\sim D_2} \right) = \partial D_1 \cup \partial D_2,$$

which also is a union of two sets of measure zero so is a set of measure zero, and that suffices for the proof.

A Jordan measurable set $D \subset \mathbb{R}^n$ is a bounded subset of $\mathbb{R}^n$ by definition, so $D \subset \Delta$ for some cell $\Delta \subset \mathbb{R}^n$. A Riemann integrable function f in the Jordan measurable set D can be extended to a function $\widetilde{f}$ in the cell $\overline{\Delta}$ by setting

$$\widetilde{f}(\mathbf{x}) = \begin{cases} f(\mathbf{x}) & \text{if} \quad \mathbf{x} \in D, \\ 0 & \text{if} \quad \mathbf{x} \in \overline{\Delta} \sim D. \end{cases} \tag{6.25}$$

The extended function $\widetilde{f}$ of course is bounded in $\overline{\Delta}$. It is continuous almost everywhere in the interior of D, since the function f is continuous almost everywhere in D by assumption, and it is also continuous in the complement $\overline{\Delta} \sim \overline{D}$ of the closure of D, where it vanishes identically by definition; and since $\overline{\Delta} = D^o \cup (\overline{\Delta} \sim D)^o \cup \partial D$ where ∂D is a set of measure 0 it follows that the function $\widetilde{f}$ is also continuous almost everywhere in $\overline{\Delta}$. Consequently the function $\widetilde{f}$ is Riemann integrable in the cell Δ, so it is possible to define the integral of the initial function f over the Jordan measurable set D by

$$\int_D f = \int_{\overline{\Delta}} \widetilde{f}. \tag{6.26}$$

It is evident from the definition of the Riemann integral that the value of the integral $\int_D f$ is independent of the choice of the cell $\overline{\Delta}$; for if $\overline{\Delta} \subset \overline{\Delta}'$ then in calculating the integral $\int_{\overline{\Delta}'} \widetilde{f}$ it is always possible to take partitions such that each cell $\overline{\Delta}'_i$ in the partition is either contained in $\overline{\Delta}$ or is disjoint from $\overline{\Delta}$ or is one of a finite number of cells containing $\partial \Delta$ and of total content less than any given ϵ.

As a special case, if f is a Riemann integrable function in a closed cell $\overline{\Delta}$ its restriction $f|D$ to a Jordan measurable set $D \subset \overline{\Delta}$ is a Riemann integrable function in D and the extension $\widetilde{f|D}$ can be identified with the product $\chi_D f$ where χ_D is the **characteristic function** of the set D, the function defined by

$$\chi_D(\mathbf{x}) = \begin{cases} 1 & \text{if} \quad \mathbf{x} \in D, \\ 0 & \text{if} \quad \mathbf{x} \notin D. \end{cases} \tag{6.27}$$

Thus if f is a Riemann integrable function in a closed cell $\overline{\Delta}$ then its integral over any Jordan measurable set $D \subset \overline{\Delta}$ can be defined equivalently by

$$\int_D f = \int_\Delta \chi_D f. \tag{6.28}$$

In particular the **content** of a Jordan measurable set $D \subset \Delta$ can be defined as the integral of the constant function 1 over the set D, so by the integral

$$|D| = \int_{\overline{\Delta}} \chi_D. \tag{6.29}$$

When $D = \overline{\Delta}$ this definition coincides with the previous definition of the content of a cell, since for any partition $\mathcal{P}$ of a cell $\overline{\Delta}$ it follows immediately from the definition of the upper sum that $S^*(\chi_{\overline{\Delta}}, \mathcal{P}) = \sum_i |\overline{\Delta}_i| = |\overline{\Delta}|$ and consequently that $\int_{\overline{\Delta}} \chi_{\overline{\Delta}} = |\overline{\Delta}|$.

The extension of the Riemann integral to integrals over Jordan measurable sets provides the basic tool for integration in a good deal of analysis. The properties of this extension are the next set of topics to be discussed here.

Theorem 6.8 (Linearity of the Integral). *If f_1 and f_2 are Riemann integrable functions in a Jordan measurable set $D \subset \mathbb{R}^n$ then so is any linear combination $f = c_1 f_1 + c_2 f_2$ for real constants c_1, c_2 and*

$$\int_D f = c_1 \int_D f_1 + c_2 \int_D f_2. \tag{6.30}$$

Proof: It was noted earlier that the sums and products of Riemann integrable functions are also Riemann integrable; and it is clear that constants are Riemann integrable, so any linear combination of Riemann integrable functions is Riemann integrable. The integrals over D of Riemann integrable functions are defined as the integrals over any cell Δ containing D of the extended functions (6.25); so it is sufficient just to prove the theorem for linear combinations of Riemann integrable functions over a cell $\overline{\Delta}$. If f is Riemann integrable in $\overline{\Delta}$ then for any partition $\mathcal{P}$ of that cell and any constant $c > 0$

$$S^*(cf, \mathcal{P}) = \sum_i \left(\sup_{\mathbf{x} \in \overline{\Delta}_i} cf(\mathbf{x}) \right) |\overline{\Delta}_i| = c \sum_i \left(\sup_{\mathbf{x} \in \overline{\Delta}_i} f(\mathbf{x}) \right) |\overline{\Delta}_i| = cS^*(f, \mathcal{P})$$

and consequently

$$\int_{\overline{\Delta}} cf = \int_{\overline{\Delta}}^* cf = c \int_{\overline{\Delta}}^* f = c \int_{\overline{\Delta}} f.$$

On the other hand

$$S^*(-f, \mathcal{P}) = \sum_i \left(\sup_{\mathbf{x} \in \overline{\Delta}_i} -f(\mathbf{x}) \right) |\overline{\Delta}_i| = -\sum_i \left(\inf_{\mathbf{x} \in \overline{\Delta}_i} f(\mathbf{x}) \right) |\overline{\Delta}_i| = -S_*(f, \mathcal{P})$$

and consequently

$$\int_{\overline{\Delta}} -f = \int_{\overline{\Delta}}^* -f = \inf_{\mathcal{P}} S^*(-f, \mathcal{P}) = \inf_{\mathcal{P}} -S_*(f, \mathcal{P})$$

$$= -\sup_{\mathcal{P}} S_*(f, \mathcal{P}) = -\int_{*\overline{\Delta}} f = -\int_{\overline{\Delta}} f.$$

If f_1 and f_2 are two Riemann integrable functions in $\overline{\Delta}$ then

$$\sup_{\mathbf{x} \in \overline{\Delta}_i} (f_1(\mathbf{x}) + f_2(\mathbf{x})) \le \sup_{\mathbf{x} \in \overline{\Delta}_i} f_1(\mathbf{x}) + \sup_{\mathbf{x} \in \overline{\Delta}_i} f_2(\mathbf{x})$$

hence

$$\int_{\overline{\Delta}} (f_1 + f_2) = \int_{\overline{\Delta}}^* (f_1 + f_2) \le S^*(f_1 + f_2, \mathcal{P}) \le S^*(f_1, \mathcal{P}) + S^*(f_2, \mathcal{P});$$

the corresponding argument shows that the reverse inequality holds for the lower sums, so

$$S_*(f_1, \mathcal{P}) + S_*(f_2, \mathcal{P}) \le \int_{\overline{\Delta}} (f_1 + f_2) \le S^*(f_1, \mathcal{P}) + S^*(f_2, \mathcal{P}). \tag{6.31}$$

For any $\epsilon > 0$ there exists a partition $\mathcal{P}$ such that

$$S_*(f_i, \mathcal{P}) \le \int_{\overline{\Delta}} f_i \le S^*(f_i, \mathcal{P}) \quad \text{and} \le S^*(f_i, \mathcal{P}) - S_*(f_i, \mathcal{P}) < \epsilon$$

for $i = 1, 2$; and substituting these inequalities in (6.31) shows that

$$\left| \int_{\overline{\Delta}} (f_1 + f_2) - \int_{\overline{\Delta}} f_1 - \int_{\overline{\Delta}} f_2 \right| < 2\epsilon.$$

Since that is the case for any $\epsilon > 0$ it follows that $\int_{\overline{\Delta}} (f_1 + f_2) = \int_{\overline{\Delta}} f_1 + \int_{\overline{\Delta}} f_2$, which suffices for the proof.

Integrability is really required for the preceding result to hold. For the function (6.10) for example $\int_{\overline{\Delta}}^* f = |\overline{\Delta}|$ and $\int_{*\overline{\Delta}} f = 0$, and it is clear that the same argument applied to the function $g(\mathbf{x}) = 1 - f(\mathbf{x})$ shows that $\int_{\overline{\Delta}}^* g = |\overline{\Delta}|$ and $\int_{*\overline{\Delta}} g = 0$; consequently $\int_{\overline{\Delta}}^* f + \int_{\overline{\Delta}}^* g = 2|\Delta|$ while $\int_{*\overline{\Delta}} f + \int_{*\overline{\Delta}} g = 0$. On the other hand $f + g = 1$ so $\int_{\overline{\Delta}}^* (f + g) = \int_{*\overline{\Delta}} (f + g) = \int_{\overline{\Delta}} (f + g) = |\Delta|$. Thus it is neither true that $\int_{\overline{\Delta}}^* (f + g) = \int_{\overline{\Delta}}^* f + \int_{\overline{\Delta}}^* g$ nor that $\int_{*\overline{\Delta}} (f + g) = \int_{*\overline{\Delta}} f + \int_{*\overline{\Delta}} g$.

Theorem 6.9 (Positivity of the Integral). *Let f and g be Riemann integrable functions over a Jordan measurable set $D \subset \mathbb{R}^n$.*
(i) *If $f(\mathbf{x}) \ge 0$ for all $\mathbf{x} \in D$ then $\int_D f \ge 0$.*
(ii) *If $f(\mathbf{x}) \ge g(\mathbf{x})$ for all $\mathbf{x} \in D$ then $\int_D f \ge \int_D g$.*
(iii) $\left| \int_D f \right| \le \int_D |f|$.

Proof: Again it is sufficient just to prove these assertions for Riemann integrable functions over a cell $\overline{\Delta}$. If $f(\mathbf{x}) \geq 0$ for all $\mathbf{x} \in \overline{\Delta}$ then $S^*(f, \mathcal{P}) \geq 0$ for any partition $\mathcal{P}$ of the cell $\overline{\Delta}$ so (i) holds. Then (ii) follows by applying (i) to the difference $f - g$ and using the preceding Theorem 6.8, while (iii) follows by applying (ii) to the functions $|f|$ and $\pm f$ since $|f(\mathbf{x})| \geq f(\mathbf{x})$ and $|f(\mathbf{x})| \geq -f(\mathbf{x})$ for all $\mathbf{x} \in \overline{\Delta}$; and that suffices for the proof.

The Riemann integral of course can be extended to integrals of vector-valued functions by integrating each component, so that the integral $\int_D \mathbf{f}$ of a vector-valued function $\mathbf{f} = \{f_i\}$ is defined to be the vector with i-th component the integral $\int_D f_i$. It is useful for many purposes to have available the following extension of part (iii) of the preceding theorem.

Corollary 6.10. *If $\mathbf{f}$ is a vector-valued function the components of which are Riemann integrable in a Jordan measurable set $D \subset \mathbb{R}^n$ then*

$$\left\| \int_D \mathbf{f} \right\|_2 \leq \int_D \|\mathbf{f}\|_2. \tag{6.32}$$

Proof: If $\mathbf{u}$ is a unit vector in the direction of the vector $\int_D \mathbf{f}$ then from the Cauchy-Schwarz inequality it follows that $|\mathbf{u} \cdot \int_D \mathbf{f}| = \| \int_D \mathbf{f} \|_2$ and $|\mathbf{u} \cdot \mathbf{f}| \leq \|\mathbf{f}\|_2$. By part (iii) of the preceding theorem

$$\left\| \int_D \mathbf{f} \right\|_2 = \left| \mathbf{u} \cdot \int_D \mathbf{f} \right| = \left| \int_D \mathbf{u} \cdot \mathbf{f} \right| \leq \int_D |\mathbf{u} \cdot \mathbf{f}| \leq \int_D \|\mathbf{f}\|_2$$

which suffices for the proof.

Lemma 6.11. *If f is a bounded function in a Jordan measurable set $D \subset \mathbb{R}^n$ and if $f(\mathbf{x}) = 0$ for all points $\mathbf{x} \in (D \sim E)$, where $E \subset D$ is a set of content 0, then f is Riemann integrable over D and $\int_D f = 0$.*

Proof: The Jordan measurable set D is bounded so $D \subset \overline{\Delta}$ for some cell $\overline{\Delta}$. The extended function $\widetilde{f}$ in $\overline{\Delta}$ is zero in $\overline{\Delta} \sim E$, so is Riemann integrable in Δ since E is of content zero. For any $\epsilon > 0$ it is possible to find a finite number of open cells Δ'_i such that $E \subset \cup_i \Delta'_i$ and $\sum_i |\Delta'_i| < \epsilon$. These cells can be taken to be cells in a suitably fine partition $\mathcal{P}$ of the cell $\overline{\Delta}$; the function $\widetilde{f}$ thus is bounded on each of the cells $\overline{\Delta}'_i$, with a bound M, and vanishes in the other cells $\overline{\Delta}''_j$ of that partition. Therefore

$$S^*(\widetilde{f}, \mathcal{P}) = \sum_i \left(\sup_{\mathbf{x} \in \overline{\Delta}'_i} \widetilde{f}(\mathbf{x}) \right) \left| \overline{\Delta}'_i \right| + \sum_j \left(\sup_{\mathbf{x} \in \overline{\Delta}''_j} \widetilde{f}(\mathbf{x}) \right) \left| \overline{\Delta}''_j \right|$$

$$\leq \sum_{i \in I'} M |\overline{\Delta}_i| + 0 \leq M\epsilon;$$

and since that is the case for any $\epsilon > 0$ it follows that $\int_D f = \int_{\bar{\Delta}}^* \tilde{f} \leq 0$. The corresponding argument shows that $\int_D f = \int_{*\bar{\Delta}} \tilde{f} \geq 0$, and consequently $\int_D f = 0$, which suffices for the proof.

The preceding result fails if the exceptional set is of measure 0 rather than of content 0; for a function f such that $f(\mathbf{x}) = 1$ at a countable dense set of points of a Jordan measurable set D and $f(\mathbf{x}) = -1$ at another countable dense set of points of D while $f(\mathbf{x}) = 0$ otherwise does have the property that it vanishes outside of a set of measure 0, but it is not Riemann integrable over D.

Theorem 6.12 (Invariance of the Integral). *If f is a Riemann integrable function over a Jordan measurable set $D \subset \mathbb{R}^n$, and if g is a bounded function in D such that $g(\mathbf{x}) = f(\mathbf{x})$ at all points $\mathbf{x} \in (D \sim E)$ where $E \subset D$ is a set of content zero, then g is also Riemann integrable over D and $\int_D g = \int_D f$.*

Proof: The function $f - g$ is bounded in D and vanishes on the set $D \sim E$ where $E \subset D$ is a subset of content zero, so by the preceding lemma this function is Riemann integrable in D and $\int_D (f - g) = 0$. The difference $g = f - (f - g)$ of these two Riemann integrable functions is then also Riemann integrable, and it follows from Theorem 6.8 that $\int_D g = \int_D f - \int_D (f - g) = \int_D f$, which suffices for the proof.

The preceding theorem shows that a Riemann integrable function f in a Jordan measurable set D can be modified on any subset of content 0 in D in any arbitrary way, so long as it remains bounded, and the modified function remains Riemann integrable with the same integral as the function f. The integral of a function f thus is not so much a property of the individual function f, but rather is a property of an equivalence class of functions, where two Riemann integrable functions in D are equivalent if and only if they are equal outside a set of content 0 in D. In particular when considering integrals or integrability of functions over cells or Jordan measurable sets, the values of a function at the boundary of the set are immaterial, provided of course that the function remains bounded; so as far as integration is concerned, it is immaterial whether closed or open Jordan measurable sets are considered.

Theorem 6.13 (Additivity of the Integral). *If $D_1, D_2 \subset \mathbb{R}^n$ are Jordan measurable sets where the intersection $E = D_1 \cap D_2$ is a set of content zero and if f is a Riemann integrable function in the union $D_1 \cup D_2$ then*

$$\int_{D_1 \cup D_2} f = \int_{D_1} f + \int_{D_2} f; \tag{6.33}$$

in particular

$$|D_1 \cup D_2| = |D_1| + |D_2|. \tag{6.34}$$

Proof: The union $D_1 \cup D_2$ of two Jordan measurable sets D_1, D_2 is also a Jordan measurable set by Lemma 6.7, so it is contained in a cell $\overline{\Delta}$; and then $\int_{D_1 \cup D_2} f = \int_{\overline{\Delta}} \chi_{D_1 \cup D_2} f$. The characteristic function of the union of the two Jordan measurable sets is $\chi_{D_1 \cup D_2}(\mathbf{x}) = \chi_{D_1}(\mathbf{x}) + \chi_{D_2}(\mathbf{x}) - \chi_E(\mathbf{x})$, where the intersection $E = D_1 \cap D_2$ is also a Jordan measurable set; consequently

$$\int_{D_1 \cup D_2} f = \int_{\Delta} \left(\chi_{D_1} + \chi_{D_2} - \chi_E \right) f$$

$$= \int_{\Delta} \chi_{D_1} f + \int_{\Delta} \chi_{D_2} f - \int_{\Delta} \chi_E f$$

$$= \int_{D_1} f + \int_{D_2} f$$

since E is a subset of content zero so $\int_{\Delta} \chi_E f = 0$ by Theorem 6.11. That demonstrates (6.33), which holds in particular for $f = 1$ and therefore implies (6.34), thereby concluding the proof.

A subset $D \subset \overline{\Delta}$ has content zero if and only if $\int_{\overline{\Delta}}^* \chi_D = 0$, since for any partition $\mathcal{P}$ of the cell $\overline{\Delta}$ as the union $\overline{\Delta} = \bigcup_{i \in I} \overline{\Delta}_i$ it is clear that

$$S^*(\chi_D, \mathcal{P}) = \sum_{\{ i \in I \mid D \cap \overline{\Delta}_i \neq \emptyset \}} |\overline{\Delta}_i|.$$

That suggests that the notion of the content of a Jordan measurable set might be extended to the notion of the content of an arbitrary subset $D \subset \overline{\Delta}$ by setting $|D| = \int_{\overline{\Delta}}^* \chi_D$; but for more general disjoint sets[3] it is not necessarily the case that $|D_1 \cup D_2| = |D_1| + |D_2|$. The Lebesgue integral does provide an extension of the notion of the content to some more general sets than Jordan measurable sets for which this is true though.

Problems, Group I

1.   (i) Suppose that f is a real-valued function on $\mathbb{R}$. Show that if f is Riemann integrable on $\mathbb{R}$ then so is $|f|$.
     (ii) If $|f|$ is Riemann integrable on $\mathbb{R}$ is f Riemann integrable on $\mathbb{R}$? Why?

2.   Find an example of a closed subset of $\mathbb{R}$ that has measure 0 but that does not have content 0.

---

[3] The Banach-Tarski paradox is the result that the ordinary solid ball of radius 1 in $\mathbb{R}^3$ can be decomposed into a union of finitely many disjoint sets such that when these sets are suitably rotated and translated in $\mathbb{R}^3$ and put back together again the union is the solid ball of radius 2; this rather surprising result was demonstrated by Banach and Tarski in a joint paper in *Fundamenta Mathematicae* 6 (1924): 244-277, and is discussed in the book *The Banach-Tarski Paradox* by Stan Wagon (Cambridge University Press, 1994).

3.   Evaluate the following integrals explicitly:

   (i) $\int_{[a,b]} e^{\sin x} \cos x$,
   (ii) $\int_{[a,b]} \frac{1}{\sin x} \cos x$ where $0 < a < b < \pi/2$,
   (iii) $\int_{[a,b]} e^x (\sin x + \cos x)$.

4.   (i) If $f(x)$ is a monotonically decreasing continuous nonnegative function on
      the set $\{ x \in \mathbb{R} \mid x \geq 1 \}$ show that for any integer $N \geq 2$

$$\sum_{n=2}^{N} f(n) \leq \int_{[1,N]} f(x)dx \leq \sum_{n=1}^{N-1} f(n).$$

   (ii) Use this to derive necessary and sufficient conditions on r ensuring that
       the sum $\sum_{n=1}^{\infty} 1/n^r$ converges.

5.   Either prove or find a counterexample to the assertion that if $f(\mathbf{x})$ and $g(\mathbf{x})$ are
      real-valued Riemann integrable functions in the interior of the unit ball
      $U = \left\{ \mathbf{x} \in \mathbb{R}^n \mid \|\mathbf{x}\|_2 < 1 \right\}$ in $\mathbb{R}^n$ and if $f(\mathbf{x}) = g(\mathbf{x})$ whenever $\mathbf{x} \in U$ and $\|\mathbf{x}\|_2$ is
      rational then $\int_U f = \int_U g$.

Problems, Group II

6.   If f and g are continuous real-valued functions in $[a, b] \subset \mathbb{R}$ and $g(x) > 0$ for all
      $x \in [a, b]$ show that there is a point $c \in (a, b)$ such that

$$\int_{[a,b]} fg = f(c) \int_{[a,b]} g.$$

7.   Show that for any integer $n \geq 0$

$$\int_{[0,x]} t^n e^{-t} = n! \left(1 - e^{-x} \sum_{k=0}^{n} \frac{x^k}{k!} \right).$$

8.   Show that a monotonically increasing real-valued function on an interval
      $[0, 1] \subset \mathbb{R}$ is Riemann integrable.

9.   Show that if f is a nonnegative integrable function in a cell $\overline{\Delta} \subset \mathbb{R}^n$ such that
      $\int_{\overline{\Delta}} f = 0$ then $\{ \mathbf{x} \in \overline{\Delta} \mid f(\mathbf{x}) \neq 0 \}$ is a set of measure 0.

10.   (i) Show that the Cantor set $C \subset [0, 1]$ is a set of measure 0. Is it a set of
      content 0? Why?

(ii) Show that the characteristic function χ_C of the Cantor set $C \subset [0,1]$ is Riemann integrable and find $\int_{[0,1]} \chi_C$.

11. Show that if $E \subset \mathbb{R}^n$ has content 0 then its boundary ∂E also has content 0, but that if $E \subset \mathbb{R}^n$ has measure 0 its boundary ∂E does not necessarily have measure 0.

12. If $E \subset \Delta \subset \mathbb{R}^n$ is a Jordan measurable set then for any $\epsilon > 0$ there is a compact Jordan measurable set $F \subset E$ for which $|E \sim F| < \epsilon$.

13. (i) If $A, B \subset \mathbb{R}$ are sets of content zero is their sum

$$A + B = \left\{ a + b \,\middle|\, a \in A, \ b \in B \right\} \subset \mathbb{R}$$

also a set of content zero? Why?
(ii) If $A, B \subset \mathbb{R}$ are sets of measure zero is their sum $A + B$ also a set of measure zero? Why?
[Suggestion: Consider the Cantor set.]

6.2 Calculation of Integrals

The explicit calculation of integrals in several variables can be reduced to a succession of calculations of integrals in a single variable. A cell $\overline{\Delta} \in \mathbb{R}^n = \mathbb{R}^{n'} \times \mathbb{R}^{n''}$ where $n = n' + n''$ can be written as a product $\overline{\Delta} = \overline{\Delta}' \times \overline{\Delta}''$ of cells $\overline{\Delta}' \in \mathbb{R}^{n'}$ and $\overline{\Delta}'' \in \mathbb{R}^{n''}$, so a function f in the cell $\overline{\Delta} \in \mathbb{R}^n$ can be written as a function $f(\mathbf{x}', \mathbf{x}'')$ of variables $\mathbf{x}' \in \overline{\Delta}' \in \mathbb{R}^{n'}$ and $\mathbf{x}'' \in \overline{\Delta}'' \in \mathbb{R}^{n''}$. For any fixed point $\mathbf{x}'' \in \overline{\Delta}''$ the function $f(\mathbf{x}', \mathbf{x}'')$ can be considered as a function of the variable $\mathbf{x}' \in \mathbb{R}^{n'}$ alone; so if f is bounded it is possible to consider the upper integral of this function of the variable $\mathbf{x}'$, denoted by $\int_{\mathbf{x}' \in \overline{\Delta}'}^{*} f(\mathbf{x}', \mathbf{x}'')$; this upper integral then is a well-defined bounded function of the variable $\mathbf{x}'' \in \overline{\Delta}''$, so it is possible to consider the upper integral of this function of the variable $\mathbf{x}''$, denoted by $\int_{\mathbf{x}'' \in \overline{\Delta}''}^{*} \left(\int_{\mathbf{x}' \in \overline{\Delta}'}^{*} f(\mathbf{x}', \mathbf{x}'') \right)$ and called an **iterated integral**. Of course it is possible to use the lower rather than upper integral for one or another or both of these integrals, leading to iterated integrals such as $\int_{\mathbf{x}'' \in \overline{\Delta}''}^{*} \left(\int_{* \, \mathbf{x}' \in \overline{\Delta}'} f(\mathbf{x}', \mathbf{x}'') \right)$; and it is possible to proceed in the other order, integrating first with respect to the variable $\mathbf{x}''$ and then the variable $\mathbf{x}'$, leading to iterated integrals such as $\int_{\mathbf{x}' \in \overline{\Delta}'}^{*} \left(\int_{\mathbf{x}'' \in \overline{\Delta}''}^{*} f(\mathbf{x}', \mathbf{x}'') \right)$.

Theorem 6.14 (Fubini's Theorem). *If f is a Riemann integrable function in a product cell $\overline{\Delta} = \overline{\Delta}' \times \overline{\Delta}''$ the integrals $\int_{\mathbf{x}' \in \overline{\Delta}'}^{*} f(\mathbf{x}', \mathbf{x}'')$ and $\int_{* \, \mathbf{x}' \in \overline{\Delta}'} f(\mathbf{x}', \mathbf{x}'')$ are*

Riemann integrable functions of the variable $\mathbf{x}'' \in \overline{\Delta}''$ *and*

$$\int_{\overline{\Delta}} f = \int_{\mathbf{x}'' \in \overline{\Delta}''}^* \left(\int_{\mathbf{x}' \in \overline{\Delta}'}^* f(\mathbf{x}', \mathbf{x}'') \right) = \int_{\mathbf{x}'' \in \overline{\Delta}''} \left(\int_{* \, \mathbf{x}' \in \overline{\Delta}'} f(\mathbf{x}', \mathbf{x}'') \right). \quad (6.35)$$

Proof: If $\mathcal{P}'$ is a partition $\overline{\Delta}' = \bigcup_i \overline{\Delta}'_i$ of the cell $\overline{\Delta}'$ and $\mathcal{P}''$ is a partition $\overline{\Delta}'' = \bigcup_i \overline{\Delta}''_i$ of the cell $\overline{\Delta}''$ these two separate partitions determine a partition $\mathcal{P} = \mathcal{P}' \times \mathcal{P}''$ of the cell $\overline{\Delta}$ and

$$S^*(f, \mathcal{P}) = \sum_{ij} \sup_{\substack{\mathbf{x}' \in \overline{\Delta}'_i \\ \mathbf{x}'' \in \overline{\Delta}''_j}} f(\mathbf{x}', \mathbf{x}'') \left| \overline{\Delta}'_i \times \overline{\Delta}''_j \right|$$

$$= \sum_j \left(\sum_i \sup_{\substack{\mathbf{x}' \in \overline{\Delta}'_i \\ \mathbf{x}'' \in \overline{\Delta}''_j}} f(\mathbf{x}', \mathbf{x}'') \left| \overline{\Delta}'_i \right| \right) \left| \overline{\Delta}''_j \right|.$$

For any fixed point $\mathbf{x}''_0 \in \overline{\Delta}''_j$

$$\sum_i \sup_{\substack{\mathbf{x}' \in \overline{\Delta}'_i \\ \mathbf{x}'' \in \overline{\Delta}''_j}} f(\mathbf{x}', \mathbf{x}'') \left| \overline{\Delta}'_i \right| \geq \sum_i \sup_{\mathbf{x}' \in \overline{\Delta}'_i} f(\mathbf{x}', \mathbf{x}''_0) \left| \overline{\Delta}'_i \right| = S^* \left(f(\mathbf{x}', \mathbf{x}''_0), \mathcal{P}' \right) \geq F(\mathbf{x}''_0)$$

where $F(\mathbf{x}''_0) = \int_{\mathbf{x}' \in \overline{\Delta}'}^* f(\mathbf{x}', \mathbf{x}''_0)$. That is the case for any fixed point $\mathbf{x}''_0 \in \overline{\Delta}''_j$ so the same inequalities hold for the supremum over all such points; hence

$$S^*(f, \mathcal{P}) \geq \sum_j \sup_{\mathbf{x}''_0 \in \overline{\Delta}''_j} F(\mathbf{x}''_0) \left| \overline{\Delta}''_j \right| = S^*(F, \mathcal{P}'') \geq \int_{\mathbf{x}'' \in \overline{\Delta}''}^* F(\mathbf{x}'')$$

and consequently for the upper integral

$$\int_{\overline{\Delta}}^* f \geq \int_{\mathbf{x}'' \in \overline{\Delta}''}^* \left(\int_{\mathbf{x}' \in \overline{\Delta}'}^* f(\mathbf{x}', \mathbf{x}'') \right).$$

The corresponding argument for the lower integral leads to the reversed inequality, and combining these two inequalities with the inequalities between the upper and lower Riemann integrals yields the chain of inequalities

$$\int_{\overline{\Delta}}^* f \geq \int_{\mathbf{x}'' \in \overline{\Delta}''}^* \left(\int_{\mathbf{x}' \in \overline{\Delta}'}^* f(\mathbf{x}', \mathbf{x}'') \right) \geq \int_{* \, \mathbf{x}'' \in \overline{\Delta}''} \left(\int_{\mathbf{x}' \in \overline{\Delta}'}^* f(\mathbf{x}', \mathbf{x}'') \right)$$

$$\geq \int_{* \, \mathbf{x}'' \in \overline{\Delta}''} \left(\int_{* \, \mathbf{x}' \in \overline{\Delta}'} f(\mathbf{x}', \mathbf{x}'') \right) \geq \int_{* \, \overline{\Delta}} f. \quad (6.36)$$

Since the function f is Riemann integrable $\int_{\overline{\Delta}}^* f = \int_{* \, \overline{\Delta}} f$ and consequently all of the inequalities in (6.36) must be equalities. In particular from the first line of

(6.36) it follows that

$$\int_{\mathbf{x}''\in\overline{\Delta}''}^* \left(\int_{\mathbf{x}'\in\overline{\Delta}'}^* f(\mathbf{x}',\mathbf{x}'') \right) = \int_* \, {}_{\mathbf{x}''\in\overline{\Delta}''} \left(\int_{\mathbf{x}'\in\overline{\Delta}'}^* f(\mathbf{x}',\mathbf{x}'') \right),$$

so the upper and lower integrals of the function $F(\mathbf{x}'') = \int_{\mathbf{x}'\in\overline{\Delta}'}^* f(\mathbf{x}',\mathbf{x}'')$ are equal; therefore the function $F(\mathbf{x}'')$ is a Riemann integrable function in the cell $\overline{\Delta}''$ so its upper Riemann integral is just its Riemann integral. Then from (6.36) it follows further that $\int_{\overline{\Delta}} f = \int_{\overline{\Delta}''} F$, which is the first equality in (6.35). The second inequality in (6.35) follows from (6.36) by replacing the middle term $\int_* \, {}_{\mathbf{x}''\in\overline{\Delta}''}(\int_{\mathbf{x}'\in\overline{\Delta}'}^* f(\mathbf{x}',\mathbf{x}''))$ by $\int_* \, {}_{\mathbf{x}''\in\overline{\Delta}''}(\int_{*\mathbf{x}'\in\overline{\Delta}'} f(\mathbf{x}',\mathbf{x}''))$, thereby concluding the proof.

Corollary 6.15. *If f is a continuous function in a product cell $\overline{\Delta} = \overline{\Delta}' \times \overline{\Delta}''$ then*

$$\int_{\overline{\Delta}} f(\mathbf{x}) = \int_{\mathbf{x}''\in\overline{\Delta}''} \left(\int_{\mathbf{x}'\in\overline{\Delta}'} f(\mathbf{x}',\mathbf{x}'') \right) = \int_{\mathbf{x}'\in\overline{\Delta}'} \left(\int_{\mathbf{x}''\in\overline{\Delta}''} f(\mathbf{x}',\mathbf{x}'') \right). \tag{6.37}$$

Proof: If f is continuous in $\overline{\Delta}$ then for each fixed point $\mathbf{x}'' \in \overline{\Delta}''$ the function $f(\mathbf{x}',\mathbf{x}'')$ is a continuous function of $\mathbf{x}' \in \overline{\Delta}'$ so the upper integral or lower integral in (6.35) can be replaced by the ordinary integral; and the same holds for iterated integrals in the reverse order, so that suffices for the proof.

It is customary also to set

$$\int_{\mathbf{x}''\in\overline{\Delta}''} \left(\int_{\mathbf{x}'\in\overline{\Delta}'} f(\mathbf{x}',\mathbf{x}'') \right) = \int_{\overline{\Delta}''} \int_{\overline{\Delta}'} f(\mathbf{x}',\mathbf{x}'') d\mathbf{x}' d\mathbf{x}'',$$

another commonly used notation for an iterated integral, where the inner integral, that with respect to the inner variable $\mathbf{x}'$, is calculated first, and the result is then integrated over the outer integral, that with respect to the outer variable $\mathbf{x}''$. Integrals over Jordan measurable sets $D \subset \overline{\Delta}$ are calculated similarly, by reducing them to integrals over cells; but some care often is required to keep the regions of integration straight. The process can be repeated, with more iterations; so an integral over a Jordan measurable set in $\mathbb{R}^n$ can be reduced to the calculation of n separate integrals of functions of a single variable.

For examples of the application of Fubini's Theorem, first let $D \subset \mathbb{R}^2$ be the region bounded by the lines $x_1 + x_2 = 0$, $x_1 - 2x_2 = 0$, and $x_2 = 1$ as in the accompanying Figure 6.2. An application of Fubini's Theorem shows that for any continuous function $f(\mathbf{x})$ in D

$$\int_D f(\mathbf{x}) = \int_{[0,1]} \int_{[-x_2,2x_2]} f(x_1,x_2) dx_1 dx_2,$$

thus reducing the integral to an iteration of integrals of functions of a single variable. The region D can be split into region D_1 consisting of those points

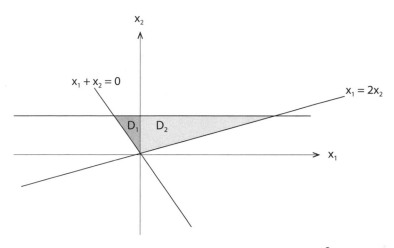

Figure 6.2. An example of Fubini's Theorem in $\mathbb{R}^2$

$\mathbf{x} \in D$ for which $-1 \le x_1 \le 0$ and the region D_2 consisting of those points $\mathbf{x} \in D$ for which $0 \le x_1 \le 2$, and then an application of Fubini's Theorem shows that for any continuous function $f(\mathbf{x})$ in D

$$\int_D f(\mathbf{x}) = \int_{[-1,0]} \int_{[-x_1,1]} f(x_1, x_2) dx_2 dx_1 + \int_{[0,2]} \int_{[x_1/2,\, 1]} f(x_1, x_2) dx_2 dx_1.$$

The order in which Fubini's Theorem is applied can make a difference in the calculation. The iterated integral

$$\int_{[0,1]} \int_{[\sqrt{x_1},\, 1]} \cos(x_2^3 + 1) dx_2 dx_1$$

involves a rather complicated trigonometric integral with respect to the variable x_2, but the equal iterated integral involves only simple integrations, since with the change of variables $t = x_2^3$

$$\int_{[0,1]} \int_{[\sqrt{x_1},\, 1]} \cos(x_2^3 + 1) dx_2 dx_1 = \int_{[0,1]} \int_{[0,\, x_2^2]} \cos(x_2^3 + 1) dx_1 dx_2$$

$$= \int_{[0,1]} x_2^2 \cos(x_2^3 + 1) dx_2 = \int_{[0,1]} \cos(t + 1) \frac{1}{3} dt = \frac{1}{3} \Big(\sin 2 - \sin 1 \Big).$$

There are some cases in which a Riemann integrable function $f(\mathbf{x}', \mathbf{x}'')$ in a product $\overline{\Delta} = \overline{\Delta}' \times \overline{\Delta}''$ is not a Riemann integrable function of one variable when the other variable is held fixed. For instance the function $f(x_1, x_2)$ of variables

$x_1, x_2 \in \mathbb{R}^1$ defined in a cell $\overline{\Delta} = \overline{\Delta'} \times \overline{\Delta''} \subset \mathbb{R}^2$ by

$$
f(x', x'') = \begin{cases} 0 & \text{if either } x' \text{ or } x'' \text{ is irrational,} \\ \frac{1}{q} & \text{if both } x' \text{ and } x'' \text{ are rational and } x'' = \frac{p}{q}, \end{cases}
$$

for coprime integers p and q actually is independent of the variable x', and as a function of the variable x'' it is continuous whenever x'' is irrational but discontinuous whenever x'' is rational. As a function of two variables its points of discontinuity are the products $\overline{\Delta'} \times \frac{p}{q}$ for all rational points $\frac{p}{q}$, a countable union of intervals; and since intervals are sets of measure zero in $\mathbb{R}^2$ the points of discontinuity form a set of measure zero, so the function f is Riemann integrable. Of course it is also Riemann integrable as a function of the variable x'' for each fixed point $x' \in \overline{\Delta'}$; but for any fixed point x'' the function takes the values 0 or $\frac{1}{q}$ on dense subsets of the interval $\overline{\Delta'}$ so the function is not Riemann integrable as a function of the variable x'. Fortunately in this case the integration can be taken first with respect to the variable x'', and the simpler version of Fubini's Theorem holds.

In Fubini's Theorem it is necessary to assume that a function f on the product cell $\overline{\Delta} = \overline{\Delta'} \times \overline{\Delta''}$ is Riemann integrable; there are functions that are Riemann integrable in the variable $\mathbf{x}' \in \overline{\Delta'}$ for each fixed point $\mathbf{x}'' \in \overline{\Delta''}$, and conversely are Riemann integrable in the variable $\mathbf{x}'' \in \overline{\Delta''}$ for each fixed point $\mathbf{x}' \in \overline{\Delta'}$, but are not Riemann integrable in the product cell $\overline{\Delta}$. For example let $f(x', x'')$ be a function on the cell $\overline{\Delta'} \times \overline{\Delta''} \in \mathbb{R}^2$ that takes the value 1 at the center of the square, that takes the value 1 at each of four points in the subsquares when the square is split in half on each line, but where these points are chosen so that no two lie on the same horizontal or vertical line, and so on; and takes the value 0 otherwise. Thus this function takes each of the values 1 and 0 on dense subsets of the square, so it is not Riemann integrable; but on each horizontal or vertical line the function takes the value 1 at a single point and is 0 otherwise, so is Riemann integrable.

As discussed in Section 3.1, pointwise limits of continuous functions are not necessarily continuous, but uniform limits of continuous functions are continuous. The pointwise limits (3.17) of Riemann integrable functions hence are not necessarily Riemann integrable, since such limits are not even necessarily continuous; moreover even if a sequence of Riemann integrable functions converges pointwise to a Riemann integrable function, the integral of the limit is not necessarily equal to the limit of the integral. For instance consider the sequence of continuous functions $f_\nu(x)$ on the unit interval $[0, 1]$ where $f_\nu(0) = 0, f_\nu(1/2\nu) = 2\nu, f_\nu(1/\nu) = 0, f_\nu(1) = 0$, and the function f_ν is extended to be linear in the intervals between these points, as in the accompanying

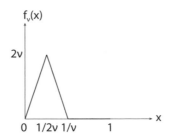

Figure 6.3. Example of a limit of integrals not equal to the integral of the limit

Figure 6.3. It is easy to see that $\lim_{\nu \to \infty} f_\nu(x) = 0$ for all points $x \in [0,1]$ but that $\int_0^1 f_\nu = 1$.

Theorem 6.16 (Convergence of Riemann Integrals). *If f_ν is a uniformly convergent sequence of Riemann integrable functions in a cell $\overline{\Delta} \subset \mathbb{R}^n$ then its limit f is a Riemann integrable function in $\overline{\Delta}$ and*

$$\lim_{\nu \to \infty} \int_{\overline{\Delta}} f_\nu = \int_{\overline{\Delta}} f.$$

Proof: Since the functions f_ν converge uniformly to f there is an integer N such that $|f(\mathbf{x}) - f_N(\mathbf{x})| < 1$ for all $\mathbf{x} \in \overline{\Delta}$; the function f_N is Riemann integrable by hypothesis so it is bounded by some constant M in $\overline{\Delta}$, and then

$$|f(\mathbf{x})| \le |f(\mathbf{x}) - f_N(\mathbf{x})| + |f_N(\mathbf{x})| \le 1 + M$$

for all $\mathbf{x} \in \overline{\Delta}$ so the limit function is bounded in $\overline{\Delta}$. The set $E_\nu \subset \overline{\Delta}$ of points at which the function f_ν fails to be continuous is a set of measure zero, since the functions f_ν are Riemann integrable by hypothesis; so the union $E = \bigcup_{\nu=1}^\infty E_\nu$ is also a set of measure zero, and all the functions f_ν are continuous in $\overline{\Delta} \sim E$. The functions f_ν converge uniformly to f in $\overline{\Delta} \sim E$ so it follows from Theorem 3.20 that the limit function f is also continuous in $\overline{\Delta} \sim E$, hence that f is Riemann integrable. For any $\epsilon > 0$ there is an N such that $|f(\mathbf{x}) - f_\nu(\mathbf{x})| < \epsilon$ for all $\mathbf{x} \in \overline{\Delta}$ whenever $\nu > N$; since the functions f and f_ν are integrable it then follows from Theorem 6.9 that for any $\nu > N$

$$\left| \int_{\overline{\Delta}} f - \int_{\overline{\Delta}} f_\nu \right| = \left| \int_{\overline{\Delta}} (f - f_\nu) \right| \le \int_{\overline{\Delta}} |f - f_\nu| \le \int_{\overline{\Delta}} \epsilon = \epsilon |\overline{\Delta}|$$

and consequently that $\lim_{\nu \to \infty} \int_{\overline{\Delta}} f_\nu = \int_{\overline{\Delta}} f$, which concludes the proof.

Another limiting process is the integral transformation defined by a kernel function $k(\mathbf{x}, \mathbf{y})$ of the variables $\mathbf{x} \in D \subset \mathbb{R}^m$ and $\mathbf{y} \in E \subset \mathbb{R}^n$ for compact Jordan measurable sets D and E. A Riemann integrable function $f(\mathbf{y})$ of the variable

$\mathbf{y} \in E$ can be transformed to the function

$$F(\mathbf{x}) = \int_{\mathbf{y} \in E} k(\mathbf{x}, \mathbf{y}) f(\mathbf{y}) \tag{6.38}$$

of the variable $\mathbf{x} \in D$, provided that the kernel $k(\mathbf{x}, \mathbf{y})$ is a Riemann integrable function of the variable $\mathbf{y} \in E$ for each point $\mathbf{x} \in D$. Of course the properties of the transform $F(\mathbf{x})$ depend on the properties of the kernel $k(\mathbf{x}, \mathbf{y})$ as a function of the variable $\mathbf{x} \in D$; a sufficiently regular kernel $k(\mathbf{x}, \mathbf{y})$ can transform a merely Riemann integrable function $f(\mathbf{y})$ of the variable $\mathbf{y} \in E$ to a continuous or differentiable function $F(\mathbf{x})$ of the variable $\mathbf{x} \in D$.

Theorem 6.17. (i) *If $D \subset \mathbb{R}^m$ and $E \subset \mathbb{R}^n$ are compact Jordan measurable sets, $k(\mathbf{x}, \mathbf{y})$ is a continuous function of the variables $\mathbf{x} \in D$, $\mathbf{y} \in E$, and $f(\mathbf{y})$ is a Riemann integrable function of the variable $\mathbf{y} \in E$, then the transform $F(\mathbf{x}) = \int_{\mathbf{y} \in E} k(\mathbf{x}, \mathbf{y}) f(\mathbf{y})$ is a continuous function of the variable $\mathbf{x} \in D$.*
(ii) If in addition there is an open neighborhood U of $D \times E$ such that the function $k(\mathbf{x}, \mathbf{y})$ is continuous in the variables x, y in U and is differentiable in the variable $\mathbf{x}$ in U with derivatives that are continuous in both variables in U then the transform $F(\mathbf{x}) = \int_{\mathbf{y} \in E} k(\mathbf{x}, \mathbf{y}) f(\mathbf{y})$ is a continuously differentiable function of the variable $\mathbf{x} \in D$ and

$$\nabla F(\mathbf{x}) = \int_{\mathbf{y} \in E} \nabla_{\mathbf{x}} k(\mathbf{x}, \mathbf{y}) f(\mathbf{y}). \tag{6.39}$$

Proof: (i) The function $f(\mathbf{y})$ is bounded, so there is a constant M such that $|f(\mathbf{y})| \leq M$ for all $\mathbf{y} \in E$. The continuous function $k(\mathbf{x}, \mathbf{y})$ in the compact set $D \times E$ is uniformly continuous, so for any $\epsilon > 0$ there is a $\delta > 0$ such that $|k(\mathbf{x}_1, \mathbf{y}_1) - k(\mathbf{x}_2, \mathbf{y}_2)| < \epsilon$ whenever $\|(\mathbf{x}_1, \mathbf{y}_1) - (\mathbf{x}_2, \mathbf{y}_2)\|_2 < \delta$. Therefore if $\mathbf{x} \in D$ and $\mathbf{h} \in \mathbb{R}^m$ satisfies $\|\mathbf{h}\|_2 < \delta$ then

$$|F(\mathbf{x} + \mathbf{h}) - F(\mathbf{x})| = \left| \int_{\mathbf{y} \in E} \left(k(\mathbf{x} + \mathbf{h}, \mathbf{y}) - k(\mathbf{x}, \mathbf{y}) \right) f(\mathbf{y}) \right|$$

$$\leq \int_{\mathbf{y} \in E} |k(\mathbf{x} + \mathbf{h}, \mathbf{y}) - k(\mathbf{x}, \mathbf{y})| \, |f(\mathbf{y})|$$

$$\leq \int_{\mathbf{y} \in E} \epsilon M = \epsilon M |E|,$$

so $F(\mathbf{x})$ is a continuous function of the variable $\mathbf{x} \in D$.
(ii) By hypothesis the kernel $k(\mathbf{x}, \mathbf{y})$ is continuous in the open neighborhood U of the set $D \times E$ and is a differentiable function of the variable $\mathbf{x}$ in U; and since D and E are compact, then by shrinking the neighborhood U as necessary it can be assumed that $\overline{U}$ is compact and the function $k(\mathbf{x}, \mathbf{y})$ is continuous in $\overline{U}$.

Then by the Mean Value Theorem, Theorem 3.36, for any point $\mathbf{x} \in D$ and for any $\mathbf{h} \in \mathbb{R}^m$ sufficiently small

$$k(\mathbf{x} + \mathbf{h}, \mathbf{y}) - k(\mathbf{x}, \mathbf{y}) = \nabla_\mathbf{x} k(\mathbf{z}, \mathbf{y}) \cdot \mathbf{h}$$

for some point $\mathbf{z} \in U$ between $\mathbf{x}$ and $\mathbf{x} + \mathbf{h}$ on the line joining these two points. Therefore

$$F(\mathbf{x} + \mathbf{h}) - F(\mathbf{x}) - \int_{\mathbf{y} \in E} \nabla_\mathbf{x} k(\mathbf{x}, \mathbf{y}) \cdot \mathbf{h} f(\mathbf{y})$$

$$= \int_{\mathbf{y} \in E} \big(k(\mathbf{x} + \mathbf{h}, \mathbf{y}) - k(\mathbf{x}, \mathbf{y}) - \nabla_\mathbf{x} k(\mathbf{x}, \mathbf{y}) \cdot \mathbf{h} \big) f(\mathbf{y})$$

$$= \int_{\mathbf{y} \in E} \big(\nabla_\mathbf{x} k(\mathbf{z}, \mathbf{y}) - \nabla_\mathbf{x} k(\mathbf{x}, \mathbf{y}) \big) \cdot \mathbf{h} f(\mathbf{y}).$$

Since the function $\nabla_\mathbf{x} k(\mathbf{z}, \mathbf{y})$ is assumed to be continuous on the compact set $\overline{U}$ it is uniformly continuous, and consequently for any $\epsilon > 0$ there is a $\delta > 0$ sufficiently small that whenever $\|\mathbf{h}\|_2 < \delta$

$$\|\nabla_\mathbf{x} k(\mathbf{z}, \mathbf{y}) - \nabla_\mathbf{x} k(\mathbf{x}, \mathbf{y})\|_2 < \epsilon;$$

hence by the Cauchy-Schwarz inequality

$$\left| \big(\nabla_\mathbf{x} k(\mathbf{z}, \mathbf{y}) - \nabla_\mathbf{x} k(\mathbf{x}, \mathbf{y}) \big) \cdot \mathbf{h} \right| < \epsilon \|\mathbf{h}\|_2.$$

Thus if $|f(\mathbf{y})| \leq M$ for all $\mathbf{y} \in E$ then

$$\left| F(\mathbf{x} + \mathbf{h}) - F(\mathbf{x}) - \int_{\mathbf{y} \in E} \nabla_\mathbf{x} k(\mathbf{x}, \mathbf{y}) \cdot \mathbf{h} f(\mathbf{y}) \right| \leq \int_{\mathbf{y} \in E} \epsilon M \|\mathbf{h}\|_2 \leq \epsilon M |E| \|\mathbf{h}\|_2,$$

which shows that $F(\mathbf{x})$ is differentiable and that its derivative is given by (6.39), which suffices for the proof.

The integral has been defined so far just for Riemann integrable functions in a Jordan measurable set in $\mathbb{R}^n$. It is possible to extend the Riemann integral in order to integrate a wider class of functions over a wider class of subsets of $\mathbb{R}^n$. The extension however does not necessarily satisfy all of the properties of integrals that have been considered in the preceding discussion; that is one of the reasons that the extension is called an **improper Riemann integral**. Some care must be taken when working with these integrals, for as with the Cheshire cat some perplexing results can appear unexpectedly, even though nothing but the grin remains of the improper integral. However there are many advantages in being able to use integrals of more general functions, such as unbounded continuous functions, over more general sets, such as unbounded sets and general open subsets of $\mathbb{R}^n$; for not every open subset of $\mathbb{R}^n$ is a Jordan measurable set. Indeed the rational neighborhood S, the open subset $S \subset \mathbb{R}^1$ containing all the rationals but not all of $\mathbb{R}^1$, is not a Jordan measurable set. To demonstrate that, if r_j is a list of the rational numbers in the open unit interval $(0, 1)$ for $j \geq 1$ and if Δ_j is an open interval of length at most $|\Delta_j| = \frac{\epsilon}{2^j}$ contained in $(0, 1)$ and containing the point r_j, where $\epsilon < \frac{1}{2}$, let $S = \bigcup \Delta_j$. Clearly $\overline{S} = [0, 1]$, and it follows that $\partial S = [0, 1] \sim S$. If S were a Jordan measurable set its boundary would be a compact set of content 0, which could be covered by finitely many open intervals of total length less than ϵ; thus there would be finitely many open intervals $I_j \subset \mathbb{R}^1$ such that $([0, 1] \sim S) \subset \bigcup_{j=1}^{N} I_j$ and $\sum_{j=1}^{N} |I_j| < \epsilon$, and consequently

$$[0, 1] \subset \left(\bigcup_{j=1}^{N} I_j \cup \bigcup_{i=1}^{\infty} \Delta_i \right).$$

However the compact set $[0, 1]$ would be covered by finitely many of these open intervals, so that actually

$$[0, 1] \subset \left(\bigcup_{j=1}^{N} I_j \cup \bigcup_{i=1}^{M} \Delta_i \right),$$

which is impossible since $\sum_{j=1}^{N} |I_j| + \sum_{i=1}^{M} |\Delta_i| < 2\epsilon < 1$; consequently S cannot be a Jordan measurable set. However arbitrary open sets can be expressed as natural limits of Jordan measurable sets.

Theorem 6.18. *Any open subset $U \subset \mathbb{R}^n$ can be written as a union $U = \bigcup_{i=1}^{\infty} D_i$ where D_i are open Jordan measurable sets such that $\overline{D_i} \subset D_{i+1}$ for all indices i. Moreover these Jordan measurable sets can be assumed to be unions of cells in a partition of $\mathbb{R}^n$.*

Proof: For any natural number r let $\mathcal{P}_r$ be a partition of the entire space $\mathbb{R}^n$ into cells having as their sides intervals of length 2^{-r}, where $\mathcal{P}_{r+1}$ is a refinement of the partition $\mathcal{P}_r$; and let Δ_r be the open cell consisting of those points $\mathbf{x} \in \mathbb{R}^n$ such that $\|\mathbf{x}\|_\infty < r$. For a sufficiently large index r_1 the partition $\mathcal{P}_{r_1}$ will be sufficiently fine that some closed cells in the partition will be contained in the open set $U \cap \Delta_{r_1}$; the interior of the union of those cells then is an open Jordan measurable set D_1 such that $\overline{D_1} \subset U \cap \Delta_{r_1}$. For a sufficiently large index $r_2 > r_1$ each point on the compact boundary ∂D_1 will be contained in the interior of the union of all closed cells $\overline{\Delta_i}$ of the partition $\mathcal{P}_{r_2}$ for which $\overline{\Delta_i} \subset U \cap \Delta_{r_2}$; the interior of the union of those closed cells in the partition $\mathcal{P}_{r_2}$ that are contained in $U \cap \Delta_{r_2}$ then is an open Jordan measurable set D_2 such that $\overline{D_1} \subset D_2$ and $\overline{D_2} \subset U \cap \Delta_{r_2}$. The process can be continued, providing a sequence of open Jordan measurable sets D_i contained in U such that $\overline{D}_i \subset D_{i+1}$ for each index i. Since the cells of partition $\mathcal{P}_{r_i}$ have sides of length 2^{-r_i} and $\bigcup_r \Delta_r = \mathbb{R}^n$ it follows that for any point $\mathbf{x} \in U$ there will be some index r_i sufficiently large that for some closed cell $\overline{\Delta}$ of the partition $\mathcal{P}_{r_i}$ it is the case that $\mathbf{x} \in \overline{\Delta} \subset U \cap \Delta_{r_i}$. Then either $\mathbf{x} \in D_i$ or $\mathbf{x} \in \partial D_i$ hence $\mathbf{x} \in D_{i+1}$, and that suffices for the proof.

Consider then an open subset $U \subset \mathbb{R}^n$ that is a union $U = \bigcup_{i=1}^{\infty} D_i$ of open Jordan measurable sets D_i such that $\overline{D}_i \subset D_{i+1}$ for all indices i. If f is a function that is continuous almost everywhere in U set

$$f_+(\mathbf{x}) = \max(f(\mathbf{x}), 0) \quad \text{and} \quad f_-(\mathbf{x}) = \max(-f(\mathbf{x}), 0) \tag{6.40}$$

for any point $\mathbf{x} \in U$. Clearly $f_+(\mathbf{x}) \geq 0$ and $f_-(\mathbf{x}) \geq 0$ at all points $\mathbf{x} \in U$, and $f(\mathbf{x}) = f_+(\mathbf{x}) - f_-(\mathbf{x})$ is a unique decomposition of the function f as a difference of two nonnegative functions, in the sense that if $f = g - h$, where $g \geq 0$ and $h \geq 0$, then $f_+ \leq g$ and $f_- \leq h$; and both functions f_+ and f_- are also continuous almost everywhere in U. For any integer $i > 0$ the functions $\min(f_+(\mathbf{x}), i)$ and $\min(f_-(\mathbf{x}), i)$ are bounded and continuous almost everywhere in U, and consequently they are Riemann integrable over the Jordan measurable sets D_j. Moreover since these are positive functions and $\min(f_+(\mathbf{x}), i) \leq \min(f_+(\mathbf{x}), i+1)$ while $D_i \subset D_{i+1}$ it follows that

$$\int_{D_i} \min(f_+(\mathbf{x}), i) \leq \int_{D_{i+1}} \min(f_+(\mathbf{x}), i+1);$$

consequently the sequence of real numbers $\int_{D_i} \min(f_+(\mathbf{x}), i)$ for $i = 1, 2 \ldots$ has a well-defined limit as $i \to \infty$, either a real number or $+\infty$, so it is possible to define

$$\int_{U, \{D_i\}} f_+ = \lim_{i \to \infty} \int_{D_i} \min(f_+(\mathbf{x}), i) \tag{6.41}$$

and correspondingly of course

$$\int_{U, \{D_i\}} f_- = \lim_{i \to \infty} \int_{D_i} \min(f_-(\mathbf{x}), i). \tag{6.42}$$

These expressions actually are independent of the choice of the sequence of Jordan measurable sets D_i. Indeed if U also is the union $U = \bigcup_{j=1}^{\infty} E_j$ for some Jordan measurable sets $E_j \subset \mathbb{R}^n$ such that $\overline{E}_j \subset E_{j+1}$ then the same construction leads to the integrals $\int_{U, \{E_j\}} f_+$ and $\int_{U, \{E_j\}} f_-$. Since each set $\overline{D}_i$ is compact and $\overline{D}_i \subset \bigcup_{j=1}^{\infty} E_j$ for the open sets E_j then D_i is actually contained in a finite union of the sets E_j, and consequently in the largest of the sets E_j in this finite union, say the set E_{j_i}; and since D_i is contained in E_j whenever $j > j_i$ it is possible to assume that $j_i \geq i$. Therefore

$$\int_{D_i} \min(f_+, i) \leq \int_{E_{j_i}} \min(f_+, j_i) \leq \int_{U, \{E_j\}} f_+,$$

and since that is the case for any i it follows that

$$\int_{U, \{D_i\}} f_+ \leq \int_{U, \{E_j\}} f_+;$$

and of course the same holds for the integrals of the function f_-. On the other hand the argument can be reversed, so that the opposite inequality $\int_{U, \{E_j\}} f_+ \leq \int_{U, \{D_i\}} f_+$ also holds, and consequently

$$\int_{U, \{E_j\}} f_+ = \int_{U, \{D_i\}} f_+$$

and correspondingly $\int_{U, \{E_j\}} f_- = \int_{U, \{D_i\}} f_-$. If at least one of these two integrals $\int_{U, \{D_i\}} f_+$ or $\int_{U, \{D_i\}} f_-$ is finite the **improper integral** of the function f over the open set U is defined to be the difference

$$\widetilde{\int}_U f = \int_{U, \{D_i\}} f_+ - \int_{U, \{D_i\}} f_-, \tag{6.43}$$

which is either a real number or $+\infty$ or $-\infty$.

A possible difficulty with improper Riemann integrals is that some of the integrals may have infinite values. For instance if f is an unbounded function that is continuous almost everywhere in an open subset U but $\widetilde{\int}_U f = \infty$ then $\widetilde{\int}_U (f + g) = 0$ where $g = -f$ while $\widetilde{\int}_U g = -\infty$ so the equation $\widetilde{\int}_U (f + g) = \widetilde{\int}_U f + \widetilde{\int}_U g$ does not really make any sense. On the other hand if f and g are functions that are continuous almost everywhere in an open subset and have finite improper integrals then since $(f + g)_+(\mathbf{x}) \leq f_+(\mathbf{x}) + g_+(\mathbf{x})$ and $(f + g)_-(\mathbf{x}) \leq f_-(\mathbf{x}) + g_-(\mathbf{x})$ at all points $\mathbf{x} \in U$ it follows fairly readily that $\widetilde{\int}_U (f + g) = \widetilde{\int}_U f + \widetilde{\int}_U g$. Thus the basic identities of Theorem 6.8 hold if the improper integrals are actually finite. It is possible to define the content of an arbitrary open subset U as the improper integral $|U| = \widetilde{\int}_U 1$, which is always well defined although possibly $|U| = +\infty$.

There are various alternative ways of extending the class of functions and domains for integration that arise in various applications. The advantage of the procedure discussed here is that it is intrinsic: it attaches to any function f that is continuous almost everywhere in an open subset U a unique value for the integral $\widetilde{\int}_U f$, unless that value would be the indeterminate form $\infty - \infty$.

There are situations in which it is possible to attach to an integral that would otherwise be of the indeterminate form $\infty - \infty$ a well-defined value by specifying a particular way of taking the limit. The classical example is that of the Cauchy principal value of the integral of a function of a single real variable that is unbounded in a neighborhood of a point, as for example

$$\lim_{\substack{\epsilon \to 0 \\ \epsilon > 0}} \left(\int_{[-1,-\epsilon]} \frac{1}{x} + \int_{[\epsilon,1]} \frac{1}{x} \right) = 0,$$

$$\lim_{\substack{\epsilon \to 0 \\ \epsilon > 0}} \left(\int_{[-1,-\epsilon]} \frac{1}{x} + \int_{[2\epsilon,1]} \frac{1}{x} \right) = -\log 2.$$

In this way it is possible to assign a definite value to the integral $\int_{-1}^{1} \frac{1}{x}$ that would otherwise be of the indeterminate form $\infty - \infty$; but the actual value depends upon the particular choice of the way in which the limit is taken.

Another important and useful technique for the actual calculation of integrals is the change of variables in an integration. For a simple special case, a C^1 diffeomorphism $\phi : [a, b] \longrightarrow [c, d]$ between two nontrivial segments of the real line has a nonzero derivative at all points of $[a, b]$, so either $\phi'(x) > 0$ for all points $x \in (a, b)$ and the mapping ϕ is orientation preserving so $\phi(a) = c$ and $\phi(b) = d$, or $\phi'(x) < 0$ for all points $x \in (a, b)$ and the mapping ϕ is orientation reversing so $\phi(a) = d$ and $\phi(b) = c$. If f is a continuous function in the interval $[c, d]$ the composition $f \circ \phi$ is a continuous function in the interval $[a, b]$. By the First Fundamental Theorem of Calculus, Theorem 6.6, there is a C^1 function F in $[c, d]$ for which $F'(x) = f(x)$ at all points $x \in (c, d)$; and by the Second Fundamental Theorem of Calculus, Theorem 6.4,

$$\int_{[c,d]} f = F(d) - F(c). \tag{6.44}$$

The composition $F \circ \phi$ then is a C^1 function in $[a, b]$ such that $(F \circ \phi)' = (F' \circ \phi) \cdot \phi' = (f \circ \phi) \cdot \phi'$; so by the Second Fundamental Theorem of Calculus again

$$\int_{[a,b]} (f \circ \phi) \cdot \phi' = (F \circ \phi)(b) - (F \circ \phi)(a). \tag{6.45}$$

If $\phi'(x) > 0$ then $\phi(b) = d$ and $\phi(a) = c$ so (6.45) takes the form

$$\int_{[a,b]} (f \circ \phi) \cdot \phi' = F(d) - F(c) = \int_{[c,d]} f, \tag{6.46}$$

while if $\phi'(x) < 0$ then $\phi(b) = c$ and $\phi(a) = d$ so (6.45) takes the form

$$\int_{[a,b]} (f \circ \phi) \cdot \phi' = F(c) - F(d) = -\int_{[c,d]} f. \tag{6.47}$$

Since $\phi'(x) > 0$ in the first of the two preceding equations, and $\phi'(x) < 0$ in the second, these two equations can be combined into the change of variables formula

$$\int_{[a,b]} (f \circ \phi) \cdot |\phi'| = \int_{[c,d]} f. \qquad (6.48)$$

For a diffeomorphism $\phi : D \longrightarrow E$ between two open subsets of $\mathbb{R}^n$ the derivative ϕ' is a matrix; and if D is connected either $\det \phi'(x) > 0$ for all points $x \in D$ and the mapping ϕ is orientation preserving or $\det \phi'(x) < 0$ for all points $x \in D$ and the mapping ϕ is orientation reversing. Under this change of variables the integral changes in the following general form of the formula (6.48) for the case $n = 1$.

Theorem 6.19 (Change of Variables Theorem). *If $\phi : D \longrightarrow E$ is a C^1 homeomorphism between two open subsets in $\mathbb{R}^n$ and if f is a function that is continuous almost everywhere in the open set E and has a finite improper integral in E then the function $(f \circ \phi) \, | \det \phi' |$ is continuous almost everywhere in D and has a finite improper integral in D, and*

$$\breve{\int}_E f = \breve{\int}_D (f \circ \phi) \, | \det \phi' |. \qquad (6.49)$$

Proof: As a preliminary observation, if $\phi : D \longrightarrow E$ is a C^1 homeomorphism between two open subsets $D, E \subset \mathbb{R}^n$ then for any cell $\Delta \subset \overline{\Delta} \subset D$ the boundary of the image of Δ is the image of the boundary of Δ, that is, $\partial \phi(\Delta) = \phi(\partial \Delta)$. Since each face of Δ is locally a submanifold of $\mathbb{R}^n$ of dimension $n - 1$ that is also the case for the image under a C^1 homeomorphism, so the boundary of the image $\phi(\Delta)$ consists of a collection of pieces that are also locally submanifolds of $\mathbb{R}^n$ of dimension $n - 1$ hence are subsets of measure zero; consequently $\partial \phi(\Delta)$ is a set of measure zero so the image $\phi(\Delta)$ is a Jordan measurable set in $\mathbb{R}^n$. A set that is the image of a cell Δ under a C^1 homeomorphism of an open neighborhood of the closure $\overline{\Delta}$ will be called a **special Jordan measurable set**, although only for the duration of this proof. It is evident from this definition that the image of a special Jordan measurable set under a C^1 homeomorphism defined in an open neighborhood of its closure is also a special Jordan measurable set.

To turn to the proof of the theorem itself, first it will be demonstrated that for any cell $\Delta \subset \overline{\Delta} \subset E$ and for any constant c

$$\int_\Delta c = \int_{\phi^{-1}(\Delta)} c \, | \det \phi' | \qquad (6.50)$$

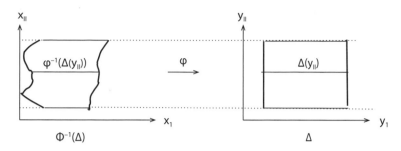

Figure 6.4. Mappings that change a single variable

if $\phi : D \longrightarrow E$ is a $\mathcal{C}^1$ homeomorphism that changes only a single variable. To simplify the notation, relabel the variables so that only the first variable is changed. When points $\mathbf{x} \in D$ and $\mathbf{y} \in E$ are written

$$\mathbf{x} = \begin{pmatrix} x_1 \\ \mathbf{x}_{II} \end{pmatrix} \quad \text{where} \quad \begin{cases} x_1 \in \mathbb{R}^1 \\ \mathbf{x}_{II} \in \mathbb{R}^{n-1} \end{cases} \quad \text{and} \quad \mathbf{y} = \begin{pmatrix} y_1 \\ \mathbf{y}_{II} \end{pmatrix} \quad \text{where} \quad \begin{cases} y_1 \in \mathbb{R}^1 \\ \mathbf{y}_{II} \in \mathbb{R}^{n-1} \end{cases}$$

the homeomorphism ϕ thus has the form

$$\phi(\mathbf{x}) = \begin{pmatrix} \phi_1(\mathbf{x}) \\ \mathbf{x}_{II} \end{pmatrix} \quad \text{where} \quad \begin{cases} \phi_1(\mathbf{x}) \in \mathbb{R}^1 \\ \mathbf{x}_{II} \in \mathbb{R}^{n-1} \end{cases} ;$$

therefore

$$\phi'(\mathbf{x}) = \begin{pmatrix} \partial_1 \phi_1(\mathbf{x}) & \partial_{II} \phi_1(\mathbf{x}) \\ 0 & I_{n-1} \end{pmatrix}$$

where I_{n-1} is the $(n-1) \times (n-1)$ identity matrix, so $\det \phi'(\mathbf{x}) = \partial_1 \phi_1(\mathbf{x})$. The cell $\Delta \subset E$ is the product of an interval Δ_1 in the y_1-axis and a cell Δ_{II} in the space $\mathbb{R}^{n-1}$ of the remaining variables $\mathbf{y}_{II}$, so it is the union of the one-dimensional slices $\Delta(\mathbf{y}_{II}) = \Delta_1 \times \mathbf{y}_{II}$ for points $\mathbf{y}_{II} \in \Delta_{II}$. The image of the cell Δ under the inverse homeomorphism $\phi^{-1} : E \longrightarrow D$ is the special Jordan measurable set $\phi^{-1}(\Delta) \subset D$, which correspondingly is the union of the one-dimensional slices

$$\phi^{-1}\big(\Delta(\mathbf{y}_{II})\big) = \Big\{ \mathbf{x} = \{x_1, \mathbf{x}_{II}\} \in \phi^{-1}(\Delta) \,\Big|\, \mathbf{x}_{II} = \mathbf{y}_{II} \Big\}$$

as sketched in the accompanying Figure 6.4. It follows from Fubini's Theorem and the change of variables formula for the restriction

$$\phi : \phi^{-1}\big(\Delta(\mathbf{y}_{II})\big) \longrightarrow \Delta(\mathbf{y}_{II})$$

of the homeomorphism ϕ to the one-dimensional slices that

$$\int_\Delta c = \int_{\mathbf{y}_{II} \in \Delta_{II}} \int_{y_1 \in \Delta(\mathbf{y}_{II})} c = \int_{\mathbf{x}_{II} \in \Delta_{II}} \int_{x_1 \in \phi^{-1}(\Delta(\mathbf{y}_{II}))} c \, |\phi_1'|$$

$$= \int_{\mathbf{x}_{II} \in \Delta_{II}} \int_{x_1 \in \phi^{-1}(\Delta(\mathbf{y}_{II}))} c \, |\det \phi'| = \int_{\phi^{-1}(\Delta)} c \, |\det \phi'|,$$

so the formula (6.50) for the effect of a change of variables in integration holds in this case.

Next it will be demonstrated that for any cell $\Delta \subset \overline{\Delta} \subset E$ and for any function f that is Riemann integrable in Δ the function $(f \circ \phi)|\det \phi'|$ is Riemann integrable in $\phi^{-1}(\Delta)$ and

$$\int_\Delta f = \int_{\phi^{-1}(\Delta)} (f \circ \phi) \, |\det \phi'| \tag{6.51}$$

if $\phi : D \longrightarrow E$ is a $\mathcal{C}^1$ homeomorphism that changes only a single variable. Of course it is clear that $(f \circ \phi)|\det \phi'|$ is bounded in $\phi^{-1}(\Delta)$; but the remainder of the preceding assertions remains to be proved. For any partition $\mathcal{P}$ of the cell $\overline{\Delta}$ as the union $\overline{\Delta} = \bigcup_i \overline{\Delta}_i$, introduce the function f^* in $\overline{\Delta}$ defined by

$$f^*(\mathbf{y}) = \begin{cases} \sup_{\mathbf{z} \in \overline{\Delta}_i} f(\mathbf{z}) & \text{if} \quad \mathbf{y} \in \Delta_i, \\ f(\mathbf{y}) & \text{if} \quad \mathbf{y} \in \bigcup_j \partial \Delta_j. \end{cases} \tag{6.52}$$

The function f^* is constant in each open cell Δ_i of the partition $\mathcal{P}$, so by what has just been proved it follows that

$$\int_{\Delta_i} f^* = \int_{\phi^{-1}(\Delta_i)} (f^* \circ \phi) \, |\det \phi'|. \tag{6.53}$$

The function f^* is continuous in Δ except for the union of the boundaries of the cells Δ_i, and that union is a set of measure 0; hence f^* is Riemann integrable over Δ, as is $(f^* \circ \phi) \, |\det \phi'|$. The special Jordan measurable sets Δ_i are disjoint, as are the special Jordan measurable sets $\phi^{-1}(\Delta_i)$, so it follows from (6.53) and Theorem 6.13 that

$$\int_\Delta f^* = \sum_i \int_{\Delta_i} f^* = \sum_i \int_{\phi^{-1}(\Delta_i)} (f^* \circ \phi) \, |\det \phi'| = \int_{\phi^{-1}(\Delta)} (f^* \circ \phi) \, |\det \phi'|. \tag{6.54}$$

It is clear from the definition (6.52) that $\int_{\Delta_i} f^* = (\sup_{\mathbf{y} \in \Delta_i} f(\mathbf{y})) \, |\Delta_i|$, hence

$$\int_\Delta f^* = \sum_i \int_{\Delta_i} f^* = \sum_i \left(\sup_{\mathbf{y} \in \Delta_i} f(\mathbf{y}) \right) |\Delta_i| = S^*(f, \mathcal{P}). \tag{6.55}$$

It is also clear from (6.52) that $(f^* \circ \phi)(\mathbf{x}) \, |\det \phi'(\mathbf{x})| \geq (f \circ \phi)(\mathbf{x}) \, |\det \phi'(\mathbf{x})|$ at all points $\mathbf{x} \in \phi^{-1}(\Delta)$, so the upper sums of these two functions for any partition of a cell containing the special Jordan measurable set $\phi^{-1}(\Delta)$ satisfy the corresponding inequality and consequently so do the upper integrals; therefore

$$\int_{\phi^{-1}(\Delta)} (f^* \circ \phi) \, |\det \phi'| = \int_{\phi^{-1}(\Delta)}^* (f^* \circ \phi) \, |\det \phi'| \geq \int_{\phi^{-1}(\Delta)}^* (f \circ \phi) \, |\det \phi'|. \quad (6.56)$$

Combining (6.54), (6.55), and (6.56) shows that

$$S^*(f, \mathcal{P}) \geq \int_{\phi^{-1}(\Delta)}^* (f \circ \phi) \, |\det \phi'|; \quad (6.57)$$

and since that is true for any partition $\mathcal{P}$ of the cell Δ

$$\int_\Delta^* f \geq \int_{\phi^{-1}(\Delta)}^* (f \circ \phi) \, |\det \phi'|. \quad (6.58)$$

The corresponding argument for the lower integrals, for which the inequalities are reversed, shows that

$$\int_{*\Delta} f \leq \int_{*\phi^{-1}(\Delta)} (f \circ \phi) \, |\det \phi'|, \quad (6.59)$$

and combining (6.58) and (6.59) yields the inequalities

$$\int_\Delta^* f \geq \int_{\phi^{-1}(\Delta)}^* (f \circ \phi) \, |\det \phi'| \geq \int_{*\phi^{-1}(\Delta)} (f \circ \phi) \, |\det \phi'| \geq \int_{*\Delta} f. \quad (6.60)$$

However $\int_\Delta^* f = \int_{*\Delta} f = \int_\Delta f$ since the function f is Riemann integrable by hypothesis, so the inequalities in (6.60) are equalities; hence the function $(f \circ \phi) \, |\det \phi'|$ is Riemann integrable over $\phi^{-1}(\Delta)$ and the formula (6.51) for the effect of a change of variables in integration holds in this case.

Actually the preceding result holds for somewhat more general domains than a cell Δ. Indeed if $\phi : D \longrightarrow E$ is still a $\mathcal{C}^1$ homeomorphism that changes only a single variable then for any point $\mathbf{b} \in E$, any sufficiently small special Jordan measurable set $U \subset E$ for which $\mathbf{b} \in U^o$, and any Riemann integrable function f in U,

$$\int_U f = \int_{\phi^{-1}(U)} (f \circ \phi) \, |\det \phi'|. \quad (6.61)$$

Indeed if U is sufficiently small there is a cell Δ such that $U \subset \overline{\Delta} \subset E$; then (6.51) holds for the extension $\tilde{f}$ of the function f to the entire cell Δ, and (6.61) follows immediately since

$$\int_U f = \int_\Delta \tilde{f} \quad \text{and} \quad \int_{\phi^{-1}(U)} (f \circ \phi)|\phi'| = \int_{\phi^{-1}(\Delta)} (\tilde{f} \circ \phi)|\phi'|.$$

From this it will be demonsrated that if $\phi : D \longrightarrow E$ is an arbitrary $\mathcal{C}^1$ homeomorphism and if f is a function that is Riemann integrable in E then for any point $\mathbf{b}_0 \in E$ and for any sufficiently small cell Δ such that $\mathbf{b}_0 \in \Delta \subset \overline{\Delta} \subset E$

$$\int_\Delta f = \int_{\phi^{-1}(\Delta)} (f \circ \phi) \, |\det \phi'|. \tag{6.62}$$

Indeed since $\det \phi'(\phi^{-1}(\mathbf{b}_0)) \neq 0$ for a $\mathcal{C}^1$ homeomorphism $\phi : D \longrightarrow E$ it follows from Corollary 5.4 that for a sufficiently small open cell $\Delta \subset \overline{\Delta} \subset E$ containing $\mathbf{b}_0$ the homeomorphism ϕ can be written as the composition

$$\phi = \phi_r \circ \phi_{r-1} \circ \cdots \circ \phi_2 \circ \phi_1 \tag{6.63}$$

of $\mathcal{C}^1$ homeomorphisms that change only a single variable. Each of these separate mappings ϕ_j will be a mapping between special Jordan measurable sets, so that (6.61) holds for each. To demonstrate (6.62) it suffices then merely to show that if (6.61) holds for two $\mathcal{C}^1$ homeomorphisms $\phi : D \longrightarrow E$ and $\psi : E \longrightarrow F$ then it also holds for the composition $\psi \circ \phi : D \longrightarrow F$. For that purpose, if f is a Riemann integrable function in F then

$$\int_F f = \int_E (f \circ \psi)|\det \psi'| = \int_D ((f \circ \psi) \circ \phi)|(\det \psi') \circ (\phi)| \, |\det \phi'|$$
$$= \int_D (f \circ (\psi \circ \phi))|\det(\psi \circ \phi)'|$$

since $(\psi \circ \phi)'(\mathbf{x}) = \psi'(\phi(\mathbf{x}))\phi'(\mathbf{x})$ by the chain rule for differentiation.

Finally to complete the proof of the theorem itself, by Theorem 6.18 any open subset $E \in \mathbb{R}^n$ can be written as the union of an increasing sequence of open Jordan measurable sets E_i, each of which is a union $E_i = \bigcup_j \Delta_{ij}$ of cells Δ_{ij} that are disjoint aside from their boundaries; and D is correspondingly the union of the increasing sequence of open subsets $D_i = \phi^{-1}(E_i)$ each of which is a union $D_i = \bigcup \phi^{-1}(\Delta_{ij})$ of special Jordan measurable sets that are disjoint aside from their boundaries. By passing to refinements of these cells it can be assumed by what already has been demonstrated that $\int_{\Delta_{ij}} f = \int_{\phi^{-}(\Delta_{ij})}(f \circ \phi)|\det \phi'|$ for any Riemann integrable function in E; and it then follows from Theorerm 6.13 that

$$\int_{E_i} f = \int_{D_i} (f \circ \phi)|\det \phi'|$$

for each set E_i. The improper integrals are defined as in (6.43) in terms of the limits of the integrals over the sets D_i and E_i, so (6.49) holds in the limit and that suffices for the proof.

For an explicit example, consider the integral $\int_E \log(x_1^2 + x_2^2)$ where

$$E = \left\{ (x_1, x_2) \in \mathbb{R}^2 \,\middle|\, a \le x_1^2 + x_2^2 \le b, \ x_1 \ge 0, \ x_2 \ge 0 \right\}.$$

This is most conveniently handled by introducing polar coordinates through the change of coordinates

$$\phi : \mathbb{R}^2(r, \theta) \longrightarrow \mathbb{R}^2(x_1, x_2)$$

where $x_1 = r \cos \theta$, $x_2 = r \sin \theta$. The mapping ϕ establishes a one-to-one mapping $\phi : D \longrightarrow E$ where

$$D = \left\{ (r, \theta) \in \mathbb{R}^2 \,\middle|\, a \le r \le b, \ 0 \le \theta \le \tfrac{\pi}{2} \right\};$$

and

$$\phi'(x, y) = \frac{\partial(x_1, x_2)}{\partial(r, \theta)} = \begin{pmatrix} \dfrac{\partial x_1}{\partial r} & \dfrac{\partial x_1}{\partial \theta} \\ \dfrac{\partial x_2}{\partial r} & \dfrac{\partial x_2}{\partial \theta} \end{pmatrix} = \begin{pmatrix} \cos \theta & -r \sin \theta \\ \sin \theta & r \cos \theta \end{pmatrix}$$

so $|\det \phi'(r, \theta)| = r$. The formula for the change of variables and an application of Fubini's Theorem show that

$$\begin{aligned}
\int_E \log(x_1^2 + x_2^2) &= \int_D \log(r^2)\, r \\
&= \int_a^b \int_0^{\pi/2} 2r \log r \; d\theta \, dr \\
&= \frac{\pi}{2} \int_a^b 2r \log r = \frac{\pi}{2} r^2 \left(\log r - \frac{1}{2} \right) \Big|_a^b \\
&= \frac{\pi}{2} (b^2 \log b - a^2 \log a) - \frac{\pi}{4}(b^2 - a^2).
\end{aligned}$$

Problems, Group I

1. When the unit cell $\Delta \subset \mathbb{R}^2$ is viewed as a product of unit intervals $\Delta = I_1 \times I_2$ in terms of coordinates $(x_1, x_2) \in \mathbb{R}^2$ show that

$$\int_\Delta f_1(x_1) f_2(x_2) = \left(\int_{I_1} f_1 \right) \left(\int_{I_2} f_2 \right)$$

where f_1, f_2 are Riemann integrable functions on I_1 and I_2, respectively.

2.   Calculate $\int_\Delta \log((1 + x_1)(1 + x_2))$ where Δ is the unit cell in $\mathbb{R}^2$.

3.   Calculate

$$\int_{[0,\,1]} \int_{[0,\,x_1^2]} (x_1^2 + x_1 x_2 - x_2^2).$$

4.   Calculate $\int_D e^{x_1 - x_2}$ where D is the triangular region in the plane having vertices $(0,0), (1,3), (2,2)$.

5.   Calculate $\breve{\int}_{\mathbb{R}} e^{-x^2}$ by noting that $(\breve{\int}_{\mathbb{R}} e^{-x^2})^2 = \breve{\int}_{\mathbb{R}^2} e^{-x_1^2 - x_2^2}$ and using polar coordinates.

Problems, Group II

6.   Find the content of the plane region bounded by the curve $r = 1 + \sin\theta$ in polar coordinates in $\mathbb{R}^2$.

7.   Let f be the real-valued function in the unit cell in $\mathbb{R}^2$ defined by

$$f(x_1, x_2) = \begin{cases} 1 & \text{if } x_1 \text{ is rational,} \\ 2x_2 & \text{if } x_1 \text{ is irrational.} \end{cases}$$

Show that the function $f(x_1, x_2)$ is not Riemann integrable in the unit cell but that nonetheless the iterated integral $\int_{x_1 \in [0,1]} \int_{x_2 \in [0,1]} f(x_1, x_2) dx_2 dx_1$ exists.

8.   Show that if g is integrable then $\int_{[0,x]} \int_{[0,y]} g(t) dt dy = \int_{[0,x]} (x - t) g(t) dt$, thus reducing a double integral to a single integral.

9.   If f is a Riemann integrable nonnegative function on the unit interval $[0, 1]$ show that the subset

$$E = \left\{ (x_1, x_2) \in \mathbb{R}^2 \,\middle|\, 0 \le x_1 \le 1,\ 0 \le x_2 \le f(x_1) \right\} \subset \mathbb{R}^2$$

is a Jordan measurable set and that its content is $|E| = \int_{[0,1]} f$.

10.   A **step function** ϕ in a cell $\Delta \subset \mathbb{R}^n$ is a real-valued function that is constant in each open cell Δ_i of a finite partition $\Delta = \bigcup_i \Delta_i$ of Δ; thus a step function is a function that can be written as the sum $\phi = \sum_i c_i \chi_{\Delta_i}$ in terms of the characteristic functions of the cells Δ_i of the partition.
       (i) Show that the step function $\phi = \sum_i c_i \chi_{\Delta_i}$ is Riemann integrable in Δ and that $\int_\Delta \phi = \sum_i c_i |\Delta_i|$.
       (ii) Show that any continuous real-valued function f on $\overline{\Delta}$ can be written as a uniform limit $f = \lim_\nu \phi_\nu$ of step functions ϕ_ν on Δ, hence that the integral

of f can be expressed as the limit $\int_\Delta f = \lim_\nu \int_\Delta \phi_\nu$ for some sequence ϕ_ν of step functions on Δ.

11. Show that if the characteristic function χ_E of a subset $E \subset \mathbb{R}^2$ is Riemann integrable then E is a set of measure zero if and only if for almost every $x_1 \in \mathbb{R}$ the set

$$E_{x_1} = \left\{ x_2 \in \mathbb{R} \,\middle|\, (x_1, x_2) \in E \right\}$$

is a set of measure zero in $\mathbb{R}$.

12. If D is a Jordan measurable set in $\mathbb{R}^n$ and $\phi : U \longrightarrow \mathbb{R}^n$ is a mapping that is C^1 on an open neighborhood U of $\overline{D}$ show that $\phi(D)$ is also a Jordan measurable set.

7

Differential Forms

7.1 Line Integrals

The Riemann integral of a function over a $\mathcal{C}^1$ compact submanifold of $\mathbb{R}^n$ is zero, since such a submanifold is a set of content 0; nonetheless there are variants of the integral that are defined nontrivially over submanifolds and similar subsets of $\mathbb{R}^n$, and they play a very important role in mathematics and its applications. The simplest case is probably that of integrals over curves in $\mathbb{R}^n$. A **parametrized curve** in an open subset $U \subset \mathbb{R}^n$ is a continuous mapping $\phi : [a, b] \longrightarrow U$ from a closed interval $[a, b] \subset \mathbb{R}^1$ into $\mathbb{R}^n$. It is necessary to distinguish between a mapping and its image, that is, to distinguish between a parametrized curve and the subset of $\mathbb{R}^n$ that is the image of the parametrized curve; a great many distinct parametrized curves may have the same image, but they are considered as distinct parametrized curves. It is customary to consider the parameter space as an oriented interval $[a, b] \subset \mathbb{R}^1$ with $a \leq b$; hence a parametrized curve also has a natural orientation, an order corresponding to the order of the parameter in $[a, b]$. Much of the subsequent discussion will focus on $\mathcal{C}^1$ parametrized curves ϕ, those for which the mapping extends to a $\mathcal{C}^1$ mapping $\phi : (a - \epsilon, b + \epsilon) \longrightarrow U$ for some $\epsilon > 0$; but it is common also to consider $\mathcal{C}^k$ parametrized curves for any integer $k > 0$, or even $\mathcal{C}^\infty$ parametrized curves, those that are $\mathcal{C}^k$ parametrized curves for all $k > 0$. Unless explicitly assumed otherwise, however, parametrized curves are merely continuous mappings. Also unless explicitly assumed otherwise, parametrized curves are not required to be one-to-one mappings, nor are $\mathcal{C}^1$ parametrized curves required to have a nonvanishing derivative; thus a parametrized curve may cover its image repeatedly or with a variety of self-intersections, or even be stationary at some points, as sketched in the accompanying Figure 7.1.

Two parametrized curves $\phi : [a, b] \longrightarrow U$ and $\psi : [c, d] \longrightarrow U$ are said to be **equivalent** parametrized curves if there is an orientation preserving homeomorphism $h : [a, b] \longrightarrow [c, d]$ such that $\phi(t) = \psi\big(h(t)\big)$ for all $t \in [a, b]$. It is clear that this is an equivalence relation in the usual sense; an equivalence class γ of parametrized curves is called a **curve**. It is often assumed as a standard convention that the parameter interval for a parametrized curve is the unit interval $[0, 1]$, since any closed interval is homeomorphic to $[0, 1]$. If ϕ and ψ are equivalent parametrized curves then $\phi([a, b]) = \psi([c, d])$, so all the parametrized curves in an equivalence class have the same image; this image is

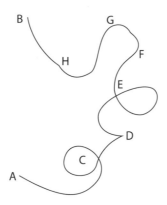

Figure 7.1. A parametrized curve from A to B; it may repeat the loop C several times, it may have a cusp as at D or a self intersection as at E, it may follow the segment from F to G then return to F and then continue back to G and beyond, or it may map an entire segment of the parameter space to the single point H; thus the image or support of a curve can be far from actually describing the curve

called the **support** of the curve γ and is denoted by $|\gamma| = |\phi|$. A parametrized curve for which the mapping $\phi : [a, b] \longrightarrow \mathbb{R}^n$ is injective is called a **simple** parametrized curve. Clearly any parametrized curve that is equivalent to a simple parametrized curve also is a simple parametrized curve; hence the entire equivalence class is called a **simple curve**. For any simple curve $\phi : [a, b] \longrightarrow \mathbb{R}^n$ the image $\phi([a, b]) \subset \mathbb{R}^n$ with the induced topology as a subset of $\mathbb{R}^n$ is a Hausdorff topological space; and since the set $[a, b]$ is compact the mapping $\phi : [a, b] \longrightarrow \phi([a, b])$ is a homeomorphism, by Theorem 3.10. It follows that a simple curve is fully determined by its image; indeed if $\phi : [a, b] \longrightarrow \mathbb{R}^n$ and $\psi : [c, d] \longrightarrow \mathbb{R}^n$ are simple parametrized curves with the same image then since ϕ and ψ are orientation preserving homeomorphisms the composition $h = \phi^{-1} \circ \psi : [c, d] \longrightarrow [a, b]$ is also an orientation preserving homeomorphism and consequently ϕ and $\psi = \phi \circ \phi^{-1} \circ \psi = \phi \circ h$ are equivalent parametrized curves so determine the same curve. Thus for a simple curve γ in $\mathbb{R}^n$ it is customary to identify the curve with its support $|\gamma| \subset \mathbb{R}^n$; but the orientation is determined only by specifying a parametrization. Some care should be taken in identifying curves other than simple curves with their images. In an extension of the notion of equivalence of curves, two $\mathcal{C}^1$ parametrized curves $\phi : [a, b] \longrightarrow U$ and $\psi : [c, d] \longrightarrow U$ are called $\mathcal{C}^1$ **equivalent** parametrized curves if the mapping $h : [a, b] \longrightarrow [c, d]$ extends to a $\mathcal{C}^1$ orientation preserving homeomorphism between open neighborhoods of $[a, b]$ and $[c, d]$, and of course correspondingly for $\mathcal{C}^k$ parametrized curves or $\mathcal{C}^\infty$ parametrized curves. An equivalence class of $\mathcal{C}^k$ parametrized curves naturally is called a $\mathcal{C}^k$ curve.

If $\phi : [0, 1] \longrightarrow U$ is a parametrized curve in an open subset $U \subset \mathbb{R}^n$ and $f : U \longrightarrow \mathbb{R}$ is a function in U the composition $f \circ \phi$, also denoted by $\phi^*(f)$, is a well-defined function on the parameter interval $[0, 1]$ called the **induced**

function or the **pullback** of the function f under the mapping ϕ. If the function f is continuous then the induced function $\phi^*(f)$ is a continuous function in $[0, 1]$; and if the parametrized curve and the function are C^1 functions then the pullback is a C^1 function. The **integral** of a continuous function f in U on the parametrized curve ϕ is defined by

$$\int_\phi f = \int_{[0,1]} \phi^*(f). \tag{7.1}$$

The value of the integral depends on the parametrization, for the integrals of a function over equivalent parametrized curves are not necessarily equal. Indeed if $\phi : [0, 1] \longrightarrow U$ and $\psi : [0, 1] \longrightarrow U$ are C^1 equivalent parametrized curves and $\phi = \psi \circ h$ for a continuously differentiable mapping $h : [0, 1] \longrightarrow [0, 1]$ then it follows from the Change of Variables Theorem, Theorem 6.19, that $\int_\psi f = \int_{[0,1]} f \circ \psi = \int_{[0,1]} (f \circ \psi) \circ h \ |h'| = \int_{[0,1]} f \circ \phi \ |h'|$, which is not necessarily equal to $\int_{[0,1]} f \circ \phi = \int_\phi f$. Consequently the notion of the integral of a function over a curve, as distinct from the integral over a parametrized curve, is not really well defined.

One way to obtain an intrinsically defined integral over a curve is to use an intrinsically defined parametrization of the curve; perhaps the only natural candidate for an intrinsically defined parametrization is by the arc length of a curve. If $\phi : [0, 1] \longrightarrow U$ is a parametrized curve then for any partition $\mathcal{P}$ of the interval $[0, 1]$ by points of subdivision $0 = t_0 < t_1 < \cdots < t_m = 1$ the parametrized curve ϕ can be approximated by the **piecewise linear** parametrized curve $\phi_\mathcal{P}$ for which $\phi_\mathcal{P}(t_i) = \phi(t_i)$ for each point of subdivision and the mapping $\phi_\mathcal{P}$ is extended to be a linear mapping between the points $\phi(t_{i-1})$ and $\phi(t_i)$, so that if $t_{i-1} \leq t \leq t_i$ then

$$\phi_\mathcal{P}(t) = \frac{t - t_i}{t_{i-1} - t_i} \phi(t_{i-1}) + \frac{t - t_{i-1}}{t_i - t_{i-1}} \phi(t_i). \tag{7.2}$$

This is something like approximating the image $\phi([0, 1]) \subset U$ by the polygon joining the successive points $\phi(t_i)$; but the mapping ϕ is not necessarily a one-to-one mapping, and consequently the approximation of the mapping ϕ by the piecewise linear mapping $\phi_\mathcal{P}$ may differ significantly from a polygonal approximation of the point set $\phi([0, 1]) \subset U$. The **arc length** of the piecewise linear curve $\phi_\mathcal{P}$ is defined to be the sum

$$L(\phi, \mathcal{P}) = \sum_{i=1}^{m} \|\phi(t_i) - \phi(t_{i-1})\|_2, \tag{7.3}$$

in terms of the ℓ_2 norm in $\mathbb{R}^n$. If $\mathcal{P}'$ is a refinement of the partition $\mathcal{P}$, obtained by adding a point of subdivision t' between t_{i-1} and t_i, the length $\|\phi(t_i) - \phi(t_{i-1})\|_2$ of the linear segment of the piecewise linear approximation $\phi_\mathcal{P}$ between the points $\phi(t_i)$ and $\phi(t_{i-1})$ is replaced by the larger sum $\|\phi(t_i) - \phi(t')\|_2 + \|\phi(t') - \phi(t_{i-1})\|_2$ in the piecewise linear approximation $\phi_{\mathcal{P}'}$, so that $L(\phi, \mathcal{P}') \geq L(\phi, \mathcal{P})$; iterating this observation shows that the same inequality holds for any refinement $\mathcal{P}'$ of the partition $\mathcal{P}$. The **arc length** of the parametrized curve ϕ is defined to be

$$L(\phi) = \sup_{\mathcal{P}} L(\phi, \mathcal{P}), \qquad (7.4)$$

where the supremum is extended over all partitions $\mathcal{P}$ of the parameter interval $[0, 1]$. If the arc length is finite the curve is said to be a **rectifiable parametrized curve**; so the arc length of a rectifiable curve is a well-defined real number. There are parametrized curves that are not rectifiable, such as the parametrized curve in $\mathbb{R}^2$ given by

$$\phi(t) = \begin{cases} \{t, t \sin \frac{\pi}{2t}\} & \text{for} \quad 0 < t \leq 1, \\ \{0, 0\} & \text{for} \quad t = 0. \end{cases} \qquad (7.5)$$

If $\mathcal{P}_n$ is the partition given by $0 < \frac{1}{n} < \frac{1}{n-1} < \cdots < \frac{1}{2} < 1$ then since

$$\left\| \phi\left(\tfrac{1}{i}\right) - \phi\left(\tfrac{1}{i+1}\right) \right\|_2 \geq \left| \tfrac{1}{i} \sin \tfrac{\pi i}{2} - \tfrac{1}{i+1} \sin \tfrac{\pi(i+1)}{2} \right| = \begin{cases} \frac{1}{i+1} & \text{if } i \text{ is even,} \\ \frac{1}{i} & \text{if } i \text{ is odd,} \end{cases}$$

it is apparent that $\lim_{n \to \infty} L(\phi, \mathcal{P}_n) = \infty$, hence that this parametrized curve is not rectifiable.

Lemma 7.1. *If $\phi : [a, b] \longrightarrow U$ and $\psi : [c, d] \longrightarrow U$ are two equivalent parametrized curves and if ϕ is rectifiable then ψ also is rectifiable and $L(\phi) = L(\psi)$.*

Proof: Since the parametrized curves ϕ and ψ are equivalent there is an orientation preserving homeomorphism $h : [a, b] \longrightarrow [c, d]$ for which $\phi = \psi \circ h$, so the homeomorphism h is monotonically increasing. Any partition

$$\mathcal{P} : a = t_0 < t_1 < \cdots < t_m = b$$

of the interval $[a, b]$ determines a partition

$$h(\mathcal{P}) : c = h(a) = h(t_0) < h(t_1) < \cdots < h(t_m) = h(b) = d$$

of the interval $[c, d]$, and conversely any partition of the interval $[c, d]$ arises in this way from a partition of the interval $[a, b]$. Since $\psi(h(t_i)) = \phi(t_i)$ it is clear from the definition (7.3) that $L(\psi, h(\mathcal{P})) = L(\phi, \mathcal{P})$, hence that $L(\psi) = \sup_{\mathcal{P}} L(\psi, h(\mathcal{P})) = \sup_{\mathcal{P}} L(\phi, \mathcal{P}) = L(\phi)$ and consequently that ψ also is rectifiable. That suffices for the proof.

The preceding lemma shows that if a parametrized curve is rectifiable then so is any equivalent parametrized curve, so the equivalence class can be called a **rectifiable curve**; and equivalent rectifiable parametrized curves have the same arc length, which can be called the arc length of the curve. For any parametrized curve $\phi : [0, 1] \longrightarrow \mathbb{R}^n$ the restriction of the mapping ϕ to a segment $[a, b] \subset [0, 1]$ is another parametrized curve $\phi|[a, b] : [a, b] \longrightarrow \mathbb{R}^n$ called a **segment** of the initial parametrized curve. The arc length of the segment $\phi|[a, b]$ of a parametrized curve ϕ is denoted by $L(\phi|[a, b])$.

Lemma 7.2. *For any rectifiable parametrized curve $\phi : [0, 1] \longrightarrow \mathbb{R}^n$ and any subinterval $[a, b] \subset [0, 1]$*
(i) $L(\phi|[a, b]) \geq \|\phi(b) - \phi(a)\|_2$; *and*
(ii) $L(\phi|[a, b]) = L(\phi|[a, c]) + L(\phi|[c, b])$ *if* $0 \leq a \leq c \leq b \leq 1$.

Proof: (i) If $0 \leq a \leq b \leq 1$ then $a \leq b$ is a particular partition $\mathcal{P}$ of the interval $[a, b]$ so it follows from (7.3) and (7.4) that

$$\|\phi(b) - \phi(a)\|_2 = L(\phi, \mathcal{P}) \leq L(\phi|[a, b]).$$

(ii) For any partition $\mathcal{P}$ of the interval $[a, b]$ the length $L(\phi, \mathcal{P})$ is increased by adjoining the point c to the other points of subdivision; so when examining the supremum of the lengths $L(\phi, \mathcal{P})$ for partitions $\mathcal{P}$ of the interval $[a, b]$ it always can be assumed that c is one of the points of subdivision of the partition $\mathcal{P}$, and consequently that the partition $\mathcal{P}$ consists of a partition $\mathcal{P}_1$ of the interval $[a, c]$ and a partition $\mathcal{P}_2$ of the interval $[c, b]$. Therefore $L(\phi, \mathcal{P}) = L(\phi, \mathcal{P}_1) + L(\phi, \mathcal{P}_2)$ for all partitions, so $L(\phi, [a, b]) = L(\phi|[a, c]) + L(\phi|[c, b])$, which suffices for the proof.

Lemma 7.3. *If $\phi : [0, 1] \longrightarrow \mathbb{R}^n$ is a rectifiable parametrized curve and $s(t) = L(\phi|[0, t])$ for any $t \in [0, 1]$ then*
(i) *$s(t)$ is a monotonically increasing continuous function for which $s(0) = 0$ and $s(1) = l = L(\phi)$; and*
(ii) *$s(t_1) = s(t_2)$ for some parameter values $0 \leq t_1 < t_2 \leq 1$ if and only if the mapping ϕ is constant in the interval $[t_1, t_2]$.*

Proof: (i) It follows immediately from Lemma 7.2 (ii) that if $0 \leq t' \leq t'' \leq 1$ then $s(t'') = s(t') + L(\phi|[t', t'']) \geq s(t')$, so the function $s(t)$ is monotonically increasing. Consider a fixed point $t_0 \in [0, 1]$ and a value $\epsilon > 0$. Since the mapping ϕ is uniformly continuous on the compact interval $[0, 1]$ there is a $\delta > 0$ such that

$$\|\phi(t') - \phi(t'')\|_2 < \epsilon \quad \text{whenever } |t' - t''| < \delta. \tag{7.6}$$

Furthermore by definition of the arc length of a parametrized curve there is a partition $\mathcal{P}$ of the interval $[0, 1]$ such that

$$0 \leq L(\boldsymbol{\phi}) - L(\boldsymbol{\phi}, \mathcal{P}) < \epsilon. \tag{7.7}$$

After passing to a refinement of the partition $\mathcal{P}$ it can be assumed that t_0 is one of the points of subdivision of the partition $\mathcal{P}$, and that if $t_0' < t_0 < t_0''$ are three consecutive points of subdivision of $\mathcal{P}$ then $|t_0 - t_0'| < \delta$ and $|t_0 - t_0''| < \delta$. For any t in the interval $t_0' < t < t_0$ let $\mathcal{P}_t$ be the refinement of $\mathcal{P}$ that arises by adding t as another point of subdivision. The restriction of the partition $\mathcal{P}_t$ to the interval $[t, t_0]$ consists just of the two points $\boldsymbol{\phi}(t)$ and $\boldsymbol{\phi}(t_0)$, and since $|t_0 - t| < |t_0' - t_0| < \delta$ it follows from (7.6) that

$$L(\boldsymbol{\phi}|[t, t_0], \mathcal{P}_t) = \|\boldsymbol{\phi}(t) - \boldsymbol{\phi}(t_0)\|_2 < \epsilon. \tag{7.8}$$

On the other hand it follows from Lemma 7.2 (ii) and (7.7) that

$$(L(\boldsymbol{\phi}|[0, t]) - L(\boldsymbol{\phi}|[0, t], \mathcal{P}_t)) + (L(\boldsymbol{\phi}|[t, t_0]) - L(\boldsymbol{\phi}|[t, t_0], \mathcal{P}_t)) +$$
$$+ (L(\boldsymbol{\phi}|[t_0, 1]) - L(\boldsymbol{\phi}|[t_0, 1], \mathcal{P}_t)) \leq L(\boldsymbol{\phi}) - L(\boldsymbol{\phi}, \mathcal{P}_t) < \epsilon,$$

so since all the differences in the parentheses are nonnegative it follows that

$$L(\boldsymbol{\phi}|[t, t_0]) - L(\boldsymbol{\phi}|[t, t_0], \mathcal{P}_t) < \epsilon. \tag{7.9}$$

Then from (7.8) and (7.9) it follows that

$$|s(t) - s(t_0)| = L(\boldsymbol{\phi}|[t, t_0])$$
$$= (L(\boldsymbol{\phi}|[t, t_0]) - L(\boldsymbol{\phi}|[t, t_0], \mathcal{P}_t)) + L(\boldsymbol{\phi}|[t, t_0], \mathcal{P}_t)$$
$$< \epsilon + \epsilon = 2\epsilon.$$

The corresponding argument yields the same result for points t in the interval $t_0 < t < t_0''$, so the function $s(t)$ is continuous.

(ii) If $\boldsymbol{\phi}(t) = \boldsymbol{\phi}(t_1)$ for all parameter values $t \in [t_1, t_2]$ it is clear from the definition that $L(\boldsymbol{\phi}, \mathcal{P}) = 0$ for any partition $\mathcal{P}$ of the interval $[t_1, t_2]$ and consequently that $0 = L(\boldsymbol{\phi}|[t_1, t_2]) = L(\boldsymbol{\phi}|[0, t_2]) - L(\boldsymbol{\phi}|[0, t_1]) = s(t_2) - s(t_1)$. On the other hand if $s(t_1) = s(t_2)$ then since $s(t)$ is monotonically increasing $s(t_1) = s(t)$ whenever $t_1 \leq t \leq t_2$, hence Lemma 7.2 (i) shows that $0 = s(t) - s(t_1) = L(\boldsymbol{\phi}|[t_1, t]) \geq \|\boldsymbol{\phi}(t) - \boldsymbol{\phi}(t_1)\|_2$ and consequently that $\boldsymbol{\phi}(t) = \boldsymbol{\phi}(t_1)$. That suffices for the proof.

If $\boldsymbol{\phi} : [0, 1] \longrightarrow \mathbb{R}^n$ is a parametrized curve for which $s(t) = L(\boldsymbol{\phi}|[0, t])$ is strictly increasing then the mapping $s : [0, 1] \longrightarrow [0, l]$ is a homeomorphism, since it is a continuous one-to-one mapping between two compact sets; and if $t : [0, l] \longrightarrow [0, 1]$ is the inverse homeomorphism then the composition $\sigma(s) = \phi(t(s))$ is a parametrized curve that is equivalent to the curve ϕ. From Lemma 7.1 it follows that

$$L(\sigma|[0, s]) = L(\phi|[0, t(s)]) = s, \tag{7.10}$$

so the parametrized curve σ is parametrized by arc length. On the other hand if the function $s(t)$ fails to be strictly increasing then by Lemma 7.3 (ii) that is because the mapping ϕ is constant on some subintervals of $[0, 1]$; and the mapping ϕ can be modified, essentially by shrinking such subintervals to points, so that $s(t)$ becomes strictly increasing. In more detail, $s : [0, 1] \longrightarrow [0, l]$ is a surjective mapping, since the image of the connected set $[0, 1]$ under the continuous mapping s is also connected and contains the points $s(0) = 0$ and $s(1) = l$. Therefore for any point $s_0 \in [0, l]$ there is some point $t_0 \in [0, 1]$ for which $s(t_0) = s_0$; and if there is another point t_0' for which $s(t_0') = s_0$ it follows from Lemma 7.3 (ii) that $\phi(t_0) = \phi(t_0')$, so setting $\psi(s_0) = \phi(t_0)$ yields a well-defined mapping $\psi : [0, l] \longrightarrow \mathbb{R}^n$, independent of the choice of the particular point t_0, such that $\psi(s(t)) = \phi(t)$. This mapping ψ is a continuous mapping, since if $0 \le s_1 \le s_2 \le l$ and if $t_1, t_2 \in [0, 1]$ are any parameter values s for which $s(t_1) = s_1$ and $s(t_2) = s_2$ then by Lemma 7.2 (i)

$$\|\psi(s_2) - \psi(s_1)\|_2 = \|\phi(t_2) - \phi(t_1)\|_2$$
$$\le L(\phi|[t_1, t_2]) = s(t_2) - s(t_1) = s_2 - s_1.$$

Furthermore the function $L(\psi|[0, s])$ is a strictly increasing function of s. To see that, if $L(\psi|[0, s_1]) = L(\psi|[0, s_2])$ where $s_1 < s_2$ then by Lemma 7.3 (ii) the mapping ψ must be constant in the interval $[s_1, s_2]$ so in particular $\psi(s_1) = \psi(s_2)$. If $s_1 = s(t_1)$ and $s_2 = s(t_2)$ then $t_1 < t_2$ since s is monotone increasing by Lemma 7.3 (i). If $t \in [t_1, t_2]$ then $s(t) \in [s_1, s_2]$ hence $\phi(t) = \psi(s(t)) = \psi(s_1)$ and consequently $s_1 = s(t_1) = s(t_2) = s_2$ by Lemma 7.3 (ii) again, a contradiction.

Thus for any rectifiable parametrized curve ϕ it is possible to modify the parametrization by deleting subintervals of the parameter space on which the mapping ϕ is constant, so that the parametrization is not constant in subintervals of the parameter set; this will be called a **regular** parametrization of the curve, and the process of modifying the parameter space in this way is called the **regularization** of the parametrized curve. Any regular parametrization is equivalent to the parametrization by arc length; and a parametrized curve with arc length as the parameter is said to be a curve **parametrized by arc length**. If f is a continuous function in an open neighborhood of a regularly parametrized curve ϕ its integral over ϕ can be defined uniquely as the integral with respect to arc length, so by definition

$$\int_{\phi} f = \int_{s \in [0,l]} f(s) \tag{7.11}$$

where the curve ϕ is parametrized by arc length; traditionally that is denoted by

$$\int_{\phi} f = \int_{s \in [0,l]} f(s)\, ds \tag{7.12}$$

following the convention that the symbol ds indicates integration with respect to parametrization by arc length. With this notation the arc length of the curve ϕ can be defined by $L(\phi) = \int_{\phi} 1 \, ds$. As an additional word of caution, the integral is still the integral of an equivalence class of parametrized curves ϕ, and is not the integral just over the point set $|\phi|$, the support of the curve ϕ; for instance a parametrized curve may cover its image repeatedly, as in the case of integrating twice along a circle.

Theorem 7.4. *A C^1 parametrized curve $\phi : [0,1] \longrightarrow U$ in an open subset $U \subset \mathbb{R}^n$ is rectifiable and its arc length is*

$$L(\phi) = \int_{[0,1]} \|\phi'\|_2. \tag{7.13}$$

Proof: For any partition $\mathcal{P}$ of the interval $[0,1]$ it follows from the second Fundamental Theorem of Calculus (Theorem 6.4) in one variable applied to each coordinate function of the mapping ϕ and the inequality of Corollary 6.10 that

$$L(\phi, \mathcal{P}) = \sum_{i=1}^{m} \|\phi(t_i) - \phi(t_{i-1})\|_2 = \sum_{i=1}^{m} \left\| \int_{[t_{i-1}, t_i]} \phi' \right\|_2$$

$$\leq \sum_{i=1}^{m} \int_{[t_{i-1}, t_i]} \|\phi'\|_2 = \int_{[0,1]} \|\phi'\|_2; \tag{7.14}$$

so since $\|\phi'\|_2$ is continuous on the compact set $[0,1]$ this integral is finite and the curve ϕ is rectifiable. Moreover since $L(\phi) = \sup_{\mathcal{P}} L(\phi, \mathcal{P})$ it is also the case that

$$L(\phi) \leq \int_{[0,1]} \|\phi'\|_2. \tag{7.15}$$

For any point $t \in [0,1)$ and any $h > 0$ sufficiently small that $t + h \leq 1$ it follows from (7.15) and Lemma 7.2 (ii) that

$$s(t+h) - s(t) = L\big(\phi|[t, t+h]\big) \leq \int_{[t, t+h]} \|\phi'\|_2. \tag{7.16}$$

On the other hand for any unit vector $\mathbf{u} \in \mathbb{R}^n$ it follows from Lemma 7.2 (i), the Cauchy-Schwarz inequality, and the Mean Value Theorem applied to the real-valued function $\mathbf{u} \cdot \boldsymbol{\phi}(t)$ that

$$L\big(\boldsymbol{\phi}\big|[t, t+h]\big) \geq \|\boldsymbol{\phi}(t+h) - \boldsymbol{\phi}(t)\|_2 \geq |\mathbf{u} \cdot \big(\boldsymbol{\phi}(t+h) - \boldsymbol{\phi}(t)\big)| = |\mathbf{u} \cdot \boldsymbol{\phi}'(\tau)| \, h \quad (7.17)$$

for some point τ in the interval $(t, t+h)$. Since $L\big(\boldsymbol{\phi}\big|[t, t+h]\big) = s(t+h) - s(t)$ by Lemma 7.2 (ii) again the preceding inequality (7.17) can be rewritten

$$\frac{1}{h}\big(s(t+h) - s(t)\big) \geq |\mathbf{u} \cdot \boldsymbol{\phi}'(\tau)|, \quad (7.18)$$

and when combined with the inequality (7.16) that leads to the inequality

$$|\mathbf{u} \cdot \boldsymbol{\phi}'(\tau)| \leq \frac{1}{h}\big(s(t+h) - s(t)\big) \leq \frac{1}{h}\int_{[t,t+h]} \|\boldsymbol{\phi}'\|_2. \quad (7.19)$$

The same inequality holds for the interval $(t - h, t)$ in place of $(t, t+h)$, which is readily seen to amount to the inequality (7.19) for a negative value of h and a point τ' in the interval $(t - h, t)$. If $\mathbf{u}$ is taken to be the unit vector in the direction of the vector $\boldsymbol{\phi}'(t)$, then since τ and τ' tend to t as h tends to 0 through either positive or negative values

$$\lim_{h \to 0} |\mathbf{u} \cdot \boldsymbol{\phi}'(\tau)| = |\mathbf{u} \cdot \boldsymbol{\phi}'(t)| = \|\boldsymbol{\phi}'(t)\|_2. \quad (7.20)$$

On the other hand it follows from the First Fundamental Theorem of Calculus, Theorem 6.6, that

$$\lim_{h \to 0} \frac{1}{h}\int_{[t,t+h]} \|\boldsymbol{\phi}'\|_2 = \frac{d}{dt}\int_0^t \|\boldsymbol{\phi}'\|_2 = \|\boldsymbol{\phi}'(t)\|_2. \quad (7.21)$$

Therefore taking the limit in (7.19) as h tends to zero through positive or negative values and applying (7.20) and (7.21) shows that the function $s(t)$ is differentiable and

$$s'(t) = \|\boldsymbol{\phi}'(t)\|_2. \quad (7.22)$$

It finally follows from the First Fundamental Theorem of Calculus again that

$$L(\boldsymbol{\phi}) = s(1) = \int_{[0,1]} s'(t) = \int_{[0,1]} \|\boldsymbol{\phi}'\|_2,$$

and that concludes the proof.

For any $\mathcal{C}^1$ parametrized curve $\boldsymbol{\phi}$ it follows from (7.22) that its arc length $s(t) = L(\boldsymbol{\phi}|[0, t])$ is a $\mathcal{C}^1$ mapping $s : [0, 1] \longrightarrow [0, l]$ where $s(1) = l = L(\boldsymbol{\phi})$ is the arc length of ϕ; and if it is also assumed that $\phi'(t) \neq \mathbf{0}$ at all points $t \in (0, 1)$ then the function $s(t)$ is strictly increasing so the curve ϕ is $\mathcal{C}^1$ equivalent to a curve that is parametrized by arc length.

Corollary 7.5. *A $\mathcal{C}^1$ mapping $\phi : [0, l] \longrightarrow \mathbb{R}^n$ is the parametrization of a curve by arc length if and only if $\|\phi'(s)\|_2 = 1$ for all points $s \in [0, l]$.*

Proof: If $\phi : [0, l] \longrightarrow \mathbb{R}^n$ is a $\mathcal{C}^1$ parametrization by arc length then $s(t) = L(\phi|[0, t]) = t$ and it follows from (7.22) that $\|\phi'(t)\|_2 = s'(t) = 1$. Conversely if $\phi : [0, l] \longrightarrow \mathbb{R}^n$ is a parametrized curve for which $\|\phi'(s)\|_2 = 1$ then it follows from (7.13) that $L(\phi|[0, t]) = \int_{[0, t]} \|\phi'\|_2 = \int_{[0, t]} 1 = t$, so the parameter t is arc length, which suffices for the proof.

If $\phi : [0, l] \longrightarrow U$ is a $\mathcal{C}^1$ curve parametrized by arc length then since arc length is strictly increasing $\phi'(t) \neq \mathbf{0}$ at all points $t \in [0, l]$; hence by Theorem 5.11, the corollary to the Rank Theorem, the image of an open neighborhood of any parameter value $s_0 \in [0, l]$ is a $\mathcal{C}^1$ submanifold of an open neighborhood of the point $\phi(s_0) \in U$. The vector $\phi'(s_0)$ is a tangent vector to that submanifold and has length $\|\phi'(s_0)\|_2 = 1$, by the preceding Corollary 7.5; thus it is the **unit tangent vector** to the local piece of the curve ϕ at the point $\phi(s_0)$, so it is denoted by $\tau(s_0) = \phi'(s_0)$. The curve ϕ may intersect itself at some points, at which there may be different unit tangent vectors to the various local components of the curve ϕ. If $\mathbf{f}$ is a continuous vector field in an open neighborhood of $|\phi|$ the dot product $\mathbf{f}(\phi(s)) \cdot \tau(s) = \mathbf{f}(\phi(s)) \cdot \phi'(s)$ is a well-defined continuous function of the variable $s \in [0, l]$ so it has a well-defined integral

$$\int_\phi \mathbf{f} \cdot \tau \, ds = \int_{[0, l]} (\mathbf{f} \circ \phi) \cdot \phi', \tag{7.23}$$

where $\int_\phi \mathbf{f} \cdot \tau \, ds$ again indicates integration of the function $\mathbf{f} \cdot \tau$ on the curve ϕ parametrized by arc length. This is called the **line integral** of the vector field $\mathbf{f}$ along the curve ϕ. What is particularly interesting is that this integral can be calculated in terms of any parametrization of the curve ϕ that is equivalent to the parametrization by arc length; in this sense the integration of vector fields along curves is a more intrinsic construction than the integration of functions along parametrized curves.

Theorem 7.6. *If $\psi : [a, b] \longrightarrow \mathbb{R}^n$ is a $\mathcal{C}^1$ parametrized curve and $\psi'(t) \neq \mathbf{0}$ for $a \leq t \leq b$, and if ϕ is the parametrization of that curve by arc length, then*

$$\int_\phi \mathbf{f} \cdot \tau \, ds = \int_{[a, b]} (\mathbf{f} \circ \psi) \cdot \psi' \tag{7.24}$$

for any continuous vector field $\mathbf{f}$ in an open neighborhood of $\psi([a, b])$.

Proof: The parametrized curve ψ is equivalent to a curve ϕ parametrized by arc length, since $\psi'(t) \neq \mathbf{0}$ at all points $t \in [a, b]$; and the line integral of the vector field $\mathbf{f}$ along the parametrized curve ψ is defined in terms of the parametrized curve ϕ. If $h : [a, b] \longrightarrow [0, l]$ is a C^1 homeomorphism such that $\psi(t) = \phi(h(t))$ then $h'(t) > 0$ and by the chain rule $\psi'(t) = \phi'(h(t))h'(t)$, so by the formula for the change of variables in integration for functions of a single variable it follows as in (7.23) that

$$\int_\phi \mathbf{f} \cdot \tau \, ds = \int_{[0,l]} \mathbf{f}(\phi(s)) \cdot \phi'(s) = \int_{[a,b]} \mathbf{f}(\phi(h(t))) \cdot \phi'(h(t)) \, h'(t)$$

$$= \int_{[a,b]} \mathbf{f}(\psi(t)) \cdot \psi'(t),$$

which suffices for the proof.

For example, consider the integral of the vector field $\mathbf{f}(\mathbf{x}) = \{x_1^2, x_1 x_2\}$ in terms of the variables x_1, x_2 in $\mathbb{R}^2$ over the parabola described parametrically by the mapping $\phi : [0, 1] \longrightarrow \mathbb{R}^2$ for which $\phi(t) = \{t, t^2\}$. Since $\phi'(t) = \{1, 2t\}$ then $\mathbf{f}(\phi(t)) = \{t^2, t^3\}$ so $\mathbf{f}(\phi(t)) \cdot \phi'(t) = t^2 + 2t^4$ and consequently

$$\int_\gamma \mathbf{f} \cdot \tau \, ds = \int_{[0,1]} \mathbf{f}(\phi(t)) \cdot \phi'(t)$$

$$= \int_{[0,1]} (t^2 + 2t^4) = \left(\frac{1}{3}t^3 + \frac{2}{5}t^5 \right) \bigg|_0^1 = \frac{11}{15}.$$

A more interesting example is the line integral of a vector field that is the gradient $\mathbf{f}(\mathbf{x}) = \nabla h(\mathbf{x})$ of a function h in an open set $U \subset \mathbb{R}^n$, a vector field called a **conservative** vector field with the **potential** h.

Theorem 7.7. *If $\mathbf{f}(\mathbf{x}) = \nabla h(\mathbf{x})$ is a conservative vector field in a connected open subset $U \subset \mathbb{R}^n$ with a C^1 potential h, and if ϕ is a C^1 oriented curve in U from a point $\mathbf{a} \in U$ to a point $\mathbf{b} \in U$ then*

$$\int_\phi \mathbf{f} \cdot \tau \, ds = h(\mathbf{b}) - h(\mathbf{a}). \tag{7.25}$$

Proof: In view of the preceding theorem it is sufficient to demonstrate this result just for a curve parametrized by arc length; so let $\phi : [0, l] \longrightarrow U$ be the parametrization of a curve by arc length, where $\phi(s) = \{\phi_j(s)\}$ and $\phi(0) = \mathbf{a}$, $\phi(l) = \mathbf{b}$. By the chain rule

$$\frac{d}{ds} h(\phi(s)) = h'(\phi(s)) \, \phi'(s) = \sum_{j=1}^n \partial_j h(\phi(s)) \phi'_j(s) = \sum_{j=1}^n f_j(\phi(s)) \phi'_j(s)$$

$$= \mathbf{f}(\phi(s)) \cdot \phi'(s),$$

and consequently by the second Fundamental Theorem of Calculus

$$\int_\gamma \mathbf{f} \cdot \boldsymbol{\tau} \, ds = \int_{[0,l]} \mathbf{f}(\boldsymbol{\phi}(s)) \cdot \boldsymbol{\phi}'(s) = \int_{[0,l]} h'(\boldsymbol{\phi}(s))$$

$$= h(\boldsymbol{\phi}(l)) - h(\boldsymbol{\phi}(0)) = h(\mathbf{b}) - h(\mathbf{a}),$$

which suffices for the proof.

Thus the integral of a conservative vector field along a curve really depends only on the beginning and end points of the curve, and is quite independent of the particular path taken between these two points. There are vector fields that are not conservative, and for which the line integral between two points depends on the path chosen between those two points. For example for the vector field $\mathbf{f}(\mathbf{x}) = \{x_2, -x_1\}$ in $\mathbb{R}^2$, and the parametrized curves $\boldsymbol{\psi}_1(t) = \{t, -t\}$ and $\boldsymbol{\psi}_2(t) = \{t, -t^2\}$ for $t \in [0, 1]$

$$\int_{\psi_1} \mathbf{f} \cdot \boldsymbol{\tau} \, ds = \int_{[0,1]} \{-t, -t\} \cdot \{1, -1\} \, dt = \int_{[0,1]} 0 \, dt = 0,$$

$$\int_{\psi_2} \mathbf{f} \cdot \boldsymbol{\tau} \, ds = \int_{[0,1]} \{-t^2, -t\} \cdot \{1, -2t\} \, dt = \int_{[0,1]} t^2 \, dt = \frac{1}{3}.$$

Actually the condition that the line integral of a vector field is independent of the path characterizes conservative vector fields.

Theorem 7.8. *A C^1 vector field $\mathbf{f} : U \longrightarrow \mathbb{R}^n$ in an open subset $U \subset \mathbb{R}^n$ is conservative if and only if the line integrals $\int_\phi \mathbf{f} \cdot \boldsymbol{\tau} \, ds$ depend only on the end points of C^1 parametrized curves γ in U.*

Proof: The preceding theorem demonstrated that if $\mathbf{f}$ is a conservative vector field and $\boldsymbol{\psi} : [0, 1] \longrightarrow U$ is a C^1 parametrized curve in U then the line integral $\int_\psi \mathbf{f} \cdot \boldsymbol{\tau} \, ds$ depends just on the beginning and end points of $\boldsymbol{\psi}$. Conversely suppose that $\mathbf{f}$ is a C^1 vector field in a connected open subset $U \subset \mathbb{R}^n$ such that the line integrals $\int_\psi \mathbf{f} \cdot \boldsymbol{\tau} \, ds$ depend only on the beginning and end points of all C^1 parametrized curves $\boldsymbol{\psi} : [0, 1] \longrightarrow U$. For a fixed base point $\mathbf{a} \in U$ the integral $h(\mathbf{x}) = \int_{\psi_\mathbf{x}} \mathbf{f} \cdot \boldsymbol{\tau} \, ds$ along any C^1 parametrized curve $\boldsymbol{\psi}_\mathbf{x}$ from the point $\mathbf{a}$ to the point $\mathbf{x}$ is a well-defined function of the point $\mathbf{x} = \{x_1, \ldots, x_n\} \in U$, independent of the particular choice of the parametrized curve $\boldsymbol{\psi}_\mathbf{x}$, provided of course that $\mathbf{x}$ is in the same connected component of U as the point $\mathbf{a}$. In particular consider a parametrized curve $\boldsymbol{\psi}_\mathbf{x}$ that describes a path from $\mathbf{a}$ to the beginning of a segment of the straight line λ_i through the point $\mathbf{x}$ parallel to the coordinate axis of the variable x_i and then along λ_i to the point $\mathbf{x}$, near the point $\mathbf{x}$, as in

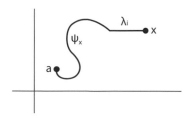

Figure 7.2. A path of integration λ_i

the accompanying figure. The parametrization has the form

$$\boldsymbol{\psi}_{\mathbf{x}}(t) = \{x_1, \ldots, x_{i-1}, t, x_{i+1}, \ldots, x_n\} \quad \text{for } b_i \leq t \leq x_i$$

so $\boldsymbol{\psi}'_{\mathbf{x}} = \delta_i$ is the unit vector parallel to λ_i. The function $h(\mathbf{x})$ then can be written as the sum of the value A of the integral up to a point $\boldsymbol{\psi}_{\mathbf{x}}(b_i)$ on the line λ_i and the integral

$$h(\mathbf{x}) - A = \int_{t \in [b_i, x_i]} \mathbf{f}(\mathbf{x}) \cdot \tau \, ds = \int_{t \in [b_i, x_i]} f_i(x_1, \ldots, x_{i-1}, t, x_{i+1}, \ldots, x_n)$$

along λ_i; and it then follows from the First Fundamental Theorem of Calculus (Theorem 6.6) in one variable that $\partial_i h(\mathbf{x}) = f_i(\mathbf{x})$. The corresponding argument for paths ending parallel to other coordinate axes in $\mathbb{R}^n$ shows that $\nabla h(\mathbf{x}) = \mathbf{f}(\mathbf{x})$, hence that the vector field $\mathbf{f}$ is conservative, thus concluding the proof.

Problems, Group I

1. Let γ_1 be the straight line segment from the point $(0, 3)$ to the point $(3, 0)$ in the plane $\mathbb{R}^2$, and γ_2 be the arc of the circle of radius 3 centered at the origin from the point $(0, 3)$ to the point $(3, 0)$. For each of these paths γ_i calculate the integral $\int_{\gamma_i} x_1 x_2^2 ds$, with respect to arc length.

2. Find the arc length of the curve in $\mathbb{R}^2$ described parametrically by $\phi(t) = (\frac{1}{3}t^3 - t, \ t^2)$ for $0 \leq t \leq 2$.

3. Evaluate the integral $\int_\phi \mathbf{f}(\mathbf{x}) \cdot \tau \, ds$ with respect to arc length for the vector field

    $$\mathbf{f}(\mathbf{x}) = \begin{pmatrix} x_1 + x_2 \\ 1 \end{pmatrix} \text{ where } \phi(t) = (t, \ t^2) \text{ for } 0 \leq t \leq 1.$$

4. Let γ be the oriented curve from the origin $(0, 0, 0)$ to the point $(1, 1, 1)$ in $\mathbb{R}^3$ described by the equations $x_2 = x_1$, $x_3 = x_1^2$.

i) Find the arc length of this curve.
ii) Calculate the integral $\int_\gamma \mathbf{f} \cdot \boldsymbol{\tau} \, ds$ with respect to the arc length, where $\boldsymbol{\tau}$ is the unit tangent vector to the curve γ and $\mathbf{f}$ is the vector field defined by
$\mathbf{f} = {}^t(x_1, -x_2, x_3)$.
iii) Calculate the integral $\int_\gamma x_1 ds$ with respect to arc length.

Problems, Group II

5. Find a curve γ in the plane that has arc length 1 and has the property that
$\int_\gamma \mathbf{f} \cdot \boldsymbol{\tau} \, ds = 0$ for the vector field $\mathbf{f}(x) = \begin{pmatrix} x_1 \\ x_2 \end{pmatrix}$.

6. Let γ be a curve in $\mathbb{R}^3$ parametrized by arc length for a C^∞ mapping
$\mathbf{f} : [0, l] \longrightarrow \mathbb{R}^3$, and let $\boldsymbol{\tau}(s) = \mathbf{f}'(s)$ be the unit tangent vector to γ at the point $\mathbf{f}(s)$.

i) Show that if $\boldsymbol{\tau}'(s) \neq 0$ then that vector is perpendicular to $\boldsymbol{\tau}(s)$.
The length $\|\boldsymbol{\tau}'(s)\|_2 = \kappa(s)$ of this vector is called the **curvature** of γ at the point $\mathbf{f}(s)$, and its inverse $1/\kappa(s)$ is called the **radius of curvature** of γ at the point $\mathbf{f}(s)$.
ii) Show that $\kappa(s) = 0$ for all $s \in [0, L]$ if and only if γ is a straight line in $\mathbb{R}^3$.
iii) Show that the radius of curvature of a circle in a plane in $\mathbb{R}^3$ is equal to the radius of the circle. If $\boldsymbol{\tau}'(s) \neq 0$ the unit vector $\boldsymbol{v}(s) = \frac{1}{\kappa(s)} \boldsymbol{\tau}'(s)$ is called the **principal normal vector** to the curve γ. The vector $\boldsymbol{\beta}(s) = \boldsymbol{\tau}(s) \times \boldsymbol{v}(s)$, called the **binormal vector** to the curve γ at the point $\mathbf{f}(s)$, is a unit vector that is perpendicular both to the tangent vector $\boldsymbol{\tau}(s)$ and to the principal normal vector $\boldsymbol{v}(s)$.
iv) Show that $\boldsymbol{\beta}'(s) \cdot \boldsymbol{\beta}(s) = \boldsymbol{\beta}'(s) \cdot \mathbf{f}'(s) = 0$ and consequently that
$\boldsymbol{\beta}'(s) = -\tau(s)\boldsymbol{v}(s)$ for a real-valued function $\tau(s)$ and that
$\boldsymbol{v}'(s) = -\kappa(s)\boldsymbol{\tau}(s) + \tau(s)\boldsymbol{\beta}(s)$.
The function $\tau(s)$ is called the **torsion** of the curve γ. The set of differential equations in (iv), called the **Serret-Frenet equations**, can be written equivalently

$$\begin{pmatrix} \boldsymbol{\tau}'(s) & \boldsymbol{v}'(s) & \boldsymbol{\beta}'(s) \end{pmatrix} = \begin{pmatrix} \boldsymbol{\tau}(s) & \boldsymbol{v}(s) & \boldsymbol{\beta}(s) \end{pmatrix} \begin{pmatrix} 0 & -\kappa(s) & 0 \\ \kappa(s) & 0 & -\tau(s) \\ 0 & \tau(s) & 0 \end{pmatrix}$$

v) Show that $\tau(s) = 0$ if and only if the curve γ is contained in a two-dimensional linear subspace of $\mathbb{R}^3$.
vi) Assuming the general result from the elementary theory of ordinary differential equations that all parametrized curves for which the parametrization $\mathbf{f}(t)$ satisfies $f_i'(t) = \sum_{j=1}^n m_{ij}(t)f_j(t)$ for some given

functions $m_{ij}(t)$ differ by a linear transformation of $\mathbb{R}^n$, show that any two curves with the same curvature and torsion can be transformed into one another by a linear transformation and translation in $\mathbb{R}^3$. For this reason curvature and torsion are sometimes called the intrinsic equations of space curves.

(For more information about the geometry of curves see, for example, volume 2 of Spivak's *Differential Geometry*.)

7.2 Differential Forms

The further discussion of integration in n-dimensional spaces can be simplified considerably by using another real algebra, the exterior algebra $\Lambda(V)$ over a real vector space V. This algebra is often introduced rather generally and abstractly; but for the purposes here it is easier and quicker to introduce it through a simple explicit construction. To an n-dimensional real vector space V with a basis $\mathbf{v}_1, \ldots, \mathbf{v}_n$ there is associated another finite dimensional real vector space $\Lambda(V)$ with a basis consisting of the formal symbols

$$\mathbf{v}_{i_1} \wedge \mathbf{v}_{i_2} \wedge \cdots \wedge \mathbf{v}_{i_r} \tag{7.26}$$

for arbitrary integers r and indices $1 \le i_1, i_2, \ldots, i_r \le n$, with the understanding that

$$\mathbf{v}_{i_1} \wedge \mathbf{v}_{i_2} \wedge \cdots \wedge \mathbf{v}_{i_r} = \text{sign} \begin{pmatrix} i_1 & i_2 & \cdots & i_r \\ j_1 & j_2 & \cdots & j_r \end{pmatrix} \mathbf{v}_{j_1} \wedge \mathbf{v}_{j_2} \wedge \cdots \wedge \mathbf{v}_{j_r} \tag{7.27}$$

if the set of indices $i_1, i_2, \ldots, i_r$ is a permutation of the set of indices $j_1, j_2, \ldots, j_r$, where $\text{sign} \begin{pmatrix} i_1 & i_2 & \cdots & i_r \\ j_1 & j_2 & \cdots & j_r \end{pmatrix}$ is the sign of that permutation; thus two of the basic vectors (7.26) are equal if the ordered sets of indices are an even permutation of one another, and are the negatives of one another if the ordered sets of indices are an odd permutation of one another. In particular if two of the indices $i_1, i_2, \ldots, i_r$ in (7.26) are equal then that basis vector is the same as the basis vector for which those two terms are interchanged; but by the understanding of (7.27) those two basis vectors are also the negatives of one another, and consequently that basis vector is necessarily the zero vector. Thus the basis of $\Lambda(V)$ really consists just of those symbols (7.26) described by distinct sets of indices $1 \le i_1 < i_2 < \cdots < i_r \le n$. For example if dim $V = 3$ then the basis vectors for the vector space $\Lambda(V)$ are the seven symbols

$$\mathbf{v}_1, \quad \mathbf{v}_2, \quad \mathbf{v}_3, \quad \mathbf{v}_1 \wedge \mathbf{v}_2, \quad \mathbf{v}_1 \wedge \mathbf{v}_3, \quad \mathbf{v}_2 \wedge \mathbf{v}_3, \quad \mathbf{v}_1 \wedge \mathbf{v}_2 \wedge \mathbf{v}_3, \tag{7.28}$$

since for instance $\mathbf{v}_1 \wedge \mathbf{v}_1 = -\mathbf{v}_1 \wedge \mathbf{v}_1 = \mathbf{0}$ and $\mathbf{v}_2 \wedge \mathbf{v}_1 = -\mathbf{v}_1 \wedge \mathbf{v}_2$ is a nonzero basis vector, while any two symbols that involve 3 distinct basic vectors $\mathbf{v}_i$ differ only by a sign and any symbols involving 4 or more of the basic vectors must have two indices that are the same so must necessarily vanish. It is evident that the vector space $\Lambda(V)$ is just another finite dimensional real vector space; but it is defined in terms of a specific basis so that it is possible to introduce quite simply a multiplication on that vector space.

The subspace of the vector space $\Lambda(V)$ spanned by the symbols involving r basic vectors is denoted by $\Lambda^r(V)$. In particular then there is the natural identification $V = \Lambda^1(V)$, since the vector space V is spanned by the basis vectors $\mathbf{v}_j$; and it is convenient and traditional to set $\Lambda^0(V) = \mathbb{R}$, the one-dimensional real vector space spanned by the identity 1, while $\Lambda^r(V) = \mathbf{0}$ if $r > n = \dim V$. Thus if $\dim V = n$ the vector space $\Lambda(V)$ really is the direct sum

$$\Lambda(V) = \Lambda^0(V) \oplus \Lambda^1(V) \oplus \cdots \oplus \Lambda^n(V) \tag{7.29}$$

where $\Lambda^r(V)$ is the vector subspace of $\Lambda(V)$ with a basis consisting of the symbols $\mathbf{v}_{i_1} \wedge \mathbf{v}_{i_2} \wedge \cdots \wedge \mathbf{v}_{i_r}$ where $1 \leq i_1 < i_2 < \ldots < i_r \leq n$. The number of distinct basis vectors thus is the number of sets of r distinct integers from among the first n integers so

$$\dim \Lambda^r(V) = \binom{n}{r}; \tag{7.30}$$

that also holds for $r = 0$, since $\Lambda^0(V) = \mathbb{R}$ and by convention $\binom{n}{0} = 1$. It follows that

$$\dim \Lambda(V) = \sum_{r=0}^{n} \dim \Lambda^r(V) = \sum_{r=0}^{n} \binom{n}{r} = 2^n, \tag{7.31}$$

in view of the binomial expansion $(1+1)^n = 2^n$. A vector $\omega \in \Lambda^r(V)$ can be written as the sum

$$\omega = \sum_{1 \leq i_1 < i_2 < \cdots < i_r \leq n} a_{i_1 i_2 \ldots i_r} \mathbf{v}_{i_1} \wedge \mathbf{v}_{i_2} \wedge \cdots \wedge \mathbf{v}_{i_r} \tag{7.32}$$

for some real numbers $a_{i_1 i_2 \ldots i_r} \in \mathbb{R}$; this is called the **reduced form** for the vector ω, and the coefficients $a_{i_1 i_2 \ldots i_r}$ are uniquely determined. For some purposes it is more convenient to write a vector $\omega \in \Lambda^r(V)$ as a sum

$$\omega = \sum_{i_1, i_2, \cdots, i_r = 1}^{n} b_{i_1 i_2 \ldots i_r} \mathbf{v}_{i_1} \wedge \mathbf{v}_{i_2} \wedge \cdots \wedge \mathbf{v}_{i_r} \tag{7.33}$$

over all orders of the indices $i_1, i_2, \ldots, i_r$; this is called the **unreduced form** for the vector ω. In this form the coefficients $b_{i_1 i_2 \ldots i_r}$ are not uniquely determined; but ω can be rewritten in the reduced form by combining the coefficients of vectors $\mathbf{v}_{i_1} \wedge \mathbf{v}_{i_2} \wedge \cdots \wedge \mathbf{v}_{i_r}$ that differ only in sign. For example if $\dim V = 3$ a vector ω in the unreduced form

$$\omega = b_{12}\mathbf{v}_1 \wedge \mathbf{v}_2 + b_{13}\mathbf{v}_1 \wedge \mathbf{v}_3 + b_{23}\mathbf{v}_2 \wedge \mathbf{v}_3$$
$$+ b_{21}\mathbf{v}_2 \wedge \mathbf{v}_1 + b_{31}\mathbf{v}_3 \wedge \mathbf{v}_1 + b_{32}\mathbf{v}_3 \wedge \mathbf{v}_2 \qquad (7.34)$$

can be rewritten in the reduced form

$$\omega = (b_{12} - b_{21})\mathbf{v}_1 \wedge \mathbf{v}_2 + (b_{13} - b_{31})\mathbf{v}_1 \wedge \mathbf{v}_3 + (b_{23} - b_{32})\mathbf{v}_2 \wedge \mathbf{v}_3, \qquad (7.35)$$

and correspondingly for any of the vector spaces $\Lambda^r(V)$. The expressions (7.32) and (7.33) are a bit unwieldy at times, so it is often convenient to use **multi-index notation**, writing the unreduced form (7.33) as

$$\omega = \sum_I b_I \mathbf{v}_I \qquad (7.36)$$

where $I = (i_1, i_2, \ldots, i_r)$ and b_I is an abbreviation for $b_{i_1 \ldots i_r}$, while $\mathbf{v}_I$ is an abbreviation for $v_{i_1} \wedge \cdots \wedge v_{i_r}$, and writing the reduced form (7.32) as

$$\omega = \sum_I' a_I \mathbf{v}_I \qquad (7.37)$$

with the corresponding meaning of the terms; thus $\sum_I$ denotes a sum over all indices $i_1, \ldots, i_r$ in the range $1 \le i_1, \ldots, i_r \le n$ while $\sum_I'$ denotes a sum over all ordered sets of indices $1 \le i_1 < \cdots < i_r \le n$.

A multiplication operation in $\Lambda(V)$ is defined by associating to two basis vectors $\omega = \mathbf{v}_{i_1} \wedge \mathbf{v}_{i_2} \wedge \cdots \wedge \mathbf{v}_{i_r} \in \Lambda^r(V)$ and $\sigma = \mathbf{v}_{j_1} \wedge \mathbf{v}_{j_2} \wedge \cdots \wedge \mathbf{v}_{j_s} \in \Lambda^s(V)$ their **exterior product** or **wedge product**, defined as the vector

$$\omega \wedge \sigma = \mathbf{v}_{i_1} \wedge \mathbf{v}_{i_2} \wedge \cdots \wedge \mathbf{v}_{i_r} \wedge \mathbf{v}_{j_1} \wedge \mathbf{v}_{j_2} \wedge \cdots \wedge \mathbf{v}_{j_s} \in \Lambda^{r+s}(V);$$

this is the zero vector if $i_p = j_q$ for any p, q, but otherwise is one of the basis vectors for the subspace $\Lambda^{r+s}(V)$. In general then for any vectors $\omega \in \Lambda^r(V)$ and $\sigma \in \Lambda^s(V)$ written out in unreduced form as

$$\omega = \sum_{i_1, \ldots, i_r = 1}^{n} a_{i_1 \ldots i_r} \mathbf{v}_{i_1} \wedge \cdots \wedge \mathbf{v}_{i_r} \in \Lambda^r(V) \qquad (7.38)$$

and

$$\sigma = \sum_{j_1, \ldots, j_s = 1}^{n} b_{j_1 \ldots j_s} \mathbf{v}_{j_1} \wedge \cdots \wedge \mathbf{v}_{j_s} \in \Lambda^s(V) \qquad (7.39)$$

their wedge product is the vector

$$\omega \wedge \sigma = \sum_{i_1,\ldots,i_r,j_1,\ldots,j_s=1}^{n} a_{i_1\cdots i_r} b_{j_1\cdots j_s} \mathbf{v}_{i_1} \wedge \cdots \wedge \mathbf{v}_{i_r} \wedge \mathbf{v}_{j_1} \wedge \cdots \wedge \mathbf{v}_{j_s}, \qquad (7.40)$$

which also is in the unreduced form. Even if the vectors ω and σ are in the reduced form their wedge product is in the unreduced form, which of course can be rewritten in the reduced form. Note that the exterior product (7.40) is a bilinear function of the vectors ω and σ. The vector space $\Lambda(V)$ with this multiplication is a real algebra called an **exterior algebra**.

It is clear that the wedge product is associative, in the sense that

$$\omega \wedge (\sigma \wedge \tau) = (\omega \wedge \sigma) \wedge \tau; \qquad (7.41)$$

this product hence is usually written just as $\omega \wedge \sigma \wedge \tau$, and correspondingly for products of more vectors. The wedge product however is not commutative, for

$$\omega \wedge \sigma = (-1)^{rs} \sigma \wedge \omega \quad \text{if } \omega \in \Lambda^r(V) \text{ and } \sigma \in \Lambda^s(V). \qquad (7.42)$$

Indeed the permutation that exchanges

$$\mathbf{v}_{i_1} \wedge \cdots \wedge \mathbf{v}_{i_r} \wedge \mathbf{v}_{j_1} \wedge \cdots \wedge \mathbf{v}_{j_s} \text{ and } \mathbf{v}_{j_1} \wedge \cdots \wedge \mathbf{v}_{j_s} \wedge \mathbf{v}_{i_1} \wedge \cdots \wedge \mathbf{v}_{i_r}$$

can be written as the composition first of the permutation that transfers $\mathbf{v}_{j_1}$ across $\mathbf{v}_{i_1} \wedge \cdots \wedge \mathbf{v}_{i_r}$, a permuation with sign $(-1)^r$, followed by the permutation that transfers $\mathbf{v}_{j_2}$ across $\mathbf{v}_{i_1} \wedge \cdots \wedge \mathbf{v}_{i_r}$, another permutation of sign $(-1)^r$, and so on; and this composition of s permutations each of sign $(-1)^r$ is a permutation of sign $(-1)^{rs}$. In the special case in which $r = 0$ a vector $\omega \in \Lambda^0(V)$ is just a real number a, and the wedge product $\omega \wedge \sigma$ is just the vector $a\,\sigma$. Of course $\omega \wedge \sigma = \mathbf{0}$ automatically if $r + s > n$.

To apply this algebraic machinery to analysis in $\mathbb{R}^n$, the differentials dx_j of the coordinate functions x_j in $\mathbb{R}^n$ are a basis for the vector space $\mathbb{R}^n$, as in Section 3.2; and in terms of this basis the vector space $\mathbb{R}^n$ can be identified with the vector space $\Lambda^1(\mathbb{R}^n)$. A vector field over an open subset $U \subset \mathbb{R}^n$ is a mapping $\mathbf{f} : U \longrightarrow \mathbb{R}^n$ with the coordinate functions $\mathbf{f}(\mathbf{x}) = \{f_j(\mathbf{x})\}$ in terms of the basis dx_j of $\mathbb{R}^n$, so it can be viewed also as a mapping $\mathbf{f} : U \longrightarrow \Lambda^1(\mathbb{R}^n)$; and when it is written in terms of the basis vectors dx_j as the sum

$$\omega(\mathbf{x}) = \sum_{j=1}^{n} f_j(\mathbf{x})\, dx_j \qquad (7.43)$$

this expression is called a **differential form** of degree 1 in U, or for short a differential 1-form in U. Similarly a mapping $\mathbf{f} : U \longrightarrow \mathbb{R}^N$ where $N = \binom{n}{r}$ can be viewed as a mapping $\mathbf{f} : U \longrightarrow \Lambda^r(\mathbb{R}^n)$; and when written in terms of the basis vectors $dx_{j_1} \wedge \ldots \wedge dx_{j_r}$ as the sum

$$\omega(\mathbf{x}) = \sum_{1 \leq j_1 < \cdots < j_r \leq n} f_{j_1 \cdots j_r} dx_{j_1} \wedge \cdots \wedge dx_{j_r} \tag{7.44}$$

this expression is called a **differential form** of degree r in U, or a differential r-**form** for short. In particular $\Lambda^0(U)$, the vector space of 0-forms on U, is identified with the vector space of functions on U. A continuous differential form in U is one for which the functions $f_{j_1 \cdots j_r}$ are continuous, while a $\mathcal{C}^1$ or $\mathcal{C}^\infty$ differential form is one for which the functions are $\mathcal{C}^1$ or $\mathcal{C}^\infty$ functions. The vector space of differential r-forms in an open subset $U \subset \mathbb{R}^n$ is denoted by $\Lambda^r(U)$; whether that refers to the vector space of continuous, $\mathcal{C}^1$, or $\mathcal{C}^\infty$ differential r-forms in U will usually be determined by context. In particular when considering differential operators it is easiest just to consider the $\mathcal{C}^\infty$ differential forms. If it is necessary or convenient to be explicit, $\Lambda^r(\mathcal{C}, U)$ denotes the vector space of continuous r-forms in U while $\Lambda^r(\mathcal{C}^1, U)$ denotes the space of $\mathcal{C}^1$ differential forms and $\Lambda^r(\mathcal{C}^\infty, U)$ denotes the space of $\mathcal{C}^\infty$ differential forms.

The special case of differential forms in an open set $U \subset \mathbb{R}^3$ is both a useful example and a case that arises sufficiently often that it is convenient to have an alternative description of differential form in that case. A differential 1-form can be viewed as an alternative way of writing a vector field in $\mathbb{R}^3$, a mapping $f : U \longrightarrow \mathbb{R}^3$, as already discussed; and since $\dim \Lambda^2(\mathbb{R}^3) = 3$ a vector field also can be identified with a differential 2-form. Thus to a vector field $\mathbf{f}(\mathbf{x}) = \{f_j(\mathbf{x})\}$ in $U \subset \mathbb{R}^3$ there are naturally associated the two differential forms

$$\omega_{\mathbf{f}}(\mathbf{x}) = f_1(\mathbf{x})dx_1 + f_2(\mathbf{x})dx_2 + f_3(\mathbf{x})dx_3 \in \Lambda^1(U)$$

and   (7.45)

$$\Omega_{\mathbf{f}}(\mathbf{x}) = f_1(\mathbf{x})dx_2 \wedge dx_3 + f_2(\mathbf{x})dx_3 \wedge dx_1 + f_3(\mathbf{x})dx_1 \wedge dx_2 \in \Lambda^2(U),$$

where the notation for the 2-form $\Omega_{\mathbf{f}}(\mathbf{x})$, involving cyclic permutations of the indices, is chosen to fit into the classical form for the algebraic operations. The exterior product of the 1-form associated to a vector field $\mathbf{f} : U \longrightarrow \mathbb{R}^3$ and the 2-form associated to a vector field $\mathbf{g} : U \longrightarrow \mathbb{R}^3$ is easily seen to be the 3-form

$$\omega_{\mathbf{f}}(\mathbf{x}) \wedge \Omega_{\mathbf{g}}(\mathbf{x}) = \Big(f_1(\mathbf{x})g_1(\mathbf{x}) + f_2(\mathbf{x})g_2(\mathbf{x}) + f_3(\mathbf{x})g_3(\mathbf{x})\Big)dx_1 \wedge dx_2 \wedge dx_3; \quad (7.46)$$

this is the 3-form associated to the dot product or inner product $\mathbf{f}(\mathbf{x}) \cdot \mathbf{g}(\mathbf{x})$ of the two vector fields $\mathbf{f}(\mathbf{x})$ and $\mathbf{g}(\mathbf{x})$, so that

$$\omega_{\mathbf{f}}(\mathbf{x}) \wedge \Omega_{\mathbf{g}}(\mathbf{x}) = \mathbf{f}(x) \cdot \mathbf{g}(x) \, dx_1 \wedge dx_2 \wedge dx_3, \tag{7.47}$$

and in this case the exterior product is commutative. On the other hand a straightforward calculation shows that the exterior product of the 1-forms associated to two vector fields $\mathbf{f} : U \longrightarrow \mathbb{R}^3$ and $\mathbf{g} : U \longrightarrow \mathbb{R}^3$ is the 2-form

$$\omega_{\mathbf{f}}(\mathbf{x}) \wedge \omega_{\mathbf{g}}(\mathbf{x}) = \det \begin{pmatrix} f_2(\mathbf{x}) & f_3(\mathbf{x}) \\ g_2(\mathbf{x}) & g_3(\mathbf{x}) \end{pmatrix} dx_2 \wedge dx_3 + \det \begin{pmatrix} f_3(\mathbf{x}) & f_1(\mathbf{x}) \\ g_3(\mathbf{x}) & g_1(\mathbf{x}) \end{pmatrix} dx_3 \wedge dx_1$$

$$+ \det \begin{pmatrix} f_1(\mathbf{x}) & f_2(\mathbf{x}) \\ g_1(\mathbf{x}) & g_2(\mathbf{x}) \end{pmatrix} dx_1 \wedge dx_2, \tag{7.48}$$

which is the 2-form associated to the vector field $\mathbf{f}(\mathbf{x}) \times \mathbf{g}(\mathbf{x})$, traditionally called **cross-product** of the two vector fields $\mathbf{f}(\mathbf{x})$ and $\mathbf{g}(\mathbf{x})$, defined by

$$\mathbf{f}(\mathbf{x}) \times \mathbf{g}(\mathbf{x}) = \left\{ \det \begin{pmatrix} f_2(\mathbf{x}) & f_3(\mathbf{x}) \\ g_2(\mathbf{x}) & g_3(\mathbf{x}) \end{pmatrix}, \det \begin{pmatrix} f_3(\mathbf{x}) & f_1(\mathbf{x}) \\ g_3(\mathbf{x}) & g_1(\mathbf{x}) \end{pmatrix}, \right.$$

$$\left. \det \begin{pmatrix} f_1(\mathbf{x}) & f_2(\mathbf{x}) \\ g_1(\mathbf{x}) & g_2(\mathbf{x}) \end{pmatrix} \right\}; \tag{7.49}$$

the easiest way to remember this definition is to write it symbolically as

$$\mathbf{f}(\mathbf{x}) \times \mathbf{g}(\mathbf{x}) = \det \begin{pmatrix} \epsilon_1 & \epsilon_2 & \epsilon_3 \\ f_1(\mathbf{x}) & f_2(\mathbf{x}) & f_3(\mathbf{x}) \\ g_1(\mathbf{x}) & g_2(\mathbf{x}) & g_3(\mathbf{x}) \end{pmatrix}, \tag{7.50}$$

where ϵ_1, ϵ_2, ϵ_3 is a basis for $\mathbb{R}^3$. In these terms then

$$\omega_{\mathbf{f}}(\mathbf{x}) \wedge \omega_{\mathbf{g}}(\mathbf{x}) = \Omega_{\mathbf{f} \times \mathbf{g}}(\mathbf{x}); \tag{7.51}$$

the exterior product of 1-forms is skew-symmetric, as is the cross-product, so

$$\omega_{\mathbf{g}}(\mathbf{x}) \wedge \omega_{\mathbf{f}}(\mathbf{x}) = -\omega_{\mathbf{f}}(\mathbf{x}) \wedge \omega_{\mathbf{g}}(\mathbf{x}) \quad \text{and} \quad \mathbf{f}(\mathbf{x}) \times \mathbf{g}(\mathbf{x}) = -\mathbf{g}(\mathbf{x}) \times \mathbf{f}(\mathbf{x}). \tag{7.52}$$

It is clear from (7.49) that the cross-product $\mathbf{f}(\mathbf{x}) \times \mathbf{g}(\mathbf{x})$ is trivial at a point $\mathbf{x}$ precisely when the two vectors $\mathbf{f}(\mathbf{x})$ and $\mathbf{g}(\mathbf{x})$ are linearly dependent. It is also clear from (7.49) that for any other vector field $\mathbf{h}(\mathbf{x})$ in U

$$\mathbf{h}(\mathbf{x}) \cdot \left(\mathbf{f}(\mathbf{x}) \times \mathbf{g}(\mathbf{x})\right) = \det \begin{pmatrix} h_1(\mathbf{x}) & h_2(\mathbf{x}) & h_3(\mathbf{x}) \\ f_1(\mathbf{x}) & f_2(\mathbf{x}) & f_3(\mathbf{x}) \\ g_1(\mathbf{x}) & g_2(\mathbf{x}) & g_3(\mathbf{x}) \end{pmatrix}, \tag{7.53}$$

which provides an alternative characterization of the cross-product. An evident consequence of (7.53) is that $\mathbf{f}(\mathbf{x}) \cdot \left(\mathbf{f}(\mathbf{x}) \times \mathbf{g}(\mathbf{x})\right) = \mathbf{g}(\mathbf{x}) \cdot \left(\mathbf{f}(\mathbf{x}) \times \mathbf{g}(\mathbf{x})\right) = 0$; hence the cross-product $\mathbf{f}(\mathbf{x}) \times \mathbf{g}(\mathbf{x})$ is a vector in $\mathbb{R}^3$ that is perpendicular to both $\mathbf{f}(\mathbf{x})$ and $\mathbf{g}(\mathbf{x})$ if these vectors are linearly independent, so it is determined uniquely up to sign. As for the sign, it follows from (7.49) then that if $\mathbf{f}(\mathbf{x}_0) = \{1, 0, 0\}$ and $\mathbf{g}(\mathbf{x}_0) = \{0, 1, 0\}$ at a point $\mathbf{x}_0 \in U$ then $\mathbf{f}(\mathbf{x}_0) \times \mathbf{g}(\mathbf{x}_0) = \{0, 0, 1\}$; consequently the three vectors $\mathbf{f}(\mathbf{x}_0), \mathbf{g}(\mathbf{x}_0), \mathbf{f}(\mathbf{x}_0) \times \mathbf{g}(\mathbf{x}_0)$ have the same orientation as the three basis vectors $dx_1, dx_2, dx_3 \in \mathbb{R}^3$, the natural orientation associated of the vector space $\mathbb{R}^3$ with this choice of coordinates.

In $\mathbb{R}^n$ the vector space of differential r-forms and the vector space of differential $(n - r)$-forms also have the same dimension, so as in the preceding special case $n = 3$ these differential forms can be viewed as differential forms associated to vector fields $\mathbf{f} : U \longrightarrow \mathbb{R}^N$ where $N = \binom{n}{n-r}$. Alternatively there is the Hodge **duality** mapping, the explicit linear isomorphism

$$* : \Lambda^r(U) \longrightarrow \Lambda^{n-r}(U) \quad \text{for open sets } U \subset \mathbb{R}^n \tag{7.54}$$

that associates to a differential form $\omega = \sum_I' f_I \, d\mathbf{x}_I \in \Lambda^r(U)$ in the reduced form the differential form $*\omega$ in reduced form

$$*\omega = \sum_{I,J}' \epsilon(I\, J) f_I \, d\mathbf{x}_J \in \Lambda^{n-r}(U) \tag{7.55}$$

where

$$\epsilon(I\, J) = \text{sign} \begin{pmatrix} I & J \\ 1 \cdots n \end{pmatrix} \tag{7.56}$$

is the sign of the permutation of the sequence $(1, \dots, n)$ leading to the order (I, J) of these integers, and is set equal to 0 if the sequence (I, J) is not a permutation of the sequence $(1, \dots, n)$. Thus for example in $\mathbb{R}^n$

$$*f = f dx_1 \wedge dx_2 \wedge \cdots \wedge dx_n \tag{7.57}$$

while in particular in $\mathbb{R}^2$

$$*(f_1 dx_1 + f_2 dx_2) = \sum_{1 \leq i,j \leq 2} \epsilon(i\, j) f_i \, dx_j = f_1 \, dx_2 - f_2 \, dx_1 \tag{7.58}$$

and in $\mathbb{R}^3$

$$*(f_1\,dx_1 + f_2\,dx_2 + f_3\,dx_3) = \sum_{1 \leq i \leq 3} \sum_{1 \leq j_1 < j_2 \leq 3} \epsilon(i\,j_1 j_2)f_i\,dx_{j_1} \wedge dx_{j_2}$$

$$= f_1\,dx_2 \wedge dx_3 - f_2 dx_1 \wedge dx_3 + f_3 dx_1 \wedge dx_2 \quad (7.59)$$

or equivalently $*\omega_{\mathbf{f}} = \Omega_{\mathbf{f}}$ for the differential forms associated to a vector field $\mathbf{f}$. The general Hodge duality operator satisfies

$$* * \omega = (-1)^{r(n-r)}\omega \quad \text{for any differential } r\text{-form } \omega \text{ in } \mathbb{R}^n. \quad (7.60)$$

To demonstrate this, if $\omega = \sum_I' f_I\,d\mathbf{x}_I$ is an r-form in $\mathbb{R}^n$ then by definition

$$*(*\omega) = *\left(\sum_{I,J}' \epsilon(I\,J)f_I\,d\mathbf{x}_J\right)$$

$$= \sum_{I,J,K}' \epsilon(J\,K)\epsilon(I\,J)f_I\,d\mathbf{x}_K. \quad (7.61)$$

For any ordered set of r indices $1 \leq k_1 < \cdots < k_r \leq n$ there is a unique complementary ordered set of $n - r$ indices $1 \leq j_1 < \cdots < j_{n-r} \leq n$ such that (J, K) a permutation of the integers $(1, 2, \ldots, n)$, and then there is a further unique complementary ordered set of r indices $1 \leq i_1 < \cdots < i_r \leq n$ such that (I, J) also is a permutation of the integers $(1, 2, \ldots, n)$, for which clearly $I = K$; thus the only nontrivial terms in the sum (7.61) are those for which the indices J and I are uniquely determined by the indices K, indeed for which $I = K$, and since $\epsilon(IJ)\epsilon(JK) = \epsilon(KJ)\epsilon(JK) = (-1)^{r(n-r)}$ just as in the earlier examination of exterior products it follows that (7.61) reduces to (7.60) as desired. For instance since it was already noted in (7.59) that $*\omega_{\mathbf{f}} = \Omega_{\mathbf{f}}$ for any vector field $\mathbf{f}$ in $\mathbb{R}^3$, it follows from (7.60) that $*\Omega_{\mathbf{f}} = * * \omega_{\mathbf{f}} = \omega_{\mathbf{f}}$.

The characteristic property of Hodge duality[1] though is that if $\omega = \sum_I' f_I(\mathbf{x})d\mathbf{x}_I$ and $\sigma = \sum_J' g_J(\mathbf{x})d\mathbf{x}_J$ are two r-forms in $\mathbb{R}^n$ then

$$\omega \wedge *\sigma = \sum_I' f_I(\mathbf{x})g_I(\mathbf{x})dx_1 \wedge \cdots \wedge dx_n. \quad (7.62)$$

Indeed by definition

$$\omega \wedge *\sigma = \sum_{I,J,K}' f_I(\mathbf{x})d\mathbf{x}_I \wedge \epsilon(J\,K)g_J(\mathbf{x})d\mathbf{x}_K$$

$$= \sum_{I,J,K}' \epsilon(J\,K)f_I(\mathbf{x})g_J(\mathbf{x})d\mathbf{x}_I \wedge d\mathbf{x}_K. \quad (7.63)$$

---

[1]The version of Hodge duality discussed here is a special case of the general Hodge duality, for which the characteristic property is that

$$\omega \wedge *\sigma = \sum_{I,J}' h_{I\,J} f_I(\mathbf{x})g_J(\mathbf{x})dx_1 \wedge \cdots \wedge dx_n$$

where $h_{IJ} = h_{i_1 j_1} h_{i_2 j_2} \cdots h_{i_n j_n}$ for some positive definite symmetric matrix h_{ij}; actually for some applications in physics the matrix h_{ij} is not even necessarily positive definite, but is a Lorentz metric, having one negative and $n - 1$ positive eigenvalues.

For any ordered set of r indices $1 \leq i_1 < \cdots < i_r \leq n$ there is a unique ordered set of $n - r$ indices $1 \leq k_1 < \cdots < k_{n-r} \leq n$ such that $d\mathbf{x}_I \wedge d\mathbf{x}_K$ is nontrivial, and for that ordered set of indices $d\mathbf{x}_I \wedge d\mathbf{x}_K = \epsilon(IK)dx_1 \wedge dx_2 \wedge \cdots \wedge dx_n$; then there is a unique ordered set of r indices $1 \leq j_1 < \cdots < j_r \leq n$ such that $\epsilon(JK) \neq 0$, and clearly $J = I$. Thus the choice of an ordered set of indices I determines unique ordered sets of indices J and K for which the associated term in the sum (7.63) is nontrivial, and since $\epsilon(IK)d\mathbf{x}_I \wedge d\mathbf{x}_K = dx_1 \wedge dx_2 \wedge \cdots \wedge dx_n$ that sum reduces to (7.62).

There are other operators on differential forms that arise in analysis. If $\boldsymbol{\phi}$: $D \longrightarrow E$ is a mapping from an open subset $D \subset \mathbb{R}^m$ into an open subset $E \subset \mathbb{R}^n$ then to any function f in the set E there can be associated the **induced function** $\boldsymbol{\phi}^*(f)$ in the set D, the function defined by

$$\boldsymbol{\phi}^*(f)(\mathbf{x}) = f(\boldsymbol{\phi}(\mathbf{x})) \quad \text{for all points } \mathbf{x} \in D. \tag{7.64}$$

It is clear that $\boldsymbol{\phi}^*(c_1 f_1 + c_2 f_2) = c_1 \boldsymbol{\phi}^*(f_1) + c_2 \boldsymbol{\phi}^*(f_2)$ for any functions f_1, f_2 in D and any constants $c_1, c_2 \in \mathbb{R}$. To emphasize the main point it will be assumed that the changes of coordinates and the differential forms are $\mathcal{C}^\infty$; the results also hold for $\mathcal{C}^r$ differential forms with a bit of care in tracking the differentiations involved. To any function $f \in \Lambda^0(E)$ there can be associated the induced function $\boldsymbol{\phi}^*(f) \in \Lambda^0(D)$, and that is a well-defined linear mapping

$$\boldsymbol{\phi}^* : \Lambda^0(E) \longrightarrow \Lambda^0(D). \tag{7.65}$$

Similarly to any mapping $\mathbf{f} : E \longrightarrow \mathbb{R}^N$ there can be associated the **induced mapping** $\boldsymbol{\phi}^*(\mathbf{f})$ in the set D, the mapping defined by

$$\boldsymbol{\phi}^*(\mathbf{f})(\mathbf{x}) = \mathbf{f}(\boldsymbol{\phi}(\mathbf{x})) \quad \text{for all points } \mathbf{x} \in D. \tag{7.66}$$

Differential forms $\omega \in \Lambda^r(E)$ in an open set $E \subset \mathbb{R}^n$ are just mappings from E into a vector space $\mathbb{R}^N$, but are expressed in terms of a basis for $\mathbb{R}^N$ that is determined by the basis $dy_1, \ldots, dy_n$ for the vector space $\mathbb{R}^n$ if $y_1, \ldots, y_n$ are the coordinates in $\mathbb{R}^n$. On the other hand differential forms $\omega \in \Lambda^r(D)$ in an open set $D \subset \mathbb{R}^m$ are mappings from D into a vector space $\mathbb{R}^M$, expressed in terms of a basis for $\mathbb{R}^M$ that is determined by the basis $dx_1, \ldots, dx_m$ for the vector space $\mathbb{R}^m$ if $x_1, \ldots, x_m$ are the coordinates in $\mathbb{R}^m$. The induced differential form under the mapping $\boldsymbol{\phi} : D \longrightarrow E$ must also reflect the change in the bases for the two vector spaces $\mathbb{R}^m$ and $\mathbb{R}^n$ defined by the mapping $\boldsymbol{\phi}$. By the chain rule the derivatives of these functions are the row vectors

$$y_i'(\mathbf{x}) = \left(\frac{\partial y_i}{\partial x_1}, \quad \frac{\partial y_i}{\partial x_2}, \quad \cdots, \quad \frac{\partial y_i}{\partial x_m} \right),$$

which can be written equivalently in terms of the bases dx_i of the space $\mathbb{R}^m$ and dy_i of the space $\mathbb{R}^n$ as the linear combination

$$dy_i = \sum_{j=1}^{m} \frac{\partial y_i}{\partial x_j} dx_j \tag{7.67}$$

of the basis vectors dx_j. The **induced differential form** $\phi^*(\omega) \in \Lambda^r(D)$ on D consequently is defined separately for any multi-index I by

$$\phi^*\left(f_I(\mathbf{y})d\mathbf{y}_I\right) = \sum_J f_I(\phi(\mathbf{x})) \frac{\partial y_{i_1}}{\partial x_{j_1}} dx_{j_1} \wedge \cdots \wedge \frac{\partial y_{i_r}}{\partial x_{j_r}} dx_{j_r}$$

$$= \sum_J f_I(\phi(\mathbf{x})) \frac{\partial y_{i_1}}{\partial x_{j_1}} \cdots \frac{\partial y_{i_r}}{\partial x_{j_r}} dx_{j_1} \wedge \cdots \wedge dx_{j_r}, \tag{7.68}$$

which in an abbreviated form is sometimes written

$$\phi^*\left(f_I(\mathbf{y})d\mathbf{y}_I\right) = f_I(\phi(\mathbf{x})) \sum_J \frac{\partial Y_I}{\partial X_J} d\mathbf{x}_J. \tag{7.69}$$

This is sometimes also referred to as the **pullback** of the differential form ω from E to D. It is clear that $\phi^*(\Lambda^r(E)) \subset \Lambda^r(D)$ and that

$$\phi^*(c_1\omega_1 + c_2\omega_2) = c_1\phi^*(\omega_1) + c_2\phi^*(\omega_2) \quad \text{for any } \omega_1, \omega_2 \in \Lambda^r(E) \text{ and } c_1, c_2 \in \mathbb{R}, \tag{7.70}$$

so the mapping ϕ^* is a linear mapping

$$\phi^* : \Lambda^r(E) \longrightarrow \Lambda^r(D); \tag{7.71}$$

and it is also clear that

$$\phi^*(\omega \wedge \sigma) = \phi^*(\omega) \wedge \phi^*(\sigma) \quad \text{for any } \omega \in \Lambda^r(E) \text{ and } \sigma \in \Lambda^s(E). \tag{7.72}$$

In particular if $D, E \subset \mathbb{R}^n$ are both subsets of vector spaces of the same dimension and $\omega \in \Lambda^n(E)$ is a differential n-form in E then $\omega(\mathbf{y}) = f(\mathbf{y})dy_1 \wedge \cdots \wedge dy_n$ for some function f in E and (7.68) reduces to

$$\phi^*\left(f(\mathbf{y})dy_1 \wedge \cdots \wedge dy_n\right) = \sum_J f(\phi(\mathbf{x})) \frac{\partial y_1}{\partial x_{j_1}} \cdots \frac{\partial y_n}{\partial x_{j_n}} dx_{j_1} \wedge \cdots \wedge dx_{j_n}$$

$$= f(\phi(\mathbf{x})) \det \phi'(\mathbf{x}) dx_1 \wedge \cdots \wedge dx_n. \tag{7.73}$$

This illustrates the role of exterior algebras as convenient tools in handling determinants, and in a sense underlies all the applications of differential forms in analysis and geometry.

Theorem 7.9. *If $\phi : D \longrightarrow E$ is a C^∞ mapping between two open subsets $D \subset \mathbb{R}^m$ and $E \subset \mathbb{R}^n$ while $\psi : E \longrightarrow F$ is a C^∞ mapping between open subsets $E \subset \mathbb{R}^n$ and $F \subset \mathbb{R}^p$ then $(\psi \circ \phi)^* : \Lambda^r(F) \longrightarrow \Lambda^r(D)$ is the composition $(\psi \circ \phi)^* = \phi^* \circ \psi^*$ of the mapping $\phi^* : \Lambda^r(E) \longrightarrow \Lambda^r(D)$ and the mapping $\psi^* : \Lambda^r(F) \longrightarrow \Lambda^r(E)$.*

Proof: If the coordinates in D are $t_1, \ldots, t_m$, the coordinates in E are $x_1, \ldots, x_n$, and the coordinates in F are $y_1, \ldots, y_p$, then for any differential form $\omega(\mathbf{y}) = \sum_I f_I(\mathbf{y}) d\mathbf{y}_I$ it follows that

$$\psi^*(\omega)(\mathbf{x}) = \sum_{I,J} f_I\big(\psi(\mathbf{x})\big) \frac{\partial \mathbf{y}_I}{\partial \mathbf{x}_J}(\mathbf{x}) d\mathbf{x}_J$$

and

$$\phi^*\big(\psi^*(\omega)\big)(\mathbf{t}) = \sum_{I,J,K} f_I\big(\psi(\phi(\mathbf{t}))\big) \frac{\partial \mathbf{y}_I}{\partial \mathbf{x}_J}\big(\phi(\mathbf{t})\big) \frac{\partial \mathbf{x}_J}{\partial \mathbf{t}_K}(\mathbf{t}) d\mathbf{t}_K.$$

On the other hand for the composite mapping $\psi \circ \phi$

$$(\psi \circ \phi)^*(\omega)(\mathbf{t}) = \sum_{I,K} f_I\big(\psi \circ \phi(\mathbf{t})\big) \frac{\partial \mathbf{y}_I}{\partial \mathbf{t}_K}(\mathbf{t}) d\mathbf{t}_K,$$

and it follows from the chain rule that

$$\frac{\partial \mathbf{y}_I}{\partial \mathbf{t}_K} = \sum_J \frac{\partial \mathbf{y}_I}{\partial \mathbf{x}_J}(\phi(\mathbf{t})) \frac{\partial \mathbf{x}_J}{\partial \mathbf{t}_K}(\mathbf{t}).$$

Consequently $(\psi \circ \phi)^*(\omega) = \phi^*\big(\psi^*(\omega)\big)$, which suffices for the proof. $\blacksquare$

The **exterior derivative** of a differential form $\omega(\mathbf{x}) \in \Lambda^r(U)$ of degree r is defined to be a differential form $d\omega(\mathbf{x}) \in \Lambda^{r+1}(U)$ of degree $r + 1$. Explicitly, if $f(\mathbf{x}) \in \Lambda^0(U)$ is a differential form of degree 0, a function defined in the subset $U \subset \mathbb{R}^n$, its exterior derivative is defined to be the derivative of that function expressed as a differential form, so

$$df(\mathbf{x}) = \sum_{j=1}^n \partial_j f(\mathbf{x}) dx_j \quad \text{if } f \in \Lambda^0(U). \tag{7.74}$$

If $\omega(\mathbf{x}) = \sum_I f_I(\mathbf{x}) d\mathbf{x}_I \in \Lambda^r(U)$ is a differential form of degree $r > 0$ its exterior derivative is defined by

$$d\omega(\mathbf{x}) = \sum_I df_I(\mathbf{x}) \wedge d\mathbf{x}_I, \tag{7.75}$$

so that

$$d\left(\sum_{i_1,\ldots,i_r=1}^{n} f_{i_1\cdots i_r}(\mathbf{x})\, dx_{i_1} \wedge \cdots \wedge dx_{i_r} \right) = \sum_{j,i_1,\ldots,i_r=1}^{n} \partial_j f_{i_1\cdots i_r}(\mathbf{x})\, dx_j \wedge dx_{i_1} \wedge \cdots \wedge dx_{i_r}.$$

$$(7.76)$$

In particular of course $d\omega = 0$ for any n-form ω in an open subset $U \subset \mathbb{R}^n$.

Theorem 7.10. *Exterior differentiation is a linear mapping*

$$d : \Lambda^r(U) \longrightarrow \Lambda^{r+1}(U) \tag{7.77}$$

of C^∞ differential r-forms in an open subset $U \subset \mathbb{R}^n$ such that
(i) $d\left(\omega(\mathbf{x}) \wedge \sigma(\mathbf{x})\right) = d\omega(\mathbf{x}) \wedge \sigma(\mathbf{x}) + (-1)^r \omega(\mathbf{x}) \wedge d\sigma(\mathbf{x})$
    if $\omega(\mathbf{x}) \in \Lambda^r(U)$ and $\sigma(\mathbf{x}) \in \Lambda^s(U)$ while
(ii) $d\, d\,\omega(\mathbf{x}) = 0$ *for any differential form $\omega \in \Lambda^r(U)$.*

Proof: That exterior differentiation is a linear mapping on r-forms is an immediate consequence of the definitions (7.74) and (7.75); and the exterior derivative of a C^∞ differential form also is a C^∞ differential form.
(i) For functions it follows from the usual rules for differentiation of a product that

$$d\big(f(\mathbf{x})g(\mathbf{x})\big) = f(\mathbf{x})dg(\mathbf{x}) + g(\mathbf{x})df(\mathbf{x}). \tag{7.78}$$

In general if $\omega = \sum_I f_I(\mathbf{x})d\mathbf{x}_I \in \Lambda^r(U)$ and $\sigma = \sum_J g_J(\mathbf{x})d\mathbf{x}_J \in \Lambda^s(U)$ it follows from (7.78) that

$$
\begin{aligned}
d(\omega(\mathbf{x}) \wedge \sigma(\mathbf{x})) &= d\Big(\sum_{I,J} f_I(\mathbf{x})g_J(\mathbf{x})d\mathbf{x}_I \wedge d\mathbf{x}_J \Big) \\
&= \sum_{I,J} d(f_I(\mathbf{x})g_J(\mathbf{x})) \wedge d\mathbf{x}_I \wedge d\mathbf{x}_J \\
&= \sum_{I,J} (g_J(\mathbf{x})df_I(\mathbf{x}) + f_I(\mathbf{x})dg_J(\mathbf{x})) \wedge d\mathbf{x}_I \wedge d\mathbf{x}_J \\
&= \sum_{I,J} df_I(\mathbf{x}) \wedge d\mathbf{x}_I \wedge g_J(\mathbf{x})d\mathbf{x}_J \\
&\quad + (-1)^r \sum_{I,J} f_I(\mathbf{x})d\mathbf{x}_I \wedge dg_J(\mathbf{x}) \wedge d\mathbf{x}_J,
\end{aligned}
$$

where the sign $(-1)^r$ arises since the 1-form $dg_J(\mathbf{x})$ must be moved across the r differentials in $d\mathbf{x}_I$.

(ii) If $\omega(\mathbf{x}) = \sum_I f_I(\mathbf{x}) d\mathbf{x}_I \in \Lambda^r(U)$ then

$$d\big(d\omega(\mathbf{x})\big) = d\Big(\sum_I df_I(\mathbf{x}) \wedge d\mathbf{x}_I \Big) = d\Big(\sum_{I,j} \frac{\partial f_I(\mathbf{x})}{\partial x_j} dx_j \wedge d\mathbf{x}_I \Big)$$

$$= \sum_{I,j,k} \frac{\partial^2 f_I(\mathbf{x})}{\partial x_j \partial x_k} dx_k \wedge dx_j \wedge d\mathbf{x}_I,$$

which is zero since $\partial_{jk} f_I(\mathbf{x})$ is symmetric in the indices j, k by Theorem 3.40 while $dx_k \wedge dx_j$ is skew-symmetric in these indices. That suffices for the proof.

A differential form $\omega(\mathbf{x})$ is said to be **closed** if $d\omega(\mathbf{x}) = 0$, and it is said to be **exact** if $\omega(\mathbf{x}) = d\sigma(\mathbf{x})$ for some differential form $\sigma(\mathbf{x})$. A consequence of Theorem 7.10 (ii) is that any exact differential form is necessarily a closed differential form; but the converse is not true in general. For example, in the complement of the origin in the plane $\mathbb{R}^2$ of the variables x_1, x_2 the function $\theta(x_1, x_2) = \tan^{-1}(x_2/x_1)$, the angle in polar coordinates, is well defined in any half-plane bounded by a line through the origin, such as the half-planes $x_1 > 0$ and $x_1 < 0$, but it cannot be defined in the entire complement of the origin as a single-valued function. However a simple calculation shows that

$$\omega(\mathbf{x}) = d\theta(x_1, x_2) = -\frac{x_2}{x_1^2 + x_2^2} dx_1 + \frac{x_1}{x_1^2 + x_2^2} dx_2,$$

so this is a well-defined closed differential 1-form in the complement of the origin in $\mathbb{R}^2$. If $\omega(\mathbf{x})$ were an exact differential form in the full complement of the origin it would be the exterior derivative $\omega(\mathbf{x}) = df(\mathbf{x})$ of a well-defined function $f(\mathbf{x})$ in the complement of the origin; but then $df(\mathbf{x}) = d\theta(\mathbf{x})$ so that $d\big(f(\mathbf{x}) - \theta(\mathbf{x})\big) = 0$ hence $f(\mathbf{x}) - \theta(\mathbf{x}) = c$ is a constant in the complement of the origin, which would imply that $\theta(\mathbf{x}) = f(\mathbf{x}) - c$ is a well-defined function in the complement of the origin, a contradiction. However it is at least true locally that any closed form is exact.

Theorem 7.11 (Poincaré's Lemma). *If $\omega(\mathbf{x}) \in \Lambda^r(\Delta)$ is a closed C^∞ differential form of degree $r > 0$ in the standard open cell*

$$\Delta = \Big\{ \mathbf{x} \in \mathbb{R}^n \,\Big|\, 0 < x_i < 1 \ \text{for} \ 1 \le i \le n \Big\} \subset \mathbb{R}^n \tag{7.79}$$

then there is a C^∞ differential form $\sigma(\mathbf{x}) \in \Lambda^{r-1}(\Delta)$ of degree $r - 1$ in Δ such that $\omega(\mathbf{x}) = d\sigma(\mathbf{x})$.

Proof: Consider differential r-forms $\omega(\mathbf{x}) = \sum_I' f(\mathbf{x}) d\mathbf{x}_I$ that involve only the differentials $dx_1, dx_2, \ldots, dx_k$ for some index $1 \le k \le n$, so that

$$\omega(\mathbf{x}) = \sum_{1 \le i_1 < \cdots < i_r \le k} f_{i_1, \ldots, i_r}(\mathbf{x}) dx_{i_1} \wedge \cdots \wedge dx_{i_r}.$$

The theorem will be demonstrated by induction on the index k. If $k = 1$ then $\omega(\mathbf{x}) = f(\mathbf{x})dx_1$ and $0 = d\omega(\mathbf{x}) = \sum_{i=2}^n \partial_i f(\mathbf{x})dx_i \wedge dx_1$ so $\partial_i f(\mathbf{x}) = 0$ for $i > 1$ and consequently $f(\mathbf{x}) = f(x_1)$ is a function of the variable x_1 alone, where $x_1 \in (0, 1)$. The integral $g(x_1) = \int_a^{x_1} f$ for a point $a \in (0, 1)$ is a function of the variable $x_1 \in (0, 1)$ for which $dg(\mathbf{x}) = dg(x_1) = f(x_1)dx_1 = \omega(\mathbf{x})$, which establishes the desired result in this special case. Assume then that the theorem has been demonstrated for an index k, and consider a differential r-form $\omega(\mathbf{x})$ that involves only the differentials $dx_1, \ldots, dx_k, dx_{k+1}$. This differential form can be written

$$\omega(\mathbf{x}) = \omega_1(\mathbf{x}) \wedge dx_{k+1} + \omega_2(\mathbf{x})$$

where the differential $(r-1)$-form $\omega_1(\mathbf{x})$ and the differential r-form $\omega_2(\mathbf{x})$ are in reduced form and involve only the differentials $dx_1, \ldots, dx_k$, so that

$$\omega_1(\mathbf{x}) = \sideset{}{'}\sum_I f_I(\mathbf{x})d\mathbf{x}_I \quad \text{and} \quad \omega_2(\mathbf{x}) = \sideset{}{'}\sum_J g_J(\mathbf{x})d\mathbf{x}_J$$

for indices

$$1 \le i_1 < i_2 < \ldots < i_{r-1} \le k \quad \text{and} \quad 1 \le j_1 < j_2 < \cdots < j_r \le k.$$

By assumption

$$0 = d\omega(\mathbf{x}) = d\omega_1(\mathbf{x}) \wedge dx_{k+1} + d\omega_2(\mathbf{x})$$
$$= \sideset{}{'}\sum_I \sum_{i=1}^n \partial_i f_I(\mathbf{x})dx_i \wedge d\mathbf{x}_I \wedge dx_{k+1} + \sideset{}{'}\sum_J \sum_{j=1}^n \partial_j g_J(\mathbf{x})dx_j \wedge d\mathbf{x}_J.$$

The only terms in this expression that involve dx_{k+1} and dx_l for $l > k+1$ are just in the first sum and just for the index $i = l$; and the basis vectors $dx_l \wedge d\mathbf{x}_I \wedge dx_{k+1}$ are linearly independent since $1 \le i_1 < i_1 < \cdots < i_{r-1} \le k$ and $k < k+1 < l$. Consequently $\partial_l f_I(\mathbf{x}) = 0$ for $l > k+1$, so $f_I(\mathbf{x}) = f_I(x_1, \cdots, x_{k+1})$ is a function just of the variables $x_1, \ldots, x_{k+1}$, where $0 < x_i < 1$. The integrals

$$h_I(x_1, \ldots, x_{k+1}) = \int_a^{x_{k+1}} f_I(x_1, \ldots, x_k, t)dt$$

then are functions in the cell Δ for which $\partial_{k+1} h_I(\mathbf{x}) = f_I(\mathbf{x})$, by the Fundamental Theorem of Calculus for functions of a single variable; hence the differential $(r-1)$-form

$$\omega_3(\mathbf{x}) = \sideset{}{'}\sum_I h_I(\mathbf{x})d\mathbf{x}_I,$$

which involves just the differentials $dx_1, \ldots, dx_k$, satisfies

$$
\begin{aligned}
d\omega_3(\mathbf{x}) &= {\sum_I}' \left(\sum_{i=1}^{k} \partial_i h_I(\mathbf{x}) dx_i \wedge d\mathbf{x}_I + \partial_{k+1} h_I(\mathbf{x}) dx_{k+1} \wedge d\mathbf{x}_I \right) \\
&= {\sum_I}' \left(\sum_{i=1}^{k} \partial_i h_I(\mathbf{x}) dx_i \wedge d\mathbf{x}_I + f_I(\mathbf{x}) dx_{k+1} \wedge d\mathbf{x}_I \right) \\
&= \omega_4(\mathbf{x}) + (-1)^{r-1} \omega_1(\mathbf{x}) \wedge dx_{k+1}
\end{aligned}
$$

for a differential form $\omega_4(\mathbf{x})$ that involves only the differentials $dx_1, \ldots, dx_k$. Then

$$
\begin{aligned}
\omega(\mathbf{x}) - (-1)^{r-1} d\omega_3(\mathbf{x}) &= \omega_1(\mathbf{x}) \wedge dx_{k+1} + \omega_2(\mathbf{x}) - (-1)^{r-1} \omega_4(\mathbf{x}) - \omega_1(\mathbf{x}) \wedge dx_{k+1} \\
&= \omega_2(\mathbf{x}) - (-1)^{r-1} \omega_4(\mathbf{x})
\end{aligned}
$$

involves only the differentials $dx_1, \ldots, dx_k$ and

$$
d\big(\omega(\mathbf{x}) - (-1)^{r-1} d\omega_3(\mathbf{x}) \big) = 0;
$$

by the induction hypothesis then there is a differential form $\omega_5(\mathbf{x})$ in the cell Δ such that

$$
\omega(\mathbf{x}) - (-1)^{r-1} d\omega_3(\mathbf{x}) = d\omega_5(\mathbf{x}),
$$

hence such that $\omega(\mathbf{x}) = d\sigma(\mathbf{x})$ where $\sigma(\mathbf{x}) = (-1)^{r-1} \omega_3(\mathbf{x}) + \omega_5(\mathbf{x})$. That establishes the induction step and thereby concludes the proof.

The preceding theorem also holds for C^k differential forms, by following through the construction keeping track of the differentiability of the terms involved. The theorem is an example of an integrability theorem for a linear system of partial differential equations. For example, the existence of a function h such that $dh = \sigma$ for a 1-form $\sigma = \sum_{j=1}^{n} f_j(\mathbf{x}) dx_j$ in $\mathbb{R}^n$ amounts to the existence of a solution of the system of partial differential equations $\partial_i h(\mathbf{x}) = f_i(\mathbf{x})$ for $1 \le i \le n$. It is clear that if there is a solution $h(\mathbf{x})$ then $\partial_j f_i(\mathbf{x}) = \partial_j \partial_i h(\mathbf{x}) = \partial_i \partial_j h(\mathbf{x}) = \partial_i f_j(\mathbf{x})$; that is just the condition that $d\sigma(\mathbf{x}) = 0$, which by Poincaré's Lemma is also a sufficient condition for the existence of the solution $h(\mathbf{x})$. There are corresponding interpretations for Poincaré's Lemma for differential forms of higher degree, although they are a bit more complicated to state without using the machinery of differential forms. Another important property of exterior differentiation is its invariance under the pullback of differential forms.

Theorem 7.12. *If $\phi : D \longrightarrow E$ is a C^∞ mapping from an open subset $D \subset \mathbb{R}^m$ into an open subset $E \subset \mathbb{R}^n$ then for any C^∞ differential form $\omega \in \Lambda^r(E)$*

$$
d\phi^*(\omega) = \phi^*(d\omega) \in \Lambda^{r+1}(D). \tag{7.80}
$$

Proof: For a differential form $\omega \in \Lambda^r(E)$ written explicitly as $\omega = \sum_I f_I(\mathbf{y})d\mathbf{y}_I$ it follows from the definition (7.68) that

$$\boldsymbol{\phi}^*(\omega) = \sum_{I,J} f_I(\boldsymbol{\phi}(\mathbf{x})) \frac{\partial y_{i_1}}{\partial x_{j_1}} \cdots \frac{\partial y_{i_r}}{\partial x_{j_r}} dx_{j_1} \wedge \cdots \wedge dx_{j_r}.$$

The exterior derivative of ω is

$$d\omega = \sum_{I,k} \frac{\partial f_I}{\partial y_k} dy_k \wedge d\mathbf{y}_I$$

and it follows from the definition (7.68) that

$$\boldsymbol{\phi}^*(d\omega) = \sum_{I,J,k,l} \frac{\partial f_I}{\partial y_k}(\boldsymbol{\phi}(\mathbf{x})) \frac{\partial y_k}{\partial x_l} \frac{\partial y_{i_1}}{\partial x_{j_1}} \cdots \frac{\partial y_{i_r}}{\partial x_{j_r}} dx_l \wedge dx_{j_1} \wedge \cdots \wedge dx_{j_r}$$

On the other hand

$$d\boldsymbol{\phi}^*(\omega) = \sum_{I,J,l} \frac{\partial}{\partial x_l} \left(f_I(\boldsymbol{\phi}(\mathbf{x})) \frac{\partial y_{i_1}}{\partial x_{j_1}} \cdots \frac{\partial y_{i_r}}{\partial x_{j_r}} \right) dx_l \wedge dx_{j_1} \wedge \cdots \wedge dx_{j_r}$$

$$= \sum_{I,J,k,l} \frac{\partial f_I}{\partial y_k}(\boldsymbol{\phi}(\mathbf{x})) \frac{\partial y_k}{\partial x_l} \frac{\partial y_{i_1}}{\partial x_{j_1}} \cdots \frac{\partial y_{i_r}}{\partial x_{j_r}} dx_l \wedge dx_{j_1} \wedge \cdots \wedge dx_{j_r}$$

$$+ \sum_{I,J,k,l} f_I(\boldsymbol{\phi}(\mathbf{x})) \frac{\partial^2 y_{i_1}}{\partial x_l \partial x_{j_1}} \frac{\partial y_{i_2}}{\partial x_{j_2}} \cdots \frac{\partial y_{i_r}}{\partial x_{j_r}} dx_l \wedge dx_{j_1} \wedge \cdots \wedge dx_{j_r}$$

$$+ \cdots$$

where the remaining terms involve successively the second derivatives of the partial derivatives $\frac{\partial^2 y_{i_2}}{\partial x_l \partial x_{j_2}}$, $\frac{\partial^2 y_{i_3}}{\partial x_l \partial x_{j_3}}$, and so on. These terms that involve the second derivatives all vanish, since for example $\frac{\partial^2 y_{i_1}}{\partial x_l \partial x_{j_1}}$ is symmetric in the indices l and j_1 while $dx_l \wedge dx_{j_1}$ is skew-symmetric in these indices. That suffices for the proof.

The exterior derivatives of differential forms in $\mathbb{R}^3$ have classical forms, which are possibly quite familiar from physics. For the 2-form $\Omega_\mathbf{f}(\mathbf{x})$ associated to a vector field $\mathbf{f}(\mathbf{x})$ in an open set $U \subset \mathbb{R}^3$ as in (7.45) it follows quite directly from the definition of exterior derivative that

$$d\Omega_\mathbf{f}(\mathbf{x}) = \left(\frac{\partial f_1(\mathbf{x})}{\partial x_1} + \frac{\partial f_2(\mathbf{x})}{\partial x_2} + \frac{\partial f_3(\mathbf{x})}{\partial x_3} \right) dx_1 \wedge dx_2 \wedge dx_3. \tag{7.81}$$

The coefficient of this differential 3-form is known as the **divergence** of the vector field $\mathbf{f}$ and is often denoted by $\mathbf{div}\, \mathbf{f}(\mathbf{x})$, so by definition it is the function

$$\mathbf{div}\, \mathbf{f}(\mathbf{x}) = \frac{\partial f_1(\mathbf{x})}{\partial x_1} + \frac{\partial f_2(\mathbf{x})}{\partial x_2} + \frac{\partial f_3(\mathbf{x})}{\partial x_3}. \tag{7.82}$$

A traditional alternative notation is expressed in terms of the vector differential operator

$$\nabla = \left\{ \frac{\partial}{\partial x_1},\quad \frac{\partial}{\partial x_2},\quad \frac{\partial}{\partial x_3} \right\}. \tag{7.83}$$

When applied to functions $\nabla f(\mathbf{x})$ is the gradient of the function $f(\mathbf{x})$; and the dot product of this operator with a vector field is the function

$$\nabla \cdot \mathbf{f}(\mathbf{x}) = \frac{\partial}{\partial x_1} f_1(\mathbf{x}) + \frac{\partial}{\partial x_2} f_2(\mathbf{x}) + \frac{\partial}{\partial x_3} f\mathbf{x}) = \mathbf{div}\, \mathbf{f}(\mathbf{x}). \tag{7.84}$$

In terms of this operator (7.81) can be rewritten

$$d\Omega_{\mathbf{f}}(\mathbf{x}) = \begin{cases} \mathbf{div}\, \mathbf{f}(\mathbf{x})\ dx_1 \wedge dx_2 \wedge dx_3 \\ \text{or} \\ \nabla \cdot \mathbf{f}(\mathbf{x})\ dx_1 \wedge dx_2 \wedge dx_3. \end{cases} \tag{7.85}$$

For the 1-form $\omega_{\mathbf{f}}(\mathbf{x})$ associated to a vector field $\mathbf{f}(\mathbf{x})$ in an open set $U \subset \mathbb{R}^3$ as in (7.45) it follows by a straightforward calculation that

$$d\omega_{\mathbf{f}}(\mathbf{x}) = \big(\partial_2 f_3(\mathbf{x}) - \partial_3 f_2(\mathbf{x})\big) dx_2 \wedge dx_3 + \big(\partial_3 f_1(\mathbf{x}) - \partial_1 f_3(\mathbf{x})\big) dx_3 \wedge dx_1$$
$$+ \big(\partial_1 f_2(\mathbf{x}) - \partial_2 f_1(\mathbf{x})\big) dx_1 \wedge dx_2; \tag{7.86}$$

this is the 2-form associated to the vector field called the **curl** of the vector field $\mathbf{f}(\mathbf{x})$, defined as the vector field

$$\mathbf{curl}\, \mathbf{f}(\mathbf{x}) = \left\{ \partial_2 f_3(\mathbf{x}) - \partial_3 f_2(\mathbf{x}),\quad \partial_3 f_1(\mathbf{x}) - \partial_1 f_3(\mathbf{x}),\quad \partial_1 f_2(\mathbf{x}) - \partial_2 f_1(\mathbf{x}) \right\}. \tag{7.87}$$

It follows from (7.49) that the curl of a vector field can be written in terms of the operator ∇ as

$$\mathbf{curl}\, \mathbf{f}(\mathbf{x}) = \nabla \times \mathbf{f}(\mathbf{x}), \tag{7.88}$$

so that

$$d\omega_{\mathbf{f}}(\mathbf{x}) = \begin{cases} \Omega_{\mathbf{curl}\, \mathbf{f}(\mathbf{x})} \\ \text{or} \\ \Omega_{\nabla \times \mathbf{f}(\mathbf{x})}. \end{cases} \tag{7.89}$$

The condition that $dd\omega = 0$ for any differential form ω, as in Theorem 7.10 (ii), has classical interpretations as well. For any $\mathcal{C}^\infty$ function $f(\mathbf{x})$ in $\mathbb{R}^3$ since $df(\mathbf{x}) = \omega_{\nabla f(\mathbf{x})}$, another interpretation of the first line of (7.45), it follows from (7.89) that $0 = d\,df(\mathbf{x}) = d\omega_{\nabla f(\mathbf{x})} = \Omega_{\nabla \times \nabla f(\mathbf{x})}$ and consequently that

$$\nabla \times \nabla f(\mathbf{x}) = \mathbf{0} \quad \text{or alternatively} \quad \textbf{curl grad} f(\mathbf{x}) = \mathbf{0}. \qquad (7.90)$$

On the other hand for any $\mathcal{C}^\infty$ vector field $\mathbf{f}(\mathbf{x}) = \{f_1(\mathbf{x}), f_2(\mathbf{x}), f_3(\mathbf{x})\}$ since $d\omega_{\mathbf{f}(\mathbf{x})} = \Omega_{\nabla \times \mathbf{f}(\mathbf{x})}$ by (7.89) it follows from (7.81) that $0 = d\,d\omega_{\mathbf{f}(\mathbf{x})} = d\Omega_{\nabla \times \mathbf{f}(\mathbf{x})} = \nabla \cdot (\nabla \times \mathbf{f}(\mathbf{x}))\, dx_1 \wedge dx_2 \wedge dx_3$ and consequently that

$$\nabla \cdot (\nabla \times \mathbf{f}(\mathbf{x})) = 0 \quad \text{or alternatively} \quad \textbf{div curl } \mathbf{f}(\mathbf{x}) = 0. \qquad (7.91)$$

Both (7.90) and (7.91) of course can be demonstrated by direct calculation from the definition of the differential operator ∇. However Poincaré's Lemma is a far less trivial result, so its classical consequences cannot be proved merely by a direct calculation, as should be expected. If $\mathbf{f}(\mathbf{x})$ is a $\mathcal{C}^\infty$ vector field in $\mathbb{R}^3$ the condition that $\mathbf{f}(\mathbf{x}) = \nabla g(\mathbf{x})$ for a function $g(\mathbf{x})$ in Δ means in terms of differential forms that $\omega_{\mathbf{f}}(\mathbf{x}) = dg(\mathbf{x})$; and by Poincaré's Lemma that is the case for a vector field in a cell $\Delta \subset \mathbb{R}^3$ if $0 = d\omega_{\mathbf{f}(\mathbf{x})} = \Omega_{\nabla \times \mathbf{f}(\mathbf{x})}$, consequently

$$\mathbf{f}(\mathbf{x}) = \nabla g(\mathbf{x}) \quad \text{for a function } g(\mathbf{x}) \text{ in } \Delta \text{ if } \nabla \times \mathbf{f}(\mathbf{x}) = \mathbf{0}, \qquad (7.92)$$

or alternatively in the even more classical notation

$$\mathbf{f}(\mathbf{x}) = \textbf{grad } g(\mathbf{x}) \quad \text{for a function } g(\mathbf{x}) \text{ in } \Delta \text{ if } \textbf{curl } \mathbf{f}(\mathbf{x}) = \mathbf{0}. \qquad (7.93)$$

On the other hand that $\mathbf{f}(\mathbf{x}) = \nabla \times \mathbf{g}(\mathbf{x})$ for a vector field $\mathbf{g}(\mathbf{x})$ in Δ means in terms of differential forms that $\Omega_{\mathbf{f}(\mathbf{x})} = d\omega_{\mathbf{g}(\mathbf{x})}$; and by Poincaré's Lemma that is the case for differential forms in a cell $\Delta \subset \mathbb{R}^3$ if $0 = d\Omega_{\mathbf{f}(\mathbf{x})} = \nabla \cdot \mathbf{f}(\mathbf{x})\, dx_1 \wedge dx_2 \wedge dx_3$, and consequently

$$\mathbf{f}(\mathbf{x}) = \nabla \times \mathbf{g}(\mathbf{x}) \quad \text{for a vector field } \mathbf{g}(\mathbf{x}) \text{ in } \Delta \text{ if } \nabla \cdot \mathbf{f}(\mathbf{x}) = 0 \qquad (7.94)$$

or alternatively

$$\mathbf{f}(\mathbf{x}) = \textbf{curl } \mathbf{g}(\mathbf{x}) \quad \text{for a vector field } \mathbf{g}(\mathbf{x}) \text{ in } \Delta \text{ if } \textbf{div } \mathbf{f}(\mathbf{x}) = 0. \qquad (7.95)$$

Problems, Group I

1. If ω_1 is a closed C^∞ differential form and ω_2 is an exact C^∞ differential form in an open subset $U \subset \mathbb{R}^n$ is the product $\omega_1 \wedge \omega_2$ closed? Why? Is the product $\omega_1 \wedge \omega_2$ exact? Why?

2. Calculate the exterior derivatives of the following differential forms:

   i) $\omega_1 = x_1 dx_1 + x_2^2 dx_2 + x_3^3 dx_3$

   ii) $\omega_2 = \dfrac{-x_2 dx_1 + x_1 dx_2}{x_1^2 + x_2^2}$

3. In $\mathbb{R}^4$ consider the differential forms

   $$\omega_1 = x_1 dx_2 + x_3 dx_4,$$
   $$\omega_2 = (x_1^2 + x_3^2)dx_1 \wedge dx_2 + (x_2^2 + x_4^2)dx_3 \wedge dx_4,$$
   $$\omega_3 = dx_1 \wedge dx_3 + x_2 dx_1 \wedge dx_4 + x_1 dx_2 \wedge dx_4.$$

   i) Find the exterior products $\omega_1 \wedge \omega_1$, $\omega_1 \wedge \omega_2$, and $\omega_2 \wedge \omega_2$ in reduced form.
   ii) Find the exterior derivatives $d\omega_1$, $d\omega_2$, and $d\omega_3$ in reduced form.
   iii) Find the Hodge duals $*\omega_1$ and $*\omega_3$ in reduced form.
   iv) Which of these three differential forms are closed?
   v) For each of the differentials ω_i that is closed find a differential form σ such that $\omega = d\sigma$. Having found one such form σ, what is the most general such form? (Answer this part of the question without giving explicit formulas for all possible solutions, but just by characterizing all possible solutions using your general knowledge of the properties of differential forms.)

4. i) For the mapping $\phi : \mathbb{R}^2(r, \theta) \longrightarrow \mathbb{R}^2(x, y)$ defined by $x = r \cos \theta$ and $y = r \sin \theta$ calculate the differential forms $\phi^*(\omega_i)$ where $\omega_1 = dx + dy$ and $\omega_2 = dx \wedge dy$.

   ii) If ψ is a local inverse of the mapping ϕ calculate the differential forms $\psi^*(\sigma_i)$ where $\sigma_1 = dr + d\theta$ and $\sigma_2 = dr \wedge d\theta$.

5. Show that if $\phi : U \longrightarrow V$ is a C^∞ homeomorphism between two open subsets of $\mathbb{R}^n$ and if every closed differential form in U is exact then every closed differential form in V is exact.

Problems, Group II

6. Find a smooth two-dimensional piece of a surface $S \subset \mathbb{R}^3$ such that the differential form on S induced by the differential 2-form

   $$\omega = x_1 dx_2 \wedge dx_3 + x_2 dx_3 \wedge dx_1 + x_3 dx_1 \wedge dx_2$$

   on $\mathbb{R}^3$ is identically zero.

7. Show that for any function $f(\mathbf{x})$ in $\mathbb{R}^n$ the differential operator $\Delta f = *d*d f$ is just the classical Laplacian operator on functions, the differential operator $\Delta f = \sum_{i=1}^{n} \frac{\partial^2 f}{\partial x_i^2}$.

8. Hyperspherical coordinates in $\mathbb{R}^4$ are defined by

$$x_1 = r\cos\theta_1,$$
$$x_2 = r\sin\theta_1\cos\theta_2,$$
$$x_3 = r\sin\theta_1\sin\theta_2\cos\theta_3,$$
$$x_4 = r\sin\theta_1\sin\theta_2\sin\theta_3.$$

Find the explicit formula for the change of variables expressing integration in $\mathbb{R}^4$ in terms of hyperspherical coordinates. Use this to calculate the content of the unit hyperball in $\mathbb{R}^4$, the set of points at which $r \le 1$ where $r = \sqrt{x_1^2 + x_2^2 + x_3^2 + x_4^2}$.

9. To a vector field $\mathbf{f} = \{f_i\}$ in $\mathbb{R}^4$ there can be associated the differential forms

$$\omega_{\mathbf{f}} = f_1 dx_1 + f_2 dx_2 + f_3 dx_3 + f_4 dx_4,$$
$$\Omega_{\mathbf{f}} = f_1 dx_2 \wedge dx_3 \wedge dx_4 - f_2 dx_3 \wedge dx_4 \wedge dx_1$$
$$+ f_3 dx_4 \wedge dx_1 \wedge dx_2 - f_4 dx_1 \wedge dx_2 \wedge dx_3.$$

Show that for any three vector fields $\mathbf{f}_1$, $\mathbf{f}_2$, $\mathbf{f}_3$ in $\mathbb{R}^4$ the vector field $\mathbf{F}$ for which $\Omega_{\mathbf{F}} = \omega_{\mathbf{f}_1} \wedge \omega_{\mathbf{f}_2} \wedge \omega_{\mathbf{f}_3}$ can be interpreted as an analogue $\mathbf{F} = \mathbf{f}_1 \times \mathbf{f}_2 \times \mathbf{f}_3$ of the cross-product in $\mathbb{R}^3$, that is, as a vector field perpendicular to the vector fields $\mathbf{f}_i$ with the appropriate orientation.

10. Show that an exterior 2-form $\omega = \sum_{1 \le i < j \le n} a_{ij} \mathbf{u}_i \wedge \mathbf{u}_j$ in an n-dimensional vector space V in terms of a basis $\mathbf{u}_i$ also can be written as the sum $\omega = (\mathbf{v}_1 \wedge \mathbf{v}_2) + (\mathbf{v}_3 \wedge \mathbf{v}_4) + \cdots + (\mathbf{v}_{2r-1} \wedge \mathbf{v}_{2r})$ for an appropriate basis $\mathbf{v}_1, \ldots, \mathbf{v}_n$ of V, where r is the largest integer such that the product $\omega \wedge \omega \wedge \cdots \wedge \omega \ne 0$ of r copies of the form ω is nonzero.

11. An exterior r-form $\omega \in \Lambda^r(V)$ for an n-dimensional real vector space V is said to be **decomposable** if it can be written in the form $\omega = \mathbf{v}_1 \wedge \cdots \wedge \mathbf{v}_r$ for a suitable basis $\mathbf{v}_1, \ldots, \mathbf{v}_n$ for V, and it is said to be **indecomposable** if it is not decomposable.

   i) Show that if $\dim V \le 3$ then every differential form $\omega \in \Lambda^2(V)$ is decomposable.
   ii) Show that if $\mathbf{v}_1, \mathbf{v}_2, \mathbf{v}_3, \mathbf{v}_4$ are linearly independent vectors in v then the exterior form $\omega = (\mathbf{v}_1 \wedge \mathbf{v}_2) + (\mathbf{v}_3 \wedge \mathbf{v}_4)$ is indecomposable. [Suggestion: Look at $\omega \wedge \omega$.]

7.3 Integrals of Differential Forms

Integrals of differential n-forms in open subsets $E \subset \mathbb{R}^n$ can be defined, but only in terms of the orientation of the vector space $\mathbb{R}^n$; these integrals are **oriented integrals**, extending that notion from spaces of dimension 1 as defined in (6.22) to spaces of dimensions $n > 1$. The integral of a continuous n-form $\omega(\mathbf{y}) = f(\mathbf{y})dy_1 \wedge \cdots \wedge dy_n$ over an open subset $E \subset \mathbb{R}^n$ of an oriented vector space $\mathbb{R}^n$ with the orientation $dy_1 \wedge \cdots \wedge dy_n$ in terms of coordinates $y_1, \ldots, y_n$ is defined by

$$\int_E f(\mathbf{y})dy_1 \wedge \cdots \wedge dy_n = \int_E f, \tag{7.96}$$

so that

$$\int_E f(\mathbf{y})dy_{i_1} \wedge \cdots \wedge dy_{i_n} = \mathrm{sgn}\begin{pmatrix} i_2 & i_2 & \cdots & i_n \\ 1 & 2 & \cdots & n \end{pmatrix} \int_E f(\mathbf{y}) \tag{7.97}$$

where $\mathrm{sgn}\begin{pmatrix} i_1 & i_2 & \cdots & i_n \\ 1 & 2 & \cdots & n \end{pmatrix}$ is the sign of the permutation $(i_1\ i_2\ \cdots\ i_n)$ of the integers $(1\ 2\ \cdots\ n)$; thus under a permutation of the variables the integral of a differential form changes by the sign of the permutation.

Theorem 7.13. *If $\phi : D \longrightarrow E$ is a C^1 homeomorphism between two connected open subsets $D, E \subset \mathbb{R}^n$ then for any continuous differential n-form ω in E*

$$\int_E \omega = \pm \int_D \phi^*(\omega), \tag{7.98}$$

where the sign is $+$ if the mapping ϕ preserves the orientation and $-$ if the mapping ϕ reverses the orientation.

Proof: Suppose that the vector space $\mathbb{R}^n$ containing D is oriented by the differential form $dx_1 \wedge \cdots \wedge dx_n$ in terms of coordinates $x_1, \ldots, x_n$, and that the vector space $\mathbb{R}^n$ containing E is oriented by the differential form $dy_1 \wedge \cdots \wedge dy_n$ in terms of coordinates $y_1, \ldots, y_n$. The integral of a continuous differential n-form $\omega(\mathbf{y}) = f(\mathbf{y})dy_1 \wedge \cdots \wedge dy_n$ in E is

$$\int_E \omega = \int_E f, \tag{7.99}$$

as in (7.96); and the integral of the induced differential form $\phi^*(\omega)$ in D, which by (7.73) has the explicit form $\phi^*(\omega)(\mathbf{x}) = f(\phi(\mathbf{x})) \det \phi'(\mathbf{x})dx_1 \wedge \cdots \wedge dx_n$, is

$$\int_D \phi^*(\omega) = \int_D (f \circ \phi) \det \phi'. \tag{7.100}$$

If the mapping ϕ preserves orientation then $\det \phi'(\mathbf{x}) > 0$ whenever $\mathbf{x} \in D$ and consequently $\det \phi'(\mathbf{x}) = |\det \phi'(\mathbf{x})|$, while if the mapping ϕ reverses orientation then $\det \phi'(\mathbf{x}) < 0$ at all points $\mathbf{x} \in D$ and consequently $\det \phi'(\mathbf{x}) = -|\det \phi'(\mathbf{x})|$; therefore (7.100) can be rewritten

$$\int_D \phi^*(\omega) = \pm \int_D (f \circ \phi) |\det \phi'| \qquad (7.101)$$

where the sign is $+$ if the mapping ϕ preserves orientation and $-$ if the mapping ϕ reverses orientation. By the change of variables formula (6.49) in Theorem 6.19

$$\int_E f = \int_D (f \circ \phi) |\det \phi'|, \qquad (7.102)$$

and (7.98) follows immediately by substituting (7.99) and (7.101) into (7.102), to conclude the proof.

There also are oriented integrals of m-forms over suitable lower-dimensional subsets of the spaces $\mathbb{R}^n$. An m-dimensional **singular cell** in an open subset $U \subset \mathbb{R}^n$ is a continuous mapping $\phi : \overline{\Delta} \longrightarrow U$ from an oriented closed cell $\overline{\Delta} \subset \mathbb{R}^m$ into the open subset $U \subset \mathbb{R}^n$. It is often assumed that the parameter cell is the **standard unit cell**

$$\overline{\Delta}_m = \left\{ \{x_1, \ldots, x_m\} \in \mathbb{R}^m \,\middle|\, 0 \leq x_j \leq 1 \quad \text{for} \quad 1 \leq j \leq m \right\} \qquad (7.103)$$

with the orientation indicated by the order of the variables, although other normalizations are sometimes more appropriate. A $\mathcal{C}^1$ **singular cell** is a singular cell for which the mapping ϕ extends to a $\mathcal{C}^1$ mapping from an open neighborhood of the closed cell $\overline{\Delta}$ into U; of course a $\mathcal{C}^k$ singular cell for any integer $k > 0$ is defined correspondingly, and a $\mathcal{C}^\infty$ singular cell is defined as one for which the mapping ϕ is of class $\mathcal{C}^k$ for all integers $k > 0$. For convenience much of the further discussion will focus on $\mathcal{C}^\infty$ singular cells, to avoid the necessity of keeping track of the degree of differentiabilitiy of singular cells and to ensure that the singular cells are sufficiently smooth for the arguments in later proofs. A one-dimensional singular cell is a parametrized curve since an interval $[0, 1]$ is a one-dimensional cell that is naturally oriented. For singular cells of dimension $m > 1$ however the orientation always must be specified explicitly. As in the case of curves, the mappings describing singular cells are not required to be one-to-one mappings, nor for $\mathcal{C}^k$ singular cells is it required that the derivative ϕ' of the mapping ϕ have maximal rank; so the images of singular cells are not necessarily submanifolds of U even locally, and that is the reason for the adjective singular. It is even possible for the image of a singular m-cell to be a submanifold of $\mathbb{R}^n$ of a dimension lower than m. Two m-dimensional singular $\mathcal{C}^1$ cells $\phi_1 : \overline{\Delta}_1 \longrightarrow \mathbb{R}^n$ and $\phi_2 : \overline{\Delta}_2 \longrightarrow \mathbb{R}^n$ are **equivalent** singular cells if there is an orientation preserving $\mathcal{C}^1$ homeomorphism $\mathbf{h} : \overline{\Delta}_1 \longrightarrow \overline{\Delta}_2$ such that $\phi_1(\mathbf{t}) = \phi_2(\mathbf{h}(\mathbf{t}))$; this is easily seen to be an equivalence relation between singular cells.

Equivalent singular cells of course are of the same dimension and have the same image; but in general singular cells having the same image are not necessarily equivalent singular cells and are not necessarily of the same dimension. The corresponding notion and results hold for C^k or C^∞ singular cells.

If $\phi : \overline{\Delta} \longrightarrow U$ is an m-dimensional C^∞ singular cell in an open subset $U \subset \mathbb{R}^n$, and if $\omega \in \Lambda^m(U)$ is a continuous differential m-form in the open set U, the **integral** of the differential form ω over the singular cell ϕ is defined by

$$\int_\phi \omega = \int_{\overline{\Delta}} \phi^*(\omega). \tag{7.104}$$

If $\phi_1 : \overline{\Delta_1} \longrightarrow U$ and $\phi_2 : \overline{\Delta_2} \longrightarrow U$ are C^∞ equivalent singular cells, so that $\phi_1(\mathbf{t}) = \phi_2(\mathbf{h}(\mathbf{t}))$ where $\mathbf{h} : V_1 \longrightarrow V_2$ is an orientation preserving C^∞ homeomorphism between open neighborhoods V_i of the cells $\overline{\Delta}_i$, then by Theorems 7.9 and 7.13

$$\int_{\overline{\Delta_1}} \phi_1^*(\omega) = \int_{\overline{\Delta_1}} \mathbf{h}^*(\phi_2^*(\omega)) = \int_{\overline{\Delta_2}} \phi_2^*(\omega); \tag{7.105}$$

thus the integrals of a differential form over C^∞ equivalent singular m-cells are equal.

Line integrals in $\mathbb{R}^n$ are special cases of integrals of differential forms over singular cells, since a parametrized curve is a special case of a singular cell. If $\phi : [0, 1] \longrightarrow U$ is a C^∞ parametrized 1-cell in an open subset $U \subset \mathbb{R}^n$, where the parameter in $[0, 1]$ is t and the parameters in $\mathbb{R}^n$ are $x_1, \ldots, x_n$, and if $\omega(\mathbf{x}) = \sum_{j=1}^n f_j(\mathbf{x})dx_j$ is a continuous differential form in U, the induced differential form is $\phi^*(\omega)(t) = \sum_{j=1}^n f_j(\phi(t))\dfrac{dx_j}{dt}dt$; and if $\mathbf{f}(\mathbf{x}) = \{f_j(\mathbf{x})\}$ is the vector field with component functions $f_j(\mathbf{x})$ this induced differential form can be written as the dot product $\phi^*(\omega)(t) = \mathbf{f}(\phi(t)) \cdot \phi'(t)dt$. The integral of the differential 1-form $\omega(\mathbf{x})$ over the singular 1-cell ϕ hence is just the familiar line integral

$$\int_\phi \omega = \int_\phi \mathbf{f} \cdot \tau \, ds \tag{7.106}$$

as in (7.23); and for simple parametrized curves the integral is determined by the image curve $\gamma = \phi([0, 1])$ so it is often denoted just by $\int_\gamma \omega = \int_\gamma \mathbf{f} \cdot \tau \, ds$. For example if $\gamma \subset \mathbb{R}^3$ is the curve from the origin $\mathbf{a} = \{0, 0, 0\}$ to the point $\mathbf{b} = \{1, 1, 1\}$ that is the image of the mapping $\phi : [0, 1] \longrightarrow \mathbb{R}^3$ given by

$\phi(t) = \{t, t^2, t^3\}$, and if $\mathbf{f}$ is the vector field $\mathbf{f}(\mathbf{x}) = \{x_1^2, x_2, x_1 x_2 x_3\}$, then

$$
\begin{aligned}
\int_\gamma \mathbf{f} \cdot \boldsymbol{\tau} \; ds &= \int_\gamma \left(x_1^2 dx_1 + x_2 dx_2 + x_1 x_2 x_3 dx_3 \right) \\
&= \int_{[0,1]} \left(t^2 \cdot dt + t^2 \cdot 2t dt + t^6 \cdot 3t^2 dt \right) \\
&= \int_{[0,1]} \left(t^2 + 2t^3 + 3t^8 \right) dt = \frac{7}{6}.
\end{aligned}
$$

As this example illustrates, the explicit calculation of line integrals is often simpler when these integrals are expressed in terms of differential 1-forms.

Another commonly considered special case is that of the integration of differential 2-forms over two-dimensional singular cells in $\mathbb{R}^3$. Let $\phi : \overline{\Delta} \longrightarrow U$ be a C^∞ singular 2-cell in an open subset $U \subset \mathbb{R}^3$, where the parameters in $\mathbb{R}^2$ are t_1, t_2 with the natural orientation $dt_1 \wedge dt_2$ and the parameters in $\mathbb{R}^3$ are x_1, x_2, x_3 with the natural orientation $dx_1 \wedge dx_2 \wedge dx_3$. A continuous 2-form in $\mathbb{R}^3$ can be viewed as the differential form

$$
\Omega_{\mathbf{f}}(\mathbf{x}) = f_1(\mathbf{x}) dx_2 \wedge dx_3 + f_2(\mathbf{x}) dx_3 \wedge dx_1 + f_3(\mathbf{x}) dx_1 \wedge dx_2
$$

associated to the vector field $\mathbf{f}(\mathbf{x})$, with the usual cyclic notation as in (7.45). The induced differential form under the mapping ϕ is the differential 2-form

$$
\begin{aligned}
\phi^*(\Omega_{\mathbf{f}})(\mathbf{t}) = \sum_{j_1, j_2 = 1}^{2} & \left(f_1(\phi(\mathbf{t})) \frac{\partial x_2}{\partial t_{j_1}} \frac{\partial x_3}{\partial t_{j_2}} + f_2(\phi(\mathbf{t})) \frac{\partial x_3}{\partial t_{j_1}} \frac{\partial x_1}{\partial t_{j_2}} \right. \\
& \left. + f_3(\phi(\mathbf{t})) \frac{\partial x_1}{\partial t_{j_1}} \frac{\partial x_2}{\partial t_{j_2}} \right) dt_{j_1} \wedge dt_{j_2} \\
= & \left(f_1(\phi(\mathbf{t})) \left(\frac{\partial x_2}{\partial t_1} \frac{\partial x_3}{\partial t_2} - \frac{\partial x_2}{\partial t_2} \frac{\partial x_3}{\partial t_1} \right) + f_2(\phi(\mathbf{t})) \left(\frac{\partial x_3}{\partial t_1} \frac{\partial x_1}{\partial t_2} - \frac{\partial x_3}{\partial t_2} \frac{\partial x_1}{\partial t_1} \right) \right. \\
& \left. + f_3(\phi(\mathbf{t})) \left(\frac{\partial x_1}{\partial t_1} \frac{\partial x_2}{\partial t_2} - \frac{\partial x_1}{\partial t_2} \frac{\partial x_2}{\partial t_1} \right) \right) dt_1 \wedge dt_2,
\end{aligned}
$$

which can be written more succinctly as

$$
\phi^*(\Omega_{\mathbf{f}})(\mathbf{t}) = \det \begin{pmatrix} f_1(\phi(\mathbf{t})) & f_2(\phi(\mathbf{t})) & f_3(\phi(\mathbf{t})) \\[6pt] \dfrac{\partial x_1}{\partial t_1} & \dfrac{\partial x_2}{\partial t_1} & \dfrac{\partial x_3}{\partial t_1} \\[10pt] \dfrac{\partial x_1}{\partial t_2} & \dfrac{\partial x_2}{\partial t_2} & \dfrac{\partial x_3}{\partial t_2} \end{pmatrix} dt_1 \wedge dt_2, \tag{7.107}
$$

or using the cross-product introduced in (7.49) as

$$\boldsymbol{\phi}^*(\Omega_{\mathbf{f}})(\mathbf{t}) = \mathbf{f}(\boldsymbol{\phi}(\mathbf{t})) \cdot \left(\frac{\partial \mathbf{x}}{\partial t_1} \times \frac{\partial \mathbf{x}}{\partial t_2} \right) dt_1 \wedge dt_2 \tag{7.108}$$

in terms of the vector fields

$$\frac{\partial \mathbf{x}}{\partial t_j} = \left\{ \frac{\partial x_1}{\partial t_j}, \frac{\partial x_2}{\partial t_j}, \frac{\partial x_3}{\partial t_j} \right\} \quad \text{for } j = 1, 2.$$

Thus the integral of the differential 2-form $\Omega_{\mathbf{f}}$ on the singular 2-cell $\boldsymbol{\phi}$ is given by

$$\int_{\boldsymbol{\phi}} \Omega_{\mathbf{f}} = \int_{\Delta} \mathbf{f}(\boldsymbol{\phi}(t)) \cdot \left(\frac{\partial \mathbf{x}}{\partial t_1} \times \frac{\partial \mathbf{x}}{\partial t_2} \right) dt_1 \wedge dt_2. \tag{7.109}$$

The value of this integral depends on both the orientation of the space $\mathbb{R}^3$, which determines the order of the columns in (7.107), and the orientation of the space $\mathbb{R}^2$, which determines the order of the rows in (7.107). If the two vectors $\partial \mathbf{x}/\partial t_1$ and $\partial \mathbf{x}/\partial t_2$ are linearly independent at a point $\mathbf{t} \in \Delta$ then rank $\boldsymbol{\phi}'(\mathbf{t}) = 2$ and it follows from the Rank Theorem that the image of an open neighborhood of the point $\mathbf{t} \in \Delta$ is a two-dimensional submanifold of an open neighborhood of $\boldsymbol{\phi}(\mathbf{t})$ in U; but of course this may not be all of the image Γ near the point $\boldsymbol{\phi}(\mathbf{t}) \in U$. The three vectors

$$\frac{\partial \mathbf{x}}{\partial t_1}, \quad \frac{\partial \mathbf{x}}{\partial t_2}, \quad \frac{\partial \mathbf{x}}{\partial t_1} \times \frac{\partial \mathbf{x}}{\partial t_2}$$

have the orientation compatible with the chosen orientation of $\mathbb{R}^3$, described by the order $dx_1, \ dx_2, \ dx_3$ of the basis vectors along the coordinate axes. The unit vector in the direction of the cross-product $\dfrac{\partial \mathbf{x}}{\partial t_1} \times \dfrac{\partial \mathbf{x}}{\partial t_2}$ is the **unit normal vector** to that oriented submanifold, which is denoted by $\boldsymbol{v}(\mathbf{t})$; and in these terms

$$\frac{\partial \mathbf{x}}{\partial t_1} \times \frac{\partial \mathbf{x}}{\partial t_2} = \left\| \frac{\partial \mathbf{x}}{\partial t_1} \times \frac{\partial \mathbf{x}}{\partial t_2} \right\|_2 \boldsymbol{v}(\mathbf{t}). \tag{7.110}$$

In parallel to the interpretation of line integrals in (7.106), the integral (7.109) can be written

$$\int_{\Gamma} \Omega_{\mathbf{f}} = \int_{\Gamma} \mathbf{f} \cdot \boldsymbol{v} \, dS \tag{7.111}$$

and can be viewed as the integral over the surface Γ of the dot product of the vector field $\mathbf{f}$ and the unit normal vector $\boldsymbol{v}$ to Γ, where dS indicates integration with respect to the **surface area** of Γ where the surface area $|\Gamma|$ is defined by[2]

$$|\Gamma| = \int_{\Delta} \left\| \frac{\partial \mathbf{x}}{\partial t_1} \times \frac{\partial \mathbf{x}}{\partial t_2} \right\|_2. \tag{7.112}$$

The calculation of the surface area can be a rather complicated one, involving explicit integrals of the form

$$|\Gamma| = \int_{\Delta} \sqrt{ \det \begin{pmatrix} \frac{\partial x_2}{\partial t_1} & \frac{\partial x_3}{\partial t_1} \\ \frac{\partial x_2}{\partial t_2} & \frac{\partial x_3}{\partial t_2} \end{pmatrix}^2 + \det \begin{pmatrix} \frac{\partial x_3}{\partial t_1} & \frac{\partial x_1}{\partial t_1} \\ \frac{\partial x_3}{\partial t_2} & \frac{\partial x_1}{\partial t_2} \end{pmatrix}^2 + \det \begin{pmatrix} \frac{\partial x_1}{\partial t_1} & \frac{\partial x_2}{\partial t_1} \\ \frac{\partial x_1}{\partial t_2} & \frac{\partial x_2}{\partial t_2} \end{pmatrix}^2 }. \tag{7.113}$$

As in the case of line integrals it is generally much, much simpler to calculate the integral (7.109) as the integral of the differential form $\Omega_{\mathbf{f}}$ in terms of any parametrization of the singular cell Γ, without calculating the surface area.

For example the cylindrical surface $\Gamma \subset \mathbb{R}^3$ defined by $x_1^2 + x_2^2 = 1$, $0 \leq x_3 \leq 1$, can be described as the image of the singular 2-cell $\boldsymbol{\phi} : \overline{\Delta}_2 \longrightarrow \Gamma$ given by $\boldsymbol{\phi}(\mathbf{t}) = \{\cos t_1,\ \sin t_1,\ t_2\}$, where the orientation of the cell is described by the differential form $dt_1 \wedge dt_2$. At the point $\boldsymbol{\phi}(\mathbf{0}) = \boldsymbol{\phi}(\{0,0\}) = \{0,0,1\}$ on the boundary of the surface Γ

$$\frac{\partial \mathbf{x}}{\partial t_1} = \{-\sin t_1,\ \cos t_1,\ 0\}\Big|_{t_1=t_2=0} = \{0 \quad 1 \quad 0\},$$

$$\frac{\partial \mathbf{x}}{\partial t_2} = \{0 \quad 0 \quad 1\}$$

hence

$$\frac{\partial \mathbf{x}}{\partial t_1} \times \frac{\partial \mathbf{x}}{\partial t_2} = \{1,\ 0,\ 0\},$$

which points outward from the interior of the cylinder. (For the opposite orientation of the cylinder the orientation of the unit cell $\overline{\Delta}_2$ would be described by the differential form $dt_2 \wedge dt_1$.) The differential 2-form

$$\Omega_{\mathbf{f}} = x_1\ dx_2 \wedge dx_3 + x_2 dx_3 \wedge dx_1 + x_3 dx_1 \wedge dx_2$$

---

[2]The direct definition of surface area is an extremely complicated matter, as discussed in detail for instance in the classical book by T. Radó, *Length and Area* (New york: American Mathematical Society Colloquium Publications 1948). Those difficulties are finessed by using the pattern of the calculation of arc length through line integrals to define surface area, reversing the order of the discussion in Section 7.23. It is customary to attempt to justify this definition of surface area by observing that the length $\left\| \frac{\partial \mathbf{x}}{\partial t_1} \times \frac{\partial \mathbf{x}}{\partial t_2} \right\|_2$ of the cross-product of the two vectors $\frac{\partial \mathbf{x}}{\partial t_1}$ and $\frac{\partial \mathbf{x}}{\partial t_2}$ is the area of the parallelogram spanned by these two vectors, which can be viewed as a planar approximation to the surface since the vectors are tangent vectors approximating the lengths of the two curves that are the image of the t_1 and t_2 coordinate axes under the mapping $\boldsymbol{\phi} : \mathbb{R}^2 \longrightarrow \mathbb{R}^3$. What is perhaps more convincing though is that the integral $\int_{\Gamma} dS$ for some standard surfaces yields the usual values for surface areas.

is that described by the vector field $\mathbf{f}$ where $\mathbf{f}(\mathbf{x}) = \{x_1, x_2, x_3\}$, and its integral is

$$\int_\Gamma \mathbf{f} \cdot \mathbf{v} \, dS = \int_\Gamma \Omega_{\mathbf{f}} = \int_\Delta x_1 \, dx_2 \wedge dx_3 + x_2 dx_3 \wedge dx_1 + x_3 dx_1 \wedge dx_2$$

$$= \int_\Delta \det \begin{pmatrix} \cos t_1 & \sin t_1 & t_2 \\ -\sin t_1 & \cos t_1 & 0 \\ 0 & 0 & 1 \end{pmatrix} dt_1 \wedge dt_2$$

$$= \int_{t_1 \in [0, 2\pi]} \int_{t_2 \in [0,1]} dt_2 dt_1 = 2\pi.$$

Furthermore since

$$\frac{\partial \mathbf{x}}{\partial t_1} \times \frac{\partial \mathbf{x}}{\partial t_2} = \cos t_1 dx_1 + \sin t_1 dx_2 + 0 dx_3$$

when expressed in terms of the basis vectors dx_1, dx_2, dx_3 it follows that

$$\left\| \frac{\partial \mathbf{x}}{\partial t_1} \times \frac{\partial \mathbf{x}}{\partial t_2} \right\|_2 = 1$$

so the area of the cylinder Γ is

$$|\Gamma| = \int_\Gamma 1 dS = \int_{\overline{\Delta}} 1 = 2\pi.$$

That is of course a familiar result, and it should be somewhat comforting that the general formula does work in this case.

The boundary $\partial \Delta$ of an ordinary n-cell Δ consists of a collection of singular $(n-1)$-cells; to discuss this boundary in more detail it is convenient to introduce some further notation. Denote the unit cell in the space $\mathbb{R}^n$ by

$$\Delta(x_1, x_2, \ldots, x_n) = \left\{ (x_1, x_2, \ldots, x_n) \in \mathbb{R}^n \,\middle|\, 0 < x_i < 1 \quad \text{for } 1 \le i \le n \right\}, \quad (7.114)$$

with the orientation described by the differential form $dx_1 \wedge dx_2 \wedge \ldots \wedge dx_n$; and correspondingly denote the closed unit cell by

$$\overline{\Delta}(x_1, x_2, \ldots, x_n) = \left\{ (x_1, x_2, \ldots, x_n) \in \mathbb{R}^n \,\middle|\, 0 \le x_i \le 1 \quad \text{for } 1 \le i \le n \right\}, \quad (7.115)$$

also with the orientation described by the differential form $dx_1 \wedge dx_2 \wedge \ldots \wedge dx_n$. The boundary of either cell consists of those points of the closed cell for

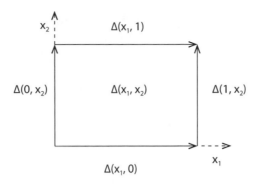

Figure 7.3. The singular cells forming the boundary of a cell $\Delta(x_1, x_2) \subset \mathbb{R}^2$

which at least one of the coordinates takes the value 0 or 1; for instance when $n = 2$ the boundary of the cell $\Delta(x_1, x_2)$ consists of the 4 segments

$$\overline{\Delta}(0, x_2) \subset \mathbb{R}^2 = \left\{ (0, x_2) \,\Big|\, 0 \le x_2 \le 1 \right\},$$
$$\overline{\Delta}(1, x_2) \subset \mathbb{R}^2 = \left\{ (1, x_2) \,\Big|\, 0 \le x_2 \le 1 \right\},$$
$$\overline{\Delta}(x_1, 0) \subset \mathbb{R}^2 = \left\{ (x_1, 0) \,\Big|\, 0 \le x_1 \le 1 \right\},$$
$$\overline{\Delta}(x_1, 1) \subset \mathbb{R}^2 = \left\{ (x_1, 1) \,\Big|\, 0 \le x_1 \le 1 \right\}.$$

(7.116)

The segment $\overline{\Delta}(0, x_2)$ can be viewed as a singular 1-cell, the image of the closed unit 1-cell $\overline{\Delta}(x_2) \subset \mathbb{R}^1$ under the mapping that takes a point $x_2 \in \overline{\Delta}(x_2)$ to the point $(0, x_2) \in \overline{\Delta}(x_1, x_2)$, and correspondingly for the other three segments. With the orientations of these singular cells defined by the natural orientation of the 1-cells $\overline{\Delta}(x_i)$, the collection of these four singular cells is sketched in the accompanying Figure 7.3. However the more natural and more traditional expression for the boundary of the cell $\overline{\Delta}(x_1, x_2)$ consists of these four singular 1-cells but with the orientations changed so that the path described by the collection of these singular cells traverses the entire boundary of the cell $\overline{\Delta}(x_1, x_2)$ in the counterclockwise direction; that involves reversing the orientations of the singular cells $\overline{\Delta}(x_1, 1)$ and $\overline{\Delta}(0, x_2)$. It is convenient to denote these singular cells with the orientation reversed by $-\overline{\Delta}(x_1, 1)$ and $-\overline{\Delta}(0, x_2)$, and to view the collection of all four of these cells, with the chosen orientations, as a formal sum

$$\partial \overline{\Delta}(x_1, x_2) = \overline{\Delta}(x_1, 0) + \overline{\Delta}(1, x_2) - \overline{\Delta}(x_1, 1) - \overline{\Delta}(0, x_2)$$

(7.117)

of singular 1-cells in $\mathbb{R}^2$. In general, a **singular n-chain** in an open subset $U \subset \mathbb{R}^m$ is defined to be a finite formal sum

$$\mathcal{C} = \sum_{j=1}^{r} v_j \cdot \boldsymbol{\phi}_j \quad \text{for integers } v_j \in \mathbb{Z} \text{ and singular } n\text{-cells } \boldsymbol{\phi}_j \text{ in } U; \tag{7.118}$$

and the **boundary** of the cell $\overline{\Delta}(x_1, x_2, \ldots, x_n)$ is defined to be the singular $(n-1)$-chain

$$\partial\overline{\Delta}(x_1, x_2, \ldots, x_n) = \sum_{i=1}^{n} \sum_{\epsilon=0}^{1} (-1)^{i+\epsilon} \overline{\Delta}(x_1, \ldots, x_{i-1}, \epsilon, x_{i+1}, \ldots, x_n). \tag{7.119}$$

The notation $\partial\overline{\Delta}$ is the same as that used for the point set boundary of $\overline{\Delta}$; generally it should be clear from the context whether $\partial\overline{\Delta}$ is to be viewed as a point set or as a singular chain, but if necessary for clarity the set will be called either the point set boundary or the boundary chain. In particular the boundary of the 1-cell $\overline{\Delta}(x_1)$ is the singular 0-chain

$$\partial\overline{\Delta}(x_1) = \overline{\Delta}(1) - \overline{\Delta}(0), \tag{7.120}$$

in which the point or 0-cell $\overline{\Delta}(1)$ at the head of the 1-cell has the positive sign and the point or 0-cell $\overline{\Delta}(0)$ at the tail of the 1-cell has the negative sign; the boundary of the 2-cell $\overline{\Delta}(x_1, x_2)$ is the singular 1-chain (7.117); and the boundary of the 3-cell $\overline{\Delta}(x_1, x_2, x_3)$ is the singular 2-chain

$$\partial\overline{\Delta}(x_1, x_2, x_3) = -\overline{\Delta}(0, x_2, x_3) + \overline{\Delta}(1, x_2, x_3) + \overline{\Delta}(x_1, 0, x_3) - \overline{\Delta}(x_1, 1, x_3)$$
$$- \overline{\Delta}(x_1, x_2, 0) + \overline{\Delta}(x_1, x_2, 1). \tag{7.121}$$

The geometric significance of the orientation of the boundary $\partial\overline{\Delta}(x_1, x_2, x_3)$ as just defined may not be all that clear, and perhaps merits a bit more discussion, for which Figure 7.4 may be helpful. The boundary segment $\overline{\Delta}(x_1, x_2, 1)$ is the top side while the boundary segment $\overline{\Delta}(x_1, x_2, 0)$ is the bottom side of the cell $\overline{\Delta}(x_1, x_2, x_3)$. By (7.121) the orientation of the top side is described by the differential form $dx_1 \wedge dx_2$ while the orientation of the bottom side is described by the differential form $-dx_1 \wedge dx_2 = dx_2 \wedge dx_1$. If $\mathbf{e}_i$ denotes the unit vector in the direction of the x_i-axis in $\mathbb{R}^3$, the orientation of the top side also can be described by the ordered pair of vectors $\mathbf{e}_1, \mathbf{e}_2$, or equivalently by their cross-product $\mathbf{e}_1 \times \mathbf{e}_2 = \mathbf{e}_3$, a vector perpendicular to the top boundary segment and in the direction pointing outward from the 3-cell $\overline{\Delta}(x_1, x_2, x_3)$; and the orientation of the bottom side is described by the ordered pair of vectors $\mathbf{e}_2, \mathbf{e}_1$, or equivalently by their cross-product $\mathbf{e}_2 \times \mathbf{e}_1 = -\mathbf{e}_3$, a vector perpendicular to the bottom boundary segment and in the direction pointing outward from the 3-cell $\overline{\Delta}(x_1, x_2, x_3)$. The boundary segment $\overline{\Delta}(1, x_2, x_3)$ is the

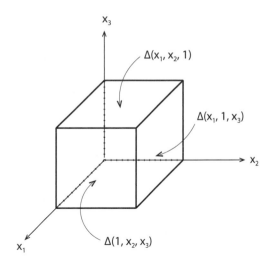

Figure 7.4. The boundary of a cell $\overline{\Delta}(x_1, x_2, x_3) \subset \mathbb{R}^3$

front side while the boundary segment $\overline{\Delta}(0, x_2, x_3)$ is the back side of the cell $\overline{\Delta}(x_1, x_2, x_3)$. The orientation of the front side is described by the differential form $dx_2 \wedge dx_3$ or the cross-product $\mathbf{e}_2 \times \mathbf{e}_3 = \mathbf{e}_1$, a vector perpendicular to that side and pointing outward, while the orientation of the back side is described by the differential form $-dx_2 \wedge dx_3$ or the cross-product $\mathbf{e}_3 \times \mathbf{e}_2 = -\mathbf{e}_1$, again a vector perpendicular to that side and pointing outward. The boundary segment $\overline{\Delta}(x_1, 1, x_3)$ is the right-hand side while the boundary segment $\overline{\Delta}(x_1, 0, x_3)$ is the left-hand side of the cell $\overline{\Delta}(x_1, x_2, x_3)$. The orientation of the right-hand side is described by the differential form $-dx_1 \wedge dx_3$ or the cross-product $\mathbf{e}_3 \times \mathbf{e}_1 = \mathbf{e}_2$, a vector perpendicular to that side and pointing outward, while the orientation of the left-hand side is described by the differential form $dx_1 \wedge dx_3$ or the cross-product $\mathbf{e}_1 \times \mathbf{e}_3 = -\mathbf{e}_2$, yet again a vector perpendicular to that side and pointing outward. Thus the orientations of the six faces of the 3-cell $\overline{\Delta}(x_1, x_2, x_3)$ are all described by the outward pointing normal vectors to these sides, an easy way to remember the orientation and to use it in explicit calculations.

The notion of the boundary chain of a standard n-cell $\overline{\Delta} \subset \mathbb{R}^n$ can be extended quite simply to singular n-cells. A singular n-cell ϕ is a continuous mapping $\phi : \overline{\Delta} \longrightarrow \mathbb{R}^n$, and its **boundary** is defined by

$$\partial \phi = \phi\big(\partial \overline{\Delta}(x_1, x_2, \ldots, x_n)\big)$$

$$= \sum_{i=1}^{n} \sum_{\epsilon=0}^{1} (-1)^{i+\epsilon} \phi\big(\overline{\Delta}(x_1, \ldots, x_{i-1}, \epsilon, x_{i+1}, \ldots, x_n)\big). \tag{7.122}$$

In this definition the singular cell $\overline{\Delta}(x_1, \ldots, x_{n-1}, 0)$ is the mapping from the standard $(n-1)$-cell $\overline{\Delta}(x_1, \ldots, x_{n-1}) \subset \mathbb{R}^{n-1}$ into $\mathbb{R}^n$ that takes a point $(x_1, \ldots, x_{n-1}) \in \mathbb{R}^{n-1}$ to the point $(x_1, \ldots, x_{n-1}, 0) \in \mathbb{R}^n$, so $\phi\big(\overline{\Delta}(x_1, \ldots, x_{n-1}, 0)\big)$

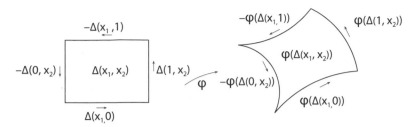

Figure 7.5. The boundary of a singular cell $\phi\left(\overline{\Delta}(x_1, x_2)\right) \subset \mathbb{R}^3$, where the mapping ϕ is orientation preserving

is just the singular $(n-1)$-cell resulting from the composition of these two mappings; the other terms in (7.122) can be interpreted correspondingly, as is illustrated in Figure 7.5. A singular n-chain $\mathcal{C}$ is said to be a **cycle** or an n-**cycle** if $\partial \mathcal{C} = 0$, and it is said to be a **boundary** if $\mathcal{C} = \partial \mathcal{C}_0$ for some singular $(n+1)$ chain $\mathcal{C}_0$. For example, in Figure 7.5 the boundary chain of the cell $\overline{\Delta}(x_1, x_2)$ is the boundary 1-chain $\partial \overline{\Delta}(x_1, x_2) = \overline{\Delta}(x_1, 0) + \overline{\Delta}(1, x_2) - \overline{\Delta}(x_1, 1) - \overline{\Delta}(0, x_2)$. Each vertex $(\epsilon_1, \epsilon_2) \in \mathbb{R}^2$ of the cell $\overline{\Delta}(x_1, x_2)$ is part of the boundary of two of these singular cells in the boundary chain $\partial \overline{\Delta}(x_1, x_2)$, at the head of one and the tail of the other; so altogether the boundary of the boundary 1-chain $\partial \overline{\Delta}(x_1, x_2)$ is the trivial 0-chain

$$\partial \partial \overline{\Delta}(x_1, x_2) = \partial \overline{\Delta}(x_1, 0) + \partial \overline{\Delta}(1, x_2) - \partial \overline{\Delta}(x_1, 1) - \partial \overline{\Delta}(0, x_2)$$

$$= \big((1,0) - (0,0)\big) + \big((1,1) - (1,0)\big) + \big((0,1) - (1,1)\big) + \big((0,0) - (0,1)\big)$$

$$= 0,$$

hence the boundary chain $\partial \overline{\Delta}(x_1, x_2)$ is a cycle. That is indeed the general situation.

Theorem 7.14. *The boundary chain of a singular chain is a cycle, or equivalently $\partial(\partial \mathcal{C}) = 0$ for any singular chain $\mathcal{C}$.*

Proof: The boundary of the standard cell $\overline{\Delta}(x_1, x_2, \ldots, x_n)$ was defined in (7.119) and the boundary of a chain was defined in (7.122), and from these definitions it follows that

$$\partial \partial \overline{\Delta}(x_1, \ldots, x_n) = \partial \left(\sum_{i=1}^{n} \sum_{\epsilon=0}^{1} (-1)^{i+\epsilon} \overline{\Delta}(x_1, \ldots, x_{i-1}, \epsilon, x_{i+1}, \ldots, x_n) \right)$$

$$= \sum_{i=1}^{n} \sum_{\epsilon=0}^{1} (-1)^{i+\epsilon} \partial \overline{\Delta}(x_1, \ldots, x_{i-1}, \epsilon, x_{i+1}, \ldots, x_n). \qquad (7.123)$$

Here $\overline{\Delta}(x_1, \ldots, x_{i-1}, \epsilon, x_{i+1}, \ldots, x_n)$ is a singular $(n-1)$-cell, the image of the standard $(n-1)$-cell $\overline{\Delta}(x_1, \ldots, x_{i-1}, x_{i+1}, \ldots, x_n) \in \mathbb{R}^{n-1}$ by the mapping

$$\phi : \overline{\Delta}(x_1, \ldots, x_{i-1}, x_{i+1}, \ldots, x_n) \longrightarrow \overline{\Delta}(x_1, \ldots, x_{i-1}, \epsilon, x_{i+1}, \ldots, x_n)$$

for which

$$\phi(x_1, \ldots, x_{i-1}, x_{i+1}, \ldots, x_n) = (x_1, \ldots, x_{i-1}, \epsilon, x_{i+1}, \ldots, x_n) \in \mathbb{R}^n;$$

it then follows from the definition (7.119) of the boundary of a cell and the definition (7.122) of the boundary of a singular cell that

$$\partial \overline{\Delta}(x_1, \ldots, x_{i-1}, \epsilon, x_{i+1}, \ldots, x_n) = \partial \phi \overline{\Delta}(x_1, \ldots, x_{i-1}, x_{i+1}, \ldots, x_n)$$

$$= \phi \partial \overline{\Delta}(x_1, \ldots, x_{i-1}, x_{i+1}, \ldots, x_n)$$

$$= \phi \Bigg(\sum_{j=1}^{i-1} \sum_{\delta=0}^{1} (-1)^{j+\delta} \overline{\Delta}(x_1, \ldots, x_{j-1}, \delta, x_{j+1}, \ldots, x_{i-1}, x_{i+1}, \ldots, x_n)$$

$$+ \sum_{j=i+1}^{n} \sum_{\delta=0}^{1} (-1)^{j-1+\delta} \overline{\Delta}(x_1, \ldots, x_{i-1}, x_{i+1}, \ldots, x_{j-1}, \delta, x_{j+1}, \ldots, x_n) \Bigg)$$

$$= \sum_{j=1}^{i-1} \sum_{\delta=0}^{1} (-1)^{j+\delta} \overline{\Delta}(x_1, \ldots, x_{j-1}, \delta, x_{j+1}, \ldots, x_{i-1}, \epsilon, x_{i+1}, \ldots, x_n)$$

$$+ \sum_{j=i+1}^{n} \sum_{\delta=0}^{1} (-1)^{j-1+\delta} \overline{\Delta}(x_1, \ldots, x_{i-1}, \epsilon, x_{i+1}, \ldots, x_{j-1}, \delta, x_{j+1}, \ldots, x_n).$$

The sign change in the preceding equation is a consequence of the indexing of the cell $\overline{\Delta}(x_1, \ldots, x_{i-1}, x_{i+1}, \ldots, x_n)$ since x_j is the j-th variable in

$$\overline{\Delta}(x_1, \ldots, x_{i-1}, x_{i+1}, \ldots, x_n)$$

if $j < i$ but is the $(j-1)$-th variable if $j > i$; and

$$\overline{\Delta}(x_1, \ldots, x_{i-1}, \epsilon, x_{i+1}, \ldots x_{j-1}, \delta, x_{j+1}, \ldots, x_n)$$

is the obvious singular $(n-2)$-cell with the natural extension of the definition of the singular $(n-1)$-cell $\overline{\Delta}(x_1, \ldots, x_{i-1}, \epsilon, x_{i+1}, \ldots, x_n)$. Substituting the preceding observation into (7.123) yields

$$\partial \partial \overline{\Delta}(x_1, \ldots, x_n)$$

$$= \sum_{\substack{i,j=1 \\ j<i}}^{n} \sum_{\epsilon,\delta=0}^{1} (-1)^{i+j+\epsilon+\delta} \overline{\Delta}(x_1, \ldots, x_{j-1}, \delta, x_{j+1}, \ldots, x_{i-1}, \epsilon, x_{i+1}, \ldots, x_n)$$

$$+ \sum_{\substack{i,j=1 \\ j>i}}^{n} \sum_{\epsilon,\delta=0}^{1} (-1)^{i+j-1+\epsilon+\delta} \overline{\Delta}(x_1, \ldots, x_{i-1}, \epsilon, x_{i+1}, \ldots, x_{j-1}, \delta, x_{j+1}, \ldots, x_n).$$

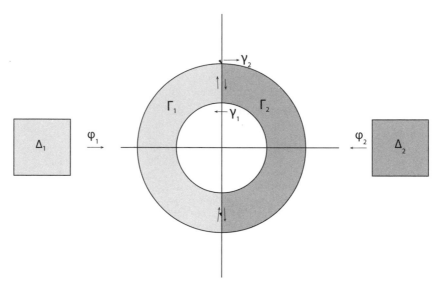

Figure 7.6. An annular region viewed as the image of a singular cell $\mathcal{C} = \phi_1 + \phi_2$

If the indices i, j are interchanged and the indices ϵ, δ are simultaneously interchanged in the first line on the right-hand side of the preceding equation the result is evidently equal to the second line in that equation, except for the sign, so that $\partial\partial\overline{\Delta}(x_1, \ldots, x_n) = 0$, which suffices for the proof.

For example the annular region in Figure 7.6, the region between two concentric circles, can be described as the image of the singular 2-chain $\mathcal{C} = \phi_1 + \phi_2$, where the images of the 2-cells $\overline{\Delta}_1$ and $\overline{\Delta}_2$ are the sets Γ_1 and Γ_2 that cover the annular region and that intersect only in the boundary segments at the top and bottom of the annular region. When the cells $\overline{\Delta}_i$ have the standard orientation, so that their boundaries are traversed in the counterclockwise order, and the mappings ϕ_i are orientation preserving, the boundary of the singular 2-chain reduces to the outer circle of the annular region, traversed in the counterclockwise direction, and the inner circle of the annular region, traversed in the clockwise direction, since the two boundary segments at the top and bottom of the annular region are traversed in opposite directions so cancel in the full boundary of the annular region.

The set of singular n-chains in an open subset $U \subset \mathbb{R}^m$ form an abelian group $C_n(U, \mathbb{Z})$, a set with an addition operation for which the sum of two singular chains $\mathcal{C}' = \sum_i v_i' \phi_i'$ and $\mathcal{C}'' = \sum_j v_j'' \phi_j''$ is defined to be the singular chain $\mathcal{C}' + \mathcal{C}'' = \sum_i v_i' \phi_i' + \sum_j v_j'' \phi_j''$. It is clear from the definition of the boundary chain that $\partial(\mathcal{C}' + \mathcal{C}'') = \partial\mathcal{C}' + \partial\mathcal{C}''$, so the set of singular n-cycles form a subgroup $Z_n(U, \mathbb{Z}) \subset C_n(U, \mathbb{Z})$, a subset closed under the operation of addition of singular chains; and it follows from Theorem 7.14 that the set of singular n-boundaries form a further subgroup $B_n(U, \mathbb{Z}) \subset Z_n(U, \mathbb{Z})$. The quotient group, defined in analogy with the quotient of two vector spaces, is a group

$$H_n(U, \mathbb{Z}) = \frac{Z_n(U, \mathbb{Z})}{B_n(U, \mathbb{Z})} \qquad (7.124)$$

called the n-dimensional **singular homology group** of the set U. These groups can be defined for more general topological spaces than just subsets of $\mathbb{R}^m$; they are algebraic structures that reflect the geometrical properties of topological spaces, and some examples of their usefulness will be included in the subsequent discussion here.[3]

The integral of a differential n-form ω over a singular n-chain is defined by

$$\int_C \omega = \sum_{j=1}^{r} v_j \int_{\phi_j} \omega \quad \text{for the chain } C = \sum_{j=1}^{r} v_j \cdot \phi_j. \qquad (7.125)$$

For example, twice the integral of a differential form ω over a singular cell ϕ can be viewed alternatively as the integral of that differential form over the singular chain $C = 2 \cdot \phi$; and the negative of the integral of ω over ϕ, or what is the same thing the integral of ω over ϕ with the opposite orientation, can be viewed alternatively as the integral of that differential form over the singular chain $C = -1 \cdot \phi = -\phi$. The Fundamental Theorem of Calculus for functions of a single variable, extends to functions of several variables in the form of Stokes's Theorem, which is expressed most naturally in terms of the integrals of differential forms over singular chains. The proof really is just a straightforward application of the classical Fundamental Theorem of Calculus in a single variable, which can be established readily in the following special case.

Theorem 7.15. *If ω is a C^∞ differential form of degree $n - 1$ defined in an open neighborhood of the cell $\overline{\Delta}(x_1, x_2, \ldots, x_n) \subset \mathbb{R}^n$ then*

$$\int_{\partial \overline{\Delta}(x_1, x_2, \ldots, x_n)} \omega = \int_{\overline{\Delta}(x_1, x_2, \ldots, x_n)} d\omega. \qquad (7.126)$$

Proof: First for a fixed index r introduce for convenience but just for the course of this proof the simplifying notation

$$\overline{\Delta} = \overline{\Delta}(x_1, x_2, \ldots, x_n) \subset \mathbb{R}^n,$$
$$\overline{\Delta}_r = \overline{\Delta}(x_1, \ldots, x_{r-1}, x_{r+1}, \ldots, x_n) \subset \mathbb{R}^{n-1}$$
$$\overline{\Delta}_{r,\epsilon} = \overline{\Delta}(x_1, \ldots, x_{r-1}, \epsilon, x_{r+1}, \ldots, x_n) \subset \mathbb{R}^n$$
$$dX_r = dx_1 \wedge \cdots \wedge dx_{r-1} \wedge dx_{r+1} \wedge \cdots \wedge dx_n,$$
$$dX = dx_1 \wedge dx_2 \wedge \cdots \wedge dx_n.$$

---

[3]Homology groups are discussed in detail in most books on algebraic topology, as for instance in the book by Allen Hatcher, *Algebraic Topology* (Cambridge University Press, 2002).

Consider then a differential $(n-1)$-form given by $\omega_r = f(\mathbf{x})\, dX_r$ for some function $f(\mathbf{x})$. For this differential form

$$d\omega_r = \partial_r f(\mathbf{x})\, dx_r \wedge dX_r = (-1)^{r-1}\partial_r f(\mathbf{x})\, dX. \tag{7.127}$$

Furthermore on a segment $\overline{\Delta}(x_1,\ldots,x_{i-1},\epsilon,x_{i+1},\ldots,x_n)$ of the boundary $\partial\overline{\Delta}$ the variable x_i is constant, so if $i \neq r$ the differential form $\omega_r = f(\mathbf{x})\, dX_r$ vanishes identically; therefore the integral of the differential form ω_r over the boundary $\partial\overline{\Delta}$ reduces to the integral just over the singular $(n-1)$-cells $(-1)^{r+1}\overline{\Delta}_{r,1}$ and $(-1)^r\overline{\Delta}_{r,0}$ of the boundary. It then follows from the definition of the integral of a differential form over a singular cell, the Fundamental Theorem of Calculus in one variable, and Fubini's Theorem that

$$\int_{\partial\overline{\Delta}} \omega_r = (-1)^{r+1}\int_{\overline{\Delta}_{r,1}} \omega_r + (-1)^r \int_{\overline{\Delta}_{r,0}} \omega_r$$

$$= (-1)^{r+1}\int_{\overline{\Delta}_r} \Big(f(x_1,\ldots,x_{r-1},1,x_{r+1},\ldots,x_n)$$

$$- f(x_1,\ldots,x_{r-1},0,x_{r+1},\ldots,x_n)\Big)dX_r$$

$$= (-1)^{r+1}\int_{\overline{\Delta}_r}\int_{\Delta(x_r)} \partial_r f(\mathbf{x})\, dx_r \wedge dX_r$$

$$= \int_{\overline{\Delta}} d\omega_r,$$

which is (7.126) for the differential form ω_r. Since any differential $(n-1)$-form ω can be written as a sum of differential forms of the special type ω_r for $r = 1, 2, \cdots, n$ and (7.126) is linear in the differential form ω it follows that (7.126) holds for arbitrary differential forms ω, which suffices for the proof.

Corollary 7.16 (Stokes's Theorem). *If $C \subset U$ is a C^∞ n-dimensional singular chain in an open subset $U \subset \mathbb{R}^m$ and ω is a C^∞ differential $(n-1)$-form in U then*

$$\int_{\partial C} \omega = \int_C d\omega. \tag{7.128}$$

Proof: First for the special case of a singular n-cell $\boldsymbol{\phi} : \overline{\Delta} \longrightarrow U$ in an open subset $U \subset \mathbb{R}^m$, if ω is an arbitrary C^1 differential $(n-1)$-form ω in U then since $\partial\boldsymbol{\phi} = \boldsymbol{\phi}\partial\overline{\Delta}$ by (7.122) it follows from the definition (7.125) of the integral of a differential form over a singular chain, the definition (7.104) of the integral of a differential form over a singular cell, and the preceding Theorem 7.15 that

$$\int_{\partial\boldsymbol{\phi}} \omega = \int_{\partial\overline{\Delta}} \boldsymbol{\phi}^*(\omega) = \int_{\overline{\Delta}} d\boldsymbol{\phi}^*(\omega) = \int_{\overline{\Delta}} \boldsymbol{\phi}^*(d\omega) = \int_{\boldsymbol{\phi}} d\omega,$$

since $d\boldsymbol{\phi}^*(\omega) = \boldsymbol{\phi}^*(d\omega)$ by Theorem 7.12; and that is (7.128) for the case of a singular cell. Then if $C = \sum_i v_i \cdot \boldsymbol{\phi}_i$ is an arbitrary singular chain where $\boldsymbol{\phi}_i$ are singular cells in U and if ω is a C^∞ differential $(n-1)$-form in U it follows from what has just been demonstrated and the definition (7.125) of the integral over a singular chain that

$$\int_{\partial C} \omega = \sum_i v_i \int_{\partial \phi_i} \omega = \sum_i v_i \int_{\phi_i} d\omega = \int_{\sum_i v_i \cdot \phi_i} d\omega = \int_C d\omega,$$

and that suffices for the proof.

Examples of the application of Stokes's Theorem abound. It can be used for instance to complement the discussion of conservative vector fields in Section 7.1. It was noted there that any closed 1-form is locally the exterior derivative of a function, indeed is the exterior derivative of a function in any cell but not necessarily so in more general regions; but it is the exterior derivative of a function in a class of regions that can be described topologically.

Corollary 7.17. [4] *If $U \subset \mathbb{R}^n$ is an open subset such that $H_1(U, \mathbb{Z}) = 0$ then any closed C^∞ differential 1-form in U is the exterior derivative of a function in U.*

Proof: It follows from Theorem 7.8 that it is sufficient to show that the integral $\int_\gamma \omega$ of any closed 1-form ω on curves γ between two points $\mathbf{a}, \mathbf{b} \in U$ depends only on the points $\mathbf{a}, \mathbf{b}$ and not on the choice of the curve γ from $\mathbf{a}$ to $\mathbf{b}$. If γ_1, γ_2 are two curves from $\mathbf{a}$ to $\mathbf{b}$, viewed as singular 1-chains, then $\gamma = \gamma_1 - \gamma_2$ is a singular 1-cycle in U so as a consequence of the hypothesis it is the boundary $\gamma = \partial C$ of a singular 2-chain $C \subset U$; then by Stokes's Theorem

$$\int_{\gamma_1} \omega - \int_{\gamma_2} \omega = \int_\gamma \omega = \int_{\partial C} \omega = \int_C d\omega = \int_C 0 = 0,$$

which suffices for the proof.

A number of applications of Stokes's Theorem can be restated in more classical terms.

Theorem 7.18. *If C is a C^∞ singular n-chain C in an open subset $U \subset \mathbb{R}^n$ and if*

$$\omega = \sum_{i=1}^n f_i dx_1 \wedge \cdots \wedge dx_{i-1} \wedge dx_{i+1} \wedge \cdots \wedge dx_n \tag{7.129}$$

is a C^∞ differential $(n-1)$-form in U then

$$\int_{\partial C} \omega = \int_{\|calC} \sum_{i=1}^n (-1)^{i-1} \frac{\partial f_i}{\partial x_i}. \tag{7.130}$$

---

[4]For extensions of this corollary see for instance the discussion of De Rham's theorem in *From Calculus to Cohomology: De Rham Cohomology and Characteristic Classes* by Ib H. Madsen and Jürgen Tornehave (Cambridge University Press, 1997), among other places.

Proof: This is simply Stokes's Theorem $\int_{\partial C} \omega = \int_C d\omega$ for the differential form (7.129), written out explicitly, so no further proof is required.

Corollary 7.19 (Green's Theorem). *If C is a singular 2-chain in an open subset $U \subset \mathbb{R}^2$ then*

$$\int_{\partial C} \left(f_1 dx_1 + f_2 dx_2 \right) = \int_D \left(\frac{\partial f_2}{\partial x_1} - \frac{\partial f_1}{\partial x_2} \right) \tag{7.131}$$

for any C^1 vector field $\mathbf{f} = \{f_1, f_2\}$ in U.

Proof: This is just the special case of the preceding theorem for the differential form $\omega_{\mathbf{f}}$ in $\mathbb{R}^2$, so no further proof is required.

Green's Theorem can be applied for example to the annular region D in Figure 7.6, which is the singular 2-chain $C = \phi_1 + \phi_2$; the boundary ∂C consists of the outer circle γ_1 oriented in the counterclockwise direction and the inner circle γ_2 oriented in the clockwise direction. If $\omega = f_1 dx_1 + f_2 dx_2$ is a C^1 differential 1-form in an open neighborhood of D then Green's Theorem asserts that $\int_{\gamma_1} \omega - \int_{\gamma_2} \omega = \int_D \left(\frac{\partial f_2}{\partial x_1} - \frac{\partial f_1}{\partial x_2} \right)$. In particular it follows that

$$\int_{\gamma_1} x_1 dx_2 - \int_{\gamma_2} x_1 dx_2 = \int_D 1 = |D|, \tag{7.132}$$

where $|D|$ is the area of the annular region D. Of course the same formula can be applied to any set that is the image of a singular chain, showing that the area of any such set can be calculated from its boundary alone.[5] As a more interesting example of this technique consider the hypocycloid curve in Figure 7.7, defined by the equation

$$x_1^{2/3} + x_2^{2/3} = a^{2/3}$$

for some constant $a > 0$. This curve can be viewed as the boundary of a singular chain C consisting of four singular 2-cells, the images of which are the segments in the four quadrants in Figure 7.7. The boundary chain reduces to the union of the four curved arcs, traversed in the counterclockwise direction, so is the image of the singular 1-cell

$$\phi(t) = \{a \cos^3 t, \, a \sin^3 t\} \quad \text{for} \quad 0 \le t \le 2\pi;$$

---

[5]This method for calculating the area of a region from its boundary is embodied in mechanical and electronic machines, called planimeters, that have long been used in engineering and surveying; these machines are discussed for example in the book by John Bryant and Christopher J. Sangwin, *How Round Is Your Circle? Where Engineering and Mathematics Meet.* (Princeton University Press, 2008).

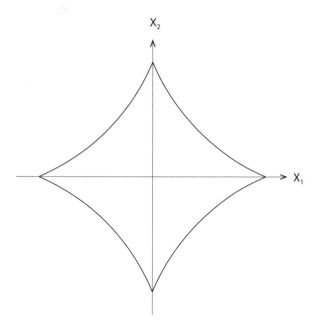

X_2

X_1

Figure 7.7. A hypocycloid

consequently the area $|D|$ of the region enclosed by the hypocycloid is

$$|D| = \int_{[0,2\pi]} x_1 dx_2 = \int_{[0,2\pi]} a\cos^3 t \cdot 3a\sin^2 t\cos t = \int_{[0,2\pi]} 3a^2 \sin^2 t\cos^4 t$$

$$= \int_{[0,2\pi]} 3a^2(\cos^4 t - \cos^6 t) = 3a^2\left(\frac{3}{4}\pi - \frac{5}{8}\pi\right) = \frac{3}{8}a^2\pi.$$

Green's Theorem can be written in terms of the differential 1-form $\omega_{\mathbf{f}} = f_1 dx_1 + f_2 dx_2$ associated to a vector field $\mathbf{f}$ as the line integral

$$\int_\gamma \mathbf{f} \cdot \tau \, ds = \int_D \left(\frac{\partial f_2}{\partial x_1} - \frac{\partial f_1}{\partial x_2}\right) \tag{7.133}$$

where τ is the unit tangent vector to γ and ds indicates integration with respect to the arc length of γ. Green's Theorem also can be written in terms of the Hodge dual $*\omega_{\mathbf{f}} = f_1 dx_2 - f_2 dx_1$ of the differential form $\omega_{\mathbf{f}}$ and since $d * \omega_{\mathbf{f}} = (\partial_1 f_1 + \partial_2 f_2) dx_1 \wedge dx_2$

$$\int_\gamma *\omega_{\mathbf{f}} = \int_D \left(\frac{\partial f_1}{\partial x_1} + \frac{\partial f_2}{\partial x_2}\right). \tag{7.134}$$

If the curve γ is parametrized by a mapping $\phi : [0, 1] \longrightarrow U$ the integral of the differential form $*\omega_{\mathbf{f}}$ is given by

$$\int_{\gamma} *\omega_{\mathbf{f}} = \int_{t \in [0,1]} \left(f_1 \frac{dx_2}{dt} - f_2 \frac{dx_1}{dt} \right). \qquad (7.135)$$

This can be viewed as the integral over the curve γ of the inner product $\mathbf{f} \cdot \mathbf{v}$ of the vector field $\mathbf{f}$ and the vector field

$$\mathbf{v} = \left\{ \frac{dx_2}{dt}, \ -\frac{dx_1}{dt} \right\}; \qquad (7.136)$$

and since $\|\mathbf{v}\|_2 = \sqrt{\left(\frac{dx_2}{dt}\right)^2 + \left(\frac{dx_1}{dt}\right)^2}$ then with the conventions adopted in the discussion of line integrals the integral (7.135) can be rewritten in terms of the unit vector $v = \mathbf{v}/\|\mathbf{v}\|_2$ as the line integral

$$\int_{\gamma} *\omega_{\mathbf{f}} = \int_{\gamma} \mathbf{f} \cdot v \, ds \qquad (7.137)$$

with respect to the arc length of the curve γ. The vector field v is orthogonal to the tangent vector field τ along the curve since

$$\mathbf{v} \cdot \tau = \left\{ \frac{dx_2}{dt}, -\frac{dx_1}{dt} \right\} \cdot \left\{ \frac{dx_1}{dt}, \frac{dx_2}{dt} \right\} = 0;$$

and when $\tau = \{1, 0\}$ then $v = \{0, -1\}$, so the pair of orthogonal vectors v, τ has the orientation of the vector space $\mathbb{R}^2$. In these terms the version (7.134) of Green's Theorem takes the form

$$\int_{\gamma} \mathbf{f} \cdot v \, ds = \int_{D} \left(\frac{\partial f_1}{\partial x_1} + \frac{\partial f_2}{\partial x_2} \right). \qquad (7.138)$$

While the line integral $\int_{\gamma} \mathbf{f} \cdot \tau \, ds$ measures the component of the vector field $\mathbf{f}$ in the direction of the tangent vector to γ, the integral $\int_{\gamma} \mathbf{f} \cdot v \, ds$ is the component of the vector field $\mathbf{f}$ in the direction of the outward normal vector to γ. The first integral sometimes is interpreted as the work done by the vector field $\mathbf{f}$ in pushing a particle along the curve γ, while the second integral sometimes is interpreted as the total flow across the curve γ of a substance moving with the velocity at any point $\mathbf{x}$ described by the vector $\mathbf{f}(\mathbf{x})$. For a conservative vector field $\mathbf{f} = \nabla h$ with the potential function h Green's Theorem in these two cases takes the forms

$$\int_{\gamma} \nabla h \cdot \boldsymbol{\tau} \, ds = \int_{D} \left(\frac{\partial^2 h}{\partial x_1 \partial x_2} - \frac{\partial^2 h}{\partial x_2 \partial x_1} \right) = 0,$$

(7.139)

$$\int_{\gamma} \nabla h \cdot \boldsymbol{v} \, ds = \int_{D} \left(\frac{\partial^2 h}{\partial x_1^2} + \frac{\partial^2 h}{\partial x_2^2} \right) = \int_{D} \Delta h$$

where Δh is the Laplacian of the function h, defined as $\Delta h(\mathbf{x}) = \frac{\partial^2 h(\mathbf{x})}{\partial x_1^2} + \frac{\partial^2 h(\mathbf{x})}{\partial x_2^2}$.

Another classical special case of Theorem 7.18 arises in three dimensions, and has a great number of applications in physics.

Corollary 7.20 (Gauss's Theorem). *If C is a singular 3-chain in an open subset $U \subset \mathbb{R}^3$ then*

$$\int_{\partial C} \left(f_1 dx_2 \wedge dx_3 + f_2 dx_3 \wedge dx_1 + f_3 dx_1 \wedge dx_2 \right) = \int_{D} \left(\sum_{i=1}^{3} \frac{\partial f_i}{\partial x_i} \right)$$

(7.140)

for any C^{∞} vector field $\mathbf{f} = \{f_1, f_2, f_3\}$ in U.

Proof: This is just the special case of the Theorem 7.18 for the differential form $\Omega_{\mathbf{f}}$ in $\mathbb{R}^3$, so no further proof is required.

For the differential form $\Omega_{\mathbf{f}}$ associated to a vector field $\mathbf{f}$ the surface integral in (7.140) can be written as in (7.111), while the integrand in the integral over the set D can be written as in (7.84), so Gauss's Theorem can be written in the alternative form

$$\int_{\Gamma} \mathbf{f} \cdot \boldsymbol{v} \, dS = \int_{D} \nabla \cdot \mathbf{f} = \int_{D} \mathbf{div} \, \mathbf{f}(\mathbf{x})$$

(7.141)

where Γ is the boundary chain of the singular 3-chain C and $D = |C|$ is the image of the chain C. As in the discussion of Green's Theorem, the surface integral can be interpreted as the total flow across the boundary Γ of a substance moving with the velocity at a point $\mathbf{x} \in \mathbb{R}^3$ described by the vector $\mathbf{f}(\mathbf{x})$. With this interpretation, $\nabla \cdot \mathbf{f}(\mathbf{x}) = \mathbf{div} \, \mathbf{f}(\mathbf{x})$ describes the amount of this mysterious substance created at the point $\mathbf{x}$, so the volume integral can be interpreted as the total amount of the substance being created within D. Gauss's Theorem asserts that the amount of substance being created in D is equal to the amount flowing outward across the boundary Γ of D, the amount of material diverging from the domain D and measured locally by $\mathbf{div} \, \mathbf{f}(\mathbf{x})$. For the particular case in which the vector field $\mathbf{f}$ is conservative, so is the gradient $\mathbf{f} = \nabla h$ of a potential function,

Gauss's Theorem takes the form

$$\int_\Gamma \nabla h(\mathbf{x}) \cdot \nu \, dS = \int_D \nabla \cdot \nabla h(\mathbf{x}) = \int_D \Delta h(\mathbf{x}) \tag{7.142}$$

in terms of the harmonic differential operator $\Delta h = \sum_{i=1}^3 \frac{\partial^2 h}{\partial x_i^2}$ in $\mathbb{R}^3$. For a simple example of Gauss's Theorem, to integrate the normal component of the vector field $\mathbf{f} = \{2x_1, x_2^2, x_3^2\}$ over the boundary Γ of the unit sphere $D = \{ \mathbf{x} \in \mathbb{R}^3 \mid \|\mathbf{x}\|_2^2 \le 1 \}$, since $\nabla \cdot \mathbf{f} = 2 + 2x_2 + 2x_3$ it follows from (7.141) that $\int_\Gamma \mathbf{f} \cdot \nu \, dS = 2 \int_D (1 + x_2 + x_3)$. The unit ball is symmetric with respect to the variables x_2 and x_3 so the second two terms in the integral yield 0 and consequently $\int_\Gamma \mathbf{f} \cdot \nu \, dS = 2 \int_D 1 = \frac{8}{3}\pi$.

A final classical interpretation of the general Stokes's Theorem of Corollary 7.16, which was traditionally called Stokes's Theorem before that term was adopted for the general case, involves line integrals in $\mathbb{R}^3$ that are the boundaries of singular 2-chains.

Corollary 7.21. *If $C \subset U$ is a continuously differentiable singular 2-chain in an open subset $U \subset \mathbb{R}^3$ then*

$$\int_{\partial C} \mathbf{f}(\mathbf{x}) \cdot \tau \, ds = \int_C (\nabla \times \mathbf{f}(\mathbf{x})) \cdot \nu \, dS = \int_C \mathbf{curl} \, \mathbf{f}(\mathbf{x}) \cdot \nu \, dS \tag{7.143}$$

for any C^∞ vector field $\mathbf{f}(\mathbf{x})$ in U.

Proof: This is just the special case of Corollary 7.16 for the differential form $\omega_\mathbf{f}$, so no further proof is necessary.

For a simple example of the classical Stokes's Theorem, let γ be the intersection of the unit sphere $S \subset \mathbb{R}^3$, the set $S = \{ (x_1, x_2, x_3) \mid x_1^2 + x_2^2 + x_3^2 = 1 \}$, with the plane $x_3 = 0$ and $\mathbf{f}(\mathbf{x})$ be the vector field $\mathbf{f}(\mathbf{x}) = \{2x_1, x_2^2, x_3^2\}$; since the curve γ is the boundary of the disc $D = \{ (x_1, x_2, x_3) \mid x_1^2 + x_2^2 \le 1, \ x_3 = 0 \}$ when the orientation of the disc is described by the differential form $dx_2 \wedge dx_1$, and $\nabla \times \mathbf{f}(\mathbf{x}) = \mathbf{0}$, Stokes's Theorem asserts that $\int_\gamma \mathbf{f}(\mathbf{x}) \cdot \tau \, ds = \int_D (\nabla \times \mathbf{f}(\mathbf{x})) \cdot \nu \, dS = 0$. For a perhaps more interesting observation though, if Ω is a closed 2-form in $\mathbb{R}^3$ and S is the unit sphere in $\mathbb{R}^3$, so that S is the boundary $S = \partial D$ of the unit ball, then by Gauss's Theorem $\int_S \Omega = \int_D d\Omega = 0$; but that does require that the differential form Ω is defined and closed in the entire ball D, just as with the corresponding example in $\mathbb{R}^2$. However if $\Omega = d\omega$ is an exact differential form defined just in an open neighborhood of the unit sphere S, and not necessarily in the entire ball D, it is actually the case that $\int_S \Omega = 0$. Indeed if γ is a circle on S, the intersection of S with a plane in $\mathbb{R}^3$, then γ splits the sphere S into two pieces, $S = S_1 \cup S_2$, where γ is the boundary of both S_1 and S_2; but if S_1 and S_2 have the same orientation as S and if γ is oriented so that $\partial S_1 = \gamma$ then clearly $\partial S_2 = -\gamma$. Then by the classical Stokes's Theorem $\int_S \Omega = \int_{S_1} \Omega + \int_{S_2} \Omega = \int_\gamma \omega - \int_\gamma \omega = 0$.

The classical Stokes's Theorem was also used to give a geometrical interpretation of the curl of a vector field. If $\mathbf{f}(\mathbf{x})$ is a $\mathcal{C}^1$ vector field in an open neighborhood U of a point $\mathbf{a} \in \mathbb{R}^3$, if D_ϵ is a disc of radius ϵ centered at the point $\mathbf{a}$ in a plane $L \subset \mathbb{R}^3$ through the point $\mathbf{a}$, with boundary $\gamma_\epsilon = \partial D_\epsilon$ in L, and if the plane L is chosen so that the unit normal vector v to the plane L is parallel to the vector $\mathbf{curl}\ \mathbf{f}(\mathbf{a})$, then by Stokes's Theorem $\int_{\gamma_\epsilon} \mathbf{f}(\mathbf{x}) \cdot \tau\,ds = \int_{D_\epsilon} \mathbf{curl}\ \mathbf{f}(\mathbf{x}) \cdot v\,dS$; and since the area of the disc D_ϵ is $|D_\epsilon| = \int_{D_\epsilon} dS$ it follows that

$$\frac{1}{\pi\epsilon^2} \int_{\gamma_\epsilon} \mathbf{f}(\mathbf{x}) \cdot \tau\,ds = \frac{\int_{D_\epsilon} \mathbf{curl}\ \mathbf{f}(\mathbf{x}) \cdot v\,dS}{\int_{D_\epsilon} dS}.$$

The function $\mathbf{curl}\ \mathbf{f}(\mathbf{x}) \cdot v$ is a continuous function of the point $\mathbf{x}$, so its average over the disc D_ϵ approaches the value $\mathbf{curl}\ \mathbf{f}(\mathbf{a}) \cdot v = \|\mathbf{curl}\ \mathbf{f}(\mathbf{a})\|_2$ in the limit as ϵ tends to zero; consequently

$$\|\mathbf{curl}\ \mathbf{f}(\mathbf{a})\|_2 = \lim_{\epsilon \to 0} \frac{1}{\pi\epsilon^2} \int_{\gamma_\epsilon} \mathbf{f}(\mathbf{x}) \cdot \tau\,ds, \qquad (7.144)$$

so that the length of the vector $\mathbf{curl}\ \mathbf{f}(\mathbf{a})$ is the limit of the line integral of the vector $\mathbf{f}(\mathbf{x})$ around the curve γ_ϵ as ϵ tends to zero, where this integral can be viewed as describing the curl of the vector field $\mathbf{f}(\mathbf{x})$ at the point $\mathbf{a}$.

Problems, Group I

1. Compute the surface integral

$$\int_S (x_1^2 + x_2^2)dx_1 \wedge dx_2 + x_1 dx_2 \wedge dx_3$$

where S is the curved part of the boundary of the cylinder of height 1 with base the unit circle at the origin in the plane of the variables x_1, x_2 oriented so that the normal vector points outward.

2. Express the surface integral $\int_S \mathbf{f} \cdot v\,dS$ of the vector-valued function

$$\mathbf{f}(x_1, x_2, x_3) = {}^t(x_1^2 + x_2^2,\ x_2 x_3,\ x_3^2)$$

on the piece S of the surface of the cylinder $x_2^2 + x_3^2 = 4$ in $\mathbb{R}^3$ for which $0 \le x_1 \le 1\ x_2 \ge 0, x_3 \ge 0$, oriented so that v is the outward normal, as the integral of a differential form and evaluate this integral.

3. Express the surface integral $\int_S \mathbf{f} \cdot v\,dS$ of the vector-valued function

$$\mathbf{f}(x_1, x_2, x_3) = {}^t(x_1^2, x_3, -x_2)$$

along the surface S of the unit sphere $x_1^2 + x_2^2 + x_3^2 = 1$, oriented so that v is the outward normal, as an integral of an appropriate differential form on S, and evaluate the integral by using Stokes's Theorem.

4.   In $\mathbb{R}^3$ let S be the surface $x_1^2 + x_2^2 = 1 - x_3$, $0 \leq x_3 \leq 1$, and let S_0 be the disc $x_1^2 + x_2^2 \leq 1$, $x_3 = 0$. Orient the surface segments S and S_0 so that the normal vectors point outwards from the interior of the region they bound.

   i) Compute the surface integral $\int_S \mathbf{f} \cdot v \, dS$ for the vector field

   $$\mathbf{f}(x_1, x_2.x_3) = {}^t(x_1, x_2, 2x_3 - x_1 - x_2).$$

   ii) Compute the integral in (i) but for the surface S_0.

5.   Compute $\int_X (\nabla \times \mathbf{f}) \cdot v \, dS$ for the vector field $\mathbf{f}(\mathbf{x}) = \{x_2, \ x_1 x_2 x_3, \ x_2\}$ where X is the piece of the paraboloid $x_3 = 3 - x_1^2 - x_2^2$ for which $x_3 \geq 0$, and v is the unit outward normal vector to X.

Problems, Group II

6.   If ω is a closed differential 1-form on the complement of the origin in $\mathbb{R}^2$ show that there is a unique constant c such that $\omega = c \, d\theta + df$ where θ is the angle (or argument) in polar coordinates and f is a differentiable function in the complement of the origin in $\mathbb{R}^2$.

7.   Show that the volume of a smoothly bounded Jordan measurable set $D \subset \mathbb{R}^3$ is given by the surface integral $V = \frac{1}{3} \int_{\partial D} \mathbf{f} \cdot v \, dS$ for the vector field $\mathbf{f}(\mathbf{x}) = {}^t(x_1, x_2, x_3)$, where the boundary surface is oriented so that the normal vector points outward.

8.   Demonstrate the differential form analogue of the formula for integration by parts, the assertion that if C is an n-dimensional chain in an open subset $U \subset \mathbb{R}^N$, if f is a smoothly differentiable function in U and if ω is a smoothly differentiable $(n-1)$-form in U then

   $$\int_C f d\omega = \int_{\partial C} f\omega - \int_C (df) \wedge \omega.$$

9.   If ϕ is a smooth oriented singular 2-cell in $\mathbb{R}^3$ and f, g are C^1 functions in an open neighborhood U of the image $\phi(\overline{\Delta})$ show that

   $$\int_{\partial \phi} f \nabla g \cdot \tau \, ds = \int_{\phi} (\nabla f \times \nabla g) \cdot v \, dS.$$

10.   i) Let $\Lambda_0^r(\Delta)$ be the set of C^∞ differential r-forms in the cell $\Delta \subset \mathbb{R}^n$ that are identically 0 in an open neighborhood of the boundary $\partial \Delta$. Show that the integral formula $(\omega_1, \omega_2) = \int_\Delta \omega_1 \wedge *\omega_2$, where $*\omega_2$ is the Hodge dual of ω_2, defines an inner product on the vector space $\Lambda_0^r(\Delta)$.

    ii) Show that for this inner product $(d\omega_1, \omega_2) = (\omega_1, \partial\omega_2)$ where ω_1 is an $(r-1)$-form, ω_2 is an r-form and the differential operator ∂ is defined by $\partial\omega = \pm * d * \omega$.

11.   A function f in $\mathbb{R}^n$ is said to have *compact support* if there is a compact set $S \subset \mathbb{R}^n$ such that $f(\mathbf{x}) = 0$ whenever $\mathbf{x} \notin S$.

    i) Show that if f_i for $1 \le i \le n$ are continuously differentiable functions in $\mathbb{R}^n$ and at least one of them has compact support then

$$\int_{\mathbb{R}^n} df_1 \wedge df_2 \wedge \cdots \wedge df_n = 0.$$

    ii) Find an explicit example of continuously differentiable functions f_i in $\mathbb{R}^n$, of course none of which has compact support, such that

$$\lim_{r \to \infty} \int_{B_r} df_1 \wedge df_2 \wedge \cdots \wedge df_n = \infty$$

where $B_r = \{\, \mathbf{x} \in \mathbb{R}^n \mid \|\mathbf{x}\|_2 \le r \,\}$.

    iii) Show that if f_1, f_2 are C^∞ functions in $\mathbb{R}^n$ and at least one of them has compact support then

$$\int_{\mathbb{R}^n} f_1(\mathbf{x})\, df_2(\mathbf{x}) \wedge dx_2 \cdots \wedge dx_n = - \int_{\mathbb{R}^n} f_2(\mathbf{x})\, df_1(\mathbf{x}) \wedge dx_2 \cdots \wedge dx_n.$$

12.   Let $\mathbf{f} = \{f_1, f_2, f_3\}$ be a C^∞ vector field and g be a C^∞ function in an open cell $\Delta \subset \mathbb{R}^3$; and let D be a singular cell in Δ such that its boundary ∂D is a C^∞ submanifold of Δ.

    i) Find $d * \omega_{\mathbf{f}}$, where $*$ indicates the Hodge dual form, and also find $* \omega_{\mathbf{f}} \wedge dg$.

    ii) Show that if $\mathbf{f}(\mathbf{x})$ is a tangent vector to the surface ∂D at each point $\mathbf{x} \in \partial D$ then

$$\int_D (f_1 \partial_1 g + f_2 \partial_2 g + f_3 \partial_3 g) = - \int_D (\partial_1 f_1 + \partial_2 f_2 + \partial_3 f_3) g.$$

    iii) Show that if $\nabla g(\mathbf{x})$ is a tangent vector to ∂D at each point $\mathbf{x} \in \partial D$ and if $\Delta g(\mathbf{x}) = \partial_1^2 g(\mathbf{x}) + \partial_2^2 g(\mathbf{x}) + \partial_3^2 g(\mathbf{x}) = 0$ at each point $\mathbf{x} \in D$ then $g(\mathbf{x})$ is constant in D.

Index